AF768122

Progress in Mathematical Physics

Volume 29

Editors-in-Chief
Anne Boutet de Monvel, Université Paris VII Denis Diderot
Gerald Kaiser, *The Virginia Center for Signals and Waves*

Editorial Board
D. Bao, *University of Houston*
C. Berenstein, *University of Maryland, College Park*
P. Blanchard, *Universität Bielefeld*
A.S. Fokas, *Imperial College of Science, Technology and Medicine*
C. Tracy, *University of California, Davis*
H. van den Berg, *Wageningen University*

Bruno Cordani

The Kepler Problem

Group Theoretical Aspects, Regularization and Quantization, with Application to the Study of Perturbations

Springer Basel AG

Author:

Bruno Cordani
Dipartimento di Matematica
Università degli Studi di Milano
Via Saldini 50
20133 Milano
Italy
e-mail: bruno.cordani@mat.unimi.it

2000 Mathematics Subject Classification 70F05; 70E20, 70F15, 70G45, 70G65, 70H33; 53A30

A CIP catalogue record for this book is available from the Library of Congress, Washington D.C., USA

Bibliographic information published by Die Deutsche Bibliothek
Die Deutsche Bibliothek lists this publication in the Deutsche Nationalbibliografie; detailed bibliographic data is available in the Internet at <http://dnb.ddb.de>.

Additional material to this book can be downloaded from http://extras.springer.com.
ISBN 978-3-0348-9421-0 ISBN 978-3-0348-8051-0 (eBook)
DOI 10.1007/978-3-0348-8051-0

This work is subject to copyright. All rights are reserved, whether the whole or part of the material is concerned, specifically the rights of translation, reprinting, re-use of illustrations, broadcasting, reproduction on microfilms or in other ways, and storage in data banks. For any kind of use whatsoever, permission from the copyright owner must be obtained.

© 2003 Springer Basel AG
Originally published by Birkhäuser Verlag, Basel - Boston - Berlin in 2003
Softcover reprint of the hardcover 1st edition 2003

Printed on acid-free paper produced of chlorine-free pulp. TCF ∞

ISBN 978-3-0348-9421-0

9 8 7 6 5 4 3 2 1 www.birkhauser-science. com

A Federica,

Violetta e Diletta

Preface

Because of the correspondences existing among all levels of reality, truths pertaining to a lower level can be considered as symbols of truths at a higher level and can therefore be the "foundation" or support leading by analogy to a knowledge of the latter. This confers to every science a superior or "elevating" meaning, far deeper than its own original one.

— R. GUÉNON, The Crisis of Modern World

Having been interested in the Kepler Problem for a long time, I have always found it astonishing that no book has been written yet that would address all aspects of the problem. Besides hundreds of articles, at least three books (to my knowledge) have indeed been published already on the subject, namely Englefield (1972), Stiefel & Scheifele (1971) and Guillemin & Sternberg (1990). Each of these three books deals only with one or another aspect of the problem, though. For example, Englefield (1972) treats only the quantum aspects, and that in a local way. Similarly, Stiefel & Scheifele (1971) only considers the linearization of the equations of motion with application to the perturbations of celestial mechanics. Finally, Guillemin & Sternberg (1990) is devoted to the group theoretical and geometrical structure.

My aim in writing this book has been precisely to compensate for this lack. Its "completeness" can, undoubtedly, be questioned. My hope is, however, to cover the main aspects of the Kepler Problem. The methods presented here also lead to a computer program, named KEPLER, which allows one to calculate and display graphically the effects of a perturbation.

More generally, this book is hoped to illustrate the use of differential geometric methods in solving concrete physical problems, with the Kepler Problem playing the role of a sort of "phylogenetic recapitulation" of the mechanics.

I would like to express my gratitude to Prof. S. Sternberg for his helpful criticisms, suggestions and encouragement. Many thanks also to the anonymous referee: among other things, he suggested the title. P. Horváthy has helped to revise parts of the book. I am indebted to S. Benenti and his coworkers C. Chanu and G. Rastelli for improving Chapter 8 and, in particular, for their permission to reproduce a delicate, not yet published, proof. I had interesting discussions with F. Fassò and A. Giacobbe on action-angle variables, degenerate systems and monodromy. R. Cushman has pointed out to me some errors and imprecisions. Many thanks also to P. Casati and M. Tarallo for their accurate reading of Chapters 9 and 11 respectively, as well as to L. Galgani and K. Payne for having corrected many errors of my poor English. I had helpful discussions with A. Alzati, E. Colombo, G. Gaeta, F. Magri, I. Mladenov, C. Morosi, P. Nicola, M. Palleschi, M. Pedroni, C. Reina and V. Zambelli. I have received considerable help from three of my students: C. Passoni for his work on monodromy, G. Merlini who has written the VisualC++ code of KEPLER, and S. Codegoni who has written the EULER program. I would like to thank B. Ruf for support and help in publishing this book, and, last but not least, my colleague and friend E. Barazzetti, since also nonstrictly technical discussions may be illuminating.

But, above all, this book would never have been written without the support of my family. Only the love of my wife Federica and my daughters Violetta and Diletta has given me strength and serenity to work hard and to respond to various bitter disappointments with a smile and a shrug. To them, and in particular to my wife, is this book dedicated.

Milano, June 2002 B. CORDANI

Contents

List of Figures

CHAPTER 1

Introductory Survey

The aim of this book is rather ambitious, namely to gather and order in a logical way most of the important ideas published on the Kepler Problem, i.e., on the motion of a particle under an attractive or repulsive force, decreasing with the square of the distance from a fixed point. These aspects include group theoretical treatment, regularization, quantization, variable separation, generalized Keplerian systems, perturbations, etc.

But what does it mean by "ordering in a logical way"? This book is not intended to be a sort of encyclopedic collection. On the contrary, we shall adopt (in the second part) a unified point of view and our main goal is to show how all these aspects can be organized. This point of view can be summarized as follows.

- The phase space of the regularized n-dimensional Kepler Problem is symplectomorphic to the cotangent bundle of the n-dimensional sphere with the zero section deleted.

- The cotangent bundle of the n-dimensional sphere is in turn symplectomorphic to a coadjoint orbit of the Lie group $SO(2, n + 1)$.

- The Kepler motion is obtained by applying to the geodesic motion on the sphere an extended canonical transformation, which involves time and energy as canonically conjugate variables.

The first two points are due essentially to J.-M. Souriau and J. Moser, whose theory is based in turn on the previous quantum approach of W. Pauli and V. Fock. The third point is due to the present author.

The book is divided into four parts in addition to the introductory chapter. The first part is devoted to an elementary exposition of the argument. The principal properties of the problem are recovered in a purely computational way, i.e., no attempt is made to grasp what goes on behind the scenes. In the second part a unified point of view is presented, and we will focus the exposition on the intrinsic, group-geometrical aspects. The third part exploits some results of the second part to give a geometrical description of the perturbation theory of the Kepler Problem. In the fourth part, to make the book as much as possible self-contained, some basic tools of differential geometry and analytical mechanics are reviewed.

While we hope that reading the first three parts entirely is worthwhile, the interested reader may limit himself to only one (or two) of them, since they are, to some extent, independent one from the other.

In this chapter, the basic points of the book are concisely reviewed, so that the advanced reader should be able to get a more precise idea of the entire argument.

1.1 Part I - Elementary Theory

1.1.1 Basic Facts

We begin in Chapter 2 by briefly recalling the three equivalent definitions and the geometrical and analytical properties of a conic.

Then the first integrals of the Kepler Problem, i.e., energy, angular momentum and eccentricity vector, are introduced, and two identities between them are established. This allows one to find the polar equation of the orbit (the angle being the true anomaly), showing that it is a conic, whose nature is determined by the sign of the energy.

Introducing the eccentric anomaly, which replaces time as the independent variable, regularizes the second order differential equation of motion, which is transformed, for each fixed value of the energy, into a linear regular one, whose integration is therefore trivial: the position vector is thus expressed as a function of the eccentric anomaly, with the first integrals playing the role of the integration constants. The finite relation between the physical time (or equivalently the mean anomaly) and the eccentric anomaly is given by the celebrated Kepler Equation, whose inversion (not achievable in closed form) is necessary to express the trajectory as a function of time.

The definitions of the three (true, eccentric and mean) anomalies are reviewed and geometrically interpreted for each value of the energy.

The general solution of the equation of motion clearly shows that the angular momentum and the eccentricity vector parametrize in a global way the set of orbits. The two vectors, however, are not independent, since they are in fact orthogonal to each other. One may extract from them five independent variables, the most popular ones being the semi-major axis, eccentricity, longitude of the ascending node, inclination and argument of the pericentre. But, in this way, the one-to-one correspondence between orbits and sets of parameters is inevitably lost. The five parameters plus the time of passage at the pericentre (which fixes the position of the planet on its orbit) are known as the Keplerian elements and form a local (in fact some orbits are missing from the parametrization) coordinate system on the phase space of the Kepler Problem.

Finally, the repulsive potential is considered and the general solution for this case is also given.

In the appendix to Chapter 2 the problem of inversion of the Kepler Equation is reviewed.

1.1.2 Separation of Variables and Action-Angle Variables

The four orthogonal coordinate systems for which the Hamilton-Jacobi Equation of the Kepler Problem is separable are reviewed in Chapter 3: they are the spherical, parabolic, elliptic and spheroconical coordinates. The Hamilton-Jacobi Equation is explicitly separated, and the problem is reduced to quadratures. The triplets of first integrals in involution, arising from each of the separation processes, are found.

Starting from a set of n independent first integrals in involution, one can calculate n other first integrals in involution (function of them), integrating the canonical 1-form along n independent cycles on the Liouville-Arnold tori. They are the action variables, and their conjugate variables are angles.

Applying this mechanism to the first integrals generated by the separation in spherical coordinates, one finds, after a further unimodular transformation, the Delaunay variables. These are not obviously global coordinates but, being very similar to the Keplerian elements, are plagued with the same singularities. A second unimodular transformation leads to the Poincaré variables, which are regular where the preceding fail (but have obviously other singularities). Complete and detailed formulas are given, expressing Delaunay and Poincaré variables as functions of position and velocity, and vice versa.

The variables of Delaunay and Poincaré are different coordinates on the same tori: these tori are obtained by the intersection of the level hypersurfaces of the first integrals relative to spherical coordinates. Arguing in a similar way but starting from the parabolic coordinates, we come to define another set of action-angle variables, which we call "Pauli variables", that are defined on tori intrinsically different from the preceding ones. We find that the Poisson bracket algebra of the angular momentum and Runge-Lenz-Laplace vector (i.e., the rescaled eccentricity vector) is isomorphic to the Lie algebra of the four-dimensional orthogonal group SO(4).

1.1.3 Quantization of the Kepler Problem

Three methods to quantize the Kepler Problem, those of Schrödinger, Pauli and Fock respectively, are reviewed in Chapter 4.

The Schrödinger method is applicable in general to a mechanical system with a kinetic energy quadratic in the momenta, thus defining a Riemannian structure on the configuration space. The Schrödinger Equation is basically a wave equation, where the Laplace operator is associated with the Riemannian structure.

We write the Schrödinger Equation for the Kepler Problem in the same four coordinate systems considered in the preceding chapter and verify its separability. The explicit integration in each of the four coordinate systems is performed, leading to the energy spectrum.

The Pauli quantization is then considered. It is an example of the general method of canonical quantization, due to Dirac and Heisenberg, and consists of substituting each classical observable with a Hermitian differential operator acting on a Hilbert space. The quantum differential operators corresponding to angular momentum and Runge-Lenz-Laplace vectors satisfy, in analogy with the classical case, the commutation rules of the Lie algebra of SO(4), supplemented with two identities. That leads immediately to finding once again probability distribution and energy levels of the Kepler Problem.

Pauli has been historically the first to point out the relevance of SO(4) for the Kepler Problem, but in a local way only: the Lie algebra, not the group, is involved. Fock instead gave a global treatment, marking the beginning of the modern point of view. Even if his argument is developed in the framework of quantum mechanics, his conclusions are valid in the classical context also.

Fock starts by performing a Fourier transform of the n-dimensional Schrödinger Equation. To the Laplace operator corresponds a multiplicative quadratic operator, while the potential term gives rise to a convolution integral. Applying a stereographic projection in the momentum space and rescaling suitably the wave function, the rotational symmetry is made manifest. Eigenvalues and eigenfunctions of the transformed equation (an integro-differential equation) are then found.

A mathematical appendix to Chapter 4 refreshes some of the arguments used.

1.1.4 Regularization and Symmetry

The Kepler Problem is not regular; in fact, its Hamiltonian vector field is not complete, since collision orbits reach the singularity of the potential in a finite time with a divergent velocity. The methods of Moser, Souriau and Kustaanheimo-Stiefel replace time with the eccentric anomaly, regularize the problem and exhibit, more or less explicitly, the various aspects of its symmetries. In Chapter 5 the three methods are reviewed, following the original exposition of the authors.

The Moser method is strongly reminiscent of the Fock quantization process, but at a classical level. Moser shows that the Kepler flow of the regularized n-dimensional Kepler Problem for negative energy is equivalent to the geodesic flow on the n-dimensional sphere, explaining in this way the $(n+1)$-dimensional orthogonal group symmetry. The result is proved by appropriately extending the stereographic projection to the (co)tangent bundle of the sphere.

In the Souriau method one introduces, after integration of the equation of motion using a clever method, two orthogonal four-dimensional vectors of equal norm, called Fock parameters, whose dynamical evolution is very nice; the eccentric anomaly running as "false time", the two vectors rotate uniformly in their initial plane. Two similar vectors, the Bacry-Györgyi parameters, may be defined; they rotate uniformly with the true time, but, while the dependence of the Fock parameters from position and velocity can be inverted explicitly, this is not possible for the Bacry-Györgyi parameters, due to the fact that transcendental functions are involved.

The Kustaanheimo-Stiefel transformation is very simple to describe but somewhat, at this stage, surprising. Generalizing (but this is not a trivial task) the Levi-Civita map, working in two dimensions, the three-dimensional Kepler Problem is transformed into that of a four-dimensional isotropic oscillator, subject to a constraint.

1.2 Part II - Group-Geometric Theory

1.2.1 Conformal Regularization

Chapter 6 is the main chapter of the book, and all the arguments of the subsequent chapters find their root here.

We consider the action of the conformal group on the Minkowski space $M_0 = \mathbb{R}^{1,n}$ and find the conformal vector fields that generate it. Computing their Lie algebra, one immediately verifies that it is isomorphic to that of the pseudo-orthogonal group $O(2, n+1)$.

There is a deep reason for this isomorphism. Since the action of $O(2, n+1)$ on $\mathbb{R}^{2,n+1}$ is linear, it induces an action on the projective manifold M of null rays. This manifold is diffeomorphic to the product $S^1 \times S^n$ of a circle by an n-dimensional sphere and is endowed with a class of pseudo-Riemannian conformally flat metrics, obtained by restriction of the pseudo-Euclidean metric of $\mathbb{R}^{2,n+1}$ on any section of the null cone $\mathcal{K}$. The linear action of $SO_0(2, n+1)$ (the connected component of the identity) sends in a continuous way a section of $\mathcal{K}$ into another section; all of these sections being conformally flat, every element of $SO_0(2, n+1)$ induces a conformal map of M to itself, and a homomorphism from $SO_0(2, n+1)$ to the conformal group is thereby established. The manifold $M = S^1 \times S^n$ is obtained by adding to M_0 some extra points at infinity; i.e., it is the conformal compactification of the Minkowski space.

The transitive action of $SO_0(2, n+1)$ on M induces a symplectic action on the cotangent bundle T^*M. This action however is not transitive, and it is natural to ask for its homogeneous components. It results that T^+M, i.e., $T^*M-\{\text{zero-section}\}$, is decomposed under the action of $SO_0(2, n+1)$ into five orbits: three are the $2(n+1)$-dimensional manifolds of space-like and time-like, future and past pointing, covectors; the other two are the two n-dimensional manifolds of null (but not identically zero) future and past pointing covectors. These five manifolds are then identified with coadjoint orbits of $SO_0(2, n+1)$.

Every section of the null cone $\mathcal{K}$ identifies an $(n+1)$-dimensional manifold, M_γ say, equipped with a conformally flat metric. A general method is described to obtain a moment map from T^+M_γ into the dual Lie algebra of the conformal group. Three sections are relevant for us, corresponding to the three signs of the energy of the Kepler Problem.

(i) $M_{(-)}$ having the topology and the pseudo-Riemannian metric of $S^1 \times S^n$;

(ii) $M_{(+)}$ having the topology and the pseudo-Riemannian metric of $\mathbb{R} \times H^n$ (this last is the n-dimensional hyperboloid);

(iii) M_0 having the topology and the pseudo-Euclidean metric of $\mathbb{R}^{1,n}$.

It is important to point out that, owing to the choice of the coordinatization of the sections (involving stereographic projection and inversion), some point is missing; restoring this point is exactly the regularization of the Kepler Problem.

The moment maps relative to these three sections are constructed. As noted above, the action of the conformal group on the $2(n+1)$-dimensional manifolds T^+M_γ is not transitive but divides each of them into five orbits. By restricting to the two $2n$-dimensional manifolds of null covectors, we find two equivalent moment maps (which we call Main Moment Map) from $T^+\mathbb{R}^n$ into the dual Lie algebra of the conformal group; here $\mathbb{R}^n = S^n - \{\text{North Pole}\}$ is obtained by a stereographic projection.

To this Main Moment Map a mechanical meaning can be associated, by considering each of its components as a dynamical variable defined on $T^+\mathbb{R}^n$, and conveniently choosing one of them as a Hamiltonian. Three choices will be relevant, one for each section $M_{(-)}, M_{(+)}$ and M_0, and for them we obtain the geodesic motion on S^n, H^n and $\mathbb{R}^n$, respectively. But the Main Moment Map gives more, displaying also the invariance groups of these geodesic motions, i.e., $\mathrm{SO}(n+1), \mathrm{SO}(1,n)$ and $\mathrm{SO}(n) \times_S \mathbb{R}^n$, respectively.

While this last result is somewhat trivial, the interesting thing is the way we have obtained them, i.e., basically by restricting ourselves to the submanifold of the null covectors in T^+M_γ and dividing out the kernel of the canonical two-form. This procedure is intrinsic and can be carried out in any canonical coordinate system of T^+M_γ : the above Hamiltonian systems i), ii) and iii) will preserve their dynamical properties (in particular the invariance groups) but may look very different. Notice, and this is a fundamental point, that T^+M_γ is the phase space-time, so that we have the possibility to perform an enlarged symplectic coordinate change, involving time and energy also. With a suitable choice of the canonical transformation the Kepler Problem is obtained, while the components, relative to the invariance groups, of the Main Moment Map are transformed into those of the angular momentum and the Runge-Lenz-Laplace vector. The remaining components give Fock or Bacry-Györgyi parameters.

Finally, we add some comments to the results of Chapter 6, which serve to clarify the geometrical meaning of the above canonical transformation.

1.2.2 Spinorial Regularization

For $n = 3$, the relevant symmetry group of the Kepler Problem is $\mathrm{SO}(2,4)$ which admits the pseudo-unitary group $\mathrm{SU}(2,2)$ as a Spin group, i.e., as a double covering. In Chapter 7 the regularization process is reformulated in terms of this latter group, so that the Kustaanheimo-Stiefel transformation is naturally defined.

The homomorphism $\mathrm{SU}(2,2) \to \mathrm{SO}(2,4)$, with kernel ± 1, is established first in an intrinsic way, and then explicitly, by exhibiting a matrix basis whose commutation relations reproduce the Lie algebra of SU(2, 2). Another basis, connected to the previous basis by a linear transformation, is then defined. Identifying the 2×2 unitary matrices with M and the 2×2 skew Hermitian matrices with M_0, one can see that the linear fractional action of SU(2, 2) over them is the conformal action of the previous chapter.

Let T_0 be the space of null twistors (a twistor is an element of $\mathbb{C}^{2,2}$ and a null twistor is a nontrivial one whose SU(2, 2) norm is zero) and define two twistors to be equivalent if they differ by a phase. The set $T_0/\sim$ of null twistors modulo a phase transformation is a 6-dimensional manifold that, being equipped with a natural symplectic form, turns out to be symplectomorphic to T^+S^3, the phase space of the regularized Kepler Problem with negative energy. This allows one to find explicitly the moment map from $T_0/\sim$ into the dual of the Lie algebra of SU(2, 2), from which the Kustaanheimo-Stiefel transformation follows. Most of the work done in this and in the previous chapter is encapsulated in the commutative diagram on page 155.

The possibility to extend the Kustaanheimo-Stiefel transformation to generic dimensions is investigated. We show that, while this generalization is abstractly possible for every n, explicit formulas can be given only for $n = 2, 3, 5$. These cases require the use of generalized twistors, defined over the real numbers, the complex numbers and the quaternions respectively.

1.2.3 Return to Separation of Variables

In Part I, we have verified that the Hamilton-Jacobi and Schrödinger Equations of the 3-dimensional Kepler Problem are separable in four coordinate systems. In Chapter 8, we shall show that general theorems on orthogonal separable systems, together with the knowledge of the maximal invariance group $SO(2) \times SO(4)$ of the Kepler Problem, enable us to deduce these four coordinate systems, and to prove that they are the only ones admitting separability.

The first general theorem is that of Stäckel, which gives necessary and sufficient conditions for a natural orthogonal system to be separable. In practice this theorem is not very useful, but allows one to prove a second one, namely that of Eisenhart, which suggests a strategy to search for the orthogonal systems separating some n-dimensional problem: firstly, find n first integrals in involution and quadratic in the momenta; secondly, verify that the n symmetric matrices associated with the quadratic forms in the momenta are all algebraically commuting; lastly, find their common n eigenvectors and, after n integrations, find the coordinate lines tangent to them.

A third theorem, due to Robertson, gives a necessary and sufficient condition for the separability of the Schrödinger Equation when the corresponding Hamilton-Jacobi Equation is separable. This condition is trivially satisfied for the Kepler Problem.

The strategy outlined above is then applied to find all the orthogonal coordinate systems separating the Kepler Problem. We know all its first integrals: energy, angular momentum and eccentricity vector, so that we must extract from them three independent functions in involution. Up to naturally defined equivalence, we prove that just four triplets exist, namely those of Chapter 3. Then we verify the algebraic commutativity. Lastly we find, for each of the triplets, the common eigenvectors with the relative coordinate systems, and the four Stäckel matrices and vectors.

If in one of the four separable coordinate systems some suitable term that preserves separability is added to the Kepler Hamiltonian, a new integrable system is obtained. The most celebrated example is that of two fixed attractive masses. The problem is reduced to quadratures, which involve elliptic integrals, and the resulting orbits are studied qualitatively. Also, the program EULER, written in the MAPLE language, is supplied, integrating the problem in an exact analytical way.

In the appendix to Chapter 8, Jacobian elliptic functions are briefly reviewed.

1.2.4 Geometric Quantization

The geometric quantization of a Hamiltonian system consists of two parts. First, view the phase space as a coadjoint orbit of some Lie group (the dynamical or symmetry group) in such a way that the relevant dynamical quantities (typically the Hamiltonian and first integrals) give the components of a moment map. Second, construct an irreducible unitary representation of the dynamical group by means of Hermitian operators acting on the Hilbert space L_2 of the square-integrable functions, which are defined over the base space of a polarization of the coadjoint orbit. If one is able to exhibit this representation, to every relevant dynamical quantity, i.e., to every observable, is associated a Hermitian operator, and the algebra of the observables under Poisson brackets is isomorphic to the algebra of the Hermitian operators under commutation. The canonical quantization, which is the prescription $p_k \mapsto -i\partial_k$, turns out a posteriori to be a mere recipe, which sometimes, but not always, gives the correct result.

As regards the Kepler Problem, the first task has already been fulfilled in Chapter 6, while in Chapter 9 we want to realize the second one. To this end, we shall utilize the theory of multiplier representations of the simple noncompact Lie groups, which we review in the first section. We will follow very closely the "classical" route of Chapter 6, and implement it at the "quantum" level.

Then we quantize the geodesic flow on the sphere S^n. For this last construction, we apply the general method to obtain, through operators acting on $L_2(S^1 \times S^n)$, a multiplier representation of the supplementary series. For a judicious choice of the parameter of the representation, this latter turns out to be reducible. The two (isomorphic) irreducible components are just the quantizations of the flow on S^n.

Finally, the canonical transformation of Chapter 6 is considered. The delicate point is to implement the interchange between coordinates and momenta. Obviously this is accomplished with a Fourier transform, but the wave function is defined on the sphere, not on a linear space. The "trick" is to make the representation fully unitary by introducing an intertwining operator (a standard tool in representation theory) that will surprisingly prove to be the inverse of the operator corresponding to the classical action of the Kepler Problem. This reveals that Fock quantization is just the implementation of the canonical transformation of Chapter 6.

1.2.5 Kepler Problem with a Magnetic Monopole

As in Chapter 7 we consider in Chapter 10 the twistor space, but without requiring the nullity condition. The manifold of the twistors of norm 2μ, modulo the usual phase transformation, has the topology of $(\mathbb{R}^3 - \{0\}) \times \mathbb{R}^3$ (rather than that of T^+S^3 as for $\mu = 0$), and is equipped with a natural symplectic structure differing from the canonical one in a magnetic term proportional to μ.

The Main Moment Map is then generalized. It is surjective, in contrast with the case $\mu = 0$, so that no points of the coadjoint orbit are missing and no regularization process is required. As in the case $\mu = 0$ we consider some components of the generalized moment map as Hamiltonians, and study the motion they generate. For the bound-unbound motions we find again the invariance groups SO(4)-SO(1,3), and that the orbits are conic sections. There are however some differences; for example, the angular momentum and the Runge-Lenz-Laplace vectors are no longer orthogonal to each other, the orbit plane does not contain the origin and there exists a whole sphere, centered in the origin, of equilibrium points.

As in Chapter 6 the generalized Main Moment Map may be transformed with a canonical map. Choosing the same transformation of that chapter, we find the MICZ Hamiltonian system, i.e., the Kepler Problem plus the magnetic term of the Dirac monopole and a repulsive potential.

Changing the canonical transformation (of which we exhibit a whole family, depending on two arbitrary functions) other systems are found, all sharing the symmetries of the Kepler Problem, although their Hamiltonians may look rather different.

An interesting example is provided by the Taub-NUT system, which has a whole sphere of singularities. As the motion approaches this latter, the velocity goes to infinity. A careful analysis of the canonical transformation explains in a direct way the dynamical properties of this system.

The above construction is based on the Hopf fibration $S^3 \to S^2$, and is therefore dimension dependent. However, it can be extended to the case $n = 5$, with analogous Hopf fibration $S^7 \to S^4$. In this case, the monopole is called "BPST instanton".

1.3 Part III - Perturbation Theory

1.3.1 General Perturbation Theory

In Chapter 11 the question of taking into account small perturbations of an integrable Hamiltonian system is investigated. The problem has two distinct aspects: the formal one and that of the convergence.

The basic idea of perturbation theory, i.e., to consider the perturbation in some sense "small" and thus to proceed to a series expansion, goes back to the early age of classical mechanics. Its actual form, however, is due to Poincaré, and may be summarized in this way: one must find a canonical transformation, differing from the identity by a quantity of the order of the perturbation, such that the transformed Hamiltonian is integrable up to second order terms. The procedure can obviously be iterated, pushing the perturbation to third order, and so on. If the process converges (but this is the key point), better and better higher order approximations are obtained.

In the sequel we will adopt this point of view, but with a technical variation. To simplify some calculations, use is made of the Lie series, which avoids those inversion problems which occur typically with methods involving the generating function of a canonical transformation.

The formal machinery of Lie series with their properties is reviewed in the first section. The core of the formal problem of the perturbations is the so-called homological equation. Solving it to some order (which requires basically integrations and differentiation) we are left with an approximate Hamiltonian which, if the unperturbed Hamiltonian is nondegenerate, turns out to be integrable.

The subtle problem of the convergence is then tackled. It is solved by the celebrated Kolmogorov Theorem: if the unperturbed Hamiltonian is nondegenerate, picking out a torus such that the corresponding frequencies of the unperturbed motion are "sufficiently irrational", it will only be slightly deformed but not destroyed, provided that the perturbation is sufficiently small.

Conserved and destroyed tori form, however, a complicated Cantor type set and the question if a certain orbit (whose initial conditions are never known with absolute precision) is stable or unstable can not be decided.

Nekhoroshev pointed out that this disappointing situation is due to the unphysical fact that we want answers valid for infinite time. We report a sketch of the proof of the Nekhoroshev Theorem, which shows that the variation of the actions is of the order of the perturbation value, thus very small, for time intervals increasing exponentially with the reciprocal of the perturbation value.

The precise meaning of the irrationality condition is given by a result from Diophantine theory, briefly recalled in an appendix to the chapter.

Moreover, in a second appendix, the Cauchy inequality, the basic tool in giving some estimates, is reviewed.

1.3.2 Perturbations of the Kepler Problem

To study, either analytically or numerically, the perturbations of the Kepler Problem we depart from the traditional method, which would require the use of action-angle variables (typically Delaunay or Poincaré variables), since these latter suffer from some drawbacks.

Indeed the Delaunay or Poincaré variables, being related to the Kepler Equation, are transcendental functions of the physical coordinates and momenta. They cannot be inverted in a closed form, and the expression of a given perturbation is therefore cumbersome.

Moreover, these coordinates are not global, so that some orbits are missing from the parametrization; this is related to the fact that the global topology of the phase space of the (regularized) Kepler Problem is that of T^+S^3 (the cotangent bundle of the 3-dimensional sphere without the zero-section, as explained in Chapter 6) while an action-angle variable system describes a manifold with the topology of $\mathbb{T}^3 \times \mathbb{R}^3$ (where $\mathbb{T}^3$ is the 3-dimensional torus).

To avoid these drawbacks, we prefer to adopt the global coordinatization given by the Fock parameters, paying for it with redundancy in the number of coordinates. Thus we will no longer deal with a 6-dimensional symplectic manifold, but with an 8-dimensional and of rank 6 Poisson manifold.

From the analytical point of view we show that in this case also the normalization procedure can be performed, requiring only a (non necessarily canonical) Poisson structure.

Unfortunately, the Kepler Problem is completely degenerate (an intrinsic, unavoidable fact), which entails that only one angle can be eliminated from the perturbation. So we are left with a Hamiltonian system with two degrees of freedom, whose phase space is symplectomorphic to $S^2 \times S^2$, each point representing an averaged ellipse. In general, no other first integrals are known besides the Hamiltonian, and the problem is not integrable.

For topological reasons, every vector field on $S^2 \times S^2$ admits at least four elliptic critical points, to which correspond four periodic orbits for the averaged truncated Hamiltonian. A theorem due to Moser ensures the prolongation of these four periodic orbits for the complete, i.e., non-truncated, Hamiltonian as well.

By means of the Kolmogorov Theorem, we show that almost all 3-dimensional orbits, corresponding to those orbits on $S^2 \times S^2$ surrounding one of the elliptic critical points, can be continued for the full Hamiltonian as well.

Also for numerical integration the Fock parameters, which regularize the Kepler problem, are well suited. Indeed, the problem is not regular, and the neighbourhood of the singularity in the origin must be absolutely avoided. This is highly advisable for the Enterprise and captain Kirk, but puts severe limitations on numerical integration. The regularized unperturbed motion, instead, becomes very simple (uniform rotation) and the method of variation of constants, requiring inversion of the motion, is easily applicable. Moreover, since the Fock parameters are global coordinates, the problems that arise during integration, when nonglobal action-angle coordinates are employed, are avoided. A stabilization method, keeping fixed the two constraints that must be satisfied by the Fock parameters, is used. The method is finally implemented in the KEPLER program.

1.3.3 Perturbations with Axial Symmetry

Unlike the general case, the axisymmetric perturbations lead, after normalization, to an integrable Liouville problem; in addition to the Kepler Hamiltonian and the full perturbed Hamiltonian, the projection of the angular momentum on the symmetry axis is a first integral, clearly in involution with the other two. This allows one to study the perturbed motion in a graphical and global way. The basic idea is to reduce the manifold of the orbits, which has the topology of $S^2 \times S^2$, by the rotational symmetry. One obtains a symplectic manifold homeomorphic to S^2; then the level surfaces of the twice reduced (with one degree of freedom) Hamiltonian, intersecting this symplectic manifold, give rise to a family of curves describing the perturbed motion.

The reduction requires however some care, since the axial symmetry has fixed points when the third integral is zero, so that the reduced phase space is not a smooth manifold. In every case we obtain a manifold homeomorphic to S^2, which we cover with two different parametrizations, one related to Pauli and the other to Delaunay action-angle variables, respectively.

In this framework we study three concrete examples. The first is the Lunar problem, i.e., a point, the Moon, moving around the Earth and perturbed by the Sun. The second is the combined Stark and quadratic Zeeman effect, i.e., the hydrogen atom subjected to the combined actions of constant electric and magnetic fields, orthogonal one to the other. The third is the problem of an artificial satellite moving around an oblate primary.

1.4 Part IV - Appendices

1.4.1 Differential Geometry

After recalling the rudiments of topology, the basic definitions of differentiable manifold and (co)tangent space, generalizing the intuitive idea of surface and tangent plane, are given. The concepts of pull-back and push-forward of vectors and covectors arise naturally when one considers transformations from a manifold respectively to another.

Tensors are defined as multilinear maps from Cartesian products of a vector space with itself into the real numbers. The particular, but very important case, of the completely skew symmetric tensors (or forms) is considered. The basic operations that one can perform on them are Lie (or directional) derivative, exterior derivative and integration. The Stokes and deRahm Theorems are briefly exposed.

A distribution is a continuous field of linear subspaces of tangent spaces. If an embedded submanifold generates naturally a distribution, the inverse is not assured: the Frobenius Theorem gives the pertinent integrability condition.

Giving a metric structure to the (co)tangent spaces of a manifold allows one to transform a vector into a covector and vice versa. If the metric tensor is symmetric the structure is Riemannian; if it is skew symmetric, with a closure condition, the structure is symplectic. The Riemannian manifolds are typically the configuration space of mechanical systems, while the symplectic manifolds, with their Poisson generalization, are the phase spaces.

Then, the definition of fibre bundle is given, which is locally, but not necessarily globally, the product of two manifolds. The basic tangent and cotangent bundle to a manifold are reviewed.

1.4.2 Lie Groups and Lie Algebras

A Lie group is a group whose elements are labelled by one or more continuously variable parameters; taking these parameters as coordinates, a Lie group can thus be seen also as a manifold, while the tangent space at the identity, once equipped with a natural composition law, becomes the corresponding Lie algebra. For what concerns us, Lie groups are the fundamental tool to investigate the symmetries of mechanical systems.

The properties of the group composition laws, in particular the associative one, endow the group manifold with a very rich structure. The possibility to define invariant (under the action of the group) vector fields, which are generated by right and left translations, allows one to define Lie bracket and structure constants.

The (co)adjoint representation is then defined by considering the natural linear action of a Lie group on the (co)tangent space at the identity.

In practice, for what concerns us, the Lie groups arise as transformation groups of a manifold and describe the symmetries of mechanical systems whose configuration or phase space is just the manifold on which the group acts. On this manifold some vector fields are defined, which are, in some sense, the velocity fields of the incipient motion caused by the group action. As one might expect, there is a homomorphism from the Lie algebra of the group to the algebra (under Lie bracket) of these induced vector fields.

The classification of the simple Lie groups, i.e., orthogonal, special unitary and symplectic, is recalled.

Finally, the definitions of connection on a principal bundle and of gauge transformation are given.

1.4.3 Lagrangian Dynamics

The Lagrange Equations are proved starting from the fundamental Newton law of dynamics. Then we show that they can be derived from the variational Hamilton principle.

The several conservation laws of mechanics (for example that of the linear or angular momentum and of the energy) are particular cases of the same general theorem, due to Noether: to every one-parameter group of diffeomorphisms of the configuration space, leaving invariant the Lagrangian, corresponds a first integral of the equations of motion. Moreover, a reduced Lagrangian can be defined, exhibiting a first elementary example of reduction by symmetry. A particular case is that of invariance with respect to time translation, to which corresponds energy conservation, while the reduced Lagrangian gives rise to the Maupertuis principle.

1.4.4 Hamiltonian Dynamics

If the arena of Lagrangian dynamics is the tangent bundle of the configuration space, that of Hamiltonian dynamics is the cotangent one. The passage from one formulation to the other is given by the Legendre transformation.

The classical arguments regarding canonical transformations and the Hamilton–Jacobi integration method are reviewed, and are followed by a discussion of their geometrical meaning.

After defining the moment map (a sort of generalized angular or linear momentum) the Noether Theorem is extended, in a Hamiltonian environment, to the case of a generic Lie group (not necessarily 1-dimensional) acting on a symplectic manifold. The components of the moment map turn out to be first integrals of a Hamiltonian invariant under the action of the Lie group. The notion of reduced symplectic manifold is introduced. The reduced Hamiltonian is defined, which generates a dynamics on the reduced symplectic manifold. The generalization to Poisson manifolds is given.

If n first integrals of an n degrees of freedom Hamiltonian system are known, then the search for the general solution is reduced to quadratures. This is the content of the Liouville Theorem, while the Arnold Theorem states that the topology of the level surfaces of the first integrals is, if compact and connected, that of an n-dimensional torus. The more general case of degenerate systems is considered. Lastly, monodromy is defined: it describes the twisting of the tori when one moves along a circle in the space of the first integrals.

Part I

Elementary Theory

CHAPTER 2

Basic Facts

In this chapter basic facts on the Kepler Problem, i.e., equations of motion and their solutions for various values of the energy, are recalled. But first of all we will review the geometrical and analytical properties of the conics.

2.1 Conics

Conics are plane curves for which the following three equivalent definitions can be given.

DEFINITION 2.1 *The conics are the plane section of a double-sheeted circular cone.*

DEFINITION 2.2 *The conics are loci on the plane defined by their focal properties. More exactly we have:*

(i) The ellipse is the locus of the points such that the sum of the distances from two fixed points (the foci) is constant.

(ii) The hyperbola is the locus of the points for which the difference of the distances from two fixed points (the foci) is constant.

(iii) The parabola is the locus of those points whose distances from a fixed point (the focus) and a straight line (the directrix) are equal.

DEFINITION 2.3 *The conics are the locus for which the ratio between the distances from a fixed point (the focus) and a fixed straight line (the directrix) is a constant E called eccentricity. For* $0 < E < 1$ *the conic is an ellipse, for* $E = 1$ *a parabola and for* $E > 1$ *a hyperbola. The circumference is recovered as limit case, letting the directrix go to infinity and E to zero.*

A simple geometrical argument shows that the first two definitions are equivalent; see, for example, Apostol (1969).

The third definition is the most suitable to obtain the equation of a conic and to study its properties. Let P be the generic point belonging to the conic, F the focus, $\vec{n}$ a unit vector orthogonal to the directrix and d the distance between focus and directrix.

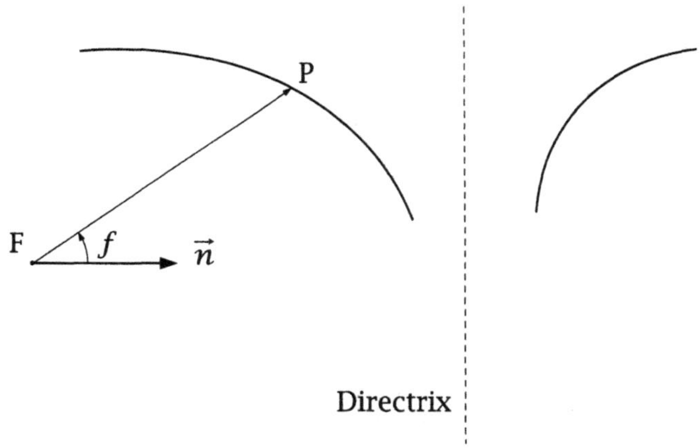

Figure 2.1: Definition of conic. F is the focus, P a generic point of the conic, $\vec{n}$ a unit vector orthogonal to the directrix.

From the definition we have

$$\left\|\overrightarrow{FP}\right\| = E\left|\overrightarrow{FP}\cdot\vec{n} - d\right|,$$

where the dot means "scalar product". Put $\left\|\overrightarrow{FP}\right\| = r$. If P lies in the left half-plane with respect to the directrix, the equation of the conic is

$$r = \frac{Ed}{E\cos f + 1},$$

while, if P lies in the right half-plane,

$$r = \frac{Ed}{E\cos f - 1}.$$

Since $r > 0$, one must exclude the values of f for which the denominator is ≤ 0. In the second case it must be $E > 1$.

Let us find (if it exists) a symmetry point O, i.e., a point such that, if $\overrightarrow{\mathrm{OP}}$ identifies a point belonging to the conic, so does the opposite vector $-\overrightarrow{\mathrm{OP}}$ also.

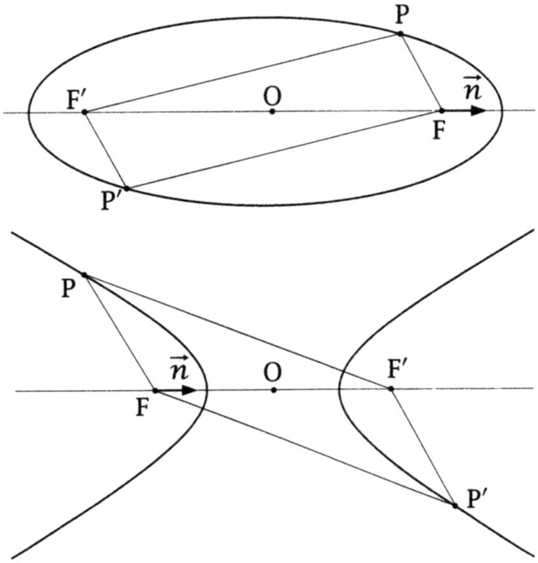

Figure 2.2: For ellipse and hyperbola a symmetry point O does exist. Their focal properties can so be deduced from the third definition.

To this end, we rewrite the conic definition as

$$\left\|\overrightarrow{\mathrm{FP}}\right\| = \left|E\,\overrightarrow{\mathrm{OP}} \cdot \vec{n} - \alpha\right|, \tag{2.1.1}$$

where

$$\alpha = E\,\overrightarrow{\mathrm{OF}} \cdot \vec{n} + Ed.$$

In Equation (2.1.1) put $\overrightarrow{\mathrm{FP}} = \overrightarrow{\mathrm{OP}} - \overrightarrow{\mathrm{OF}}$ and square both the terms. The resulting equation must be true also after the substitution $\overrightarrow{\mathrm{OP}} \to -\overrightarrow{\mathrm{OP}}$. Take the two equations so obtained and subtract one from the other. It yields

$$\overrightarrow{\mathrm{OP}} \cdot \left(\overrightarrow{\mathrm{OF}} - E\alpha\vec{n}\right) = 0.$$

In order that this equation be satisfied for every point of the curve, the point O must satisfy the relation

$$\overrightarrow{\mathrm{OF}} = E\alpha\vec{n}, \tag{2.1.2}$$

or, equivalently, $\overrightarrow{OF} \cdot \vec{n} = E\alpha$. From the definition of α we obtain

$$\alpha = \frac{Ed}{1-E^2}, \quad \overrightarrow{OF} = \frac{E^2 d}{1-E^2}\vec{n}.$$

Thus, a symmetry point exists if and only if $E \neq 1$, i.e., for ellipse and hyperbola. Notice that $\alpha > 0$ if $E < 1$. Put $a = |\alpha|$. Substituting Equation (2.1.2) into the square of Equation (2.1.1) one obtains

$$\left\|\overrightarrow{OP}\right\|^2 + E^2 a^2 = E^2 \left(\overrightarrow{OP} \cdot \vec{n}\right)^2 + a^2,$$

or, in Cartesian coordinates, the well known equation

$$\frac{x^2}{a^2} + \frac{y^2}{a^2(1-E^2)} = 1.$$

Now we can deduce the focal properties of the conics. Bearing in mind Figure 2.2, Equation (2.1.1) for the ellipse reads

$$\left\|\overrightarrow{FP}\right\| = a - E\,\overrightarrow{OP} \cdot \vec{n},$$

and, for the symmetry,

$$\left\|\overrightarrow{FP'}\right\| = a + E\,\overrightarrow{OP} \cdot \vec{n}.$$

Summing term to term, since $\left\|\overrightarrow{FP'}\right\| = \left\|\overrightarrow{F'P}\right\|$, we obtain

$$\left\|\overrightarrow{FP}\right\| + \left\|\overrightarrow{F'P}\right\| = 2a,$$

i.e., the focal defining property of the ellipse. For the hyperbola, Equation (2.1.1) reads

$$\left\|\overrightarrow{FP}\right\| = -a - E\,\overrightarrow{OP} \cdot \vec{n}$$

and, for the symmetry,

$$\left\|\overrightarrow{FP'}\right\| = a - E\,\overrightarrow{OP} \cdot \vec{n}.$$

Subtracting term by term we obtain

$$\left\|\overrightarrow{F'P}\right\| - \left\|\overrightarrow{FP}\right\| = 2a,$$

i.e., the focal defining property of the hyperbola. This shows equivalence between the second and third definitions.

2.2 Properties of the Keplerian Motion

As said in the introductory chapter, the Kepler Problem is concerned with the motion of a particle in a force field which decreases with the square of the distance from a fixed point F. The Hamiltonian is

$$H = \frac{1}{2}p^2 - \frac{1}{q}, \quad p^2 = \vec{p} \cdot \vec{p}, \quad q = \sqrt{\vec{q} \cdot \vec{q}},$$

and the equations of motion are

$$\frac{d\vec{q}}{dt} = \vec{p}, \quad \frac{d\vec{p}}{dt} = -\frac{\vec{q}}{q^3}.$$

It is a typical fact for the Hamiltonian systems that the integrals of motion play a fundamental role, so let us investigate them in our case. One can easily verify, with a direct calculation, that the integrals of motion are:

(i) the Hamiltonian, due to the fact that it is time-independent;

(ii) the angular momentum vector $\vec{G} = \vec{q} \times \vec{p}$, due to the rotational symmetry (Kepler's second law);

(iii) the eccentricity vector $\vec{E} = \vec{p} \times \vec{G} - \frac{\vec{q}}{q}$, due to ...? A simple and direct answer to this question does not exist, and in fact this is the starting point for the geometrical analysis of the Kepler Problem, as we shall see in Part II. For the moment, we limit ourself to simply ascertain the mere fact.

We have seven integrals of motion, but, since the system has six degrees of freedom in the phase space, only five can be independent. In other words, two relations must exist among them. Indeed

(i) $\vec{G} \cdot \vec{E} = 0$, which follows from: $\vec{a} \times \vec{b} \cdot \vec{c} = 0$ when two of the three vectors $\vec{a}, \vec{b}, \vec{c}$ are parallel;

(ii) $E^2 - 1 = 2HG^2$, where we adopt the convention $\left\|\vec{a}\right\| = a$ for a generic vector $\vec{a}$. Remembering that $\left(\vec{a} \times \vec{b}\right) \times \vec{c} = \left(\vec{a} \cdot \vec{c}\right) \vec{b} - \left(\vec{b} \cdot \vec{c}\right) \vec{a}$, we can verify this identity:

$$\begin{aligned} E^2 - 1 &= \left(\vec{p} \times \vec{G}\right) \cdot \vec{p} \times \vec{G} + 1 - 2\frac{\vec{q} \cdot \vec{p} \times \vec{G}}{q} - 1 \\ &= \left(\vec{p} \times \vec{G}\right) \times \vec{p} \cdot \vec{G} - \frac{2}{q}G^2 = \left[p^2\vec{G} - \left(\vec{G} \cdot \vec{p}\right) \vec{p}\right] \cdot \vec{G} - \frac{2}{q}G^2 \\ &= \left(p^2 - \frac{2}{q}\right) G^2 = 2HG^2. \end{aligned}$$

The conservation of the angular momentum ensures that the motion belongs to a plane. Let us find the equation of the orbit. To this end, take the scalar product of $\vec{E}$ and $\vec{q}$. Putting $\vec{E} \cdot \vec{q} = Eq\cos f$, where f is said to be the *true anomaly*, from the definition of $\vec{E}$ we obtain

$$q(f) = \frac{G^2}{E\cos f + 1}. \tag{2.2.1}$$

This is the well known (generalization of the) first Kepler law: the orbit is a conic. It explains why we have chosen the symbol $\vec{E}$ and its name, and clearly shows that to $f = 0$ corresponds the pericentre, where $\vec{E}$ and $\vec{q}$ have the same direction. The relation $E^2 - 1 = 2HG^2$ shows that the sign of H brings about the sign of $E^2 - 1$, thus the nature of the conic. Notice that, for the hyperbola, only the left branch is described by (2.2.1). As we shall see at the end of this chapter, the right branch corresponds to the orbit of a particle in the *repulsive* potential.

Now we integrate the Hamilton Equations of motion. Let us discuss the three cases: $H < 0$, $H > 0$, $H = 0$[1] separately. In each of the three cases we make a far reaching change in the independent variable, introducing the *eccentric anomaly* s.

2.2.1 Energy $H < 0$

Put

$$\frac{dt}{d\mathrm{s}} = \frac{q}{\sqrt{-2H}}. \tag{2.2.2}$$

We obtain

$$\frac{d^2}{d\mathrm{s}^2} = \frac{1}{-2H}\left(q\frac{dq}{dt}\frac{d}{dt} + q^2\frac{d^2}{dt^2}\right)$$

and therefore, from the equations of motion,

$$\frac{d^2\vec{q}}{d\mathrm{s}^2} = \frac{1}{-2H}\left(q\frac{dq}{dt}\frac{d\vec{q}}{dt} - \frac{\vec{q}}{q}\right).$$

Owing to the elementary vector identity already used and to the first equation of motion, the eccentricity vector can be written as

$$\vec{E} = p^2\vec{q} - \frac{\vec{q}}{q} - \left(\vec{q}\cdot\frac{d\vec{q}}{dt}\right)\frac{d\vec{q}}{dt},$$

moreover

$$\vec{q}\cdot\frac{d\vec{q}}{dt} = \frac{1}{2}\frac{d}{dt}\left(\vec{q}\cdot\vec{q}\right) = q\frac{dq}{dt}.$$

(1) Here and in the sequel, with a slight abuse of notation, we do not distinguish between a function and its numerical value.

Taking into account the definition of H yields

$$\vec{E} = 2H\frac{d^2\vec{q}}{d\mathrm{s}^2} + 2H\vec{q},$$

or equivalently

$$\frac{d^2\vec{q}}{d\mathrm{s}^2} + \vec{q} = \frac{\vec{E}}{2H}. \tag{2.2.3}$$

We have thus transformed the second order equation of motion in a *linear and regular* one: the introduction of the eccentric anomaly s allows us to eliminate the singularity in the origin.

Equation (2.2.3) is that of a harmonic oscillator with a constant forcing term and can thus be easily integrated. The general solution is

$$\vec{q}\,(\mathrm{s}) = \vec{C}_1 \cos \mathrm{s} + \vec{C}_2 \sin \mathrm{s} + \frac{\vec{E}}{2H},$$

where $\vec{C}_1$ and $\vec{C}_2$ are integration constants, which we now determine as functions of $H, \vec{G}$ and $\vec{E}$. Define t_p as the *time of passage at the pericentre*, or, in other words, $t = t_p$ when $f = 0$. Put $\mathrm{s} = 0$ when $t = t_p$. Thus

$$\vec{q}\,(0) = \vec{C}_1 + \frac{\vec{E}}{2H}, \quad \frac{d\vec{q}\,(0)}{d\mathrm{s}} = \vec{C}_2.$$

Since, from the orbit Equation (2.2.1), for $f = 0$ we have $q\,(0) = \frac{G^2}{E+1}$, we can also write

$$\vec{q}\,(0) = \frac{G^2}{E+1}\frac{\vec{E}}{E}.$$

Equating the two expressions for $\vec{q}\,(0)$ and recalling the second relation between the first integrals we obtain

$$\vec{C}_1 = \frac{1}{-2H}\frac{\vec{E}}{E}.$$

For the second constant, we notice that at the pericentre the velocity vector is orthogonal to the position vector and therefore $q\,(0)\,\frac{dq(0)}{dt} = G$. Taking into account (2.2.2), this can be written as

$$\frac{dq\,(0)}{d\mathrm{s}} = \frac{G}{\sqrt{-2H}},$$

which, in vectorial form, becomes

$$\frac{d\vec{q}\,(0)}{d\mathrm{s}} = \frac{1}{\sqrt{-2H}}\vec{G} \times \frac{\vec{E}}{E} = \vec{C}_2.$$

Therefore, the general solution of Equation (2.2.3) is

$$\vec{q}\,(\mathfrak{s}) = \frac{1}{-2H}\frac{\vec{E}}{E}\cos\mathfrak{s} + \frac{1}{\sqrt{-2H}}\vec{G}\times\frac{\vec{E}}{E}\sin\mathfrak{s} - \frac{\vec{E}}{-2H}. \tag{2.2.4}$$

Owing to regularization the collision orbits, which are characterized by $\vec{G} = 0$ and $E = 1$, are now allowed and the particle "bounces back" when it reaches the origin. Notice that the circular orbits, instead, correspond to $\vec{E} = 0$ with $\frac{\vec{E}}{E} = \vec{n}$ an arbitrary unit vector.

Equating the norm of both sides of (2.2.4), we obtain

$$q\,(\mathfrak{s}) = a\,(1 - E\cos\mathfrak{s})\,, \quad \text{where } a = \frac{1}{-2H}. \tag{2.2.5}$$

This shows that a is the semimajor axis of the ellipse. From the Definition (2.2.2) we have

$$\frac{dt}{d\mathfrak{s}} = a^{\frac{3}{2}}\,(1 - E\cos\mathfrak{s})\,,$$

and, integrating,

$$t - t_p = a^{\frac{3}{2}}\,(\mathfrak{s} - E\sin\mathfrak{s})\,, \tag{2.2.6}$$

which is the celebrated *Kepler Equation.* To express the position vector $\vec{q}$ as a function of the time (which is a natural wish) one should invert this equation and substitute $\mathfrak{s}(t)$ into the orbit Equation (2.2.4). Unfortunately, Equation (2.2.6) is a transcendental one and can not be inverted in a closed form; in the appendix to this chapter the problem of the inversion is briefly reviewed. An immediate consequence of this equation is third Kepler's law:

$$\mathfrak{T} = 2\pi a^{\frac{3}{2}},$$

where $\mathfrak{T}$ is the period. It is usual to define

$$\ell = \mathfrak{s} - E\sin\mathfrak{s},$$

and to call it *mean anomaly* .

2.2.2 Energy $H > 0$

Put

$$\frac{dt}{d\mathfrak{s}} = \frac{q}{\sqrt{2H}}.$$

Proceeding as in the case $H < 0$ we obtain

$$\frac{d^2\vec{q}}{d\mathfrak{s}^2} - \vec{q} = -\frac{\vec{E}}{2H}.$$

The general solution is

$$\vec{q}\,(\mathfrak{s}) = \vec{C}_1 \cosh \mathfrak{s} + \vec{C}_2 \sinh \mathfrak{s} + \frac{\vec{E}}{2H},$$

from which

$$\vec{q}\,(\mathfrak{s}) = -\frac{1}{2H}\frac{\vec{E}}{E}\cosh \mathfrak{s} + \frac{1}{\sqrt{2H}}\vec{G} \times \frac{\vec{E}}{E}\sinh \mathfrak{s} + \frac{\vec{E}}{2H}. \tag{2.2.7}$$

Moreover

$$q\,(\mathfrak{s}) = a\,(E \cosh \mathfrak{s} - 1)\,, \quad a = \frac{1}{2H},$$

$$t - t_p = a^{\frac{3}{2}}\,(E \sinh \mathfrak{s} - \mathfrak{s})\,,$$

$$\ell = E \sinh \mathfrak{s} - \mathfrak{s},$$

in full analogy with the previous case.

2.2.3 Energy $H = 0$

In this case the "scale factor" can not be $\frac{1}{\sqrt{|2H|}}$, but turns out to be the norm of the angular momentum. Indeed, putting

$$\frac{dt}{d\mathfrak{s}} = Gq$$

we obtain

$$\frac{d^2\vec{q}}{d\mathfrak{s}^2} = -G^2\vec{E}.$$

The general solution is

$$\vec{q}\,(\mathfrak{s}) = \vec{C}_1 + \vec{C}_2\mathfrak{s} - \frac{1}{2}G^2\vec{E}\mathfrak{s}^2.$$

Proceeding as in the previous cases and remembering that now $E = 1$ one obtains

$$\vec{q}\,(\mathfrak{s}) = -\frac{1}{2}G^2\vec{E}\mathfrak{s}^2 + G\vec{G} \times \vec{E}\mathfrak{s} + \frac{1}{2}G^2\vec{E}. \tag{2.2.8}$$

Moreover

$$q\,(\mathfrak{s}) = \frac{1}{2}G^2\left(1 + \mathfrak{s}^2\right),$$

$$t - t_p = \frac{1}{2}G^3\left(\mathfrak{s} + \frac{1}{3}\mathfrak{s}^3\right),$$

$$\ell = \mathfrak{s} + \frac{1}{3}\mathfrak{s}^3.$$

In this third case the function $t\,(\mathfrak{s})$ can be inverted exactly.

2.3 The Three Anomalies

We have therefore defined the three anomalies, namely

(i) the *true anomaly* f,

(ii) the *eccentric anomaly* $\mathfrak{s}$,

(iii) the *mean anomaly* ℓ.

Now we find the analytical and geometrical relations between them. Let us introduce a system of orthogonal coordinates with the origin in the centre of the motion, the X-axis directed as $\vec{E}$ and the Y-axis in such a way that X,Y and $\vec{G}$ form a right oriented system: the plane XY is thus the plane of the orbit.

2.3.1 Energy $H < 0$

In the plane of the orbit the 3-dimensional equation of motion (2.2.4) becomes

$$\begin{cases} X = a\cos\mathfrak{s} - aE \\ Y = a\sqrt{1-E^2}\sin\mathfrak{s}, \end{cases}$$

which shows at once the geometrical meaning of the eccentric anomaly: see Figure 2.3.

Comparing Equations (2.2.1) and (2.2.5) yields

$$\cos\mathfrak{s} = \frac{\cos f + E}{E\cos f + 1}, \quad \sin\mathfrak{s} = \frac{\sqrt{1-E^2}\sin f}{E\cos f + 1}, \tag{2.3.1}$$

or, in a more usual form,

$$\tan\frac{\mathfrak{s}}{2} = \sqrt{\frac{1-E}{1+E}}\tan\frac{f}{2}.$$

Differentiating the first expression of the (2.3.1) and taking into account the other one, we find the relation

$$\frac{df}{d\mathfrak{s}} = \frac{G}{\sqrt{-2H}}\frac{1}{q}. \tag{2.3.2}$$

2.3.2 Energy $H > 0$

The 3-dimensional equation of motion becomes

$$\begin{cases} X = aE - a\cosh\mathfrak{s} \\ Y = a\sqrt{E^2-1}\sinh\mathfrak{s}, \end{cases}$$

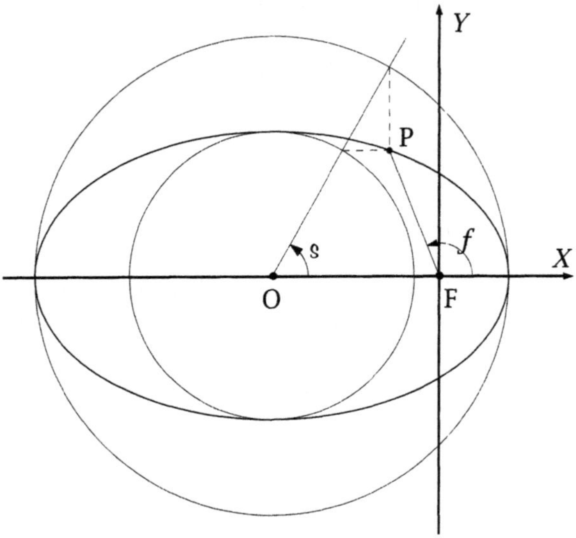

Figure 2.3: The geometrical meaning of the eccentric anomaly. The radius of the major circle is a, that of the minor is $a\sqrt{1-E^2}$ and the distance $\overline{\mathrm{OF}}$ is aE.

and, as in the previous case,

$$\tanh\frac{\mathfrak{s}}{2} = \sqrt{\frac{E-1}{E+1}}\tan\frac{f}{2}.$$

A geometrical interpretation of $\mathfrak{s}$ could be given in this case also, with equilateral hyperbolas replacing the circles of Figure 2.3.

2.3.3 Energy $H = 0$

The 3-dimensional equation of motion becomes

$$\begin{cases} X = \frac{1}{2}G^2\,(1-\mathfrak{s}^2) \\ Y = G^2\mathfrak{s}, \end{cases}$$

and, as in the previous cases,

$$\mathfrak{s} = \tan\frac{f}{2}.$$

In all three cases, we can write $\mathfrak{s}$ as a function of f, which, composed with the definition of ℓ, gives ℓ as a function of f.

2.4 The Elements of the Orbit for $H < 0$

The three Equations (2.2.4), (2.2.7) and (2.2.8), with the related Kepler Equation, completely describe the motion in the three cases $H \lesseqqgtr 0$. Let us restrict to the case $H < 0$. Notice that as integration constants we do not use the components of the initial velocity and position, but rather t_p, $\vec{G}$ and $\vec{E}$, i.e., seven quantities subjected to the constraint $\vec{G} \cdot \vec{E} = 0$. This choice is natural and presents great advantages. In fact, whereas the initial velocity and position do not suggest in any way neither the shape nor the orientation of the orbit (only the plane of the trajectory is immediately recognized), the two vectors $\vec{G}$ and $\vec{E}$ describe the orbit. Moreover, the mapping between orbits and pairs of these vectors is one-to-one.

Owing to this last property, we can immediately detect the topology of the manifold of the orbits for any fixed negative energy, once the problem has been regularized, thus allowing the case $G^2 = 0$ also. Write the second relation between the first integrals as

$$G^2 + \frac{E^2}{-2H} = \frac{1}{-2H}.$$

Then the two vectors $\vec{G} \pm \frac{\vec{E}}{\sqrt{-2H}}$ span two 2-dimensional spheres of radius $\frac{1}{\sqrt{-2H}}$, showing that the topology we seek is $S^2 \times S^2$.

But it is for the perturbation theory that this choice is particularly useful, since in this case $\vec{G}$ and $\vec{E}$ are no longer constants, and the "small" averaged action of other celestial bodies brings about a variation that is slow compared with the fast revolutionary motion. Thus we have the following picture: the planet rotates very fast on its orbit, and the perturbations cause a slow change in the geometry of this orbit. Clearly, describing the action of these perturbations, following in time the small variations of position and velocity, is much less meaningful.

One might find it disturbing that $\vec{G}$ and $\vec{E}$ are not independent. As a matter of fact, one can introduce five *independent* quantities to parametrize the set of the orbits, but in this way the one-to-one correspondence between the orbits and the sets of parameters is inevitably lost. This is a mere topological fact: the orbit manifold of the Kepler Problem is not homeomorphic to $\mathbb{R}^5$ and hence can not be covered with one coordinate system only[(2)]. The situation is, for example, very similar to that, more intuitive, of the parametrization of a sphere: a point can be characterized either with a vector from the centre or with the two local coordinates latitude and longitude. In the former case all points of the sphere are labeled, but the three components of the vector must satisfy

[(2)] The topology of the Kepler Problem will be investigated in Part II.

the constraint of having a constant norm; in the latter the two coordinates are free, but for the two poles the longitude is not unambiguously defined.

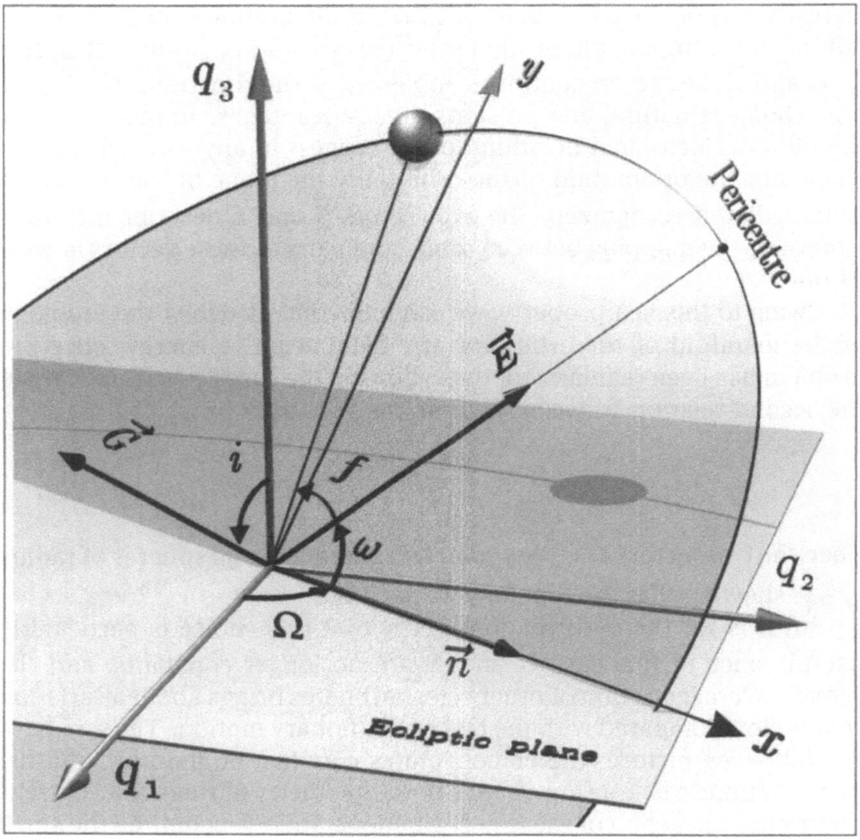

Figure 2.4: The elements of the orbit.

Let $\check{q}_1, \check{q}_2$ and $\check{q}_3$ be the orthonormal vectors of the $q_1q_2q_3$ axes: $\check{q}_i \cdot \check{q}_k = \delta_{ik}$. For $H < 0$, the most popular choice of the six independent parameters characterizing the motion is the following:

- the semimajor axis a;
- the eccentricity E;
- the longitude of the ascending node Ω;
- the inclination i;

- the argument of the pericentre ω;
- the time of passage at the pericentre t_p.

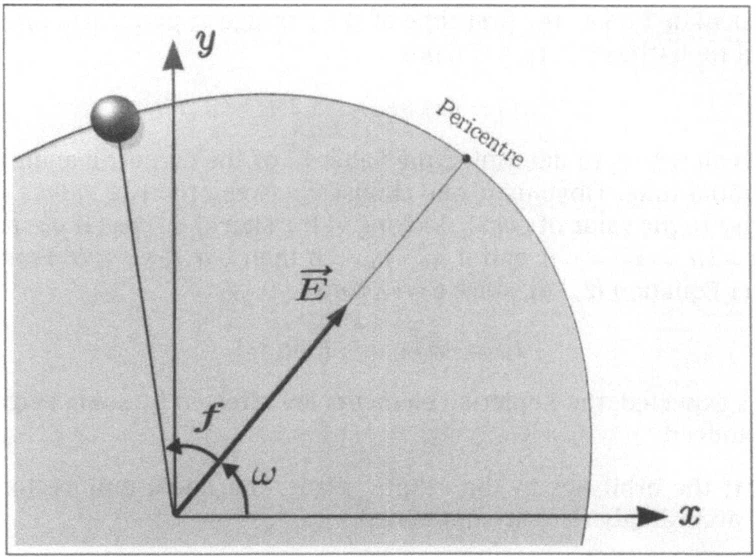

Figure 2.5: On the plane of the orbit.

See Figure 2.4 and Figure 2.5. The first two parameters fix the shape of the orbit. The following three are the Euler angles of the orbit and thus fix its orientation in space. The last one fixes the position of the point on the orbit. These six parameters are sometimes called *Kepler elements*, and can be easily expressed in terms of the initial position $\vec{q_0}$ and velocity $\vec{p_0}$ at time $t = 0$.

To this end, calculate from the very definitions firstly H, $\vec{G}$ and $\vec{E}$ in terms of the initial position and velocity. Then we can write

$$a = \frac{1}{-2H} = \frac{G^2}{1 - E^2}, \quad \cos i = \frac{G_3}{G}.$$

The *nodal unit vector* $\vec{n}$ (which determines the *node line*) is orthogonal to the projection of $\vec{G}$ onto the *ecliptic plane* $q_1 q_2$, thus

$$\vec{n} = \frac{-G_2 \check{q}_1 + G_1 \check{q}_2}{\sqrt{G_1^2 + G_2^2}},$$

from which

$$\begin{aligned} \cos\Omega &= \vec{n}\cdot\check{q}_1, \quad \sin\Omega = \vec{n}\cdot\check{q}_2, \\ \cos\omega &= \vec{n}\cdot\frac{\vec{E}}{E}, \quad \sin\omega = \vec{n}\times\frac{\vec{E}}{E}\cdot\frac{\vec{G}}{G}. \end{aligned} \tag{2.4.1}$$

To calculate t_p, i.e., the first time of the passage at pericentre *after* the initial time (thus $0 \le t_p < \mathcal{T}$), put

$$q_0 = a(1 - E\cos\mathfrak{s}_0), \quad -2\pi < \mathfrak{s}_0 \le 0,$$

which allows us to determine the value $\mathfrak{s}_0$ of the eccentric anomaly at the initial time. Obviously, one chooses between the two values of $\mathfrak{s}_0$, relative to the value of $\cos\mathfrak{s}_0$, looking at the sign of $\vec{q_0}\cdot\vec{p_0}$: if $\vec{q_0}\cdot\vec{p_0} > 0$ then $-2\pi < \mathfrak{s}_0 < -\pi$ and if $\vec{q_0}\cdot\vec{p_0} < 0$ then $-\pi < \mathfrak{s}_0 < 0$. From the Kepler Equation (2.2.6), since $\mathfrak{s} = \mathfrak{s}_0$ for $t = 0$,

$$t_p = -a^{\frac{3}{2}}(\mathfrak{s}_0 - E\sin\mathfrak{s}_0).$$

As expected, the Keplerian elements are affected by some singularities. Indeed

(i) if the orbit lies in the ecliptic plane, the nodal unit vector and accordingly Ω, ω are undefined;

(ii) if the orbit is circular, $\vec{E} = 0$ and ω, t_p are undefined;

(iii) if $\vec{G} = 0$, the motion is linear, the plane of the orbit is undefined and so are i, Ω, ω.

Studying our solar system we are very close to the cases (i) and (ii), since eccentricity and inclination of the planets are very small. To avoid the singularities one could pass to another system of elements, for example to

$$\begin{aligned} k &= E\cos(\Omega+\omega), \quad h = E\sin(\Omega+\omega), \\ r &= \sin\tfrac{i}{2}\cos\Omega, \quad\quad s = \sin\tfrac{i}{2}\sin\Omega, \end{aligned} \tag{2.4.2}$$

in addition to a and t_p. However, the radical cure is to use global, though constrained, parameters, as we show in Part III.

2.5 The Repulsive Potential

If the force acting on the particle is repulsive, the potential changes sign and the Hamiltonian becomes

$$H_+ = \frac{1}{2}p^2 + \frac{1}{q}$$

so that H_+ is always positive.

We can repeat the previous study with some minor change of sign only. Indeed, as is verified with straightforward calculations, the integrals of motion are $H_+, \vec{G}$ (obviously) and

$$\vec{E}_+ = \vec{p} \times \vec{G} + \frac{\vec{q}}{q}.$$

The two identities among $H_+, \vec{G}, \vec{E}_+$ remain unchanged but the equation of the orbit becomes

$$q(f) = \frac{G^2}{E \cos f - 1},$$

showing that the orbit is the right branch of the hyperbola. Putting

$$\frac{dt}{ds} = \frac{q}{\sqrt{2H_+}}$$

and proceeding as before, we obtain

$$\frac{d^2\vec{q}}{ds^2} - \vec{q} = -\frac{\vec{E}}{2H_+},$$

which is the same equation found for the attractive potential in the case $H > 0$. The general solution is

$$\vec{q}(s) = \frac{1}{2H_+}\frac{\vec{E}_+}{E_+}\cosh s + \frac{1}{\sqrt{2H_+}}\vec{G} \times \frac{\vec{E}_+}{E_+}\sinh s + \frac{\vec{E}_+}{2H_+}.$$

The other formulas then easily follow.

APPENDIX

2.A The Kepler Equation

It is obvious that for $E < 1$ the Kepler Equation, which we write in the form $\ell = \mathfrak{s} - E \sin \mathfrak{s}$, can be in principle unambiguously inverted to find $\mathfrak{s} = s(\ell)$, with E as parameter. Unfortunately, this can not be achieved in a closed form. The problem has a three centuries long history; see Colwell (1993) for a detailed report with an extensive bibliography.

To find the function $s(\ell)$ in a form suitable for the calculations we may proceed in two ways:

(i) expand the difference $\mathfrak{s} - \ell$ for fixed value of E in a Fourier series in the variable ℓ with coefficients depending on E;

(ii) write $s(\ell)$ as a power series in the eccentricity E with coefficients depending on ℓ.

The first approach gives

$$\mathfrak{s} = \ell + 2 \sum_{k=1}^{\infty} \frac{J_k(kE)}{k} \sin k\ell \tag{2.A.1}$$

where

$$J_k(z) = \frac{1}{\pi} \int_0^{\pi} \cos(kx - z \sin x)\, dx = \sum_{j=0}^{\infty} \frac{(-1)^j z^{k+2j}}{2^{k+2j} j!(k+j)!}, \quad k = 0, 1, \ldots$$

are the Bessel functions of order k. The proof of the expansion is a straightforward calculation. If

$$s(\ell) - \ell = \sum_{k=1}^{\infty} A_k \sin k\ell,$$

then for $k = 1, 2, 3, \ldots$

$$\begin{aligned} A_k &= \frac{2}{\pi} \int_0^{\pi} [s(\ell) - \ell] \sin k\ell \, d\ell = -\frac{2}{k\pi} \int_0^{\pi} [s(\ell) - \ell]\, d(\cos k\ell) \\ &= -\frac{2}{k\pi} [s(\ell) - \ell] \cos k\ell \Big|_0^{\pi} + \frac{2}{k\pi} \int_0^{\pi} \cos k\ell [s'(\ell) - 1]\, d\ell \\ &= \frac{2}{k\pi} \int_0^{\pi} \cos k\ell s'(\ell)\, d\ell = \frac{2}{k\pi} \int_0^{\pi} \cos k\ell \, ds(\ell). \end{aligned}$$

Since $\ell \equiv s(\ell) - E \sin s(\ell)$,

$$\begin{aligned} A_k &= \frac{2}{k\pi} \int_0^{\pi} \cos[ks(\ell) - kE \sin s(\ell)]\, ds(\ell) \\ &= \frac{2}{k} \frac{1}{\pi} \int_0^{\pi} \cos[k\mathfrak{s} - kE \sin \mathfrak{s}]\, d\mathfrak{s} = \frac{2}{k} J_k(kE), \end{aligned}$$

leading to (2.A.1).

In the second approach one uses the well known Lagrange formula for the local inversion of the holomorphic functions, which he in fact discovered precisely in his attempts to invert the Kepler Equation (see for example Goursat (1959)), to obtain

$$\mathfrak{s} = \ell + \sum_{k=1}^{\infty} a_k(\ell) E^k, \quad a_k(\ell) = \frac{d^{k-1}}{d\ell^{k-1}} \sin^k \ell. \tag{2.A.2}$$

The functions $a_k(\ell)$ are trigonometric polynomials in the mean anomaly. In Fleckenstein (1941) a sequence of operational steps leading from (2.A.2) to (2.A.1) is described.

As shown by Cauchy, series (2.A.2) converges for $E < \frac{2\sqrt{x(1-x)}}{1-2x}$, where x is the solution of the transcendental equation $\frac{1-x}{x} = \exp(\frac{2}{1-2x})$, thus for $E < 0.66274\ldots$.

CHAPTER 3

Separation of Variables and Action–Angle Coordinates

The separation of variables in the Hamilton–Jacobi Equation is the main tool for the integration of a general mechanical system. In this chapter we study all the orthogonal coordinate systems for which the Kepler Problem is separable. There exist in fact *four* coordinate systems with this property: the spherical, the parabolic, the elliptic and the sphero-conical coordinates[(1)]. In Part II we will see *why* these, and only these, coordinate systems have the separability property, but for the moment we only verify, by an explicit calculation, that the method generates the four triplets of first integrals in involution.

Taking the three integrals as new momenta in the canonical formalism, the Hamilton–Jacobi method then gives, after three integrations, the conjugate coordinates. Instead of the integrals directly generated by the separation process, one can take as new momenta a combination of them, and in particular the so-called *action coordinates*. It turns out that the corresponding conjugate coordinates are *angle* variables. The action–angle variables are, in some sense, very natural for the topology of the integrable systems, as elucidated by the Arnold Theorem. We will find them for the first two of the above coordinate systems: from the spherical coordinates we will obtain the well known *Delaunay variables*, and from the parabolic coordinates the *Pauli variables*. As will become

[(1)] These names are not universally adopted.

clear in Part III, both these action-angle coordinate systems are needed to cover, in a convenient way, the phase space of the Kepler Problem with axisymmetric perturbations.

We point out that, for the Kepler Problem with negative energy, there is a strict relation between the periodicity of all the orbits (degeneration) and its over-integrability: see Subsection D.4.2. Moreover, over-integrability involves, in general, separability in several coordinate systems of the configuration space: see the Eisenhart Theorem.

3.1 Separation of Variables

It is well known that the Hamilton-Jacobi Equation for the free particle in three-dimensional Euclidean space is separable in eleven coordinate systems (Morse & Feshbach 1953). Thus, all we have to do is to verify, for each of these coordinate systems, whether adding the Kepler potential preserves the separability. As a final result, we find just the four coordinate systems already mentioned. Let us study them in some detail.

3.1.1 Spherical Coordinates

One can pass from the Cartesian q_1, q_2, q_3 to the spherical coordinates q, ϑ, φ with

$$q_1 = q \sin\vartheta \cos\varphi, \quad q_2 = q \sin\vartheta \sin\varphi, \quad q_3 = q\cos\vartheta,$$
$$0 < q < \infty, \quad 0 \le \vartheta \le \pi, \quad 0 \le \varphi < 2\pi.$$

Let us denote, as usual, the time derivative with a dot. A straightforward calculation shows that the square of the velocity is

$$v^2 = \dot{q}^2 + q^2\dot{\vartheta}^2 + q^2 \sin^2\vartheta\, \dot{\varphi}^2$$

and the canonical momenta are

$$p_q = \dot{q}, \quad p_\vartheta = q^2\dot{\vartheta}, \quad p_\varphi = q^2 \sin^2\vartheta\, \dot{\varphi}.$$

The Hamiltonian of the Kepler Problem is

$$H = \frac{1}{2}\left(p_q^2 + \frac{p_\vartheta^2}{q^2} + \frac{p_\varphi^2}{q^2 \sin^2\vartheta}\right) - \frac{1}{q}.$$

The three first integrals in involution are:

(i) the Hamiltonian H, since it does not depend on time;

(ii) the norm of the total angular momentum $G = \sqrt{p_\vartheta^2 + \frac{p_\varphi^2}{\sin^2\vartheta}}$, since the Hamiltonian depends on the variables ϑ, p_ϑ through G only;

(iii) the third component $G_3 = p_\varphi$ of the angular momentum, since H and G do not depend on φ.

Thus H, G, G_3 can be taken as new momenta and the corresponding canonical transformation is generated by the function

$$S(q, \vartheta, \varphi, H, G, G_3) = \int p_q\, dq + p_\vartheta\, d\vartheta + p_\varphi\, d\varphi,$$

where

$$p_q = \pm\sqrt{2H + \frac{2}{q} - \frac{G^2}{q^2}}, \quad p_\vartheta = \pm\sqrt{G^2 - \frac{G_3^2}{\sin^2\vartheta}}, \quad p_\varphi = G_3. \qquad (3.1.1)$$

The problem is thus reduced to quadratures. The explicit integration is postponed to the following section.

3.1.2 Parabolic Coordinates

The parabolic coordinates ξ, η, φ are given by

$$q_1 = \xi\eta\cos\varphi, \quad q_2 = \xi\eta\sin\varphi, \quad q_3 = \frac{1}{2}\left(\xi^2 - \eta^2\right),$$
$$0 \le \xi < \infty, \quad 0 \le \eta < \infty, \quad 0 \le \varphi < 2\pi,$$

from which

$$q = \frac{1}{2}\left(\xi^2 + \eta^2\right), \quad \xi^2 = q + q_3, \quad \eta^2 = q - q_3.$$

The surfaces ξ = constant and η = constant are paraboloids of revolution with q_3 as symmetry axis and the focus placed in the origin. The square of the velocity is

$$v^2 = \left(\xi^2 + \eta^2\right)\left(\dot\xi^2 + \dot\eta^2\right) + \xi^2\eta^2\dot\varphi^2$$

and the conjugate momenta are

$$p_\xi = \left(\xi^2 + \eta^2\right)\dot\xi, \quad p_\eta = \left(\xi^2 + \eta^2\right)\dot\eta, \quad p_\varphi = \xi^2\eta^2\dot\varphi.$$

The Hamiltonian of the Kepler Problem is

$$H = \frac{1}{2}\frac{p_\xi^2 + p_\eta^2}{\xi^2 + \eta^2} + \frac{1}{2}\frac{p_\varphi^2}{\xi^2\eta^2} - \frac{2}{\xi^2 + \eta^2}.$$

As in the previous case, H and p_φ are first integrals of the motion. To separate the variables, multiply the above equation by $\xi^2 + \eta^2$ and reorder the terms, obtaining

$$\left[\frac{1}{2}\left(p_\xi^2 + \frac{p_\varphi^2}{\xi^2}\right) - H\xi^2 - 1\right] + \left[\frac{1}{2}\left(p_\eta^2 + \frac{p_\varphi^2}{\eta^2}\right) - H\eta^2 - 1\right] = 0.$$

Since the first term in the square bracket is a function of ξ, p_ξ and the second term of η, p_η only, and remembering that H and p_φ are constant, the above equation can be satisfied only if

$$\frac{1}{2}\left(p_\xi^2 + \frac{p_\varphi^2}{\xi^2}\right) - H\xi^2 - 1 = -\Gamma_1,$$

$$\frac{1}{2}\left(p_\eta^2 + \frac{p_\varphi^2}{\eta^2}\right) - H\eta^2 - 1 = \Gamma_1,$$

where Γ_1 is some integration constant. To find its dynamical meaning, subtract one of the above expressions from the other so that, taking into account the expression of the Hamiltonian, we find

$$\Gamma_1 = \frac{1}{2}\frac{1}{\xi^2 + \eta^2}\left[\xi^2\left(p_\eta^2 + \frac{p_\varphi^2}{\eta^2}\right) - \eta^2\left(p_\xi^2 + \frac{p_\varphi^2}{\xi^2}\right)\right] - \frac{\xi^2 - \eta^2}{\xi^2 + \eta^2}.$$

A straightforward calculation shows that $\Gamma_1 = E_3$, the third component of the eccentricity vector. The three first integrals in involution are:

(i) the Hamiltonian H;

(ii) the third component E_3 of the eccentricity vector;

(iii) the third component G_3 of the angular momentum.

Thus H, E_3, G_3 can be taken as new momenta and the underlying canonical transformation is generated by the function

$$S(\xi, \eta, \varphi, H, E_3, G_3) = \int p_\xi \, d\xi + p_\eta \, d\eta + p_\varphi \, d\varphi,$$

where

$$p_\xi = \pm\sqrt{2H\xi^2 + 2(1 - E_3) - \frac{p_\varphi^2}{\xi^2}},$$

$$p_\eta = \pm\sqrt{2H\eta^2 + 2(1 + E_3) - \frac{p_\varphi^2}{\eta^2}},$$

$$p_\varphi = G_3.$$

The problem is reduced to the quadratures. We postpone the explicit integrations to the following section.

3.1.3 Elliptic Coordinates

The elliptic coordinates ξ, η, φ are given by

$$\begin{aligned} q_1 &= \sqrt{(\xi^2 - 1)(1 - \eta^2)} \cos\varphi, \\ q_2 &= \sqrt{(\xi^2 - 1)(1 - \eta^2)} \sin\varphi, \\ q_3 &= \xi\eta + 1, \\ 1 \le \xi < \infty, &\quad -1 \le \eta \le 1, \quad 0 \le \varphi < 2\pi, \end{aligned}$$

from which

$$q = \xi + \eta, \quad \varrho = \sqrt{(\xi^2 - 1)(1 - \eta^2)},$$

where ϱ is the distance from the q_3-axis. Let q' be the distance from the point F′ $\equiv (0, 0, 2)$. Since

$$q' = \sqrt{\varrho^2 + (q_3 - 2)^2} = \xi - \eta,$$

the surfaces ξ = constant and η = constant are confocal ellipsoids and hyperboloids with foci F $\equiv (0, 0, 0)$, F' $\equiv (0, 0, 2)$ and q_3 as symmetry rotational axis. The square of the velocity is

$$v^2 = \frac{\xi^2 - \eta^2}{\xi^2 - 1}\dot{\xi}^2 + \frac{\xi^2 - \eta^2}{1 - \eta^2}\dot{\eta}^2 + \left(\xi^2 - 1\right)\left(1 - \eta^2\right)\dot{\varphi}^2$$

and the conjugate momenta are

$$p_\xi = \frac{\xi^2 - \eta^2}{\xi^2 - 1}\dot{\xi}, \quad p_\eta = \frac{\xi^2 - \eta^2}{1 - \eta^2}\dot{\eta}, \quad p_\varphi = \left(\xi^2 - 1\right)\left(1 - \eta^2\right)\dot{\varphi}.$$

The Hamiltonian of the Kepler Problem is

$$H = \frac{1}{2}\left(\frac{\xi^2 - 1}{\xi^2 - \eta^2}p_\xi^2 + \frac{1 - \eta^2}{\xi^2 - \eta^2}p_\eta^2 + \frac{1}{(\xi^2 - 1)(1 - \eta^2)}p_\varphi^2\right) - \frac{1}{\xi + \eta}.$$

As in the previous cases, H and p_φ are first integrals of the motion. To separate the variables, multiply the above equation by $2(\xi^2 - \eta^2)$ so that, by reordering the terms, one obtains

$$\begin{aligned} &\left[\left(\xi^2 - 1\right)p_\xi^2 + \frac{p_\varphi^2}{\xi^2 - 1} - 2H\left(\xi^2 - 1\right) - 2\xi\right] \\ &\qquad + \left[\left(1 - \eta^2\right)p_\eta^2 + \frac{p_\varphi^2}{1 - \eta^2} - 2H\left(1 - \eta^2\right) + 2\eta\right] = 0. \end{aligned}$$

The first term in the square bracket is a function of ξ, p_ξ and the second term of η, p_η only, hence, remembering that H and p_φ are constant, the above equation can be satisfied only if

$$\left(\xi^2 - 1\right) p_\xi^2 + \frac{p_\varphi^2}{\xi^2 - 1} - 2H\left(\xi^2 - 1\right) - 2\xi = -\Gamma_2,$$

$$\left(1 - \eta^2\right) p_\eta^2 + \frac{p_\varphi^2}{1 - \eta^2} - 2H\left(1 - \eta^2\right) + 2\eta = \Gamma_2,$$

where Γ_2 is some integration constant. To find its dynamical meaning, subtract one of the above expressions from the other so that, taking into account the expression of the Hamiltonian, we find

$$\Gamma_2 = -\frac{\left(\xi^2 - 1\right)\left(1 - \eta^2\right)}{\xi^2 - \eta^2}\left(p_\xi^2 - p_\eta^2\right) - \left(\frac{1}{\xi^2 - 1} - \frac{1}{1 - \eta^2}\right) p_\varphi^2 + 2\frac{\xi\eta + 1}{\xi + \eta}.$$

One can verify, with a direct calculation, that $\Gamma_2 = G^2 - 2E_3$. The three first integrals in involution are

(i) the Hamiltonian H;

(ii) the constant generated by the separation process $\Gamma_2 = G^2 - 2E_3$;

(iii) the third component G_3 of the angular momentum.

Thus H, Γ_2, G_3 can be assumed as new momenta and the canonical transformation involved is generated by the function

$$S\left(\xi, \eta, \varphi, H, \Gamma_2, G_3\right) = \int p_\xi \, d\xi + p_\eta \, d\eta + p_\varphi \, d\varphi,$$

where

$$p_\xi = \pm\sqrt{\frac{1}{\xi^2 - 1}\left(-\frac{p_\varphi^2}{\xi^2 - 1} + 2H\left(\xi^2 - 1\right) + 2\xi - \Gamma_2\right)},$$

$$p_\eta = \pm\sqrt{\frac{1}{1 - \eta^2}\left(-\frac{p_\varphi^2}{1 - \eta^2} + 2H\left(1 - \eta^2\right) - 2\eta + \Gamma_2\right)},$$

$$p_\varphi = G_3.$$

The problem is thus reduced to quadratures, and these latter involve elliptic integrals. The explicit integration is performed later, in Chapter 8, as a particular case of the more general problem of the two fixed centers. The program EULER, in MAPLE language, is supplied, which integrates and displays the orbits.

3.1.4 Spheroconical Coordinates

The spheroconical coordinates q, ξ, η are given by

$$\begin{aligned} q_1 &= \frac{q}{b}\sqrt{(b^2+\xi^2)(b^2-\eta^2)}, \\ q_2 &= \frac{q}{ab}\xi\eta, \\ q_3 &= \frac{q}{a}\sqrt{(a^2-\xi^2)(a^2+\eta^2)}, \\ a^2+b^2 &= 1, \quad \xi^2 \le a^2, \quad \eta^2 \le b^2. \end{aligned}$$

The parameters a and b are chosen to satisfy $a^2 + b^2 = 1$ and are to be regarded as constants for a given set of spheroconical coordinates. The coordinate surface q = constant is a sphere and the surfaces ξ = constant and η = constant are cones of elliptic cross-section whose equations are

$$\begin{aligned} \frac{q_1^2}{b^2+\xi^2} + \frac{q_2^2}{\xi^2} - \frac{q_3^2}{a^2-\xi^2} &= 0, \\ -\frac{q_1^2}{b^2-\eta^2} + \frac{q_2^2}{\eta^2} + \frac{q_3^2}{a^2+\eta^2} &= 0. \end{aligned} \tag{3.1.2}$$

See Figure 8.4 on page 178. The square of the velocity is

$$v^2 = \dot{q}^2 + q^2\left(\frac{\xi^2+\eta^2}{(b^2+\xi^2)(a^2-\xi^2)}\dot{\xi}^2 + \frac{\xi^2+\eta^2}{(b^2-\eta^2)(a^2+\eta^2)}\dot{\eta}^2\right)$$

and the conjugate momenta are

$$\begin{aligned} p_q &= \dot{q}, \\ p_\xi &= q^2\frac{\xi^2+\eta^2}{(b^2+\xi^2)(a^2-\xi^2)}\dot{\xi}, \\ p_\eta &= q^2\frac{\xi^2+\eta^2}{(b^2-\eta^2)(a^2+\eta^2)}\dot{\eta}. \end{aligned}$$

The Hamiltonian of the Kepler Problem is

$$\begin{aligned} H = \frac{1}{2}\Bigg[p_q^2 + \frac{1}{q^2}\Bigg(&\frac{(b^2+\xi^2)(a^2-\xi^2)}{\xi^2+\eta^2}p_\xi^2 \\ &+ \frac{(b^2-\eta^2)(a^2+\eta^2)}{\xi^2+\eta^2}p_\eta^2\Bigg)\Bigg] - \frac{1}{q}. \end{aligned}$$

The Hamiltonian H is a first integral of the motion. Another first integral is

$$G^2 = \frac{(b^2+\xi^2)(a^2-\xi^2)}{\xi^2+\eta^2}p_\xi^2 + \frac{(b^2-\eta^2)(a^2+\eta^2)}{\xi^2+\eta^2}p_\eta^2,$$

since H depends on ξ, η, p_ξ, p_η only through this expression. To obtain the third integral, multiply the above equation by $\xi^2 + \eta^2$ and reorder the terms obtaining

$$\left[\left(b^2+\xi^2\right)\left(a^2-\xi^2\right)p_\xi^2 - G^2\xi^2\right] + \left[\left(b^2-\eta^2\right)\left(a^2+\eta^2\right)p_\eta^2 - G^2\eta^2\right] = 0.$$

Since the first term in the square bracket is a function of ξ, p_ξ and the second term of p_η, η only, the above equation can be satisfied only if

$$\left(b^2+\xi^2\right)\left(a^2-\xi^2\right)p_\xi^2 - G^2\xi^2 = \Gamma_3,$$
$$\left(b^2-\eta^2\right)\left(a^2+\eta^2\right)p_\eta^2 - G^2\eta^2 = -\Gamma_3,$$

where Γ_3 is some integration constant. Subtracting the latter from the former and taking into account the expression of G, it results that

$$\Gamma_3 = \frac{1}{\xi^2+\eta^2}\left[\eta^2\left(b^2+\xi^2\right)\left(a^2-\xi^2\right)p_\xi^2 - \xi^2\left(b^2-\eta^2\right)\left(a^2+\eta^2\right)p_\eta^2\right].$$

A direct calculation shows that

$$\Gamma_3 = b^2G_1^2 - a^2G_3^2.$$

The three first integrals in involution are

(i) the Hamiltonian H;

(ii) the constant generated by the separation process $\Gamma_3 = b^2G_1^2 - a^2G_3^2$;

(iii) the norm of the total angular momentum G.

Thus H, Γ_3, G can be considered as new momenta and the canonical transformation involved is generated by the function

$$S\left(q, \xi, \eta, H, \Gamma_3, G\right) = \int p_q dq + p_\xi\, d\xi + p_\eta\, d\eta,$$

where

$$p_q = \pm\sqrt{2H + \frac{2}{q} - \frac{G^2}{q^2}},$$
$$p_\xi = \pm\sqrt{\frac{G^2\xi^2 + \Gamma_3}{(b^2+\xi^2)(a^2-\xi^2)}},$$
$$p_\eta = \pm\sqrt{\frac{G^2\eta^2 - \Gamma_3}{(b^2-\eta^2)(a^2+\eta^2)}}.$$

The problem is thus reduced to quadratures and these involve elliptic integrals.

3.2 Action-Angle Variables

Consider a Hamiltonian system with n degrees of freedom and canonical coordinates q_h, p_k, where $h, k = 1, \ldots, n$. We recall (see Appendix D) that, given a complete set of independent first integrals $\Phi_k(q, p)$ in involution, one can obtain from them another set, say I_h, integrating the canonical 1-form along n independent cycles, say γ_h, over the Liouville-Arnold tori:

$$I_h(\Phi) = \frac{1}{2\pi} \oint_{\gamma_h} \sum_{k=1}^{n} p_k(q, \Phi)\, dq_k. \tag{3.2.1}$$

The functions $p_k(q, \Phi)$ are obtained inverting the expressions of the first integrals with respect to the momenta. These new variables I_h are called *actions*.

If the system is separable (in the sense of the Stäckel theorem), each p_k depends on the respective q_k only (and on all the first integral values, in general): each cycle γ_k is thus obtained letting vary the corresponding q_k, p_k only, while the other canonical coordinates stay unchanged. The angle variables ϕ_h, canonically conjugate to the actions, are introduced via the canonical transformation defined by the generating function

$$F(q, I) = \int_{P_0}^{P(q)} \sum_{k=1}^{n} p_k(q, \Phi(I))\, dq_k,$$

P_0 being a fixed point and $P(q)$ the generic point.

3.2.1 Delaunay and Poincaré Variables

Starting from the first integrals H, G, G_3 and applying the above mechanism, one obtains the Delaunay variables. To perform more easily the integrations involved, we make a canonical transformation, the so-called *Whittaker transformation*, that allows us to pass from Cartesian coordinates in 3-dimensional space to polar coordinates in the orbital plane. It is composed of two subsequent transformations. With reference to Figure 2.4 on page 30, we first rotate the coordinates

$$\begin{pmatrix} q_1 \\ q_2 \\ q_3 \end{pmatrix} = \begin{pmatrix} \cos\Omega & -\sin\Omega & 0 \\ \sin\Omega & \cos\Omega & 0 \\ 0 & 0 & 1 \end{pmatrix} \begin{pmatrix} 1 & 0 & 0 \\ 0 & \cos i & -\sin i \\ 0 & \sin i & \cos i \end{pmatrix} \begin{pmatrix} x \\ y \\ 0 \end{pmatrix},$$

then extend the rotation to the momenta

$$\begin{pmatrix} p_1 \\ p_2 \\ p_3 \end{pmatrix} = \begin{pmatrix} \cos\Omega & -\sin\Omega & 0 \\ \sin\Omega & \cos\Omega & 0 \\ 0 & 0 & 1 \end{pmatrix} \begin{pmatrix} 1 & 0 & 0 \\ 0 & \cos i & -\sin i \\ 0 & \sin i & \cos i \end{pmatrix} \begin{pmatrix} p_x \\ p_y \\ 0 \end{pmatrix}.$$

The transformation

$$\begin{pmatrix} p_1 & p_2 & p_3 \\ q_1 & q_2 & q_3 \end{pmatrix} \mapsto \begin{pmatrix} p_x & p_y & G_3 \\ x & y & \Omega \end{pmatrix}$$

is a canonical one. In fact, as a direct calculation shows,

$$p_1 dq_1 + p_2 dq_2 + p_3 dq_3 = p_x dx + p_y dy + \left(x p_y - y p_x\right) \cos i \, d\Omega,$$

where $x p_y - y p_x = G$ and $G \cos i = G_3$. The canonical 1-form is thus invariant under this transformation. The second transformation is

$$\begin{pmatrix} p_x & p_y & G_3 \\ x & y & \Omega \end{pmatrix} \mapsto \begin{pmatrix} p_r & p_\theta & G_3 \\ r & \theta & \Omega \end{pmatrix},$$

i.e., the canonical lifting from Cartesian to polar coordinates on the orbit plane

$$\begin{aligned} x &= r \cos\theta, \quad & p_x &= p_r \cos\theta - \frac{p_\theta}{r} \sin\theta, \\ y &= r \sin\theta, \quad & p_y &= p_r \sin\theta + \frac{p_\theta}{r} \cos\theta. \end{aligned}$$

One should be careful to distinguish between ϑ (the angle between position vector and q_3-axis) and θ (the angle, in the orbit plane, between position vector and x-axis); see Figure 3.1. Notice that $\theta = \omega + f$. Obviously, p_θ equals the total angular momentum G.

Now we are ready to work out the three integrals (3.2.1), with the momenta given by (3.1.1). The first is trivial,

$$I_\varphi = \frac{1}{2\pi} \int_0^{2\pi} G_3 d\varphi = G_3.$$

The second is easily performed, noticing that the canonical 1-form can be rewritten as

$$p_q dq + p_\vartheta d\vartheta + p_\varphi d\varphi = p_r dr + p_\theta d\theta + G_3 d\Omega,$$

and that, when ϑ covers its libration cycle, θ and φ cover a complete cycle from 0 to 2π, whereas Ω stays unchanged. Therefore, since $q = r$,

$$I_\vartheta = \frac{1}{\pi} \int_{\vartheta_{\min}}^{\vartheta_{\max}} p_\vartheta d\vartheta = \frac{1}{2\pi} \int_0^{2\pi} \left(p_\theta d\theta - p_\varphi d\varphi\right) = G - G_3.$$

The third,

$$I_q = \frac{1}{\pi} \int_{q_{\min}}^{q_{\max}} \sqrt{2H + \frac{2}{q} - \frac{G^2}{q^2}} \, dq$$

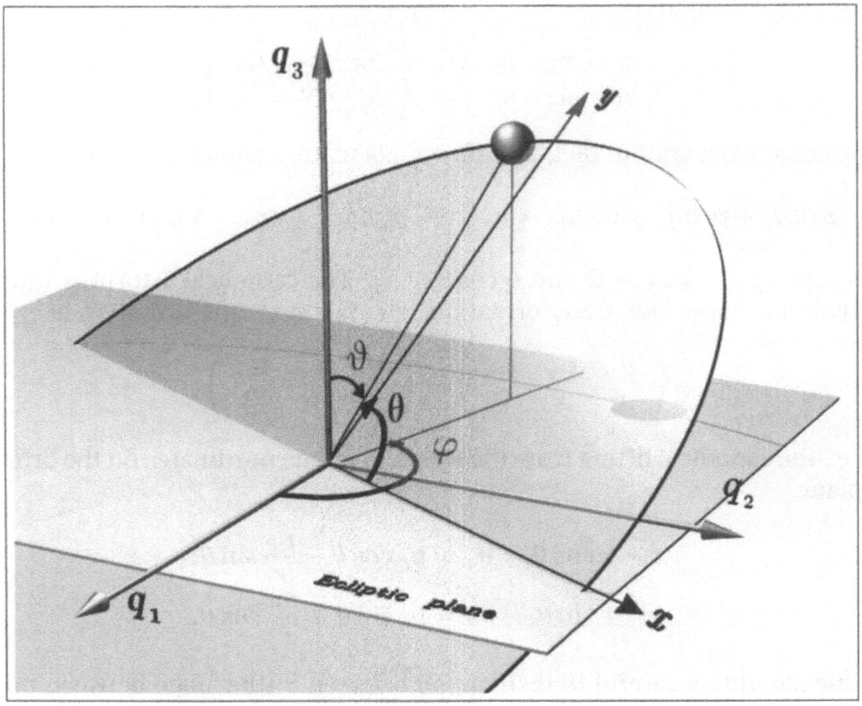

Figure 3.1: The spherical coordinates ϑ and φ, and the anomaly θ.

is reduced to elementary integrations. Indeed, substitute the expression (2.2.5) in $p_q = \dot{q}$, yielding

$$p_q = \sqrt{-2H}\frac{E\sin\mathfrak{s}}{1 - E\cos\mathfrak{s}}.$$

Then, write the first of (3.1.1) as

$$p_q = -\frac{-2H}{p_q} + \frac{2}{qp_q} - \frac{G^2}{q^2 p_q},$$

so that

$$p_q\, dq = \left[-\frac{1}{\sqrt{-2H}}(1 - E\cos\mathfrak{s}) + \frac{2}{\sqrt{-2H}} - \frac{G^2}{\sqrt{-2H}}\frac{1}{q}\right] d\mathfrak{s}.$$

Recalling (2.3.2), we find

$$I_q = \frac{1}{2\pi}\oint p_q\, dq = \frac{1}{\sqrt{-2H}} - G.$$

The actions are defined up to a *unimodular transformation*, i.e., a linear transformation whose representative matrix has integer entries and determinant ±1 (see Section D.4). The Delaunay actions are defined by

$$\begin{pmatrix} \mathcal{L} \\ \mathcal{G} \\ \mathcal{H} \end{pmatrix} = \begin{pmatrix} 1 & 1 & 1 \\ 0 & 1 & 1 \\ 0 & 0 & 1 \end{pmatrix} \begin{pmatrix} I_q \\ I_\vartheta \\ I_\varphi \end{pmatrix},$$

and, therefore,

$$\mathcal{L} = \frac{1}{\sqrt{-2H}}, \quad \mathcal{G} = G, \quad \mathcal{H} = G_3. \tag{3.2.2}$$

To find the angles canonically conjugate to the actions, let us define the function

$$F(r, \theta, \Omega, \mathcal{L}, \mathcal{G}, \mathcal{H}) = \Omega\mathcal{H} + \theta\mathcal{G} + \int_{\mathrm{P}}^{r} \sqrt{-\frac{1}{\mathcal{L}^2} + \frac{2}{r} - \frac{\mathcal{G}^2}{r^2}}\,dr,$$

where the point P must be fixed on the orbit: we choose the pericentre, one of the two roots of the integrand function. The function F generates a canonical transformation

$$\begin{pmatrix} p_r & p_\theta & G_3 \\ r & \theta & \Omega \end{pmatrix} \mapsto \begin{pmatrix} \mathcal{L} & \mathcal{G} & \mathcal{H} \\ \ell & \mathfrak{g} & \mathfrak{h} \end{pmatrix},$$

defined implicitly by the equations

$$p_r = \frac{\partial F}{\partial r}, \quad p_\theta = \frac{\partial F}{\partial \theta}, \quad G_3 = \frac{\partial F}{\partial \Omega},$$
$$\ell = \frac{\partial F}{\partial \mathcal{L}}, \quad \mathfrak{g} = \frac{\partial F}{\partial \mathcal{G}}, \quad \mathfrak{h} = \frac{\partial F}{\partial \mathcal{H}}.$$

The three equations on the first line agree with (3.2.2), as required. The first equation on the second line gives

$$\ell = \frac{1}{\mathcal{L}^3} \int_{\mathrm{P}}^{r} \left(-\frac{1}{\mathcal{L}^2} + \frac{2}{r} - \frac{\mathcal{G}^2}{r^2} \right)^{-\frac{1}{2}} dr.$$

Putting $r = a(1 - E\cos\mathfrak{s})$ and taking into account the identities

$$\mathcal{L}^2 = a, \quad \mathcal{G}^2 = a\left(1 - E^2\right),$$

one finds

$$\ell = \mathfrak{s} - E\sin\mathfrak{s},$$

showing that the angle conjugate to the action $\mathcal{L}$ is the mean anomaly, as suggested by the symbol ℓ. The second equation of the second line gives

$$\mathfrak{g} = \theta + \mathcal{G} \int_{\mathrm{P}}^{r} \left(-\frac{1}{\mathcal{L}^2} + \frac{2}{r} - \frac{\mathcal{G}^2}{r^2} \right)^{-\frac{1}{2}} d\left(\frac{1}{r}\right).$$

Substituting the orbit equation

$$\frac{1}{r} = \frac{E\cos f + 1}{a\,(1 - E^2)},$$

one finds

$$g = \theta - f = \omega,$$

showing that the angle conjugate to the action $\mathcal{G}$ is the argument of the pericentre ω. Finally, the third equation shows immediately that

$$h = \Omega,$$

i.e., the angle conjugate to the action $\mathcal{H}$ is the longitude of the ascending node.

The Delaunay variables are not well suited when the eccentricity E and the inclination i become very small, since they are plagued with the same singularities of the Keplerian elements. This implies that the variables $\mathcal{L}$, $\mathcal{G}$ and $\mathcal{H}$ differ in quantities that are of the order of the square of E and i. These three variables become identical if E and i vanish, and the angles ℓ, g and h become undetermined. In order to escape from this difficulty, Poincaré has introduced the canonical variables

$$\begin{aligned} \Lambda &= \mathcal{L}, & \lambda &= \ell + g + h, \\ \chi_1 &= \sqrt{2(\mathcal{L} - \mathcal{G})}\cos(g + h), & \eta_1 &= -\sqrt{2(\mathcal{L} - \mathcal{G})}\sin(g + h), \\ \chi_2 &= \sqrt{2(\mathcal{G} - \mathcal{H})}\cos h, & \eta_2 &= -\sqrt{2(\mathcal{G} - \mathcal{H})}\sin h, \end{aligned} \tag{3.2.3}$$

the last four being analogous (for small inclinations and eccentricities) to the variables k, h, r, s of (2.4.2) on page 32. In fact,

$$\sqrt{2(\mathcal{L} - \mathcal{G})} = \sqrt{2\mathcal{L}\left(1 - \sqrt{1 - E^2}\right)}$$

is of the same order as E and

$$\sqrt{2(\mathcal{G} - \mathcal{H})} = \sqrt{2\mathcal{L}\sqrt{1 - E^2}(1 - \cos i)}$$

is of the same order as $\sin\frac{i}{2}$.

To pass from Delaunay to Poincaré variables, one composes two canonical transformations. The first is

$$\begin{pmatrix} \Lambda \\ \rho_1 \\ \rho_2 \end{pmatrix} = \begin{pmatrix} 1 & 0 & 0 \\ 1 & -1 & 0 \\ 0 & 1 & -1 \end{pmatrix} \begin{pmatrix} \mathcal{L} \\ \mathcal{G} \\ \mathcal{H} \end{pmatrix},$$

$$\begin{pmatrix} \lambda \\ \varphi_1 \\ \varphi_2 \end{pmatrix} = \begin{pmatrix} 1 & 1 & 1 \\ 0 & -1 & -1 \\ 0 & 0 & -1 \end{pmatrix} \begin{pmatrix} \ell \\ g \\ h \end{pmatrix},$$

which is canonical since each of the two (unimodular) matrices is the transpose of the inverse of the other. The second is

$$\chi_h = \sqrt{2\rho_h}\cos\varphi_h, \quad \eta_h = \sqrt{2\rho_h}\sin\varphi_h, \quad h = 1, 2,$$

analogous to the change in the plane from polar to Cartesian coordinates, which explains why the singularity in the origin is removed.

It is useful to find explicitly the relations between the "physical" canonical variables $\vec{q}, \vec{p}$ and the Delaunay or Poincaré variables.

i) Delaunay variables

The Delaunay variables are defined when $\sqrt{G_1^2 + G_2^2} \neq 0$ and $\vec{E} \neq 0$.

Let us find the Delaunay variables as functions of $\vec{q}, \vec{p}$. For the action variables $\mathcal{L}, G, G_3$, see the definition. For the angles Ω, ω, see (2.4.1) on page 32. For the mean anomaly ℓ, recalling that

$$q(\mathfrak{s}) = a - aE\cos\mathfrak{s}, \quad \frac{dq}{d\mathfrak{s}} = \frac{dt}{d\mathfrak{s}}\frac{dq}{dt} = aE\sin\mathfrak{s}$$

one finds

$$\cos\mathfrak{s} = \frac{a-q}{aE}, \quad \sin\mathfrak{s} = \frac{\vec{q}\cdot\vec{p}}{\sqrt{a}E},$$

and, from the definition of mean anomaly,

$$\begin{pmatrix} \cos\ell \\ \sin\ell \end{pmatrix} = \begin{pmatrix} \cos\Delta & \sin\Delta \\ -\sin\Delta & \cos\Delta \end{pmatrix} \begin{pmatrix} \frac{a-q}{aE} \\ \frac{\vec{q}\cdot\vec{p}}{\sqrt{a}E} \end{pmatrix}, \quad \Delta = \frac{\vec{q}\cdot\vec{p}}{\sqrt{a}}.$$

Let us find $\vec{q}, \vec{p}$ as functions of the Delaunay variables. This, obviously, involves the inversion of the Kepler Equation, since the evolution of the canonical coordinates is naturally expressed in terms of the eccentric anomaly. The expression (2.2.4) for the orbit and that obtained differentiating it with respect to the time show that we reach our goal if, after inverting the Kepler Equation, we write $\vec{G}, \vec{E}$ as functions of $\mathcal{L}, G, G_3, \Omega, \omega$. With a little geometry, this is easily achieved. In fact, recalling Figure 2.4 on page 30, we can write the relations

$$\begin{aligned}
G_1 &= \sqrt{G^2 - G_3^2}\sin\Omega, \quad G_2 = -\sqrt{G^2 - G_3^2}\cos\Omega, \\
E_1 &= \sqrt{1 - \frac{G^2}{\mathcal{L}^2}}\left(\cos\omega\cos\Omega - \sin\omega\sin\Omega\cos i\right), \\
E_2 &= \sqrt{1 - \frac{G^2}{\mathcal{L}^2}}\left(\cos\omega\sin\Omega + \sin\omega\cos\Omega\cos i\right), \\
E_3 &= \sqrt{1 - \frac{G^2}{\mathcal{L}^2}}\sin\omega\sin i,
\end{aligned} \tag{3.2.4}$$

where $\cos i = \frac{G_3}{G}$ and $\sin i = \sqrt{1 - \frac{G_3^2}{G^2}}$.

ii) Poincaré variables

The Poincaré variables are well defined when $G_3 \neq -G$ (the retrograde motion on the q_1q_2 plane is excluded), and $G \neq 0$ (the collision orbits are excluded).

Let us find the Poincaré variables as functions of $\vec{q}, \vec{p}$. It is enough to write them as functions of $\vec{G}, \vec{E}$. Firstly

$$\chi_2 = \sqrt{2(G - G_3)}\vec{n} \cdot \check{q}_1 = -\sqrt{\frac{2}{G + G_3}} G_2,$$
$$\eta_2 = -\sqrt{2(G - G_3)}\vec{n} \cdot \check{q}_2 = -\sqrt{\frac{2}{G + G_3}} G_1.$$

To calculate χ_1, η_1, since

$$\cos(\omega + \Omega) = \frac{1}{E}\left[\left(\vec{n} \cdot \vec{E}\right)\left(\vec{n} \cdot \check{q}_1\right) - \left(\vec{n} \times \vec{E} \cdot \frac{\vec{G}}{G}\right)\left(\vec{n} \cdot \check{q}_2\right)\right],$$
$$\sin(\omega + \Omega) = \frac{1}{E}\left[\left(\vec{n} \times \vec{E} \cdot \frac{\vec{G}}{G}\right)\left(\vec{n} \cdot \check{q}_1\right) + \left(\vec{n} \cdot \vec{E}\right)\left(\vec{n} \cdot \check{q}_2\right)\right],$$
$$\sqrt{2(\mathcal{L} - G)} = \sqrt{2\mathcal{L}(1 - \sqrt{1 - E^2})} = E\sqrt{\frac{2\mathcal{L}}{1 + \sqrt{1 - E^2}}},$$

we find

$$\chi_1 = \sqrt{\frac{2\mathcal{L}}{1 + \sqrt{1 - E^2}}}\left[\frac{G_3E_1 - G_1E_3}{G} - \frac{G_2}{G + G_3}\frac{G_1E_2 - G_2E_1}{G}\right],$$
$$\eta_1 = -\sqrt{\frac{2\mathcal{L}}{1 + \sqrt{1 - E^2}}}\left[\frac{G_3E_2 - G_2E_3}{G} + \frac{G_1}{G + G_3}\frac{G_1E_2 - G_2E_1}{G}\right].$$

Notice that the singularities affecting the Delaunay variables have been eliminated. In fact, if the orbit lies in the q_1q_2 plane, then $G_1 = G_2 = 0$, hence $\chi_2 = \eta_2 = 0$, while if the orbit is a circle, then $\vec{E} = 0$, hence $\chi_1 = \eta_1 = 0$.

To calculate $\lambda = \ell + \Omega + \omega$, the so-called *mean longitude*, write it as

$$\lambda = (\mathrm{s} - f) - \Delta + (f + \Omega + \omega).$$

The angle $f + \Omega + \omega$, the so-called *true longitude*, is well defined for $E = 0$ also and can be calculated. In fact

$$\cos(f + \Omega + \omega) = \cos(f + \omega)\cos\Omega - \sin(f + \omega)\sin\Omega,$$
$$\sin(f + \Omega + \omega) = \sin(f + \omega)\cos\Omega + \cos(f + \omega)\sin\Omega,$$

and this calculation is fully similar to that, already performed, of $\cos(\omega + \Omega)$ and $\sin(\omega + \Omega)$; we must only replace ω with $f + \omega$, that is $\vec{E}$ with $\vec{q}$, obtaining

$$\cos(f + \Omega + \omega) = \frac{1}{q}\left[\frac{G_3 q_1 - G_1 q_3}{G} - \frac{G_2}{G + G_3}\frac{G_1 q_2 - G_2 q_1}{G}\right],$$
$$\sin(f + \Omega + \omega) = \frac{1}{q}\left[\frac{G_3 q_2 - G_2 q_3}{G} + \frac{G_1}{G + G_3}\frac{G_1 q_2 - G_2 q_1}{G}\right].$$

For the term $(\mathfrak{s} - f)$, remembering (2.3.1) and (2.2.1), we may write

$$\cos(\mathfrak{s} - f) = \frac{q}{G^2}\left[1 + \left(\sqrt{1 - E^2} - 1\right)\sin^2 f + E\cos f\right],$$
$$\sin(\mathfrak{s} - f) = \frac{q}{G^2}\left[\left(\sqrt{1 - E^2} - 1\right)\sin f \cos f - E\sin f\right].$$

Substituting in these last two equations the obvious relations

$$\cos f = \frac{\vec{q}}{q} \cdot \frac{\vec{E}}{E}, \quad \sin f = \frac{\vec{G}}{G} \times \frac{\vec{q}}{q} \cdot \frac{\vec{E}}{E},$$

we obtain

$$\cos(\mathfrak{s} - f) = \frac{q}{G^2}\left[1 - \frac{1}{1 + \sqrt{1 - E^2}}\left(\frac{\vec{G} \times \vec{q} \cdot \vec{E}}{Gq}\right)^2 + \frac{\vec{q} \cdot \vec{E}}{q}\right],$$

$$\sin(\mathfrak{s} - f) = -\frac{q}{G^2}\frac{\vec{G} \times \vec{q} \cdot \vec{E}}{Gq}\left[1 + \frac{1}{1 + \sqrt{1 - E^2}}\frac{\vec{q} \cdot \vec{E}}{q}\right],$$

which are well defined for $E = 0$ also. Notice that $\lambda = f + \Omega + \omega$ for $E = 0$.

Lastly, let us find $\vec{q}, \vec{p}$ as functions of the Poincaré variables. It is useful to find first $\vec{G}$ and $\vec{E}$ as functions of the Poincaré variables. From (3.2.3) we obtain

$$G = \Lambda - \frac{1}{2}(\chi_1^2 + \eta_1^2), \quad G_3 = \Lambda - \frac{1}{2}(\chi_1^2 + \eta_1^2 + \chi_2^2 + \eta_2^2),$$

and from the first line of (3.2.4)

$$G_1 = -\eta_2\sqrt{\frac{1}{2}(G + G_3)}, \quad G_2 = -\chi_2\sqrt{\frac{1}{2}(G + G_3)}.$$

Moreover, from (3.2.3) again,

$$\cos(\omega + \Omega) = \frac{\chi_1}{\sqrt{\chi_1^2 + \eta_1^2}}, \quad \sin(\omega + \Omega) = -\frac{\eta_1}{\sqrt{\chi_1^2 + \eta_1^2}},$$
$$\cos\Omega = \frac{\chi_2}{\sqrt{\chi_2^2 + \eta_2^2}}, \quad \sin\Omega = -\frac{\eta_2}{\sqrt{\chi_2^2 + \eta_2^2}},$$

from which

$$\cos\omega = \frac{\chi_1\chi_2 + \eta_1\eta_2}{\sqrt{\chi_1^2+\eta_1^2}\sqrt{\chi_2^2+\eta_2^2}}, \quad \sin\omega = \frac{\chi_1\eta_2 - \eta_1\chi_2}{\sqrt{\chi_1^2+\eta_1^2}\sqrt{\chi_2^2+\eta_2^2}}.$$

Substituting into (3.2.4) and taking into account that

$$\cos i = \frac{G_3}{G} = 1 - \frac{\chi_2^2+\eta_2^2}{2G}, \quad \sin i = \frac{\sqrt{G-G_3}\sqrt{G+G_3}}{G},$$

we obtain

$$E_1 = \sqrt{\frac{\Lambda+G}{2\Lambda^2}}\left[\chi_1 + \frac{\eta_2(\eta_1\chi_2 - \chi_1\eta_2)}{2G}\right],$$
$$E_2 = \sqrt{\frac{\Lambda+G}{2\Lambda^2}}\left[-\eta_1 + \frac{\chi_2(\eta_1\chi_2 - \chi_1\eta_2)}{2G}\right],$$
$$E_3 = \sqrt{\frac{\Lambda+G}{2\Lambda^2}}\sqrt{\frac{G+G_3}{2G^2}}\,(\chi_1\eta_2 - \eta_1\chi_2)\,.$$

The starting point to calculate $\vec{q}, \vec{p}$ as functions of the Poincaré variables is, as for the Delaunay variables, the expression (2.2.4)

$$\vec{q} = \Lambda^2\left[\frac{\vec{E}}{E}\cos\mathfrak{s} + \frac{1}{\Lambda}\vec{G}\times\frac{\vec{E}}{E}\sin\mathfrak{s} - \vec{E}\right], \tag{3.2.5}$$

where, however, the eccentric anomaly must be substituted by the *eccentric longitude* $\sigma = \mathfrak{s} + \omega + \Omega$, which is well defined also for $E = 0$ and $G = G_3$. In fact, by definition of mean longitude and from the Kepler Equation,

$$\lambda = \sigma - E\sin\mathfrak{s} = \sigma - E\cos(\omega+\Omega)\sin\sigma + E\sin(\omega+\Omega)\cos\sigma;$$

but

$$E\cos(\omega+\Omega) = \sqrt{\frac{\Lambda^2-G^2}{\Lambda^2}}\frac{\chi_1}{\sqrt{2(\Lambda-G)}} = \sqrt{\frac{\Lambda+G}{2\Lambda^2}}\chi_1,$$
$$E\sin(\omega+\Omega) = -\sqrt{\frac{\Lambda+G}{2\Lambda^2}}\eta_1,$$

so that we obtain a *modified Kepler Equation*

$$\lambda = \sigma - \sqrt{\frac{\Lambda+G}{2\Lambda^2}}(\chi_1\sin\sigma + \eta_1\cos\sigma)$$

that, given the Poincaré variables, one must invert to find σ. Excluding the collision orbits, this inversion is, in principle, always possible. Indeed $\frac{d\lambda}{d\sigma} \geq 0$, where the equality holds only when $G = 0$, as one easily verifies.

Come back to the expression (3.2.5), that we may write as

$$\frac{\vec{q}}{\Lambda^2} = \left[\frac{\vec{E}}{E}\cos(\omega+\Omega) - \frac{1}{\Lambda}\vec{G}\times\frac{\vec{E}}{E}\sin(\omega+\Omega)\right]\cos\sigma$$
$$+ \left[\frac{\vec{E}}{E}\sin(\omega+\Omega) + \frac{1}{\Lambda}\vec{G}\times\frac{\vec{E}}{E}\cos(\omega+\Omega)\right]\sin\sigma - \vec{E}.$$

Referring to Figure 2.4, let us define the orthonormal vectors $\check{x} = \vec{n}$ and $\check{y}$ of the x and y axes. We may write

$$\vec{E} = E\cos\omega\check{x} + E\sin\omega\check{y}, \quad \vec{G}\times\frac{\vec{E}}{E} = -G\sin\omega\check{x} + G\cos\omega\check{y},$$

from which

$$\frac{\vec{E}}{E}\cos(\omega+\Omega) - \frac{1}{\Lambda}\vec{G}\times\frac{\vec{E}}{E}\sin(\omega+\Omega)$$
$$= \left[\cos\omega\cos(\omega+\Omega) + \frac{G}{\Lambda}\sin\omega\sin(\omega+\Omega)\right]\check{x}$$
$$+ \left[\sin\omega\cos(\omega+\Omega) - \frac{G}{\Lambda}\cos\omega\sin(\omega+\Omega)\right]\check{y}.$$

Let us calculate

$$\cos\omega\cos(\omega+\Omega) + \frac{G}{\Lambda}\sin\omega\sin(\omega+\Omega)$$
$$= \cos\omega\cos(\omega+\Omega) + \left(1 - \frac{\chi_1^2+\eta_1^2}{2\Lambda}\right)\sin\omega\sin(\omega+\Omega)$$
$$= \cos\Omega - \frac{\chi_1^2+\eta_1^2}{2\Lambda}\sin\omega\sin(\omega+\Omega)$$
$$= \cos\Omega - \frac{\chi_1^2+\eta_1^2}{2\Lambda}\left[\sin^2(\omega+\Omega)\cos\Omega - \sin(\omega+\Omega)\cos(\omega+\Omega)\sin\Omega\right]$$
$$= \cos\Omega - \frac{\chi_1^2+\eta_1^2}{2\Lambda}\left[\frac{\eta_1^2}{\chi_1^2+\eta_1^2}\cos\Omega + \frac{\chi_1\eta_1}{\chi_1^2+\eta_1^2}\sin\Omega\right]$$
$$= \left(1 - \frac{\eta_1^2}{2\Lambda}\right)\cos\Omega - \frac{\chi_1\eta_1}{2\Lambda}\sin\Omega.$$

Similarly

$$\sin\omega\cos(\omega+\Omega) - \frac{G}{\Lambda}\cos\omega\sin(\omega+\Omega)$$
$$= -\left(1-\frac{\eta_1^2}{2\Lambda}\right)\sin\Omega - \frac{\chi_1\eta_1}{2\Lambda}\cos\Omega.$$

We can therefore write

$$\frac{\vec{E}}{E}\cos(\omega+\Omega) - \frac{1}{\Lambda}\vec{G}\times\frac{\vec{E}}{E}\sin(\omega+\Omega) = \left(1-\frac{\eta_1^2}{2\Lambda}\right)\check{x}' - \frac{\chi_1\eta_1}{2\Lambda}\check{y}',$$

where $\check{x}'$ and $\check{y}'$ are the orthonormal vectors $\check{x}$ and $\check{y}$ rotated through the angle Ω :

$$\check{x}' = \check{x}\cos\Omega - \check{y}\sin\Omega, \quad \check{y}' = \check{x}\sin\Omega + \check{y}\cos\Omega.$$

Alternatively, $\check{x}'$ and $\check{y}'$ can be obtained by rotating $\check{q}_1, \check{q}_2$ through the angle i around $\vec{n}$. They can be expressed as functions of the components of $\vec{G}$, thus of the Poincaré parameters. In fact, taking into account that $\check{y} = \frac{\vec{G}}{G}\times\vec{n}$, it turns out that

$$\check{x}\cos\Omega - \check{y}\sin\Omega = \frac{G_2\check{q}_1 - G_1\check{q}_2}{\sqrt{G_1^2+G_2^2}}\frac{G_2}{\sqrt{G_1^2+G_2^2}}$$
$$-\frac{1}{G}\left(-\frac{G_1G_3}{\sqrt{G_1^2+G_2^2}}\check{q}_1 - \frac{G_2G_3}{\sqrt{G_1^2+G_2^2}}\check{q}_2 + \sqrt{G_1^2+G_2^2}\check{q}_3\right)\frac{G_1}{\sqrt{G_1^2+G_2^2}}$$
$$= \left(\frac{G_2^2}{G(G+G_3)} + \frac{G_3}{G}\right)\check{q}_1 - \frac{G_1G_2}{G(G+G_3)}\check{q}_2 - \frac{G_1}{G}\check{q}_3,$$

and

$$\check{x}\sin\Omega + \check{y}\cos\Omega = -\frac{G_1G_2}{G(G+G_3)}\check{q}_1 + \left(\frac{G_1^2}{G(G+G_3)} + \frac{G_3}{G}\right)\check{q}_2 - \frac{G_2}{G}\check{q}_3.$$

Notice that $\check{x}'$ and $\check{y}'$ are well defined for the circular orbits and for those lying on the q_1q_2 plane also. Arguing in the same way let us calculate

$$\frac{\vec{E}}{E}\sin(\omega+\Omega) + \frac{1}{\Lambda}\vec{G}\times\frac{\vec{E}}{E}\cos(\omega+\Omega) = -\frac{\chi_1\eta_1}{2\Lambda}\check{x}' + \left(1-\frac{\chi_1^2}{2\Lambda}\right)\check{y}'.$$

Summing up, we have reached the goal to express $\vec{q}$ as a function of the Poincaré parameters, up to obviously the inversion of the modified

Kepler Equation. Indeed

$$\frac{\vec{q}}{\Lambda^2} = \left[\cos\sigma - \left(\frac{\eta_1^2}{2\Lambda}\cos\sigma + \frac{\chi_1\eta_1}{2\Lambda}\sin\sigma\right) - \sqrt{\frac{\Lambda+G}{2\Lambda^2}}\chi_1\right]\check{x}'$$
$$+\left[\sin\sigma - \left(\frac{\chi_1\eta_1}{2\Lambda}\cos\sigma + \frac{\chi_1^2}{2\Lambda}\sin\sigma\right) + \sqrt{\frac{\Lambda+G}{2\Lambda^2}}\eta_1\right]\check{y}'.$$

It is now immediate to find $\vec{p}$ as a function of the Poincaré parameters, in fact

$$\vec{p} = \frac{d}{dt}\vec{q} = \frac{1}{\Lambda}\frac{1}{q}\frac{d}{d\sigma}\vec{q}, \quad \text{since } \frac{d}{d\varsigma} = \frac{d}{d\sigma}$$

and

$$q = \Lambda^2(1 - E\cos\varsigma) = \Lambda^2\left(1 - \sqrt{\frac{\Lambda+G}{2\Lambda^2}}\chi_1\cos\sigma + \sqrt{\frac{\Lambda+G}{2\Lambda^2}}\eta_1\sin\sigma\right).$$

3.2.2 Pauli Variables

Starting from the first integrals H, E_3, G_3, which we have obtained separating the variables in parabolic coordinates, and applying, modulo some technical details, the general mechanism, we obtain a new set of action-angle variables that we call *Pauli variables*(2). Unfortunately, with the direct application of the method to the full 3-dimensional problem we get into trouble with the geometrical interpretation of the angles, so we choose to solve, first, the 2-dimensional case and, then, to extend the result. Therefore, let us consider the orbit plane, in which we introduce the parabolic coordinates

$$x = \xi\eta, \quad y = \frac{1}{2}\left(\xi^2 - \eta^2\right).$$

Equivalently

$$y + ix = \left[\frac{1}{\sqrt{2}}(\xi + i\eta)\right]^2,$$

which shows that the plane $\xi\eta$ is the double covering of the plane xy. This transformation is also known as the *Levi-Civita regularizing map*, as we shall see later in Chapter 5.

(2)The choice of the name *Pauli variables* will become clear later, when we will examine the Pauli quantization of the hydrogen atom.

Separating the variables we obtain

$$p_\xi^2 = 2\left(1 - E_y\right) + 2H\xi^2,$$
$$p_\eta^2 = 2\left(1 + E_y\right) + 2H\eta^2,$$

the new separation variable E_y being the projection of the eccentricity vector on the y-axis. Remembering the formula for the area of the ellipse, the action integrals (3.2.1) are easily calculated:

$$I_\xi = \frac{1}{\sqrt{-2H}} - \frac{E_y}{\sqrt{-2H}},$$
$$I_\eta = \frac{1}{\sqrt{-2H}} + \frac{E_y}{\sqrt{-2H}}.$$

We prefer to take, as action variables, the linear combination

$$\begin{pmatrix} \mathcal{L} \\ \mathcal{R} \end{pmatrix} = \frac{1}{2}\begin{pmatrix} 1 & 1 \\ -1 & 1 \end{pmatrix}\begin{pmatrix} I_\xi \\ I_\eta \end{pmatrix} = \begin{pmatrix} \frac{1}{\sqrt{-2H}} \\ \frac{E_y}{\sqrt{-2H}} \end{pmatrix}.$$

The matrix of the above transformation is not unimodular, and connects two discrete subgroups Γ_i, $i = 1, 2$ of $\mathbb{R}^2$, defining two *different* tori $\mathbb{R}^2/\Gamma_i$ (see Subsection D.4.1). This makes up for the double covering mentioned above, so that we expect that the coordinates, conjugate to $\mathcal{L}$ and $\mathcal{R}$, span the angles on the torus given by $\mathcal{L}$ = constant and $\mathcal{R}$ = constant and living on the cotangent bundle to the physical configuration space $\mathbb{R}^2$. We can rewrite the momenta as

$$p_\xi^2 = 2\left(1 - \frac{\mathcal{R}}{\mathcal{L}}\right) - \frac{\xi^2}{\mathcal{L}^2},$$
$$p_\eta^2 = 2\left(1 + \frac{\mathcal{R}}{\mathcal{L}}\right) - \frac{\eta^2}{\mathcal{L}^2}.$$

The function

$$F(\xi, \eta, \mathcal{L}, \mathcal{R}) = \int p_\xi d\xi + p_\eta d\eta$$

generates the canonical transformation

$$\begin{pmatrix} p_\xi & p_\eta \\ \xi & \eta \end{pmatrix} \mapsto \begin{pmatrix} \mathcal{L} & \mathcal{R} \\ \ell' & \hbar \end{pmatrix},$$

defined implicitly by the equations

$$p_\xi = \frac{\partial F}{\partial \xi}, \quad p_\eta = \frac{\partial F}{\partial \eta}, \quad \ell' = \frac{\partial F}{\partial \mathcal{L}},$$
$$\hbar = \frac{\partial F}{\partial \mathcal{R}} = -\int_{\xi_0}^{\xi} \frac{d\xi}{\sqrt{2\mathcal{L}(\mathcal{L} - \mathcal{R}) - \xi^2}} + \int_{\eta_0}^{\eta} \frac{d\eta}{\sqrt{2\mathcal{L}(\mathcal{L} + \mathcal{R}) - \eta^2}},$$

where ξ_0 and η_0 are one of the zero's of the respective denominator. We do not calculate ℓ', which differs from the mean anomaly ℓ only by an additive constant. The two integrals relative to $\mathfrak{k}$ are elementary:

$$\mathfrak{k} = -\arcsin \frac{\xi}{\sqrt{2\mathcal{L}(\mathcal{L}-\mathcal{R})}} + \arcsin \frac{\eta}{\sqrt{2\mathcal{L}(\mathcal{L}+\mathcal{R})}}.$$

To put $\mathfrak{k}$ in a better form, we express ξ and η as some functions of the eccentric anomaly $\mathfrak{s}$. To this end, recalling that

$$r = a(1 - E\cos\mathfrak{s}), \quad \sin f = \sqrt{1-E^2}\frac{\sin\mathfrak{s}}{1-E\cos\mathfrak{s}}, \quad \cos f = \frac{\cos\mathfrak{s} - E}{1 - E\cos\mathfrak{s}},$$

we obtain

$$y = r\sin(f+\omega) = a\left[\sqrt{1-E^2}\sin\mathfrak{s}\cos\omega + (\cos\mathfrak{s} - E)\sin\omega\right],$$

from which

$$\begin{aligned}
\xi^2 &= r + y \\
&= a\left[(\sin\omega - E)\cos\mathfrak{s} + \sqrt{1-E^2}\cos\omega\sin\mathfrak{s} + 1 - E\sin\omega\right] \\
&= a\left[\sqrt{(1-\sin\omega)(1+E)}\sin\frac{\mathfrak{s}}{2} + \sqrt{(1+\sin\omega)(1-E)}\cos\frac{\mathfrak{s}}{2}\right]^2 \\
&= 2a(1 - E\sin\omega)\sin^2\left[\frac{\mathfrak{s}}{2} + \arctan\sqrt{\frac{(1+\sin\omega)(1-E)}{(1-\sin\omega)(1+E)}}\right].
\end{aligned}$$

Analogously

$$\begin{aligned}
\eta^2 &= r - y \\
&= \left[-(\sin\omega + E)\cos\mathfrak{s} - \sqrt{1-E^2}\cos\omega\sin\mathfrak{s} + 1 + E\sin\omega\right] \\
&= a\left[\sqrt{(1+\sin\omega)(1+E)}\sin\frac{\mathfrak{s}}{2} - \sqrt{(1-\sin\omega)(1-E)}\cos\frac{\mathfrak{s}}{2}\right]^2 \\
&= 2a(1 + E\sin\omega)\sin^2\left[\frac{\mathfrak{s}}{2} - \arctan\sqrt{\frac{(1-\sin\omega)(1-E)}{(1+\sin\omega)(1+E)}}\right].
\end{aligned}$$

Since $2a(1 \pm E\sin\omega) = 2\mathcal{L}(\mathcal{L} \pm \mathcal{R})$, one gets

$$\mathfrak{k} = -\arctan\sqrt{\frac{(1+\sin\omega)(1-E)}{(1-\sin\omega)(1+E)}} - \arctan\sqrt{\frac{(1-\sin\omega)(1-E)}{(1+\sin\omega)(1+E)}}$$

that, remembering the trigonometric identity

$$\tan(\alpha+\beta) = \frac{\tan\alpha + \tan\beta}{1 - \tan\alpha\tan\beta},$$

can be written as

$$-\tan \hbar = \frac{\sqrt{1-E^2}}{E\cos\omega} = \frac{G}{\mathcal{L}E_x}.$$

This suggests the definition of the *Runge-Lenz-Laplace vector*(3)

$$\vec{R} = \frac{\vec{E}}{\sqrt{-2H}} = \mathcal{L}\vec{E}.$$

The above expression of $\hbar$ shows that $\mathcal{R} = R_y$, regarded as the generator of an infinitesimal canonical transformation, generates a rotation on the plane containing $\vec{G}$ and the x-axis. By symmetry, R_x will generate a rotation on the plane containing $\vec{G}$ and the y-axis. Finally, since G generates a rotation on the xy plane, we conclude that R_x, R_y, G have the same Poisson algebra of the angular momentum, isomorphic to the Lie algebra of the three-dimensional rotation group.

This is a fundamental result, which can be extended to the three-dimensional case. We recall the definition of Poisson bracket $\{f,g\}$ between two functions $f(q_h, p_k)$ and $g(q_h, p_k)$ on the phase space of an n-dimensional configuration manifold:

$$\{f,g\} = \sum_{k=1}^{n} \frac{\partial f}{\partial p_k}\frac{\partial g}{\partial q_k} - \frac{\partial f}{\partial q_k}\frac{\partial g}{\partial p_k}.$$

The wanted extension, which one can verify with a direct calculation, is

$$\{G_i, G_j\} = -\epsilon_{ijk}G_k, \quad \{G_i, R_j\} = -\epsilon_{ijk}R_k, \quad \{R_i, R_j\} = -\epsilon_{ijk}G_k, \tag{3.2.6}$$

where ϵ_{ijk} is equal to $1(-1)$ if ijk is an even (odd) permutation of 123, or zero if two indices are equal. The first two equations simply state that $\vec{G}$ and $\vec{R}$ are vectors, while the third shows that they generate the Lie algebra of the *four*-dimensional rotation group.

This (at this stage) surprising symmetry cannot obviously be fortuitous and will be explained in Part II. For the moment, it allows us to easily find the two angles conjugate to the action variables(4) G_3 and R_3. Since the components of each of the two *Pauli vectors*

$$\vec{M} = \frac{1}{2}\left(\vec{G} + \vec{R}\right), \quad \vec{N} = \frac{1}{2}\left(\vec{G} - \vec{R}\right)$$

satisfy the Poisson algebra of angular momentum, while the mutual Poisson brackets vanish,

$$\{M_i, M_j\} = -\epsilon_{ijk}M_k, \quad \{N_i, N_j\} = -\epsilon_{ijk}N_k, \quad \{M_i, N_j\} = 0,$$

(3) We stress that this definition is made in the hypothesis $H < 0$.

(4) It should be obvious that $\frac{1}{\sqrt{-2H}}, R_3, G_3$ are the action variables in the 3-dimensional case. Anyway, the relative integrations are elementary, though somewhat laborious.

put here

$$\varphi_G = \frac{1}{2}\left(\arctan\frac{M_2}{M_1} + \arctan\frac{N_2}{N_1}\right),$$
$$\varphi_R = \frac{1}{2}\left(\arctan\frac{M_2}{M_1} - \arctan\frac{N_2}{N_1}\right), \tag{3.2.7}$$

to obtain

$$\{G_3, \varphi_G\} = \{R_3, \varphi_R\} = 1,$$
$$\{G_3, \varphi_R\} = \{R_3, \varphi_G\} = 0.$$

The two Pauli vectors are not independent; indeed, they satisfy the relation

$$\left\|\vec{M}\right\|^2 = \left\|\vec{N}\right\|^2 = \frac{1}{4}\mathcal{L}^2,$$

as one easily verifies.

3.2.3 Monodromy

It is easy to verify that the monodromy (see Subsection D.4.3 for the definitions) is trivial for the Delaunay and Pauli variables. The actions, therefore, are univalued functions of the first integrals in involution.

The triviality of the monodromy is, by definition, equivalent to the triviality of a bundle, whose fibre is a torus $\mathbb{T}^3$, and the base an open in $\mathbb{R}^3$ parametrizing the set of the first integrals in involution. Obviously, the triviality of the bundle is guaranteed if the base can be continuously contracted to a point.

The Kepler Problem is completely degenerate, so that, as explained in Subsection D.4.2, whatever is the choice of the action-angle system, inevitably singularities in the parametrization appear, and one must remove from the phase space some subsets. In the Delaunay case, one is forced to take

$$H < 0, \quad G < \frac{1}{\sqrt{-2H}}, \quad G_3 < G.$$

The second inequality excludes the circular orbits, the third excludes the orbits lying on the ecliptic plane.

In the Pauli case,

$$H < 0,$$
$$-\frac{1}{\sqrt{-2H}} < G_3 + \frac{E_3}{\sqrt{-2H}} < \frac{1}{\sqrt{-2H}},$$
$$-\frac{1}{\sqrt{-2H}} < G_3 - \frac{E_3}{\sqrt{-2H}} < \frac{1}{\sqrt{-2H}},$$

the second and third inequalities exclude the orbits for which φ_G and φ_R are not defined.

It is clear that the open sets in $\mathbb{R}^3$, described by the above inequalities, are homeomorphic to open balls in $\mathbb{R}^3$, thus contractible to a point: the triviality of the bundles follows.

CHAPTER 4

Quantization of the Kepler Problem

The expression "quantization of a mechanical system" is somewhat vague and this is not the place to give an exact and precise definition (which, probably[1], can not be given at all). So we take it for granted that the reader is familiar with the fascinating subtleties of quantum mechanics and only recall that, from the physical point of view, quantizing the Kepler Problem means to find the spectrum of the energy of the hydrogen atom and the probability distribution of the electron. We will try, in Part II of this book, to make *mathematically* well defined the term "quantization", while, in this chapter, we review the historical methods for quantizing the problem, and, more precisely, the Schrödinger, the Pauli and the Fock methods.

4.1 The Schrödinger Quantization

First of all we summarize the main points of the method and subsequently we apply it to the Kepler Problem. Since the "output" of the method is a partial differential equation (the *Schrödinger Equation* of course) whose differential part is the Laplacian in $\mathbb{R}^3$, our task is greatly simplified in its integration if the variables can be separated. But, in a Euclidean space, the coordinates for which the Schrödinger Equation

[1] The very nature of the argument forces us to use this term.

is separable are the same that separate the corresponding Hamilton-Jacobi Equation: see Robertson Theorem. From the work of the previous chapter, we know all the four coordinate systems that have this property.

Let us consider a *natural* mechanical system with n degrees of freedom, i.e., a system whose Hamiltonian is of the type(2)

$$H(q^h, p_k) = \frac{1}{2} g^{ik}(q^h) p_i p_k + V(q^h) \quad h, i, k = 1, \ldots, n, \tag{4.1.1}$$

where q^h are local coordinates on the configuration manifold M. In other words, a mechanical system is natural if its energy is the sum of a kinetic energy quadratic in the velocities and thus in the momenta, and of a potential, which is a function of the coordinates only. The coefficients of the quadratic form are the local contravariant components of a *metric tensor*, and give to the configuration manifold the structure of a Riemannian manifold.

The basic experimental fact of quantum mechanics is the following: at a microscopic scale, matter manifests a wavelike behavior, giving rise to phenomena like interference or diffraction that are inconsistent with the corpuscular model. But it is well known that when the wave length is very small with respect to the spatial scale of variation of the wave amplitude, the approximation of *geometrical optics* can be used. It is thus possible to replace wave fronts with (orthogonal to them) ray paths, losing however the possibility to describe interference and diffraction . So the idea of Schrödinger can be stated in this way: classical mechanics is the limit (in the sense of geometrical optics) of a more general *wave mechanics* and the quantization of the physical observables, like energy, is caused by the boundary conditions, which force the wave length to take discrete values. We are therefore led to study the wave propagation in the manifold M, the configuration space of our mechanical system. The wave equation is

$$\Delta_M \phi - \frac{\nu^2}{c^2} \frac{\partial^2 \phi}{\partial t^2} = 0,$$

where Δ_M is the Laplacian operator in the Riemannian manifold M

$$\Delta_M \phi = \frac{1}{\sqrt{g}} \partial_h \left(\sqrt{g} g^{hk} \partial_k \phi \right), \quad g = \det(g_{hk}), \quad \partial_h = \frac{\partial}{\partial q^h},$$

ν is the *refraction index* (which is in general a function of the position) and c the speed of propagation of the wave in empty space. In the stationary case put

$$\phi(q^h, t) = \psi(q^h) e^{i\omega t}, \quad \omega = 2\pi \frac{c}{\lambda_0},$$

(2)We use for a moment the same symbol H denoting the Hamiltonian of the Kepler Problem.

where λ_0 is the wave length in empty space, so that the wave equation may be written as

$$\Delta_M \psi + 4\pi^2 \frac{\nu^2}{\lambda_0^2} \psi = 0,$$

i.e., the *Helmholtz Equation.* Since $\psi(q^h)$ is a complex-valued function, the above equation is equivalent to two real equations, which we find putting

$$\psi(q^h) = A(q^h) e^{\frac{2\pi i}{\lambda_0} W(q^h)}.$$

The real functions $A(q^h)$ and $W(q^h)$ are the *amplitude* and *phase* of the wave, respectively. Replacing the above expression into the Helmholtz Equation and separating imaginary and real part, we obtain

$$\nabla \cdot (A^2 \nabla W) = 0, \quad (\nabla W)^2 = \nu^2 + \frac{\lambda_0^2}{4\pi^2} \frac{\Delta_M A}{A}. \tag{4.1.2}$$

The former equation is the continuity equation of a fluid with density A^2 and velocity ∇W. The limit process from wave to geometrical optics entails taking the relative amplitude variations as negligible on intervals comparable with the wave length. With this assumption, the latter of the above equations becomes

$$(\nabla W_{\text{cl}})^2 = \nu^2,$$

where W_{cl} is the approximate "classical" solution. This equation is the *Eikonal Equation* and the surfaces W_{cl} = constant are the wave fronts. Notice that adding to the Helmholtz Equation a term linear in $\nabla \psi$ we obtain, by the limit process, the same Eikonal Equation. Thus, the transition from geometrical to wave optics exhibits some ambiguities, if one considers the Eikonal Equation only. In other words, merely geometrical considerations are not sufficient to determine unambiguously the second order propagation equation, if one knows only the equation of the wave fronts. If, on the contrary, to the Eikonal Equation we add the first of the two Equations (4.1.2), involving also the amplitude A, the form of the second order equation is unambiguously determined.

It is well known that the integration of the Eikonal Equation, owing to the method of Cauchy bicharacteristics, is equivalent to the integration of a system of n second order differential equations. Using the line element $ds = \sqrt{g_{hk} dq^h dq^k}$ as independent variable, this system is

$$\frac{D}{Ds} \left(\nu g_{hk} \frac{dq^k}{ds} \right) = \partial_h \nu, \tag{4.1.3}$$

where $\frac{D}{Ds}$ is the absolute derivative. The bicharacteristic equations of the wave equation describe the ray paths in the geometrical optics approximation.

Let us consider the Lagrange Equations (C.1.1), which we write in the form

$$\frac{D}{Dt}\left(g_{hk}\dot{q}^k\right) = -\partial_h V.$$

Replacing $dt = \frac{ds}{\sqrt{2(H-V)}}$, where H is the constant value of the energy, we obtain

$$\frac{D}{Ds}\left(\sqrt{2(H-V)}g_{hk}\frac{dq^k}{ds}\right) = \partial_h\sqrt{2(H-V)}.$$

Comparing these last with the bicharacteristic Equations (4.1.3), we notice that they coincide, provided that the correspondence

$$\nu \longleftrightarrow \sqrt{2\,(H-V)}$$

holds.

Equivalently, we notice the analogy between the Eikonal Equation and the Hamilton-Jacobi Equation relative to the Hamiltonian (4.1.1)

$$(\nabla S)^2 = 2\,(H-V)\,,$$

or between the Fermat variational principle

$$\delta\int \nu\, ds = 0$$

and the Maupertuis principle

$$\delta\int\sqrt{2\,(H-V)}\,ds = 0.$$

It is thus natural to introduce, in classical mechanics, the equivalent of the Helmholtz Equation. But, as said, there are many second order equations that, in the geometrical optics approximation, reduce to the Eikonal Equation. We need, for the classical mechanics, the analog of the first of (4.1.2), i.e., a sort of continuity equation. To find this latter, let us define the scalar density in the configuration manifold

$$D(q^h,\alpha_k) = \det\left(\frac{\partial^2 S(q^h,\alpha_k)}{\partial q^h\partial\alpha_k}\right),$$

where $S(q^h,\alpha_k)$ is a complete integral of the Hamilton-Jacobi Equation, hence $D(q^h,\alpha_k) \neq 0$. Define the velocity field in the configuration manifold

$$v = \frac{\partial H(q,\nabla S)}{\partial\,(\nabla S)}.$$

We can prove that

$$\nabla\cdot(Dv) = 0.$$

Indeed, differentiate the Hamilton–Jacobi Equation $H(q, \nabla S) = \text{constant}$ with respect to α_k and q^h . By putting $D^k_h = \frac{\partial^2 S}{\partial \alpha_k \partial q^h}$ and denoting with $\check{D}^i_j$ the inverse matrix (it always exists since S is a complete integral) we have $\partial_h \left(v^i D^k_i \right) = 0$. Contracting with $\check{D}^h_k$, remarking that $\partial_h D^i_k = \partial_k D^i_h$ and $dD = D\check{D}^k_h dD^h_k$, completes the proof.

Therefore, D can be interpreted as the particle density. If we consider one particle only and we know the constants of motion α_k, but not the other initial conditions, D may be interpreted as a probability density. Finally, we can write the analog of the Helmholtz Equation for classical mechanics, i.e., the stationary Schrödinger Equation

$$\Delta_M \psi + \frac{2(H - V)}{\hbar^2} \psi = 0,$$

where $\hbar$, the Planck constant divided by 2π, is the quantum analog of $\frac{\lambda_0}{2\pi}$, and plays the role of a sort of material wave length. One cannot, *a priori*, say anything about the value of this constant, all information being deferred to the experiment. Hereafter we pose for simplicity $\hbar = 1$. It is preferable to write the Schrödinger Equation as

$$\left[-\frac{1}{2} \Delta_M + V \right] \psi = H \psi,$$

to stress that we are concerned with an eigenvalue problem. The correspondence

$$\sqrt{g} \psi^* \psi \longleftrightarrow D, \quad \psi^* = \text{complex conjugate},$$

suggest of interpreting $\sqrt{g} \psi^* \psi$ as the probability density of the particle, and

$$\int_\tau \sqrt{g} \psi^* \psi \, dq^1 \ldots dq^n$$

as the probability of finding the particle in the volume τ.

We must warn the reader: from the above arguments, one could infer that the wave function ψ describes a sort of material field or that probabilistic concepts arise only from our ignorance of the initial conditions of the particle. Although some people adhere to these (in our opinion naive) interpretations, the physical reality seems to be much more subtle, as the Einstein–Podolsky–Rosen argument and the violation of the Bell inequalities indicate. But we will not go further into this mined field.

At this point, we are ready to write and solve the Schrödinger Equation for the Kepler Problem in those four coordinate systems, seen in the previous chapter, that admit separation of variables.

4.1.1 Spherical Coordinates

The metric tensor in covariant form is

$$(g_{hk}) = \begin{pmatrix} 1 & 0 & 0 \\ 0 & q^2 & 0 \\ 0 & 0 & q^2 \sin^2 \vartheta \end{pmatrix},$$

and the Schrödinger Equation for the Kepler Problem is

$$-\frac{1}{2}\left[\frac{1}{q^2}\partial_q\left(q^2\partial_q\right) + \frac{1}{q^2}\Delta_{S^2}\right]\psi - \frac{1}{q}\psi = H\psi, \tag{4.1.4}$$

where Δ_{S^2} is the Laplacian on the two-dimensional unitary sphere. Separate the variables by putting

$$\psi = R(q)Y(\zeta),$$

where ζ are coordinates on the unitary sphere. As shown in the appendix to this chapter,

$$\left[\Delta_{S^2} + \ell\left(\ell+1\right)\right] f_\ell\left(\zeta\right) = 0$$

holds, where $f_\ell\left(\zeta\right)$ is a homogeneous harmonic polynomial, of degree ℓ in the variables q_1, q_2, q_3, restricted to the unitary sphere. The homogeneous harmonic polynomials of degree ℓ, restricted to the unitary sphere, are called *spherical harmonics.* To find explicitly these eigenfunctions, write the above equation in the coordinates ϑ, φ :

$$\frac{1}{\sin\vartheta}\frac{\partial}{\partial\vartheta}\left(\sin\vartheta\frac{\partial Y(\vartheta,\varphi)}{\partial\vartheta}\right) + \frac{1}{\sin^2\vartheta}\frac{\partial^2 Y(\vartheta,\varphi)}{\partial\varphi^2} + \lambda Y(\vartheta,\varphi) = 0,$$

leaving the eigenvalue parameter λ unspecified. Separate further the variables by putting

$$Y(\vartheta,\varphi) = \Theta(\vartheta)\Phi(\varphi),$$

so that the equation breaks into the two equivalent

$$\frac{d^2\Phi}{d\varphi^2} + \mu\Phi = 0,$$

$$\frac{1}{\sin\vartheta}\frac{d}{d\vartheta}\left(\sin\vartheta\frac{d\Theta}{d\vartheta}\right) + \left(\lambda - \frac{\mu}{\sin^2\vartheta}\right)\Theta = 0.$$

The first is easily integrated: since Φ must be univalued, then $\mu = m^2$, where m is an integer, and[(3)] $\Phi = e^{im\varphi}$. The latter, putting $x = \cos\vartheta$, becomes

$$(1-x^2)\frac{d^2\Theta}{dx^2} - 2x\frac{d\Theta}{dx} + \left(\lambda - \frac{m^2}{1-x^2}\right)\Theta = 0. \tag{4.1.5}$$

[(3)] We systematically omit the normalization factor.

This equation has (see Appendix 4.A.1), three singular points: $x = \pm 1$ and $x = \infty$, and these singularities are all fuchsian. Since $-1 \leq x \leq 1$ we consider the first two points only. With the characteristic equation method (see (4.A.4) on page 89), we find that, in the neighbourhood of $x = +1$, the function Θ has the form

$$\Theta = (1-x)^{\frac{m}{2}} f_1(1-x),$$

where $f_1(0) \neq 0$. The negative values of m are to be rejected in order to avoid a singularity in $x = +1$. Similarly for the other point $x = -1$, we obtain

$$\Theta = (1+x)^{\frac{m}{2}} f_2(1+x).$$

Since these two expressions are two different developments of the same branch of the continuous function Θ, we can write

$$\Theta = (1-x^2)^{\frac{|m|}{2}} P(x)$$

with

$$P(x) = \sum_{r=0}^{\infty} a_r x^r.$$

Substituting these two expressions into the Equation (4.1.5), one finds for $P(x)$ the equation

$$(1-x^2)\frac{d^2P}{dx^2} - 2(|m|+1)x\frac{dP}{dx} + [\lambda - |m|\,(|m|+1)]\,P = 0, \quad (4.1.6)$$

and for its coefficients a_r the recurrence relation

$$(r+1)(r+2)a_{r+2} = [r(r-1) + 2(|m|+1)r - \lambda + |m|\,(|m|+1)]\,a_r.$$

One may therefore take for P a series of odd or even powers, with an arbitrary first coefficient. To have P regular in the interval we must impose that the series terminates at some point, so that P will be a polynomial. Putting $a_{y+2} = 0$ with $a_y \neq 0$, we obtain

$$\lambda = y(y-1) + 2(|m|+1)y + |m|\,(|m|+1) = (y+|m|)(y+|m|+1),$$

and with $\ell = y + |m|$ we find the known result. Clearly, $\ell \geq 0$ and $-\ell \leq m \leq \ell$. The explicit form of the function Θ is obtained by recurrence relation or invoking the *associated Legendre functions*; see the relevant appendix to this chapter. Indeed, compare (4.1.6) with (4.A.5) and (4.1.5) with (4.A.6).

Summing up, to each eigenvalue $\lambda = \ell(\ell+1)$ there correspond $2\ell+1$ eigenfunctions $Y_\ell^m(\vartheta,\varphi)$, with $m = -\ell, -\ell+1, \ldots, 0, 1, \ldots, \ell$, given by

$$Y_\ell^m(\vartheta,\varphi) = e^{im\varphi}\sin^{|m|}\vartheta\,\frac{d^{|m|}P_\ell(\cos\vartheta)}{d(\cos\vartheta)^{|m|}},$$

where we have taken into account the expression found in the appendix. Their explicit form for the first values of ℓ and m is

$$
\begin{aligned}
&Y_1^0 = \cos\vartheta, \quad Y_1^{\mp 1} = \sin\vartheta e^{\mp i\varphi},\\
&Y_2^0 = 3\cos^2\vartheta - 1, \quad Y_2^{\mp 1} = \sin\vartheta\cos\vartheta e^{\mp i\varphi}, \quad Y_2^{\mp 2} = \sin^2\vartheta e^{\mp 2i\varphi},\\
&Y_3^0 = 5\cos^3\vartheta - 3\cos\vartheta, \quad Y_3^{\mp 1} = \sin\vartheta(5\cos^2\vartheta - 1)e^{\mp i\varphi},\\
&Y_3^{\mp 2} = \sin^2\vartheta\cos\vartheta e^{\mp 2i\varphi}, \quad Y_3^{\mp 3} = \sin^3\vartheta e^{\mp 3i\varphi}.
\end{aligned}
$$

Notice that, with the substitutions

$$\sin\vartheta e^{\mp i\varphi} = \frac{q_1 \mp i q_2}{q}, \quad \cos\vartheta = \frac{q_3}{q},$$

the functions $q^\ell Y_\ell^m$ for any fixed (not summed) ℓ become $2\ell + 1$ independent homogeneous harmonic polynomials of degree ℓ, forming thus a basis for $\mathcal{H}^{2,\ell}$ (notation of Theorem 4.2 on page 91)

Having solved the spherical part, the Schrödinger Equation is reduced to the ordinary differential equation

$$\frac{d^2R}{dq^2} + \frac{2}{q}\frac{dR}{dq} + \left[2H + \frac{2}{q} - \frac{\ell(\ell+1)}{q^2}\right]R = 0,$$

which can be simplified by changing both the dependent and independent variables:

$$R(q) = \frac{y(q)}{q}, \quad q = \frac{1}{2\sqrt{2|H|}}x,$$

so that

$$\frac{d^2y}{dx^2} + \left[\pm\frac{1}{4} + \frac{\Lambda}{x} - \frac{\ell(\ell+1)}{x^2}\right]y = 0, \quad \Lambda = \frac{1}{\sqrt{2|H|}}. \tag{4.1.7}$$

The two signs takes into account the fact that H may be positive or negative. Note that, for negative energy, Λ is equal to the action $\mathcal{L}$. The equation has a fuchsian singularity for $x = 0$ and a nonfuchsian one for $x = \infty$. We must find the values of the parameter Λ in such a way that this equation has solutions satisfying the conditions: $y(0) = 0$ and $y < \infty$ for $x \to \infty$. We treat separately the two cases: $H < 0$ (negative sign in the above equation) and $H > 0$ (positive sign).

Energy $H < 0$

For $x \to \infty$ Equation (4.1.7) tends to the asymptotic form

$$\frac{d^2y}{dx^2} - \frac{1}{4}y = 0,$$

and the fundamental solutions are

$$y = e^{\pm\frac{x}{2}}.$$

The asymptotic solution with the + sign is discarded, since it tends to ∞. We are led to seek a solution of Equation (4.1.7) of the type

$$y(x) = e^{-\frac{x}{2}} v(x).$$

Replacing this expression, the equation for $v(x)$ becomes

$$\frac{d^2 v}{dx^2} - \frac{dv}{dx} + \left[\frac{\Lambda}{x} - \frac{\ell(\ell+1)}{x^2}\right] v = 0,$$

whose characteristic equation is

$$\alpha(\alpha - 1) - \ell(\ell + 1) = 0,$$

from which $\alpha = -\ell$ or $\alpha = \ell+1$. Discarding the negative solution ($v \to \infty$ for $x \to 0$), we put

$$v(x) = x^{\ell+1} w(x),$$

where $w(x)$ is of the type

$$w(x) = \sum_{k=0}^{\infty} a_k x^k.$$

Then $w(x)$ must satisfy the equation

$$\frac{d^2 w}{dx^2} + \left[\frac{2(\ell+1)}{x} - 1\right]\frac{dw}{dx} + \frac{\Lambda - (\ell+1)}{x} w = 0. \tag{4.1.8}$$

Inserting the series expansion and equating to zero the coefficients of each power, one finds, for the a's, the recurrence formula

$$a_{k+1} = a_k \frac{k + \ell + 1 - \Lambda}{(k+1)(k + 2\ell + 2)}.$$

Since we require that $w(x)$ is quadratically integrable, it must be a polynomial (of degree n', say). The series must therefore terminate, i.e., $a_{n'+1} = 0$, with $a_{n'} \neq 0$, so that

$$n' + \ell + 1 - \Lambda = 0.$$

We have reached a fundamental result: $\Lambda = \frac{1}{\sqrt{2|H|}}$ can take only integer and strictly positive values $N = n' + \ell + 1$; this implies that the energy is quantized

$$H = -\frac{1}{2N^2} \quad N = 1, 2, \ldots,$$

which is the energy spectrum. Since $\ell = 0, 1, 2, \ldots, N-1$, and for every value of ℓ the integer m spans the values between $-\ell$ and ℓ, the degeneracy[4] of the energy level labelled by N is

$$\sum_{\ell=0}^{N-1} (2\ell + 1) = N^2.$$

The explicit form of the polynomials $w(x)$ is obtained with the recurrence formula for the a's, or by noticing that Equation (4.1.8) becomes Equation (4.A.8) on page 94 of the generalized Laguerre polynomials $L_k^{(j)}$, if we agree to put

$$j = 2\ell + 1, \quad k = 2\ell + 1 + n' = N + \ell.$$

The polynomials $w(x)$ are hence proportional to $L_{N+\ell}^{(2\ell+1)}$, and the proportionality factor is determined by imposing the usual normalization conditions.

Energy $H > 0$

For $x \to \infty$, (4.1.7) has the asymptotic form

$$\frac{d^2y}{dx^2} + \frac{1}{4}y = 0,$$

whose general integral is

$$y = c_1 e^{i\frac{x}{2}} + c_2 e^{-i\frac{x}{2}}.$$

It is always bounded: thus we are not forced to impose any restriction on Λ, every value of the energy H is an eigenvalue, and the spectrum is continuous.

As in the previous case, we put

$$y = e^{\pm i\frac{x}{2}} v(x),$$

which gives for $v(x)$ the equation

$$\frac{d^2v}{dx^2} \pm i\frac{dv}{dx} + \frac{\Lambda - \ell(\ell+1)}{x^2} v = 0.$$

Putting

$$v = x^{\ell+1} w, \quad w = \sum_{s=o}^{\infty} a_s x^s,$$

[4] *Degeneracy* is a technical term of physicists, denoting the number of the independent eigenfunctions associated with the same fixed eigenvalue.

we find the recurrence formula

$$a_{s+1} = a_s \frac{-\Lambda \mp i(\ell + s + 1)}{(2\ell + s + 2)(s + 1)}.$$

Since Λ is real, the numerator of this fraction will never vanish, and the series does not truncate. The general integral is the real part of

$$y = e^{i\frac{x}{2}} x^{\ell+1} \sum_{s=o}^{\infty} a_s x^s,$$

and the series converges for every value of x.

4.1.2 Parabolic Coordinates

The integration of the Schrödinger Equation in parabolic coordinates is more direct after the change of coordinates

$$\rho_1 = q + q_3 = \eta^2, \quad \rho_2 = q - q_3 = \xi^2.$$

The metric tensor in covariant form is

$$(g_{hk}) = \begin{pmatrix} \frac{\rho_1+\rho_2}{4\rho_1} & 0 & 0 \\ 0 & \frac{\rho_1+\rho_2}{4\rho_2} & 0 \\ 0 & 0 & \rho_1\rho_2 \end{pmatrix}.$$

The wave equation is

$$-\frac{1}{2}\left[\frac{4}{\rho_1+\rho_2}\frac{\partial}{\partial\rho_1}\left(\rho_1\frac{\partial}{\partial\rho_1}\right) + \frac{4}{\rho_1+\rho_2}\frac{\partial}{\partial\rho_2}\left(\rho_2\frac{\partial}{\partial\rho_2}\right) + \frac{1}{\rho_1\rho_2}\frac{\partial^2}{\partial\varphi^2}\psi\right] - \frac{2}{\rho_1+\rho_2}\psi = H\psi.$$

In order to separate the variables put

$$\psi(\rho_1, \rho_2, \varphi) = f_1(\rho_1) f_2(\rho_2) \Phi(\varphi)$$

and obtain

$$\frac{d}{d\rho_1}\left(\rho_1\frac{df_1}{d\rho_1}\right) - \left(\frac{m^2}{4\rho_1} - \frac{H}{2}\rho_1 - \beta\right) f_1 = 0,$$

$$\frac{d}{d\rho_2}\left(\rho_2\frac{df_2}{d\rho_2}\right) - \left(\frac{m^2}{4\rho_2} - \frac{H}{2}\rho_2 + \beta - 1\right) f_2 = 0,$$

$$\Phi(\varphi) = e^{im\varphi}, \quad m = 0, \pm 1, \pm 2, \ldots,$$

where β is a separation constant. Let us suppose that $H < 0$ and put

$$\sigma_1 = \sqrt{-2H}\rho_1, \quad \sigma_2 = \sqrt{-2H}\rho_2,$$
$$\lambda_1 = \frac{\beta}{\sqrt{-2H}}, \quad \lambda_2 = \frac{1-\beta}{\sqrt{-2H}},$$

then the two differential equations become

$$\frac{d}{d\sigma_k}\left(\sigma_k \frac{df_k}{d\sigma_k}\right) + \left(\lambda_k - \frac{\sigma_k}{4} - \frac{m^2}{4\sigma_k}\right) f_k = 0, \quad k = 1, 2 \quad (k \text{ not summed}).$$

We note that, if $\sigma_k \to \infty$, then $f_k \to e^{-\frac{\sigma_k}{2}}$ and, if $\sigma_k \to 0$, then $f_k \to \sigma_k^{\frac{|m|}{2}}$. Therefore we seek two solutions of the type

$$f_k = e^{-\frac{\sigma_k}{2}} \sigma_k^{\frac{|m|}{2}} w(\sigma_k).$$

We obtain

$$\sigma_k \frac{d^2 w}{d\sigma_k^2} + (|m| + 1 - \sigma_k) \frac{dw}{d\sigma_k} + \left[\lambda_k - \frac{1}{2}(|m| + 1)\right] w = 0,$$

that is the differential equation of the generalized Laguerre polynomials. The two solutions are

$$w(\sigma_k) = L^{|m|}_{N_k+|m|}(\sigma_k), \quad N_k = \lambda_k - \frac{1}{2}(|m| + 1),$$

where N_1 and N_2 are nonnegative integers. Define the strictly positive integer

$$N = \lambda_1 + \lambda_2 = N_1 + N_2 + |m| + 1.$$

From the definitions of λ_1 and λ_2, we find the known energy spectrum:

$$H = -\frac{1}{2N^2}, \quad N = 1, 2, \dots .$$

The degeneracy of the energy level, labelled by N, is equal to the number of different ways in which one can combine N_1, N_2 and m to obtain N. For $m = 0$, there are N ways to choose N_1 and N_2; for $|m| \neq 0$, there are two ways to choose m (i.e., $\pm m$), and $N - |m|$ ways to choose N_1 and N_2. The degeneracy of the N-level is therefore

$$N + 2 \sum_{|m|=1}^{N-1} (N - |m|) = N^2,$$

which agrees with the previous result.

4.1.3 Elliptic Coordinates

The metric tensor in covariant form is

$$(g_{hk}) = \begin{pmatrix} \frac{\xi^2-\eta^2}{\xi^2-1} & 0 & 0 \\ 0 & \frac{\xi^2-\eta^2}{1-\eta^2} & 0 \\ 0 & 0 & (\xi^2-1)(1-\eta^2) \end{pmatrix},$$

and the Schrödinger Equation is

$$\frac{\partial}{\partial\xi}\left[\left(\xi^2-1\right)\frac{\partial\psi}{\partial\xi}\right]+\frac{\partial}{\partial\eta}\left[\left(1-\eta^2\right)\frac{\partial\psi}{\partial\eta}\right]$$
$$+\left(\frac{1}{\xi^2-1}+\frac{1}{1-\eta^2}\right)\frac{\partial^2\psi}{\partial\varphi^2}+\left[2H\left(\xi^2-\eta^2\right)+2\xi-2\eta\right]\psi=0.$$

In order to separate the variables[5] put $\psi(\xi,\eta,\varphi)=X(\xi)\,Y(\eta)\,\Phi(\varphi)$. The following equations are obtained

$$\frac{d}{d\xi}\left[\left(\xi^2-1\right)\frac{dX}{d\xi}\right]+\left[A+2\xi+2H\xi^2-\frac{m^2}{\xi^2-1}\right]X=0,$$
$$\frac{d}{d\eta}\left[\left(1-\eta^2\right)\frac{dY}{d\eta}\right]-\left[A+2\eta+2H\eta^2+\frac{m^2}{1-\eta^2}\right]Y=0,$$
$$\Phi(\varphi)=e^{im\varphi},\quad m=0,\pm1,\pm2,\ldots,$$

where A is a separation constant. The two equations for X and Y take exactly the same form, but we cannot conclude at once that the functions $X(\xi)$ and $Y(\eta)$ will be the same, because the ranges of ξ and η are different: however, it turns out that the arguments for Y are parallel to the following for X

Arguing as previously, we are led to searching a solution of the form

$$X(\xi)=e^{-\sqrt{-2H}\xi}\left(\xi^2-1\right)^{\frac{|m|}{2}}\sum_{s=0}^{\infty}a_s(\xi-1)^s,\quad a_0\neq0.$$

A calculation shows that the coefficients a_s must satisfy the recurrence formula

$$\alpha_s a_{s+1}+(\beta_s+A)\,a_s+\gamma_s a_{s-1}=0,$$

(5) We will follow, with some minor changes in the notation, the exposition of Coulson & Robinson (1958), to which we refer for the omitted details.

where

$$\alpha_s = 2(s+1)(s+|m|+1),$$
$$\beta_s = s^2 + s\left(2|m| + 1 - 4\sqrt{-2H}\right) + 2H + 2 - 2\sqrt{-2H}\,(|m|+1) + |m|\,(|m|+1),$$
$$\gamma_s = 2\sqrt{-2H}\left(\frac{1}{\sqrt{-2H}} - |m| - s\right).$$

For a physically acceptable behavior the series must terminate to some finite order, so suppose a_{M-1} is the last nonzero coefficient in the series, i.e., $a_M = a_{M+1} = \cdots = 0$. Then since

$$\alpha_M a_{M+1} + (\beta_M + A)\, a_M + \gamma_M a_{M-1} = 0,$$

we must have $\gamma_M = 0$, i.e., $\frac{1}{\sqrt{-2H}} = |m| + M =$ a positive integer N, say, and we find again the familiar energy spectrum.

We want to calculate the discrete values of the separation constant A and the $M = N - |m|$ coefficients a_s of the series. The recurrence formula gives the homogeneous algebraic system of M linear equations

$$(\beta_0 + A)a_0 + \alpha_0 a_1 = 0,$$
$$\gamma_1 a_0 + (\beta_1 + A)a_1 + \alpha_1 a_2 = 0,$$
$$\gamma_2 a_1 + (\beta_2 + A)a_2 + \alpha_2 a_3 = 0,$$
$$\cdots\cdots\cdots\cdots\cdots\cdots$$
$$\gamma_{M-1} a_{M-2} + (\beta_{M-1} + A)a_{M-1} = 0.$$

Defining the tridiagonal matrix **M** and the vector $\underline{a}$

$$\mathbf{M} = \begin{pmatrix} \beta_0 & \alpha_0 & & & & \\ \gamma_1 & \beta_1 & \alpha_1 & & & \\ & \gamma_2 & \beta_2 & \cdot & & \\ & & \cdot & \cdot & \cdot & \\ & & & \cdot & \cdot & \alpha_{M-2} \\ & & & & \gamma_{M-1} & \beta_{M-1} \end{pmatrix}, \quad \underline{a} = \begin{pmatrix} a_0 \\ a_1 \\ \vdots \\ a_{M-1} \end{pmatrix},$$

the linear equation system may be written as

$$\mathbf{M}\underline{a} = -A\underline{a},$$

showing that the values that $-A$ can assume are the eigenvalues of the matrix **M**. As one easily verifies, the matrix

$$\mathbf{M}' = \begin{pmatrix} \beta_0 & \sqrt{\alpha_0\gamma_1} & & & & \\ \sqrt{\alpha_0\gamma_1} & \beta_1 & \sqrt{\alpha_1\gamma_2} & & & \\ & \sqrt{\alpha_1\gamma_2} & \beta_2 & \cdot & & \\ & & \cdot & \cdot & \cdot & \\ & & & \cdot & \cdot & \sqrt{\alpha_{M-2}\gamma_{M-1}} \\ & & & & \sqrt{\alpha_{M-2}\gamma_{M-1}} & \beta_{M-1} \end{pmatrix}$$

(which is well defined since all the α_s and γ_s are positive) and the matrix **M** satisfy the same characteristic equation and therefore have the same eigenvalues. Since **M**′ is symmetric, its eigenvalues are all real and moreover, as a deeper analysis shows, are all distinct: see Coulson & Robinson (1958) for the details. The coefficients a_s are defined up to an arbitrary multiplicative constant, which can be fixed with the normalization conditions. The calculation of the degeneracy is similar to the previous case.

The integration of the Schrödinger Equation in elliptic coordinates is reduced to a *finite* eigenvalue and eigenvector problem, which is an algebraic one. Unfortunately, for $M > 3$ the characteristic equation is of degree > 4, and only numerical solutions are available.

4.1.4 Spheroconical Coordinates

The metric tensor in spheroconical coordinates is

$$(g_{hk}) = \begin{pmatrix} 1 & 0 & 0 \\ 0 & q^2 \frac{\xi^2+\eta^2}{(b^2+\xi^2)(a^2-\xi^2)} & 0 \\ 0 & 0 & q^2 \frac{\xi^2+\eta^2}{(b^2-\eta^2)(a^2+\eta^2)} \end{pmatrix}.$$

The Schrödinger Equation is similar to Equation (4.1.4), but with the Laplacian on the sphere S^2 expressed in spheroconical coordinates. Separating the radial from the spherical part, the former turns out to be exactly that already solved treating the case of the spherical coordinates, so we will consider the latter only

$$[\Delta_{S^2} + \ell\,(\ell+1)]\,\psi_\ell(\xi,\eta) = 0.$$

Putting $\psi_\ell(\xi,\eta) = U_\ell(\xi)V_\ell(\eta)$, the above equation breaks into the following two Lamé Equations[(6)]:

$$\sqrt{(b^2+\xi^2)\,(a^2-\xi^2)}\frac{d}{d\xi}\left(\sqrt{(b^2+\xi^2)\,(a^2-\xi^2)}\frac{dU_\ell}{d\xi}\right) + \left[\ell\,(\ell+1)\,\xi^2 + B\right]U_\ell = 0,$$

$$\sqrt{(b^2-\eta^2)\,(a^2+\eta^2)}\frac{d}{d\eta}\left(\sqrt{(b^2-\eta^2)\,(a^2+\eta^2)}\frac{dV_\ell}{d\eta}\right) + \left[\ell\,(\ell+1)\,\eta^2 - B\right]V_\ell = 0,$$

(6) A detailed study of the Lamé Equation and the related *ellipsoidal harmonics* can be found in Whittaker & Watson (1952).

where B is a separation constant. Since $V_\ell(\eta) = U_\ell(i\eta)$, a discussion of the former equation will suffice. It has fuchsian singularities at $\xi = \pm a$, $\xi = \pm ib$ and ∞. The solutions must be of the form[7]

$$U_\ell(\xi) = \left(b^2 + \xi^2\right)^{\frac{k_1}{2}} \left(a^2 - \xi^2\right)^{\frac{k_2}{2}} \xi^{k_3} \sum_{r=0}^{\infty} a_r \xi^{2r}, \tag{4.1.9}$$

where k_1, k_2 and k_3 can each have the value 1 or 0: the values of k_1, k_2 assure proper behavior at the singular points $\pm a, \pm ib$. Substituting the "ansatz" (4.1.9) in the first Lamé Equation, we find that the coefficients a_r have to satisfy a recurrence relation.

Let us suppose first that $k_1 = k_2 = k_3 = 0$. The recurrence relation is, as in the previous case of the elliptic coordinates,

$$\alpha_r a_{r+1} + (\beta_r + B)\, a_r + \gamma_r a_{r-1} = 0, \quad a_{-1} = 0,$$

but with different definitions:

$$\begin{aligned} \alpha_r &= 2a^2b^2(r+1)(2r+1), \\ \beta_r &= 4(a^2 - b^2)r^2, \\ \gamma_r &= (\ell - 2r + 2)(\ell + 2r - 1). \end{aligned}$$

If $U_\ell(\xi)$ is to be well behaved at infinity, the function $\sum a_r \xi^{2r}$ must be a polynomial. To this end, we notice that $\gamma_r = 0$ for $r = \frac{\ell}{2} + 1$ (we suppose thus ℓ even) and if we succeed in choosing B in such a way that $a_r = 0$, the recurrence relation shows that $a_{r+1} = 0$ too and consequently $a_{r+2} = \ldots = 0$: the last index r for which $a_r \neq 0$ is $r = \frac{\ell}{2}$. But a_r is a polynomial of degree r in B, as one can see by an inductive argument, and the problem is reduced to solving an algebraic equation of degree $\frac{\ell}{2} + 1$. It is possible to prove that this equation has all its $\frac{\ell}{2} + 1$ roots real and distinct. Another way to reach the same conclusion is to notice that we must impose consistency conditions on the eigenvalue equation

$$\mathbf{N}\underline{a} = -B\underline{a},$$

where

$$\mathbf{N} = \begin{pmatrix} \beta_0 & \alpha_0 & & & & \\ \gamma_1 & \beta_1 & \alpha_1 & & & \\ & \gamma_2 & \beta_2 & \cdot & & \\ & & \cdot & \cdot & \cdot & \\ & & & \cdot & \cdot & \alpha_{\frac{\ell}{2}-1} \\ & & & & \gamma_{\frac{\ell}{2}} & \beta_{\frac{\ell}{2}} \end{pmatrix}, \qquad \underline{a} = \begin{pmatrix} a_0 \\ a_1 \\ \vdots \\ a_{\frac{\ell}{2}} \end{pmatrix},$$

[7] Of course, the coefficients a_r have nothing to do with the parameter a occurring in the definition of the spheroconical coordinates.

then the proof follows the same line of reasoning of the previous case of the elliptic coordinates. The $\frac{\ell}{2}+1$ functions $U_\ell(\xi)$, that we have constructed, are known as *ellipsoidal harmonics of the first kind.*

Let us now suppose that $k_1 = 1$ and $k_2 = k_3 = 0$. In this case we find, after similar calculations, that $\gamma_r = 0$ for $r = \frac{1}{2}(\ell+1)$ and thus we must suppose ℓ odd. The last index for which $a_r \neq 0$ is $r = \frac{1}{2}(\ell-1)$ and we can construct $\frac{1}{2}(\ell+1)$ functions $U_\ell(\xi)$: these are the *ellipsoidal harmonics of the second kind.* The same conclusions are drawn for the other two cases: $k_2 = 1, k_1 = k_3 = 0$ and $k_3 = 1, k_1, k_2 = 0$, and so we have $\frac{3}{2}(\ell+1)$ harmonics of the second kind.

In the other three cases in which only one k is equal to zero, we find that $\gamma_r = 0$ for $r = \frac{\ell}{2}$ and thus we must suppose ℓ even. The last index for which $a_r \neq 0$ is $r = \frac{1}{2}(\ell-2)$, and we can construct $\frac{3}{2}\ell$ *ellipsoidal harmonics of the third kind.*

Lastly, if all the three k's are different from zero, we find that $\gamma_r = 0$ for $r = \frac{1}{2}(\ell-1)$ and thus we must suppose ℓ odd. The last index for which $a_r \neq 0$ is $r = \frac{1}{2}(\ell-3)$, and we can construct $\frac{1}{2}(\ell-1)$ *ellipsoidal harmonics of the fourth kind.*

Define $\sigma = k_1 + k_2 + k_3$. In order to have harmonics of the first, second, third or fourth kind, σ must assume the values $0, 1, 2$ or 3, respectively. Let us check the total number of ellipsoidal harmonics that we can construct for each fixed ℓ: if ℓ is even, then $\frac{\ell}{2}+1+\frac{3}{2}\ell = 2\ell+1$, while if ℓ is odd, then $\frac{3}{2}(\ell+1)+\frac{1}{2}(\ell-1) = 2\ell+1$, as expected.

It is interesting to find the relation that exists between ellipsoidal harmonics and harmonic homogeneous polynomials in $\mathbb{R}^3$. Since the last index for which $a_r \neq 0$ is $r = \frac{1}{2}(\ell-\sigma)$, we can write

$$\sum_{r=0}^{\frac{1}{2}(\ell-\sigma)} a_r \xi^{2r} = \prod_{s=1}^{\frac{1}{2}(\ell-\sigma)} \left(\xi^2 - \rho_s^2\right),$$

where the ρ_s's are by definition the zeros of the polynomial. The a's are the components of an eigenvector and are defined up to a multiplicative arbitrary factor, which we fix by taking $a_{\frac{1}{2}(\ell-\sigma)} = 1$. Hence

$$U_\ell(\xi) = \left(b^2+\xi^2\right)^{\frac{k_1}{2}} \left(a^2-\xi^2\right)^{\frac{k_2}{2}} \xi^{k_3} \prod_{s=1}^{\frac{1}{2}(\ell-\sigma)} \left(\xi^2-\rho_s^2\right),$$

$$V_\ell(\eta) = \left(b^2-\eta^2\right)^{\frac{k_1}{2}} \left(a^2+\eta^2\right)^{\frac{k_2}{2}} \eta^{k_3} \prod_{s=1}^{\frac{1}{2}(\ell-\sigma)} \left(\eta^2-\rho_s^2\right).$$

From the definition of the spheroconical coordinates, we have that

$$\psi_\ell = U_\ell(\xi)V_\ell(\eta) = \frac{C}{q^\sigma} q_1^{k_1} q_2^{k_3} q_3^{k_2} \prod_{s=1}^{\frac{1}{2}(\ell-\sigma)} \left(\xi^2 - \rho_s^2\right)\left(\eta^2 - \rho_s^2\right),$$

where $C = a^{k_2+k_3} b^{k_1+k_3}$. Consider the function

$$F_s = \frac{q_1^2}{b^2+\rho_s^2} + \frac{q_2^2}{\rho_s^2} - \frac{q_3^2}{a^2-\rho_s^2}.$$

It is a polynomial of the second degree in ξ and η and furthermore vanishes if $\xi^2 = \rho_s^2$ or $\eta^2 = -\rho_s^2$. Hence, we can write

$$\psi_\ell = \frac{1}{q^\ell} q_1^{k_1} q_2^{k_3} q_3^{k_2} \prod_{s=1}^{\frac{1}{2}(\ell-\sigma)} \left(\frac{q_1^2}{b^2+\rho_s^2} + \frac{q_2^2}{\rho_s^2} - \frac{q_3^2}{a^2-\rho_s^2}\right),$$

where we have omitted an inessential numerical factor. $q^\ell \psi_\ell$ (ℓ not summed) is manifestly a homogeneous polynomial of degree ℓ, and the above discussion proves moreover that it is harmonic. Unfortunately, it cannot be written explicitly, since this would require knowing all the a_r or, equivalently, all the ρ_s, i.e., solving the above algebraic equation of degree $\frac{1}{2}(\ell - \sigma) + 1$: this can be done only numerically, and only after the parameters a and b of the coordinates have been fixed.

4.2 Pauli Quantization

The *Pauli quantization* of the Kepler Problem is an instance of the general method of the *canonical quantization*, due to Dirac and Heisenberg, that we first briefly review, mainly to fix the notations. See Dirac (1958) for a presentation of the general method, and Pauli (1926), English translation in Van der Waerden (1967), for the particular application to the Kepler Problem.

4.2.1 Canonical Quantization

The basic postulates of the canonical quantization are the following.

(i) The state of a physical system is completely described by a vector of a, usually infinite-dimensional, Hilbert space, which is a vector space over the complex numbers, endowed with a scalar product and which is complete (i.e., every Cauchy sequence converges, the convergence being defined by the norm induced by the scalar product).

(ii) To every classical observable $f(q_h, p_k)$ on the phase space of a canonical Hamiltonian system, corresponds a linear Hermitian operator $\hat{f}$ acting on the Hilbert space. Defining the commutator of two operators as

$$\left[\hat{f}, \hat{g}\right] = \hat{f}\hat{g} - \hat{g}\hat{f},$$

the correspondence $f \overset{\mathcal{Q}}{\to} \hat{f}$ is called *quantization*. It must, in some sense, preserve the canonical structure, sending the Poisson bracket of two classical observables into the commutator of the corresponding quantum operators:

$$\widehat{\{f, g\}} = i\left[\hat{f}, \hat{g}\right].$$

Stating it differently, the diagram

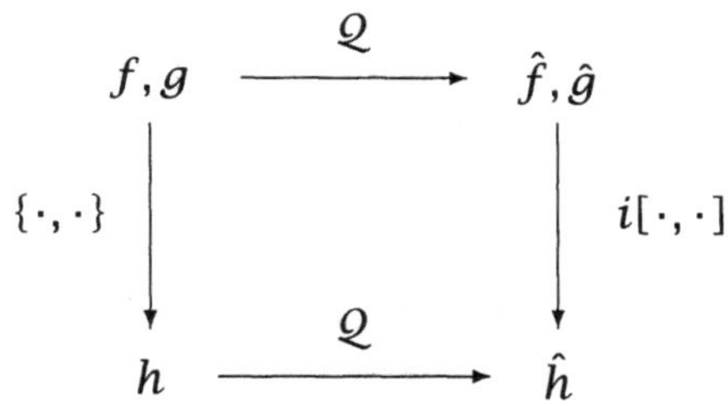

must commutate. The usual choices are

$$p_k \overset{\mathcal{Q}}{\to} \hat{p}_k = \frac{1}{i}\frac{\partial}{\partial q_k}, \quad q_k \overset{\mathcal{Q}}{\to} \hat{q}_k = q_k,$$

that are called *canonical quantization rules.*

(iii) Whereas a classical observable f can, when measured, assume every value in a continuous interval, the corresponding quantum observable can only assume the eigenvalues of the Hermitian operator $\hat{f}$. The square (calculated with the scalar product of the Hilbert space) of the eigenfunctions gives the probability distributions.

The critical point of this procedure is the second one. In some simple cases, as for a particle in Euclidean space subject to a potential, the quantization rules can be applied in an unambiguous way, and the eigenvalue equation of the energy operator agrees with the Schrödinger Equation. But, in general, these rules are plagued with the problem of the ordering: in fact, whereas, for example, there is no difference between the two classical observables fg and gf, the two operator $\hat{f}\hat{g}$ and $\hat{g}\hat{f}$ are in general different.

As shown in Groenewold (1946) and Van Hove (1951), the problem of the ordering can in general be unambiguously resolved only for the algebra, under Poisson bracket, of the polynomials that are quadratic, possibly nonhomogeneous, in the canonical variables. Thus the search for a general method to quantize unambiguously *all* the observables is hopeless and each case has to be investigated separately. Indeed, as pointed out by Guillemin & Sternberg (1984, page 104), there exist interesting groups that act as symplectic transformations and have a unitary *nonquadratic* representation, and so will have their own quantization rules. The geometric quantization of the Kepler Problem, pursued in Chapter 9, will furnish an interesting example.

4.2.2 Pauli Quantization

We recall the Poisson brackets (3.2.6) of the first integrals of the Kepler Problem, which we rewrite as follows:

$$\{G_i, G_j\} = -\epsilon_{ijk} G_k, \quad \{G_i, E_j\} = -\epsilon_{ijk} E_k, \quad \{E_i, E_j\} = 2H\epsilon_{ijk} G_k,$$

where $H < 0$ is the value of the energy. We want to quantize these first integrals. For the angular momentum, we simply pose

$$\hat{G}_i = \epsilon_{ijk} q_j \hat{p}_k.$$

For the components of the eccentricity vector, we must "symmetrize" the expressions, obtained through the application of the canonical rules, in order to make the corresponding operators Hermitian:

$$\hat{E}_i = \frac{1}{2}\epsilon_{ijk}\left(\hat{p}_j \hat{G}_k - \hat{G}_k \hat{p}_j\right) - \frac{q_i}{q}.$$

These six operators commute with the Hamiltonian $\hat{H} = \frac{1}{2}\hat{p}^2 - \frac{1}{q}$. Moreover a somewhat lengthy calculation shows that[(8)]

$$\left[\hat{G}_r, \hat{G}_j\right] = i\epsilon_{rjk}\hat{G}_k, \quad \left[\hat{G}_r, \hat{E}_j\right] = i\epsilon_{rjk}\hat{E}_k, \quad \left[\hat{E}_r, \hat{E}_j\right] = -2iH\epsilon_{rjk}\hat{G}_k,$$

where H is an eigenvalue of $\hat{H}$. These relations are to be compared with their classical version (3.2.6). They show that $\hat{G}_r$ and $\hat{R}_r = \frac{\hat{E}_r}{\sqrt{-2H}}$ generate the Lie algebra of the *four*-dimensional rotation group. Moreover

(8)Notice that historically this quantum version of the commutation rules has been found by Pauli *before* the classical version (3.2.6).

they satisfy the identities

$$\sum_{r=1}^{3} \hat{G}_r \hat{R}_r = \sum_{r=1}^{3} \hat{R}_r \hat{G}_r = 0, \quad \hat{G}^2 + \hat{R}^2 = \frac{1}{-2H} - 1,$$

which differ from the classical identities only by the term -1. Then

$$\hat{M}_r = \frac{1}{2}\left(\hat{G}_r + \hat{R}_r\right) \quad \text{and} \quad \hat{N}_r = \frac{1}{2}\left(\hat{G}_r - \hat{R}_r\right)$$

span two commuting sets of operators, each one satisfying the commutation relations of the angular momentum. Hence

$$\hat{M}^2 = \hat{N}^2 = j(j+1), \quad j = 0, \frac{1}{2}, 1, \frac{3}{2}, \ldots,$$

where the appearance of the half-integer values is due to the fact that $\hat{M}_r + \hat{N}_r = \hat{G}_r$. But

$$\hat{M}^2 = \hat{N}^2 = \frac{1}{4}\left(\hat{G}^2 + \hat{R}^2\right) = \frac{1}{4}\left(\frac{1}{-2H} - 1\right) = j(j+1),$$

therefore

$$\frac{1}{-2H} = (2j+1)^2,$$

and we find again the familiar expression for the energy levels of the hydrogen atom.

4.3 Fock Quantization

Undoubtedly the Pauli result is very interesting, but a question is left open: how do we get a four-dimensional rotation group if we live in a three-dimensional world? Or, in more mathematical terms: how do we get the *global* action of this group? The answer was given by Fock (1935) in a clever, although somewhat mysterious, way. Even if his argument is fully quantum mechanical, the conclusions go beyond the particular implementation and, as we shall see, are valid in the classical framework as well. We stress that the Fock result is a cornerstone and marks the beginning of the modern treatment of the Kepler Problem.

Another interesting property of the Fock method is that, as shown by Bander & Itzykson (1966), it can be generalized to any number $n \geq 2$ of dimensions. Let us consider the Schrödinger Equation for the n-dimensional Kepler Problem:

$$\left[-\frac{1}{2}\Delta_{\mathbb{R}^n} - \frac{1}{q}\right]\phi(q) = H\phi(q).$$

We now make a Fourier transform $\mathcal{F}$ and write the equation in momentum space. To the Laplacian operator corresponds the multiplicative operator $-p^2$, while the potential term gives rise to a convolution integral $*$, since from the definition of Fourier transform

$$\mathcal{F}(fg) = \frac{1}{(2\pi)^n}\mathcal{F}(f) * \mathcal{F}(g).$$

In, e.g., Gel'fand & Shilov (1964, page 192), the Fourier transform of a generic power is calculated as

$$\mathcal{F}(q^\lambda) = 2^{\lambda+n}\pi^{\frac{n}{2}}\frac{\Gamma\left(\frac{n+\lambda}{2}\right)}{\Gamma\left(-\frac{\lambda}{2}\right)}p^{-\lambda-n}. \tag{4.3.1}$$

Denoting, with a slight abuse of notation, the transform of $\phi(q)$ by $\phi(p)$, the Schrödinger Equation is transformed into

$$\frac{1}{2}\left(p^2 - 2H\right)\phi(p) = \frac{\Gamma\left(\frac{n-1}{2}\right)}{2\pi^{\frac{n+1}{2}}}\int_{\mathbb{R}^n}\frac{\phi(p')}{||p-p'||^{n-1}}d^np'. \tag{4.3.2}$$

Now we will replace p_k by $\sqrt{-2H}x_k$, imbed the n-dimensional space, with Cartesian coordinates x_k, into an $(n+1)$-dimensional one, with Cartesian coordinates X_α, and perform a stereographic projection from the unit sphere: see Figure 4.1.

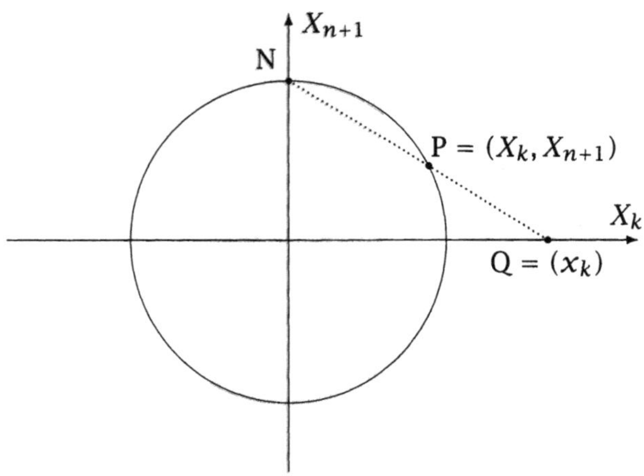

Figure 4.1: The stereographic projection σ of $P \in S^n$ on $Q \in \mathbb{R}^n$.

The x's are *local* coordinates on the sphere S^n, since, and this is a crucial point, the North pole is missing. To make explicit the stereographic mapping, notice that $x_k = \frac{X_k}{1-X_{n+1}}$ and, because of $\langle X, X\rangle = 1$,

$$x^2 = \frac{1-X_{n+1}^2}{(1-X_{n+1})^2} = \frac{1+X_{n+1}}{1-X_{n+1}}.$$

This relation can be inverted; moreover $X_k = x_k\,(1-X_{n+1})$, hence

$$X_{n+1} = \frac{x^2-1}{x^2+1}, \quad X_k = \frac{2x_k}{x^2+1}. \tag{4.3.3}$$

The Equations (4.3.3) are the explicit form of the inverse of the stereographic mapping σ. The metric tensor on S^n and volume form are easily found

$$g_{hk} = \left(\frac{2}{x^2+1}\right)^2 \delta_{hk}, \quad d\Omega_n = \left(\frac{2}{x^2+1}\right)^n d^n x.$$

We will need the expression

$$\begin{aligned}\|X-X'\|^2 &= \langle X, X\rangle + \langle X', X'\rangle - 2\,\langle X, X'\rangle \\ &= 2 - 2\frac{4\,\langle x, x'\rangle + (x^2-1)\,(x'^2-1)}{(x^2+1)\,(x'^2+1)} = 4\frac{\|x-x'\|^2}{(x^2+1)\,(x'^2+1)}.\end{aligned}$$

Replacing $x_k = \frac{p_k}{\sqrt{-2H}}$ in the above expressions, we obtain

$$\begin{aligned}X_k &= \frac{2\sqrt{-2H}}{p^2-2H}p_k, \quad X_{n+1} = \frac{p^2+2H}{p^2-2H}, \\ d\Omega_n &= \left(\frac{2\sqrt{-2H}}{p^2-2H}\right)^n d^n p, \\ \|X-X'\|^2 &= \frac{-8H}{(p^2-2H)\,(p'^2-2H)}\|p-p'\|^2.\end{aligned}$$

Moreover, let us change the wave function by defining[(9)]

$$\Phi(X) = \left[\frac{p^2-2H}{-4H}\right]^{\frac{n+1}{2}} \phi(p). \tag{4.3.4}$$

Inserting all that in Equation (4.3.2), we get

$$\frac{\Gamma\left(\frac{n-1}{2}\right)}{2\pi^{\frac{n+1}{2}}} \int_{S^n} \frac{\Phi(X')}{\|X-X'\|^{n-1}} d\Omega_n' = \sqrt{-2H}\Phi(X). \tag{4.3.5}$$

[(9)] One must admit that this is a mysterious point and, at this stage, the only explanation lies in its effectiveness. In Chapter 9 it will be better understood.

In this way, the symmetry of the $(n+1)$-dimensional orthogonal group is made manifest: in fact, the underlying manifold is now S^n instead of $\mathbb{R}^n$, while the term $\|X - X'\|^{n-1}$ and the volume form are rotationally invariant.

Define the linear operator

$$\hat{\mathcal{L}}^{-1}\Phi(X) = \frac{\Gamma\left(\frac{n-1}{2}\right)}{2\pi^{\frac{n+1}{2}}} \int_{S^n} \frac{\Phi(X')}{\|X - X'\|^{n-1}} d\Omega'_n$$

and rewrite Equation (4.3.5) as

$$\hat{\mathcal{L}}^{-1}\Phi(X) = \sqrt{-2H}\Phi(X).$$

Therefore, $\hat{\mathcal{L}}$ is the quantization of the classical action $\mathcal{L}$. Its eigenvalues and eigenfunctions can be found in a very elegant way, as shown by Bander & Itzykson (1966). The key point is the remark that in $\mathbb{R}^{n+1}$ the kernel $\frac{1}{\|X-X'\|^{n-1}}$ of the integral Equation (4.3.5) is essentially the Green function[(10)] of the Laplacian $\Delta_{\mathbb{R}^{n+1}}$, i.e.,

$$\Delta_{\mathbb{R}^{n+1}} \frac{1}{\|X - X'\|^{n-1}} = -(n-1)\Omega_n \delta(X - X'),$$

where Ω_n is the area of the unit sphere S^n (see Appendices 4.A.5 and 4.A.6 to this chapter). Let X' be a variable and $X \in S^n$ a fixed point. Define a surface Σ as the union of the two components (see Figure 4.2):

$$\Sigma \equiv \{X' : \|X'\|^2 = 1, \|X - X'\| > \varepsilon\} \cup \{X' : \|X'\|^2 \leq 1, \|X - X'\| = \varepsilon\}.$$

This surface is a sphere S^n with a small cap of radius ε, centered at the point X, removed and substituted by an internal hemisphere with the same radius and centre. Write the Green formula (see Appendix 4.A.2) for a homogeneous harmonic polynomial $h_\ell(X')$ in $\mathbb{R}^{n+1}$ and for the function $\frac{1}{\|X-X'\|^{n-1}}$, which is harmonic everywhere except at the point $X = X'$. Since the inner volume in Σ does not contain the point X, we have

$$0 = \int_\Sigma \left[h_\ell(X') \frac{d}{d\nu} \frac{1}{\|X - X'\|^{n-1}} - \frac{1}{\|X - X'\|^{n-1}} \frac{d}{d\nu} h_\ell(X') \right] d\Omega'_n,$$

where ν is the outward orthogonal direction to Σ.

The integral splits into two parts, each one relative to one of the two components of Σ. The first one, taken over the small hemisphere around the point X, tends to $\frac{n-1}{2}\Omega_n h_\ell(X)$: this is easily seen, taking

(10) Recall that the Green function $G(x,x')$ of a differential operator A is the solution of the equation: $A\left(G(x,x')\right) = \delta(x - x')$.

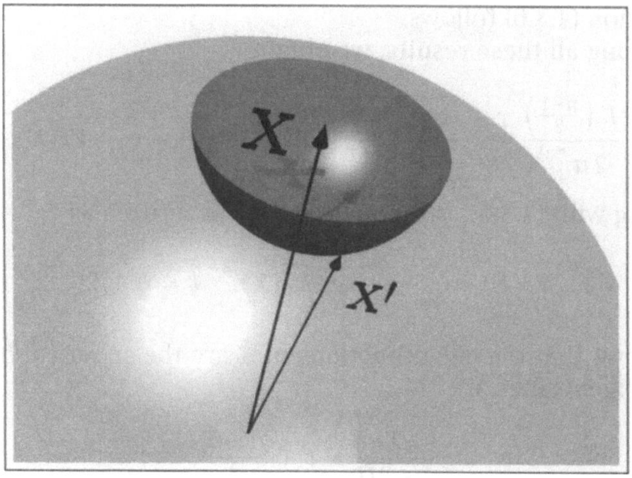

Figure 4.2: The surface Σ.

into account Appendix 4.A.6 and that the normal outward to Σ is inward to the sphere, then applying again the Green formula. Moreover, since X is on the unit sphere, $h_\ell(X)$ coincides with the spherical harmonics Y_ℓ. The second part tends smoothly, when ε goes to zero, to an integral evaluated over the entire sphere.

Let us simplify the second part, which in turn is the sum of two terms. Due to the homogeneity of h_ℓ we have

$$\frac{d}{d\nu} h_\ell(X')\Big|_{\|X'\|=1} = \ell Y_\ell(X').$$

Moreover we show that

$$\frac{d}{d\nu}\frac{1}{\|X-X'\|^{n-1}}\Big|_{\|X\|=\|X'\|=1} = -\frac{n-1}{2}\frac{1}{\|X-X'\|^{n-1}}\Big|_{\|X\|=\|X'\|=1}. \tag{4.3.6}$$

To this end, put $X' = rN$, where N is a variable unit vector $\in \mathbb{R}^{n+1}$ and compare (remembering that $X \in S^n$)

$$\langle X-X', X-X'\rangle = \|X\|^2 + \|X'\|^2 - 2\langle X, X'\rangle \overset{\text{on } S^n}{=} 2 - 2\langle X, N\rangle$$

with

$$\frac{\partial}{\partial r}\langle X-X', X-X'\rangle = -2\langle N, X-X'\rangle \overset{\text{on } S^n}{=} -2\langle N, X\rangle + 2;$$

hence

$$\frac{\partial}{\partial r}\|X-X'\|^2 \overset{\text{on } S^n}{=} \|X-X'\|^2,$$

and Equation (4.3.6) follows.

Collecting all these results, we obtain

$$\frac{\Gamma\left(\frac{n-1}{2}\right)}{2\pi^{\frac{n+1}{2}}}\int_{S^n}\frac{Y_\ell(X')}{\|X-X'\|^{n-1}}d\Omega'_n=\frac{2}{n-1+2\ell}Y_\ell(X).$$

Comparing with (4.3.5), this equation may be written as

$$\hat{\mathfrak{L}}^{-1}Y_\ell(X)=\frac{2}{n-1+2\ell}Y_\ell(X),\quad \ell=0,1,2,\ldots,$$

which shows that the eigenfunctions of $\hat{\mathfrak{L}}$ are the spherical harmonics, while its eigenvalues are

$$\frac{1}{\sqrt{-2H}}=\ell+\frac{n-1}{2},$$

with multiplicity (see Appendix 4.A.2)

$$\dim\mathcal{H}^{n,\ell}=\frac{(n+2\ell-1)\,(n+\ell-2)!}{(n-1)!\ell!}.$$

In the physical case $n=3$, we recover again the familiar formulae. $\hat{\mathfrak{L}}$ is a pseudo-differential operator, whose formal expression is

$$\hat{\mathfrak{L}}=\sqrt{-\Delta_{S^n}+\left(\frac{n-1}{2}\right)^2}.$$

This is easily seen, noticing that also the eigenfunctions of the Laplacian are the spherical harmonics, while its eigenvalues are $-\ell(\ell+n-1)$; consequently, those of $\hat{\mathfrak{L}}$ will be

$$\sqrt{\ell(\ell+n-1)+\left(\frac{n-1}{2}\right)^2}=\ell+\frac{n-1}{2},$$

as required.

APPENDIX

4.A Mathematical Review

In this appendix we briefly review some mathematical arguments used in the chapter.

4.A.1 Second Order Linear Differential Equations

Let us consider equations of the type(11)

$$\frac{d^2y}{dx^2} + P(x)\frac{dy}{dx} + Q(x)y = 0. \tag{4.A.1}$$

Every solution $y(x)$ is regular (i.e., has a power series expansion with integer and positive exponents) for those values of x for which the coefficients $P(x)$ and $Q(x)$ are regular: only in the points where one or both of them have a singularity can a singularity for $y(x)$ arise.

Consider first the regular case. If one is able to find two particular independent integrals $y_1(x)$ and $y_2(x)$, then the general solution $y(x)$ is the linear combination

$$y(x) = c_1 y_1(x) + c_2 y_2(x),$$

where c_1 and c_2 are arbitrary constants. These will be determined in such a way to satisfy some assigned conditions, e.g., at the extreme points a and b of the interval on the x-axis. In particular, we will consider the *homogeneous conditions*, for example

$$y(a) = 0, \quad y(b) = 0,$$

which have the following defining property: if they are satisfied by an integral $y(x)$, then $cy(x)$, with c an arbitrary constant, has the same property also. Note that $y(x) = 0$ on the whole interval is, trivially, a solution: of course we are interested in finding a non trivial one. In the case of the above example, c_1 and c_2 must satisfy the two conditions

$$\begin{cases} c_1 y_1(a) + c_2 y_2(a) = 0, \\ c_1 y_1(b) + c_2 y_2(b) = 0, \end{cases}$$

(11) For a more exhaustive study of the argument see for example Ince (1956, Chapter XV).

and therefore, if we want a nontrivial solution,

$$\det \begin{pmatrix} y_1(a) & y_2(a) \\ y_1(b) & y_2(b) \end{pmatrix} = 0.$$

In the cases of our interest, the coefficient $Q(x)$ is a linear function of a parameter λ, which can be determined in such a way as to have nontrivial solutions: λ is called an *eigenvalue* and the corresponding integral is an *eigenfunction.* An eigenfunction is determined modulo an arbitrary multiplicative constant that can be fixed by requiring the *normalization condition*

$$\int_a^b yy^* dx = 1,$$

whose meaning is obvious, if one thinks to the probabilistic interpretation.

If to every eigenvalue λ_h corresponds one eigenfunction y_h only, the eigenvalues are said to be *simple* and it is easy to prove that the set of the normalized eigenfunctions satisfies the orthogonality conditions

$$\int_a^b y_h y_k^* dx = \delta_{hk}.$$

But if we consider another type of homogeneous conditions

$$y(a) = y(b), \quad \left.\frac{dy}{dx}\right|_{x=a} = \left.\frac{dy}{dx}\right|_{x=b},$$

then to an eigenvalue may correspond more than one eigenfunction; in this case the eigenvalue is said to be *multiple* and, if we are in the framework of quantum mechanics, the physical system is said to be *degenerate.* These eigenfunctions do not constitute, in general, an orthonormal basis; however, one proves that it is possible to build up with them a set of eigenfunctions having this property.

Let us consider the singular case. We say that Equation (4.A.1) has, at a point x_0, a *fuchsian* (or *regular* or *nonessential*) *singularity* if for $x = x_0$ at least one of the coefficients P, Q becomes infinite, but of order one for P and two (or less) for Q. In this case, we can write the equation as

$$\frac{d^2y}{dx^2} + \frac{P_1(x - x_0)}{x - x_0}\frac{dy}{dx} + \frac{Q_1(x - x_0)}{(x - x_0)^2} y = 0, \tag{4.A.2}$$

P_1 and Q_1 being power series with integer nonnegative exponents. The reason for this definition lies in the behavior of the solutions in the neighbourhood of the point x_0: if the singularity is fuchsian, every integral of the equation is regular in x_0 or exhibits a singularity of the type

$$y = (x - x_0)^\alpha \Phi(x - x_0), \tag{4.A.3}$$

with α a real and Φ a regular function (or, in an exceptional case, with a logarithmic singularity). If, on the contrary, the singularity is not fuchsian, the solution of the equation may exhibit other types of singularities or become undetermined in $x = x_0$. We will only be concerned with fuchsian singularities.

Put (4.A.3) in the Equation (4.A.2) and equate to zero the coefficients of the powers of $x - x_0$. In this way, we formally find the coefficients of the series Φ; the first of the conditions is

$$\alpha(\alpha - 1) + \alpha P_1(0) + Q_1(0) = 0, \tag{4.A.4}$$

and gives in general two values for α, with the corresponding two independent integrals of the form (4.A.3)(12). Equation (4.A.4) is the *characteristic equation* relative to the singularity x_0.

Finally, let us consider the case of a singularity at infinity. This case is reduced to the previous one using the transformation: $x \mapsto 1/x$. The condition to have only fuchsian singularities at infinity is that the coefficients P and Q are of the type

$$P(x) = \frac{1}{x} P_1\left(\frac{1}{x}\right), \quad Q(x) = \frac{1}{x^2} Q_1\left(\frac{1}{x}\right).$$

for $x \to \infty$.

4.A.2 Laplacian on the Sphere and Homogeneous Harmonic Polynomials

Consider the n-dimensional unitary sphere

$$S^n = \{X \in \mathbb{R}^{n+1} : r^2 \equiv X_1^2 + X_2^2 + \cdots + X_{n+1}^2 = 1\}.$$

We want to solve the eigenvalue problem for the Laplacian on the sphere

$$\Delta_{S^n} \psi = \lambda \psi.$$

Some solutions are easily found with the following construction(13). Let $h_\ell(X)$ be a homogeneous harmonic polynomial of degree ℓ in $\mathbb{R}^{n+1}$, that is

$$\Delta_{\mathbb{R}^{n+1}} h_\ell = 0 \qquad \text{(harmonicity)},$$

$$\sum_{\alpha=1}^{n+1} X_\alpha \frac{\partial h_\ell}{\partial X_\alpha} = \ell h_\ell \quad \text{(homogeneity)}.$$

(12)If the two values of α differ by an integer, or, in particular, are equal, Φ will be of the type $\Phi_1 + \Phi_2 \log(x - x_0)$, with Φ_1 and Φ_2 regular functions. We will not be concerned with this exceptional case.

(13)We will follow the exposition of Vilenkin (1968, Chapter IX) .

If ζ is a point on S^n, owing to the homogeneity property we can write $h_\ell = r^\ell f_\ell(\zeta)$ (ℓ not summed), where $f_\ell(\zeta)$ is the value of the harmonic homogeneous polynomial restricted to the unit sphere. Owing to the harmonicity property we can write

$$\begin{aligned}\Delta_{\mathbb{R}^{n+1}} h_\ell(X) &= \Delta_{\mathbb{R}^{n+1}}\left(r^\ell f_\ell(\zeta)\right)\\ &= \left(\frac{\partial^2}{\partial r^2} + \frac{n}{r}\frac{\partial}{\partial r} + \frac{1}{r^2}\Delta_{S^n}\right)\left(r^\ell f_\ell(\zeta)\right) = 0,\end{aligned}$$

thus

$$[\Delta_{S^n} + \ell(\ell + n - 1)] f_\ell(\zeta) = 0.$$

In other words, a homogeneous harmonic polynomial of degree ℓ restricted to S^n is the eigenfunction of the Laplacian on the sphere relative to the eigenvalue $-\ell(\ell + n - 1)$.

Recall the Green formula:

$$\int_\tau (u\Delta w - w\Delta u)\, d\tau = \int_\sigma \left(u\frac{\partial w}{\partial \nu} - w\frac{\partial u}{\partial \nu}\right) d\sigma, \quad \forall u, w$$

where τ is a volume, σ its boundary and ν the outward orthogonal direction to σ.

PROPOSITION 4.1 *The homogeneous harmonic polynomials of different degree restricted to the unitary sphere are orthogonal:*

$$\int_{S^n} h_\ell(\zeta) h_{\ell'}(\zeta)\, d\sigma = 0 \quad \text{for } \ell \neq \ell'.$$

Proof. By Green's formula we have

$$\int_{\|X\|\le 1} [h_\ell \Delta_{\mathbb{R}^{n+1}} h_{\ell'} - h_{\ell'}\Delta_{\mathbb{R}^{n+1}} h_\ell]\, d\tau = \int_{S^n} h_\ell(\zeta) h_{\ell'}(\zeta)(\ell - \ell')\, d\sigma,$$

since the normal derivatives of h_ℓ and $h_{\ell'}$ on S^n are

$$\frac{\partial}{\partial r} h_\ell(r\zeta)\bigg|_{r=1} = \ell h_\ell(\zeta) \quad \text{and} \quad \frac{\partial}{\partial r} h_{\ell'}(r\zeta)\bigg|_{r=1} = \ell' h_{\ell'}(\zeta).$$

The proposition follows since $h_\ell(X)$ and $h_{\ell'}(X)$ are harmonic. QED

The homogeneous harmonic polynomials restricted to the sphere are called *spherical harmonics* and are denoted by Y_ℓ. We now show that, in this way, we have actually found "all" the solutions of the eigenvalue problem for the Laplacian on the sphere. More precisely, these functions form a complete set and, therefore, a continuous function on S^n can be developed into a series of spherical harmonics with arbitrary accuracy. We start by proving the

THEOREM 4.2 *The space $\mathcal{P}^{n,\ell}$ of the homogeneous polynomials in $n+1$ variables of degree ℓ is the direct sum of the sub-spaces $\mathcal{H}^{n,\ell}$ of the homogeneous harmonic polynomials in $n+1$ variables of degree ℓ and $r^2\mathcal{P}^{n,\ell-2}$ of the polynomials of the form $r^2 f_1(X)$, where $f_1(X) \in \mathcal{P}^{n,\ell-2}$:*

$$\mathcal{P}^{n,\ell} = \mathcal{H}^{n,\ell} + r^2\mathcal{P}^{n,\ell-2}.$$

Proof. First of all we prove that

$$\mathcal{H}^{n,\ell} \cap r^2\mathcal{P}^{n,\ell-2} = 0,$$

or, in other words, that if $f_1(X) \in \mathcal{P}^{n,\ell-2}$, then $\Delta_{\mathbb{R}^{n+1}}(r^2 f_1) \neq 0$. Indeed, if $f(X) \in r^2\mathcal{P}^{n,\ell-2}$, then it can be presented in the form $f(X) = r^{2k} f_k(X)$, where $f_k(X) \in \mathcal{P}^{n,\ell-2k}$, $0 < k \leq [\ell/2]$ and is not divisible by r^2. Suppose $\Delta_{\mathbb{R}^{n+1}} f = 0$. Since

$$\Delta_{\mathbb{R}^{n+1}}\left(r^{2k} f_k\right) = 2k(n+2\ell-2k-1)r^{2k-2} f_k + r^{2k}\Delta_{\mathbb{R}^{n+1}} f_k,$$

as one can verify by a direct calculation, the right-hand side would be equal to zero. But $k \geq 1$ and $\ell \geq 2k$, thus $k(n+2\ell-2k-1) > 0$. Cancelling r^{2k-2} out, we see that f_k is divisible by r^2, contrary to our assumption. This proves that $\mathcal{H}^{n,\ell} \cap r^2\mathcal{P}^{n,\ell-2} = 0$.

We now prove that $\mathcal{H}^{n,\ell}$ and $r^2\mathcal{P}^{n,\ell-2}$ make up the whole space $\mathcal{P}^{n,\ell}$. To do this, we calculate the dimensions of these subspaces. Arranging the homogeneous polynomials in powers of X_1 we see that

$$\dim \mathcal{P}^{n,\ell} = \dim \mathcal{P}^{n-1,\ell} + \dim \mathcal{P}^{n-1,\ell-1} + \cdots + \dim \mathcal{P}^{n-1,0}.$$

Moreover, it is clear that

$$\dim \mathcal{P}^{0,\ell} = 1.$$

Hence, by induction, we get

$$\dim \mathcal{P}^{n,\ell} = \frac{(n+\ell)!}{n!\ell!}.$$

Since the dimension of $r^2\mathcal{P}^{n,\ell-2}$ is equal to the dimensions of $\mathcal{P}^{n,\ell-2}$, and $\mathcal{H}^{n,\ell} \cap r^2\mathcal{P}^{n,\ell-2} = 0$, the dimension $\dim \mathcal{H}^{n,\ell}$ satisfies the inequality

$$\dim \mathcal{H}^{n,\ell} \leq \dim \mathcal{P}^{n,\ell} - \dim \mathcal{P}^{n,\ell-2}.$$

But the polynomial $\Delta_{\mathbb{R}^{n+1}} f$ with $f \in \mathcal{P}^{n,\ell}$ belongs to the space $\mathcal{P}^{n,\ell-2}$. Therefore the condition $\Delta_{\mathbb{R}^{n+1}} f = 0$ entails the existence of a number of linear relations among the coefficients of the polynomial $f(X)$ less than or equal to $\dim \mathcal{P}^{n,\ell-2}$, from which

$$\dim \mathcal{H}^{n,\ell} \geq \dim \mathcal{P}^{n,\ell} - \dim \mathcal{P}^{n,\ell-2}.$$

Thus we can infer that

$$\dim \mathcal{H}^{n,\ell} = \dim \mathcal{P}^{n,\ell} - \dim \mathcal{P}^{n,\ell-2},$$

and this concludes the proof. QED

Note that in passing the dimension of the space $\mathcal{H}^{n,\ell}$ has been obtained

$$\dim \mathcal{H}^{n,\ell} = \frac{(n+2\ell-1)\,(n+\ell-2)!}{(n-1)!\ell!}.$$

Applying the above theorem to $\mathcal{P}^{n,\ell-2}$ in the same way and continuing this process, we obtain

$$\mathcal{P}^{n,\ell} = \sum_{k=0}^{[\ell/2]} r^{2k}\mathcal{H}^{n,\ell-2k},$$

where $r^{2k}\mathcal{H}^{n,\ell-2k}$ is a subspace of $\mathcal{P}^{n,\ell}$ consisting of polynomials of the form $r^{2k}h_{\ell-2k}(X)$, with $h_{\ell-2k}(X) \in \mathcal{H}^{n,\ell-2k}$.

Thus we have proved that every homogeneous polynomial of degree ℓ in $n+1$ variables is decomposed into a sum of polynomials of the form $r^{2k}h_{\ell-2k}(X)$, where $h_{\ell-2k}(X) \in \mathcal{H}^{n,\ell-2k}$. It follows that the value of any homogeneous polynomial of degree ℓ on the unit sphere coincides with the value on this sphere of the harmonic inhomogeneous polynomial

$$h(X) = \sum_{k=0}^{[\ell/2]} h_{\ell-2k}(X).$$

Hence, in considering polynomials on the unit sphere, it is sufficient to restrict ourselves to harmonic polynomials.

By the Weierstrass Theorem any continuous function $F(\zeta)$ on S^n can be uniformly approximated, with any required degree of accuracy, by a polynomial. But every polynomial can be written as a sum of homogeneous polynomials that, in turn, can be written as a sum of harmonic polynomials. In conclusion, any continuous function $F(\zeta)$ on S^n can be uniformly approximated by a sum of functions belonging to the spaces $\mathcal{H}^{n,\ell}$. Consequently the subspaces $\mathcal{H}^{n,\ell}$ make up the whole space $L^2(S^n)$ of square-integrable functions on the unit sphere.

4.A.3 Associated Legendre Functions

Let us consider the equation

$$(1-x^2)\frac{d^2P}{dx^2} - 2x\frac{dP}{dx} + \ell(\ell+1)P = 0$$

which admits as solution the polynomial $P_\ell(x)$ of degree ℓ,

$$P_\ell(x) = \frac{1}{2^\ell \ell!}\frac{d^\ell}{dx^\ell}(x^2-1)^\ell.$$

Consider now the generalization

$$(1-x^2)\frac{d^2P}{dx^2} - 2(|m|+1)x\frac{dP}{dx} + [\ell(\ell+1) - |m|(|m|+1)]P = 0. \tag{4.A.5}$$

Differentiating it, we obtain a similar equation but with every derivative substituted by a derivative one order higher and with $|m|$ replaced by $|m|+1$: thus Equation (4.A.5) is satisfied by $\frac{d^{|m|}P_\ell}{dx^{|m|}}$ (which is a polynomial of degree $\ell - |m|$). The functions

$$P_\ell^m(x) = (1-x^2)^{\frac{|m|}{2}}\frac{d^{|m|}P_\ell}{dx^{|m|}}$$

are called *associated Legendre functions* and satisfy the equation

$$(1-x^2)\frac{d^2\Theta}{dx^2} - 2x\frac{d\Theta}{dx} + \left(\ell(\ell+1) - \frac{m^2}{1-x^2}\right)\Theta = 0. \tag{4.A.6}$$

4.A.4 Generalized Laguerre Polynomials

The Laguerre polynomials $L_k(x)$ of degree k are defined by

$$L_k(x) = e^x\frac{d^k}{dx^k}\left(x^k e^{-x}\right).$$

The explicit form of the first polynomials is

$$\begin{aligned} L_0 &= 1, \\ L_1 &= 1-x, \\ L_2 &= 2-4x+x^2, \\ L_3 &= 6-18x+9x^2-x^3. \end{aligned}$$

They satisfy the differential equation

$$\left[x\frac{d^2}{dx^2} + (1-x)\frac{d}{dx} + k\right]L_k = 0 \tag{4.A.7}$$

and are orthonormal with respect to the measure $e^{-x}dx$:

$$\int_0^\infty L_k(x)L_h(x)e^{-x}\,dx = \delta_{kh},$$

as one can prove by repeated integration by parts.

Differentiating Equation (4.A.7) and putting $L_k^{(j)} = \frac{d^j}{dx^j}L_k$, one obtains

$$\left[x\frac{d^2}{dx^2} + (2-x)\frac{d}{dx} + (k-1)\right]L_k^{(1)} = 0,$$

similar to (4.A.7), but with the second coefficient increased by 1 and the third decreased by 1, respectively. Differentiating j times we get

$$\left[x\frac{d^2}{dx^2} + (j+1-x)\frac{d}{dx} + (k-j)\right]L_k^{(j)} = 0, \qquad (4.A.8)$$

which is the differential equation satisfied by the generalized Laguerre polynomials $L_k^{(j)}$. For a fixed j they are orthogonal with respect to the measure $x^j e^{-x} dx$:

$$\int_0^\infty L_k^{(j)} L_h^{(j)} x^j e^{-x}\, dx = \begin{cases} 0 & \text{if } k \neq h \\ \frac{(k!)^3}{(k-j)!} & \text{if } k = h \end{cases}$$

and so one can normalize the generalized Laguerre polynomials.

4.A.5 Surface Measure on the Sphere and Gamma Function

The Gamma $\Gamma(z)$ function is defined as

$$\Gamma(z) = \int_0^\infty t^{z-1} e^{-t} dt :$$

see for example Goursat (1959, page 183). Integrating by parts, one obtains $\Gamma(z+1) = z\Gamma(z)$. Since $\Gamma(1) = 1$, for z integer $\Gamma(z) = (z-1)!$ holds. The value of some definite integrals is expressed as a function of Γ, for example the Eulerian integral of the first kind

$$\int_0^1 (1-x)^{p-1} x^{q-1} dx = \frac{\Gamma(p)\Gamma(q)}{\Gamma(p+q)},$$

see Goursat (1959, page 280). Setting $p = q = \frac{1}{2}$ in the above integral we find: $\Gamma(\frac{1}{2}) = \sqrt{\pi}$.

These few properties of the Γ function are useful in calculating the measure Ω_n of the n-dimensional unit sphere S^n. Let $\vec{\zeta}_k$ be a unit vector $\in S^k$, $\vec{e}$ a fixed unit vector $\in \mathbb{R}^{n+1}$ and $\vec{\zeta}_{n-1}$ a unit vector belonging to the subspace orthogonal to $\vec{e}$, then

$$\vec{\zeta}_n = t\vec{e} + \sqrt{1-t^2}\,\vec{\zeta}_{n-1}, \quad 0 \le t \le 1.$$

Let ds_k^2 be the square of the line element on S^k. We have

$$ds_n^2 = \frac{1}{1-t^2}dt^2 + (1-t^2)ds_{n-1}^2, \quad \sqrt{g_n} = (1-t^2)^{\frac{n-2}{2}}\sqrt{g_{n-1}},$$

and the area of S^n is

$$\int_{S^n} d\Omega_n = \int_{S^{n-1}}\int_{-1}^{+1}(1-t^2)^{\frac{n-2}{2}}dt\, d\Omega_{n-1}.$$

To calculate the inner integral, put $t^2 = x$, obtaining so the Eulerian integral of the first kind:

$$\int_0^1 (1-x)^{\frac{n-2}{2}}x^{-\frac{1}{2}}dx = \frac{\Gamma(\frac{n}{2})\Gamma(\frac{1}{2})}{\Gamma(\frac{n+1}{2})}.$$

Hence, by induction,

$$\Omega_n = \frac{\sqrt{\pi}\Gamma(\frac{n}{2})}{\Gamma(\frac{n+1}{2})}\Omega_{n-1} = \cdots = \frac{\sqrt{\pi}^{n-1}\Gamma(1)}{\Gamma(\frac{n+1}{2})}\Omega_1 = \frac{2\pi^{\frac{n+1}{2}}}{\Gamma(\frac{n+1}{2})},$$

since, obviously, the "area" of S^1 is 2π.

4.A.6 Green Function of the Laplacian

In this subsection let S^n be the n-dimensional sphere of generic radius ρ. We can write

$$\int_{\|X\|\leq\rho} \Delta_{\mathbb{R}^{n+1}}\frac{1}{\|X\|^{n-1}}\, d\tau = -\int_{S^n}\left\|\nabla\frac{1}{\|X\|^{n-1}}\right\|\rho^n\, d\Omega_n.$$

Since $\nabla\frac{1}{\|X\|^{n-1}} = -(n-1)\frac{X}{\|X\|^{n+1}}$, the integral on the right-hand side is equal to $-(n-1)\Omega_n$. The function $\frac{1}{\|X\|^{n-1}}$ is harmonic everywhere except at the origin. Therefore

$$\Delta_{\mathbb{R}^{n+1}}\frac{1}{\|X\|^{n-1}} = -(n-1)\Omega_n\delta(X).$$

CHAPTER 5

Regularization and Symmetry

In Chapter 2 the physical time t, i.e., the independent variable in the equation of motion, has been replaced with the eccentric anomaly $\mathfrak{s}$. This substitution has the nice property of transforming the equation of motion into a *regular* one. In fact, the Hamiltonian vector field of the Kepler Problem is not complete in its natural phase space, as collision orbits reach the singularity of the potential in a finite time and with a divergent velocity. On the contrary, all solutions of the transformed equation remain finite for every value of the eccentric anomaly.

This is an instance of a *regularization* process, which in general amounts to embedding the incomplete system into a complete one as a dense open subset, or, loosely speaking, to "adding" some orbits to those of the original system. At first sight this seems not well founded, since one may object that the topology of the phase space is changed. In Souriau (1974) the situation is clarified as follows. Define the *evolution space* as the product of the phase space by the time. Every solution of the equations of motion is described by a curve on it. Identify all points in the evolution space that belong to the same solution curve and define the *space of motions* as the quotient under this equivalence relation: each point of the space of motions is thus a trajectory and the information on the global structure of the mechanical system is encoded into its topology. A fixed $t = t_0$ section of the evolution space is the phase space at time t_0. It may happen however that the mapping: "section at $t_0 \mapsto$ space of motions" is not *onto*. An example is just the Kepler

Problem, where the domain of the mapping does not contain those motions that hit the centre at $t = t_0$, and thus does not reflect the global structure of the space of the motions, while the regularization makes the mapping one-to-one.

In this chapter we study three procedures for regularizing the Kepler Problem: namely the method of Moser, Souriau and Kustaanheimo-Stiefel. See Moser (1970), Souriau (1974) and Stiefel & Scheifele (1971), respectively. Actually, the term "regularization" is not fully adequate, since for the regularization it is sufficient, as said above, to switch from time to eccentric anomaly, but we employ it to follow the common approach. Really, these three procedures produce much more, exhibiting more or less explicitly the various facets of the symmetries of the Kepler Problem. According to the spirit of this first Part, we will limit ourselves to expose (almost verbatim and with only minor changes in notation for uniformity's sake) the original arguments, postponing to Part II the study of their intimate relations.

5.1 Moser Method

The Moser method (Moser 1970) is strongly reminiscent, at the classical level, of the Fock quantization process, described in the previous chapter. In fact, Moser shows that the Kepler flow of the regularized n-dimensional Kepler Problem for negative energy is equivalent to the geodesic flow on the n-dimensional sphere S^n, "explaining" in this way how the $(n+1)$-dimensional orthogonal symmetry arises. This result is proved by appropriately extending the stereographic projection

$$\sigma : S^n - \{\text{North Pole}\} \to \mathbb{R}^n$$

to the (co)tangent bundle of the sphere.

In order to prove the statement we begin with the discussion of the geodesic flow on S^n and transform it ultimately into the Kepler Problem. Let $X = (X_1, X_2, \ldots, X_{n+1})$ be a real vector and $\|X\|^2 = \sum_{\beta=1}^{n+1} X_\beta^2$, so that $\|X\| = 1$ represents the unit sphere S^n. The geodesic flow can be described as the solution of the variational problem

$$\delta \int \left\| \frac{dX(\alpha)}{d\alpha} \right\|^2 d\alpha = 0,$$

where $X = X(\alpha)$ is a vector function satisfying $\|X(\alpha)\|^2 = 1$. The Euler-Lagrange Equations for this problem are

$$\frac{d^2 X(\alpha)}{d\alpha^2} + \lambda^2 X(\alpha) = 0,$$

where $\lambda = \left\|\frac{dX(\alpha)}{d\alpha}\right\|$ is a Lagrange multiplier. Introducing the cotangent[(1)] vector $Y = \frac{dX(\alpha)}{d\alpha}$, we write this as a first order system

$$\frac{dX}{d\alpha} = Y, \quad \frac{dY}{d\alpha} = -\|Y\|^2 X. \tag{5.1.1}$$

One verifies that the two relations, describing the cotangent bundle T^*S^n to the sphere,

$$\|X\| = 1, \quad \langle X, Y\rangle = \sum_{\beta=1}^{n+1} X_\beta Y_\beta = 0$$

remain satisfied if they hold initially. The first order system can be written as a Hamiltonian system

$$\frac{dX}{d\alpha} = \frac{\partial F}{\partial Y}, \quad \frac{dY}{d\alpha} = -\frac{\partial F}{\partial X} \qquad \text{on } T^*S^n,$$

where

$$F(X,Y) = \frac{1}{2}\|X\|^2 \|Y\|^2 .$$

Now, as in the Fock approach, we use the stereographic projection (see Figure 4.1 on page 82)

$$X_k = \frac{2x_k}{x^2+1}, \quad X_{n+1} = \frac{x^2-1}{x^2+1},$$

and extend it to a mapping of the cotangent bundle T^*S^n into the $2n$-dimensional space $T^*\mathbb{R}^n$ with coordinates x_k, y_h. Thus we look for $Y = Y(x,y)$ such that

$$\langle Y, dX\rangle = \langle y, dx\rangle \quad \text{and} \quad \langle Y, X\rangle = 0. \tag{5.1.2}$$

The former relation says that Y is an extended covector and the latter that it is tangent to the sphere. This extension is given by

$$Y_k = \frac{1}{2}\left(\|x\|^2 + 1\right) y_k - \langle y, x\rangle x_k, \quad Y_{n+1} = \langle y, x\rangle ,$$

as one can verify directly. We briefly indicate how to calculate it. Let $\mathbf{J}$ be the $(n+1)\times n$ Jacobian matrix

$$(\mathbf{J})_{\beta k} = \frac{\partial X_\beta}{\partial x_k}, \quad \text{satisfying} \quad \mathbf{J}^T\mathbf{J} = \left(\frac{2}{\|x\|^2+1}\right)^2 \mathbf{1}_n,$$

[(1)]The Euclidean metric allows us to identify vectors with covectors.

where $\mathbf{J}^T$ is the transposed matrix and $\mathbf{1}_n$ the $n \times n$ identity matrix. Putting

$$Y = \left(\frac{\|x\|^2 + 1}{2}\right)^2 \mathbf{J}y$$

we find the required extension, in fact

$$\langle Y, dX\rangle = \left(\frac{\|x\|^2 + 1}{2}\right)^2 y^T \mathbf{J}^T \mathbf{J}\, dx = \langle y, dx\rangle\,;$$

moreover, since

$$\langle \mathbf{J}y, X\rangle = \left\langle y, \mathbf{J}^T X\right\rangle = \left\langle y, \frac{1}{2}\frac{\partial \|X\|^2}{\partial x}\right\rangle = 0$$

we find

$$\langle Y, X\rangle = \left(\frac{\|x\|^2 + 1}{2}\right)^2 \langle \mathbf{J}y, X\rangle = 0.$$

To transform the differential Equations (5.1.1) one merely has to transform the Hamiltonian, as is easily checked in virtue of (5.1.2). In other words, we define

$$F(x, y) = \frac{1}{2}\|X\|^2 \|Y\|^2 = \frac{1}{8}\|y\|^2 \left(\|x\|^2 + 1\right)^2$$

and obtain the transformed equations in the form

$$\frac{dx_k}{d\alpha} = \frac{\partial F}{\partial y_k}, \quad \frac{dy_k}{d\alpha} = -\frac{\partial F}{\partial x_k}. \tag{5.1.3}$$

Moreover, the geodesics with "unit velocity" $\|Y\| = 1$ correspond to solutions with $F = \frac{1}{2}$. Since these differential equations depend on the gradient of F only, we can replace F by $u(F)$ if $\left.\frac{du}{dF}\right|_{F=\frac{1}{2}} = 1$, because on $F = \frac{1}{2}$ the gradients of F and $u(F)$ agree. Taking in particular

$$K = u(F) = \sqrt{2F} = \frac{1}{2}\|y\|\left(\|x\|^2 + 1\right), \tag{5.1.4}$$

the system (5.1.3) on $F = \frac{1}{2}$ goes over into

$$\frac{dx_k}{d\alpha} = \frac{\partial K}{\partial y_k}, \quad \frac{dy_k}{d\alpha} = -\frac{\partial K}{\partial x_k} \quad \text{on } K = 1.$$

Changing the independent variable α to

$$dt = \|y\|\, d\alpha \tag{5.1.5}$$

results in
$$\frac{dx_k}{dt} = \frac{1}{\|y\|}\frac{\partial K}{\partial y_k}, \quad \frac{dy_k}{dt} = -\frac{1}{\|y\|}\frac{\partial K}{\partial x_k}.$$
Since on $K = 1$ we have
$$\frac{1}{\|y\|}\frac{\partial K}{\partial y_k} = \frac{\partial H}{\partial y_k}, \quad \frac{1}{\|y\|}\frac{\partial K}{\partial x_k} = \frac{\partial H}{\partial x_k},$$
where H and K are related by the *Moser-map*
$$H = \frac{K-1}{\|y\|} - \frac{1}{2}, \tag{5.1.6}$$
the resulting system is Hamiltonian, with H as the Hamiltonian function, on $H = -\frac{1}{2}$. Therefore $H = \frac{1}{2}\|x\|^2 - \frac{1}{\|y\|}$, and if we make the canonical transformation: $p_k = -x_k, q_h = y_h$, the system goes over into the Kepler Problem on the energy surface $H = -\frac{1}{2}$, with
$$H = \frac{1}{2}\|p\|^2 - \frac{1}{\|q\|}. \tag{5.1.7}$$
Definition (5.1.5) shows that on this energy surface the variable α is the eccentric anomaly $\mathfrak{s}$: see (2.2.2) on page 23.

Summarizing, we have shown that the extended stereographic transformation and (5.1.5) map the unit (indeed $F = \frac{1}{2}$ implies $\|Y\| = 1$) cotangent bundle of S^n – {North Pole} into the $2n$-dimensional phase space $T^*\mathbb{R}^n$ and the geodesic circles go into the Kepler orbits on the energy surface $H = -\frac{1}{2}$. Equivalently, the flow (with the "true" time t) of the variables q_k, p_h, given by the Hamiltonian (5.1.7) on the energy hypersurface $H = -\frac{1}{2}$, agrees with that of the variables x_k, y_h (with the "fake" time α), given by the Hamiltonian (5.1.4) on the energy hypersurface $K = 1$. So far we have excluded the North Pole $X = (0, 0, \ldots, 1)$. Now we restore this point, thereby compactifying the energy surface, and the geodesics through the North Pole are transformed into the collision orbits: we have regularized the problem.

So far we have discussed the equivalence of the Kepler Problem with the geodesic flow on the sphere only for $H = -\frac{1}{2}$. The general case of a negative energy surface $H = -\frac{1}{2\lambda^2}$ is easily reduced to this case by rescaling the variables according to
$$q \mapsto \lambda^2 q, \quad p \mapsto \frac{p}{\lambda}, \quad t \mapsto \lambda^3 t. \tag{5.1.8}$$

For the first time we encountered above the Hamiltonian K, which will play an important role in the sequel. To tell the truth, the Hamiltonian K is already "hidden" in the Fock quantization procedure, discussed in the previous chapter. To reveal it, let us return to Equation

(4.3.5) on page 83 and insert in it, instead of (4.3.4), the expression

$$\Phi(X) = \left[\frac{p^2-1}{2}\right]^{\frac{n+1}{2}} \phi(p)$$

(thus, without the rescaling $p_k \mapsto \frac{p_k}{\sqrt{-2H}}$), obtaining

$$\sqrt{-2H}\frac{1}{2}\left(p^2+1\right)\phi(p) = \frac{\Gamma\left(\frac{n-1}{2}\right)}{2\pi^{\frac{n+1}{2}}} \int_{\mathbb{R}^n} \frac{\phi(p')}{\|p-p'\|^{n-1}} d^n p',$$

instead of Equation (4.3.2). The above expression is the Fourier transform of

$$\sqrt{-2H}\frac{1}{2}\left(-\Delta_{\mathbb{R}^n}+1\right)\phi(x) = \frac{1}{\|x\|}\phi(x),$$

which may be rewritten as

$$\frac{1}{2}\|x\|\left(-\Delta_{\mathbb{R}^n}+1\right)\phi(x) = \frac{1}{\sqrt{-2H}}\phi(x). \tag{5.1.9}$$

This is the canonical quantization of the Hamiltonian K after a canonical exchange between coordinates and momenta.

To get a clearer geometrical understanding of the Moser construction, consider the 2-dimensional Kepler Problem and a geodesic circle on S^2, which we can safely rotate into the position

$$X_1 = \sin\alpha, \quad X_2 = -\cos\beta\cos\alpha, \quad X_3 = \sin\beta\cos\alpha.$$

The vector (X_1, X_2, X_3) is orthogonal to the vector $(0, \sin\beta, \cos\beta)$; β measures the angle between the equator $X_3 = 0$ and the circle in question, whereas α is the coordinate along the circle itself. The definition $Y_k = \frac{dX_k}{d\alpha}$ yields

$$Y_1 = \cos\alpha, \quad Y_2 = \cos\beta\sin\alpha, \quad Y_3 = -\sin\beta\sin\alpha.$$

Since the explicit form of the extended stereographic mapping $T^*\sigma$ is

$$x_k = \frac{X_k}{1-X_{n+1}}, \quad y_k = Y_k\left(1-X_{n+1}\right) + X_k Y_{n+1}, \tag{5.1.10}$$

the image of the circle under the extended stereographic projection is

$$x_1 = \frac{\sin\alpha}{1-\sin\beta\cos\alpha}, \quad x_2 = -\frac{\cos\beta\cos\alpha}{1-\sin\beta\cos\alpha},$$
$$y_1 = \cos\alpha - \sin\beta, \quad y_2 = \cos\beta\sin\alpha.$$

With $q_k = y_k$ and $E = \sin\beta$, this takes the form

$$q_1 = \cos\alpha - E, \quad q_2 = \sqrt{1-E^2}\sin\alpha,$$

which is the representation of an ellipse of eccentricity $E = \sin\beta$ in terms of the eccentric anomaly α. Moreover, with $p_k = -x_k$,

$$p_1^2 + (p_2 - \tan\beta)^2 = 1 + \tan^2\beta,$$

which is the representation of a circle in the Cartesian plane $p_1 p_2$, that is the *hodograph curve*. The trajectory of the planet and the trajectory of its velocity are the direct, followed by a translation, projection and the stereographic projection, respectively, of a geodesic circle onto the equatorial plane.

5.2 Souriau Method

While the Moser method works in the "unphysical" phase space with coordinates $\vec{x}, \vec{y}$, which agrees with the physical one for $H = -\frac{1}{2}$ only, the Souriau method (Souriau 1974, Souriau 1983) acts in the physical phase space with coordinates $\vec{q}, \vec{p}$ (the notation is that of Chapter 2). Adhering to the original work, we will treat here the 3-dimensional Kepler Problem, but in what follows all the formulae can be straightforwardly extended to the n-dimensional case.

Define

$$Q = \begin{pmatrix} \vec{q} \\ t \end{pmatrix}, \quad d\alpha = \frac{dt}{q},$$

then a direct calculation gives

$$\begin{aligned} \frac{dQ}{d\alpha} &= \begin{pmatrix} q\vec{p} \\ q \end{pmatrix}, & \frac{d^2Q}{d\alpha^2} &= \begin{pmatrix} \left(\vec{q}\cdot\vec{p}\right)\vec{p} - \frac{\vec{q}}{q} \\ \vec{q}\cdot\vec{p} \end{pmatrix}, \\ \frac{d^3Q}{d\alpha^3} &= \begin{pmatrix} 0 \\ 1 \end{pmatrix} + 2H\frac{dQ}{d\alpha}, & \frac{d^4Q}{d\alpha^4} &= 2H\frac{d^2Q}{d\alpha^2}. \end{aligned} \tag{5.2.1}$$

Hence the 4×4 matrix

$$\mathbf{Q} = \begin{pmatrix} Q & \frac{dQ}{d\alpha} & \frac{d^2Q}{d\alpha^2} & \frac{d^3Q}{d\alpha^3} \end{pmatrix}$$

satisfies the linear differential equation with constant coefficients

$$\frac{d\mathbf{Q}}{d\alpha} = \mathbf{QA},$$

where

$$\mathbf{A} = \begin{pmatrix} 0 & 0 & 0 & 0 \\ 1 & 0 & 0 & 0 \\ 0 & 1 & 0 & 2H \\ 0 & 0 & 1 & 0 \end{pmatrix}.$$

The integration is immediate

$$\mathbf{Q} = \mathbf{Q}_0 \exp(\alpha \mathbf{A}).$$

The exponential is easily calculated using the standard methods of linear algebra. If $H > 0$ one finds

$$\exp(\alpha\mathbf{A}) = \begin{pmatrix} 1 & 0 & 0 & 0 \\ \alpha & 1 & 0 & 0 \\ \frac{\cosh\lambda\alpha - 1}{\lambda^2} & \frac{\sinh\lambda\alpha}{\lambda} & \cosh\lambda\alpha & \lambda\sinh\lambda\alpha \\ \frac{\sinh\lambda\alpha - \lambda\alpha}{\lambda^3} & \frac{\cosh\lambda\alpha - 1}{\lambda^2} & \frac{\sinh\lambda\alpha}{\lambda} & \cosh\lambda\alpha \end{pmatrix}, \quad \lambda = \sqrt{2H}.$$

If $H < 0$ this can be rewritten

$$\exp(\alpha\mathbf{A}) = \begin{pmatrix} 1 & 0 & 0 & 0 \\ \alpha & 1 & 0 & 0 \\ \frac{1-\cos\mu\alpha}{\mu^2} & \frac{\sin\mu\alpha}{\mu} & \cos\mu\alpha & -\mu\sin\mu\alpha \\ \frac{\mu\alpha - \sin\mu\alpha}{\mu^3} & \frac{1-\cos\mu\alpha}{\mu^2} & \frac{\sin\mu\alpha}{\mu} & \cos\mu\alpha \end{pmatrix}, \quad \mu = \sqrt{-2H}.$$

Finally, if $H = 0$

$$\exp(\alpha\mathbf{A}) = \begin{pmatrix} 1 & 0 & 0 & 0 \\ \alpha & 1 & 0 & 0 \\ \frac{\alpha^2}{2} & \alpha & 1 & 0 \\ \frac{\alpha^3}{6} & \frac{\alpha^2}{2} & \alpha & 1 \end{pmatrix}.$$

These formulae give explicitly the value of the variables

$$t, q, \vec{q}, q\vec{p}, \vec{q}\cdot\vec{p}, \left(\vec{q}\cdot\vec{p}\right)\vec{p} - \frac{\vec{q}}{q}$$

as some functions of α and of the initial conditions.

Now we introduce two sets of parameters, namely the *Fock* and *Bacry-Györgyi* parameters, which will play an important role in the sequel. Hereafter we consider the case $H < 0$.

5.2.1 Fock Parameters

The Fock parameters are the two vectors U and V of $\mathbb{R}^4$ defined by[(2)]

$$U = \begin{pmatrix} \vec{U} \\ U_4 \end{pmatrix} = \begin{pmatrix} -\mathcal{L}\frac{d^2\vec{q}}{d\alpha^2} \\ -\frac{d^2t}{d\alpha^2} \end{pmatrix},$$

$$V = \begin{pmatrix} \vec{V} \\ V_4 \end{pmatrix} = \begin{pmatrix} -\mathcal{L}^2\frac{d^3\vec{q}}{d\alpha^3} \\ -\mathcal{L}\frac{d^3t}{d\alpha^3} \end{pmatrix}, \quad \mathcal{L} = \frac{1}{\sqrt{-2H}}, \tag{5.2.2}$$

which, owing to Equations (5.2.1), may be rewritten

$$U = \begin{pmatrix} \frac{1}{\sqrt{-2H}}\left[-\left(\vec{q}\cdot\vec{p}\right)\vec{p} + \frac{\vec{q}}{q}\right] \\ -\vec{q}\cdot\vec{p} \end{pmatrix}, \quad V = \begin{pmatrix} q\vec{p} \\ -\frac{1+2Hq}{\sqrt{-2H}} \end{pmatrix}.$$

As is verified by a direct calculation they satisfy the identities

$$\|U\| = \|V\| = \mathcal{L}, \quad \langle U, V\rangle = 0, \tag{5.2.3}$$

where the norm and scalar product of $\mathbb{R}^4$ are Euclidean.

The importance of these two vectors lies in the fact that, while incorporating *all* the information on the problem, they have a nice dynamical evolution. In fact, since from Definition (2.2.2) $\alpha = \mathcal{L}s$, we have

$$\frac{dU}{ds} = V, \quad \frac{dV}{ds} = -U. \tag{5.2.4}$$

The former of these two equations is an obvious consequence of Definitions (5.2.2) while the latter comes from the fourth of Equations (5.2.1). Equations (5.2.4) are easily integrated

$$U(s) = \overline{U}\cos s + \overline{V}\sin s, \quad V(s) = -\overline{U}\sin s + \overline{V}\cos s,$$

where $\overline{U}$ and $\overline{V}$ are the initial values. Thus, when the "fake time" s is running, the two vectors $U(s)$ and $V(s)$ rotate uniformly on their initial plane. Moreover, the first integrals of the Kepler Problem are easily and very elegantly obtained from the manifestly constant exterior product $U \wedge V$, in fact

$$\frac{\vec{U}\times\vec{V}}{\mathcal{L}} = \vec{G}, \quad \frac{\vec{V}U_4 - \vec{U}V_4}{\mathcal{L}} = \vec{R},$$

where $\vec{G}$ is the angular momentum vector and $\vec{R}$ the Runge-Lenz-Laplace vector. Lastly, in order to obtain the motion, notice that

$$\vec{q} = \mathcal{L}\left(\vec{U} - \vec{R}\right).$$

(2) Our definition of the Fock parameters differs from the original one in Souriau (1974) for the normalization factor and for the signs.

For initial values we may choose

$$\overline{U} = \begin{pmatrix} \mathcal{L}\frac{\vec{R}}{R} \\ 0 \end{pmatrix}, \quad \overline{V} = \begin{pmatrix} \vec{G} \times \frac{\vec{R}}{R} \\ -R \end{pmatrix}, \quad R = \left\|\vec{R}\right\|,$$

since they satisfy the identities (5.2.3), and are clearly constants of the motion; hence

$$\vec{q}\,(\mathfrak{s}) = \mathcal{L}\left(\mathcal{L}\frac{\vec{R}}{R}\cos\mathfrak{s} + \vec{G} \times \frac{\vec{R}}{R}\sin\mathfrak{s} - \vec{R}\right),$$

that is, Equation (2.2.4).

5.2.2 Bacry-Györgyi Parameters

Switching from the "fake time" $\mathfrak{s}$ to the "true" time t, or equivalently to the mean anomaly ℓ, the rotation velocity of the Fock parameters is no longer constant and become even infinite for the collision orbits. It is however possible to define two other similar vectors, namely the Bacry-Györgyi parameters[(3)], which rotate uniformly; see Bacry (1966), Györgyi (1968). Indeed, define

$$\begin{pmatrix} Z \\ W \end{pmatrix} = \begin{pmatrix} \cos\Delta & -\sin\Delta \\ \sin\Delta & \cos\Delta \end{pmatrix} \begin{pmatrix} U \\ V \end{pmatrix}, \quad \Delta = \frac{\vec{q}\cdot\vec{p}}{\mathcal{L}}.$$

Obviously, Z and W satisfy the conditions

$$\|Z\| = \|W\| = \mathcal{L}, \quad \langle Z, W\rangle = 0,$$

moreover a straightforward calculation shows that

$$\frac{dZ}{d\ell} = W, \quad \frac{dW}{d\ell} = -Z,$$

as required. Unfortunately, while the functions $U(\vec{q},\vec{p})$ and $V(\vec{q},\vec{p})$ can be inverted in closed form, this is no longer possible for Z and W, since it would require the inversion of transcendental functions, so that one can not express explicitly the motion as a function of the time.

5.3 Kustaanheimo-Stiefel Transformation

While those of Moser and Souriau are n-dimensional methods, the Kustaanheimo-Stiefel transformation $\mathbb{KS}$ is essentially related to the 3-dimensional Kepler Problem (really, it can be generalized to other dimensions, but not in a straightforward way). Basically it transforms the

(3) They are also called *Ligon-Schaaf parameters* : see Ligon & Schaaf (1976) and Cushman & Bates (1997, page 65).

Hamiltonian K in the Hamiltonian of a 4-dimensional harmonic oscillator, whose motion is subject to a constraint[(4)].

DEFINITION 5.1 *The Kustaanheimo-Stiefel transformation is the map*

$$\mathbb{KS} : \mathbb{R}^4 - \{0\} \times \mathbb{R}^4 \to \mathbb{R}^3 - \{0\} \times \mathbb{R}^3$$

given explicitly by

$$\begin{aligned} x_1 &= 2(z_1 z_3 + z_2 z_4), \\ x_2 &= 2(z_2 z_3 - z_1 z_4), \\ x_3 &= -z_1^2 - z_2^2 + z_3^2 + z_4^2, \end{aligned}$$

from which

$$\|x\| = \|z\|^2,$$

and by

$$\begin{aligned} y_1 &= -\frac{z_1 w_3 + z_2 w_4 + z_3 w_1 + z_4 w_2}{\|z\|^2}, \\ y_2 &= \frac{z_1 w_4 - z_2 w_3 - z_3 w_2 + z_4 w_1}{\|z\|^2}, \\ y_3 &= \frac{z_1 w_1 + z_2 w_2 - z_3 w_3 - z_4 w_4}{\|z\|^2}. \end{aligned}$$

The coordinates of the domain are subject to the constraint

$$z_1 w_2 - z_2 w_1 + z_3 w_4 - z_4 w_3 = 0. \tag{5.3.1}$$

Because of this constraint the domain of the $\mathbb{KS}$ map is 7-dimensional, while the target space is 6-dimensional, thus the map has a 1-dimensional kernel. In fact, as one verifies, all pairs z', w' connected to z, w by the relations

$$\begin{pmatrix} z_1' \\ z_2' \\ z_3' \\ z_4' \end{pmatrix} = \begin{pmatrix} \cos\beta & -\sin\beta & 0 & 0 \\ \sin\beta & \cos\beta & 0 & 0 \\ 0 & 0 & \cos\beta & -\sin\beta \\ 0 & 0 & \sin\beta & \cos\beta \end{pmatrix} \begin{pmatrix} z_1 \\ z_2 \\ z_3 \\ z_4 \end{pmatrix}$$

and

$$\begin{pmatrix} w_1' \\ w_2' \\ w_3' \\ w_4' \end{pmatrix} = \begin{pmatrix} \cos\beta & -\sin\beta & 0 & 0 \\ \sin\beta & \cos\beta & 0 & 0 \\ 0 & 0 & \cos\beta & -\sin\beta \\ 0 & 0 & \sin\beta & \cos\beta \end{pmatrix} \begin{pmatrix} w_1 \\ w_2 \\ w_3 \\ w_4 \end{pmatrix}$$

[(4)]In what follows do not confuse z and w with the Bacry-Györgyi parameters Z and W.

are sent by the map into the same pair x, y for every β.

Let us now suppose that the z's are coordinates and the w's canonically conjugate momenta of an 8-dimensional phase space. After a canonical exchange between the coordinates and momenta, let us compose the Moser Hamiltonian $K(x, y)$ with the KS map. A straightforward calculation shows that

$$K\left(x(z), y(z, w)\right) = \frac{1}{2}\left(\|z\|^2 + \|w\|^2\right),$$

which is the Hamiltonian of a 4-dimensional isotropic harmonic oscillator. If the constraint (5.3.1) is satisfied by the initial conditions, it will be respected by the dynamical evolution for all times.

The KS map may be restricted in particular to the 2-dimensional Kepler Problem. To this end put in its definition

$$z_2 = z_4 = w_2 = w_4 = 0,$$

so that

$$x_1 = 2z_1z_3, \quad x_3 = -z_1^2 + z_3^2,$$

and

$$y_1 = -\frac{z_1w_3 + z_3w_1}{\|z\|^2}, \quad y_3 = \frac{z_1w_1 - z_3w_3}{\|z\|^2},$$

while the constraint (5.3.1) becomes empty. This is the well known *Levi-Civita regularization map* (Levi-Civita 1956), and it is clear that the passage from 2 to 3 dimensions is not a trivial task. In fact, we will succeed in generalizing the KS to the n-dimensional case, but only after a deeper analysis of the group-geometrical structure of the Kepler Problem. And this, as we said, is the aim of Part II.

Part II

Group–Geometric Theory

CHAPTER 6

Conformal Regularization

In Part I we have discussed in detail the relevant points of the Kepler Problem in a pure computational way, without trying to understand its intrinsic structure. In this way one can work out explicitly the whole argument with relatively elementary mathematical tools. In Part II we change somewhat the presentation, which will require more sophisticated mathematical tools, including differential geometry and Lie group theory[(1)], but will provide (hopefully) a clearer understanding.

The crucial point lies in Souriau (1974) and Souriau (1983), which we have already partially reviewed: the phase space of the regularized n-dimensional Kepler Problem is symplectomorphic to a coadjoint orbit of the Lie group $SO_0(2, n + 1)$, the (double covering of the) conformal group of $(n + 1)$-dimensional Minkowski space. Moreover, all relevant quantities of the problem (the Hamiltonian, the angular momentum, the Runge-Lenz-Laplace vector and the Fock or Bacry-Györgyi parameters) fit well into this construction. They are, in fact, components of the associated moment map. This will be explained in this chapter, which is in fact intended to be the most important of the entire book. However, as proposed in Cordani (1986) and Cordani (1988), we reverse the order of the exposition with respect to Souriau: we start from the conformal group and only *a posteriori* will we recover, by a judicious choice of the coordinates, the Kepler Problem, in such a way as to focus on what is

[(1)] The background necessary to the understanding of Part II is reviewed in the appendices to this book.

intrinsic and what is coordinate dependent. This strategy offers several advantages:

1. the introduction of the regularization parameter $\mathfrak{s}$ is not postulated, but arises rather automatically, as part of a canonical transformation;
2. the relation between the Fock and Bacry-Györgyi parameters is clarified;
3. the case of the repulsive potential is automatically included;
4. the equivalence of the Fock-Moser and KS regularizations, due basically to the homomorphism $SU(2,2) \to SO(2,4)$, is proved for any value of the energy;
5. the geometric quantization is obtained, showing, in particular, that the Fock quantization method is the implementation, at a "quantum level", of our "classical" procedure presented in this chapter;
6. simply modifying the coordinates without changing the intrinsic structure, enables us to find other mechanical systems whose Hamiltonian, although seemingly different from that of the Kepler Problem, shares, nevertheless, its dynamical properties.

The first three points are discussed in this chapter; the other three are postponed to the subsequent ones.

Where not specified, the range of the indices is

$$\begin{aligned} A, B, C &= -1, 0, \dots, n+1, \\ \lambda, \mu, \nu &= 0, \dots, n, \\ \alpha, \beta, \gamma &= 1, \dots, n+1, \\ i, j, k &= 1, \dots, n, \\ a, b, c &= 2, \dots, n; \end{aligned}$$

x and y stand for the n-dimensional Euclidean norm of x_k and y_h, (x, y) for the n-dimensional scalar product, and $\langle \cdot, \cdot \rangle$ for the pairing between Lie algebra and its dual space.

6.1 The Conformal Group

Let us define the $(n+1)$-dimensional Minkowski space[(2)] M_0 as the space $\mathbb{R}^{1,n} = \mathbb{R}^1 \oplus \mathbb{R}^n$ equipped with the pseudo-Euclidean diagonal

---

(2) Later, it will be clear why we have attached the subscript "0" in this definition.

metric tensor

$$(\eta_{\mu\nu}) = \begin{pmatrix} -1 & & & \\ & 1 & & \\ & & \ddots & \\ & & & 1 \end{pmatrix}.$$

The Lorentz group $\mathfrak{L}$, with Lie algebra $\mathfrak{l}$, is the subgroup of the general linear group that leaves this metric invariant. We will only consider the *identity connected component* of the Lorentz group, that is, $\mathfrak{L}_0 = SO_0(1,n)$. Notice, in particular, that $-\mathbf{1}_{n+1}$ (i.e., minus the $(n+1)\times(n+1)$ identity matrix) does not belong to $\mathfrak{L}_0$, since it can not be generated by a hyperbolic rotation. Another group acting on M_0 and leaving its metric invariant is the $(n+1)$-dimensional abelian group $\mathfrak{T}$ of translations, with Lie algebra $\mathfrak{t}$. The semi-direct product $\mathfrak{L}_0 \times_S \mathfrak{T}$ forms the Poincaré group, which exhausts all the possibilities of "rigid" motions in M_0.

We are interested in (the identity connected component of) the *conformal group* $\mathfrak{C}$ of the Minkowski space, i.e., the group of diffeomorphisms transforming the flat metric $\eta_{\mu\nu}$ into $g_{\mu\nu}(x) = \omega^2(x)\eta_{\mu\nu}$, $x \in M_0$, where ω is some nonvanishing function. The manifolds with metric tensor $\omega^2(x)\eta_{\mu\nu}$ will be called *conformally flat* or *conformally Minkowski-related*. With the trivial choice $\omega(x) \equiv 1$ the Poincaré group is readily seen to be a subgroup of the conformal group, and the problem is thus to investigate whether different, nontrivial choices of $\omega(x)$ are possible. So we first look for all *local* conformal mappings, and then try to extend them globally. To this end, consider the most general infinitesimal conformal transformation

$$x^\mu \mapsto \overline{x}^\mu = x^\mu + \varepsilon V^\mu(x) + O(\varepsilon^2)$$

which satisfies

$$\frac{\partial \overline{x}^\mu}{\partial x^\rho}\frac{\partial \overline{x}^\nu}{\partial x^\sigma}\eta_{\mu\nu} = \omega^2(x)\eta_{\rho\sigma}, \quad \text{with } \omega(x) = 1 + \varepsilon f(x) + O(\varepsilon^2).$$

Keeping only the first order terms in ε, we obtain the flat space *conformal Killing Equation*

$$\partial_\mu V_\nu + \partial_\nu V_\mu = 2f(x)\,\eta_{\mu\nu}, \quad \partial_\mu = \frac{\partial}{\partial x^\mu}.$$

PROPOSITION 6.1 *The general solution of the conformal Killing Equation for $n \geq 2$ is a second order polynomial*

$$V_\mu(x) = t_\mu + \lambda_{\mu\nu}x^\nu + dx_\mu + 2k_\nu x^\nu x_\mu - k_\mu x_\nu x^\nu, \quad \lambda_{\mu\nu} = -\lambda_{\nu\mu}, \tag{6.1.1}$$

depending on the $\frac{1}{2}(n+2)(n+3)$ parameters $t_\mu, \lambda_{\mu\nu}, d, k_\mu$.

Proof. Contracting the indices in the Killing Equation, or else contracting with $\partial^\mu \partial^\nu$, we find

$$(n+1)f = \nabla \cdot V, \quad \Box(\nabla \cdot V) = 0, \quad \nabla \cdot V = \partial_\mu V^\mu, \quad \Box = \eta^{\mu\nu}\partial_\mu \partial_\nu.$$

Differentiating, furthermore, the Killing Equation with respect to x_ρ, permuting ρ and μ, and adding the two expressions we obtain

$$\partial_\rho \partial_\mu V_\nu = \frac{1}{n+1}\left(\eta_{\mu\nu}\partial_\rho + \eta_{\rho\nu}\partial_\mu - \eta_{\rho\mu}\partial_\nu\right)(\nabla \cdot V),$$

which gives, upon application of ∂^ν,

$$\left(1 - \frac{2}{n+1}\right)\partial_\rho \partial_\mu (\nabla \cdot V) = 0. \qquad (6.1.2)$$

Thus, for $n \geq 2$, $\nabla \cdot V$ is a linear function of x :

$$f(x) = \frac{1}{n+1}\nabla \cdot V = d + 2k_\mu x^\mu,$$

where d and k_μ are $n+2$ independent constants. The proposition is then derived in a straightforward manner. QED

Notice that for $n = 1$ condition (6.1.2) is trivially satisfied and the set of solutions depends on an infinite number of parameters. In particular, for an Euclidean 2-space with $\eta_{\mu\nu} = \delta_{\mu\nu}$, $\mu, \nu = 1, 2$, the Killing Equation reduces to the Cauchy-Riemann Equations

$$\partial_1 V_1 = \partial_2 V_2, \quad \partial_1 V_2 = -\partial_2 V_1,$$

and is therefore satisfied by the real and imaginary parts of any analytic function of $x_1 + ix_2$.

The conformal vector field (6.1.1) generates, for every value of the parameters, local conformal diffeomorphisms, which are obtained integrating the equation

$$\frac{d\overline{x}^\mu}{d\tau} = V^\mu(\overline{x}).$$

We carry out these integrations keeping only one of the four types of parameters different from zero.

- For $t^\mu \neq 0$ we find $\overline{x}^\mu = x^\mu + t^\mu \tau$, that is, the translation group $\mathfrak{T}$.
- For $\lambda^\mu{}_\nu \neq 0$ we find $\overline{x}^\mu = \Lambda^\mu{}_\nu(\tau) x^\nu$, $\boldsymbol{\Lambda}(\tau) = \exp(\boldsymbol{\lambda}\tau)$, that is, the Lorentz group $\mathfrak{L}_0$.
- For $d \neq 0$ we find $\overline{x}^\mu = x^\mu \exp(d\tau)$, the *dilation group* $\mathfrak{D}$.

- For $k_\mu \neq 0$ we find

$$\overline{x}^\mu = \frac{x^\mu - x^\nu x_\nu k^\mu \tau}{1 - 2x^\nu k_\nu \tau + k^\nu k_\nu x^\rho x_\rho \tau^2},$$

  called *conformal translation group* $\mathfrak{K}$. This latter is, as we shall see below, an abelian group.

It is clear that the action of the translation, Lorentz and dilation group is globally well defined on M_0, unlike that of the conformal translation group. For this latter, the denominator

$$1 - 2x^\nu k_\nu \tau + k^\nu k_\nu x^\rho x_\rho \tau^2 = k^\nu k_\nu \left(x^\mu \tau - \frac{k^\mu}{k^\rho k_\rho} \right) \left(x_\mu \tau - \frac{k_\mu}{k^\rho k_\rho} \right)$$

indeed vanishes at the points that belong to either the null cone

$$\left(x^\mu \tau - \frac{k^\mu}{k^\rho k_\rho} \right) \left(x_\mu \tau - \frac{k_\mu}{k^\rho k_\rho} \right) = 0,$$

or to the hyperplane $2x^\nu k_\nu \tau = 1$, if $k^\mu k_\mu = 0$. Thus, a transformation of $\mathfrak{K}$ always blows up at some point of M_0. It is possible to regularize this action by adding a null cone to M_0 (since this is the topology of the submanifold, where the action of $\mathfrak{K}$ blows up) at infinity. This provides us, as explained in the next section, with the compactified Minkowski space-time M.

The set of conformal vector fields (6.1.1) is a realization of the Lie algebra of the conformal group. The natural basis for these vector fields is $\{\partial_\mu\}$. The $\frac{1}{2}(n+2)(n+3)$ linearly independent conformal vector fields on M_0 can be labelled as follows:

- $\mathcal{P}_\mu = \partial_\mu$, spanning the Lie algebra $\mathfrak{t}$ of the translation group;
- $\mathcal{G}_{\mu\nu} = x_\nu \partial_\mu - x_\mu \partial_\nu$, spanning the Lie algebra $\mathfrak{l}$ of the Lorentz group;
- $\mathcal{D} = x^\mu \partial_\mu$, spanning the Lie algebra $\mathfrak{d}$ of the dilation group;
- $\mathcal{K}_\mu = 2x_\mu x^\nu \partial_\nu - x^\nu x_\nu \partial_\mu$, spanning the Lie algebra $\mathfrak{k}$ of the conformal translation group.

The Lie brackets $[A, B] = AB - BA$ are readily calculated, giving

$$\begin{aligned}
[\mathcal{G}_{\rho\mu}, \mathcal{G}_{\rho\nu}] &= \eta_{\rho\rho} \mathcal{G}_{\mu\nu} \quad \text{or} = 0 \quad \text{if all indices are different,} \\
[\mathcal{G}_{\mu\nu}, \mathcal{P}_\rho] &= \eta_{\mu\rho} \mathcal{P}_\nu - \eta_{\nu\rho} \mathcal{P}_\mu, \quad [\mathcal{G}_{\mu\nu}, \mathcal{K}_\rho] = \eta_{\mu\rho} \mathcal{K}_\nu - \eta_{\nu\rho} \mathcal{K}_\mu, \\
[\mathcal{G}_{\mu\nu}, \mathcal{D}] &= [\mathcal{P}_\mu, \mathcal{P}_\nu] = [\mathcal{K}_\mu, \mathcal{K}_\nu] = 0, \\
[\mathcal{P}_\mu, \mathcal{K}_\nu] &= 2\mathcal{G}_{\mu\nu} + 2\mathcal{D}\eta_{\mu\nu}, \quad [\mathcal{P}_\mu, \mathcal{D}] = \mathcal{P}_\mu, \quad [\mathcal{K}_\mu, \mathcal{D}] = -\mathcal{K}_\mu.
\end{aligned} \tag{6.1.3}$$

The, defined above, conformal group seems to have almost nothing to do with a matrix Lie group, the Lorentz subgroup only being of this type. This is a pity, since matrix groups act linearly, and this is always a nice property. But now comes a surprise.

PROPOSITION 6.2 *The conformal group is locally isomorphic to the pseudo-orthogonal group* $\mathrm{O}(2, n+1)$.

Proof. The pseudo-orthogonal group $\mathrm{O}(2, n+1)$ acts linearly on $\mathbb{R}^{2,n+1}$ with coordinates X^A and leaves the flat metric

$$(\eta_{AB}) = \left(\begin{array}{cc|ccc} -1 & & & & \\ & -1 & & & \\ \hline & & 1 & & \\ & & & \ddots & \\ & & & & 1 \end{array}\right)$$

invariant. The $\frac{1}{2}(n+2)(n+3)$ vector fields

$$\mathcal{G}_{AB} = X_B \partial_A - X_A \partial_B \tag{6.1.4}$$

span the Lie algebra of $\mathrm{O}(2, n+1)$, and satisfy the Lie brackets

$$[\mathcal{G}_{AB}, \mathcal{G}_{AC}] = \eta_{AA} \mathcal{G}_{BC}, \quad \text{or} = 0 \text{ if all indices are different.} \tag{6.1.5}$$

The subgroup acting on the linear subspace $\mathbb{R}^{1,n}$, $A, B = 0, 1, \ldots, n$ is isomorphic to the Lorentz group. The other generators are identified as follows:

$$\mathcal{G}_{\mu,n+1} = \frac{1}{2}(\mathcal{P}_\mu + \mathcal{K}_\mu), \quad \mathcal{G}_{-1,\mu} = \frac{1}{2}(\mathcal{P}_\mu - \mathcal{K}_\mu), \quad \mathcal{G}_{n+1,-1.} = \mathcal{D}.$$

As one verifies by a direct calculation, the Lie brackets (6.1.3) of the conformal group $\mathfrak{C}$ agree with the Lie brackets (6.1.5) of the pseudo-orthogonal group $\mathrm{O}(2, n+1)$: the two groups have the same Lie algebra, and are hence locally isomorphic. QED

Let us restrict ourselves to $\mathrm{SO}_0(2, n+1)$, i.e., the identity connected component. In the next section we will investigate if the two groups (we recall that $\mathfrak{C}$ is already defined as the identity connected component) are *globally* isomorphic.

6.2 The Compactified Minkowski Space

Our proof of Proposition 6.2 is a simple check, but a beautiful argument (Penrose 1974) shows the deep reason of its validity. We shall now review in detail this construction, since it will be the keystone of all that follows.

The action of $SO_0(2, n+1)$ on $\mathbb{R}^{2,n+1}$ is linear, therefore it induces an action on the projective manifold of (thus *unoriented*) rays through the origin. This action is not transitive, since the null cone

$$\mathcal{K} = \{X \in \mathbb{R}^{2,n+1} : \eta(X, X) = 0\}$$

is preserved. Define M as the manifold of the null (that is, belonging to the null cone $\mathcal{K}$) unoriented rays through the origin of $\mathbb{R}^{2,n+1}$. The action of $SO_0(2, n+1)$ on M is clearly transitive.

PROPOSITION 6.3 *The manifold M is homeomorphic to $(S^1 \times S^n)/\mathbb{Z}_2$.*

Proof. Considering the splitting $\mathbb{R}^{2,n+1} = \mathbb{R}^2 \oplus \mathbb{R}^{n+1}$ we may write for a generic vector X of $\mathbb{R}^{2,n+1}$

$$X = X_{(-)} + X_{(+)}, \quad X_{(-)} \in \mathbb{R}^2, \; X_{(+)} \in \mathbb{R}^{n+1}. \tag{6.2.1}$$

If $X \neq 0$ belongs to the null cone $\mathcal{K}$, then

$$-\eta\left(X_{(-)}, X_{(-)}\right) = \eta\left(X_{(+)}, X_{(+)}\right) = \text{some positive constant.}$$

Take X defined only up to a positive factor, so that the above positive constant can be normalized to 1. This provides us with a manifold homeomorphic to $S^1 \times S^n$, which is connected for $n \geq 1$. But the same unoriented ray is associated to both X and $-X$, implying the proposition. QED

REMARK 6.1 Due to the presence of the factor S^1, the homeomorphism

$$(S^1 \times S^n)/\mathbb{Z}_2 \simeq S^1 \times S^n$$

holds.

Define $\widetilde{M}$ as the manifold of *oriented* null lines of $\mathbb{R}^{2,n+1}$. It is a double covering of M, i.e., $M \simeq \widetilde{M}/\mathbb{Z}_2$. Let us choose a 1-codimensional submanifold of $\mathbb{R}^{2,n+1}$ that intersects the oriented generators of the null cone at most once. This provides us with a mapping

$$\gamma : \widetilde{M} \to M_\gamma$$

from the manifold of null generators, $\widetilde{M}$, to the cross-section M_γ. For regularity reasons, we choose our M_γ in such a way that its tangent space is transverse to the null line through this point. The cross-section M_γ is, at least locally, diffeomorphic to $\widetilde{M}$ (indeed, γ^{-1} may be nonsurjective) and is endowed with a natural metric tensor, namely the restriction of the flat metric of $\mathbb{R}^{2,n+1}$. Varying γ, the metric on each cross-section will also change, but not in a completely arbitrary way, as shows the

PROPOSITION 6.4 *The cross-sections M_γ are all conformally Minkowski related.*

Proof. First we verify that there exists a cross-section that we identify with M_0, whose metric is of the Minkowski type. To this end, write the quadratic line element of $\mathbb{R}^{2,n+1}$ as

$$ds^2 = (dX^{n+1} + dX^{-1})(dX^{n+1} - dX^{-1}) - (dX^0)^2 + (dX^1)^2 + \cdots + (dX^n)^2,$$

and intersect the null cone $\mathcal{K}$ with the hyperplane

$$X^{-1} + X^{n+1} = 1. \tag{6.2.2}$$

Restricted to the cross-section, this metric reduces to that of Minkowski space M_0.

Next, we verify that all cross-sections are conformally related. To see this, let us introduce generalized spherical coordinates r, ρ, x^0, x^k in both subspaces $\mathbb{R}^2$ and $\mathbb{R}^{n+1}$,

$$\begin{aligned} X^{-1} &= r\cos x^0, \\ X^0 &= r\sin x^0, \\ X^\alpha &= \rho\Gamma^\alpha(x^k), \quad \Gamma^\alpha\Gamma_\alpha = 1, \end{aligned}$$

where x^0 and x^k are local coordinates on S^1 and S^n, respectively. The restriction to the null cone $\mathcal{K}$ is obtained by putting $r = \rho$, while a cross-section of the null cone is given by specifying r as a strictly positive function of x^0, x^k on $\widetilde{M}$:

$$\gamma : \begin{cases} X^{-1} = r(x^\mu)\cos x^0, \\ X^0 = r(x^\mu)\sin x^0, \\ X^\alpha = r(x^\mu)\Gamma^\alpha(x^k). \end{cases}$$

The quadratic line element $ds^2_{(\gamma)}$ of the cross-section M_γ defined above is

$$ds^2_{(\gamma)} = r^2(x^\mu)\left[-(dx^0)^2 + ds^2_{S^n}\right],$$

where $ds^2_{S^n}$ is the line element of the unit sphere S^n. This shows clearly that the metric tensors of the various sections are all conformally related. QED

Making γ^{-1} an isometry, the manifold $\widetilde{M}$, thus M, is endowed not with a single metric alone, but rather with a whole class of metrics, whose elements are all conformally Minkowski related. The manifold M

is obtained by adding to M_0 some extra points at infinity[(3)], corresponding to those null lines of the cone that do not intersect the hyperplane (6.2.2). For this reason, M is said to be the *conformal compactification* of the Minkowski space.

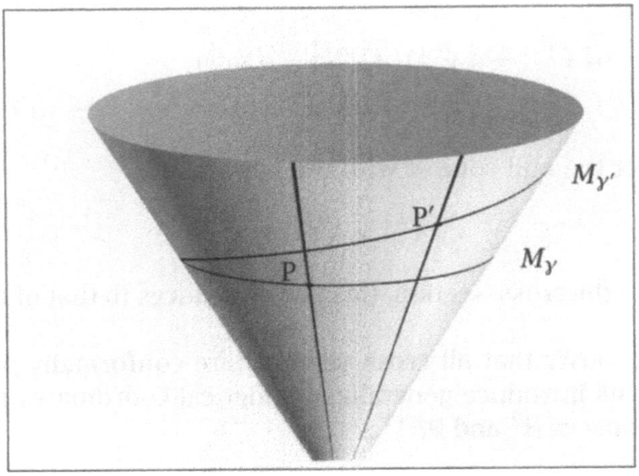

Figure 6.1: The null cone $\mathcal{K}$.

The linear action of $SO_0(2, n+1)$ on $\mathbb{R}^{2,n+1}$ sends a point P of the null cone $\mathcal{K}$ into another point P$'$ of the null cone, hence maps a section M_γ into another section $M_{\gamma'}$: see Figure 6.1. But all sections of the null cone are conformally flat, thus every element of $SO_0(2, n+1)$ induces a conformal mapping of M into itself, yielding hence a homomorphism from $SO_0(2, n+1)$ to the conformal group $\mathfrak{C}$. The two groups have the same numbers of parameters; the kernel of the homomorphism must therefore be discrete. For n odd, this kernel is $\mathbb{Z}_2$, since the element -1 is now connected to the identity of $SO_0(2, n+1)$, and acts by changing the orientations of the rays of the null cone $\mathcal{K}$. But M is the manifold of the unoriented rays, and the element -1 is therefore mapped into the identity of the conformal group $\mathfrak{C}$. For n even the kernel is trivial. Thus we have obtained the global isomorphisms

$$\mathfrak{C} = \begin{cases} \dfrac{SO_0(2, n+1)}{\mathbb{Z}_2} & n \text{ odd}, \\ SO_0(2, n+1) & n \text{ even}. \end{cases}$$

This gives us a more complete formulation of Proposition 6.2.

---

(3) The set of these points has the topology of a null cone.

The whole argument of this section is summarized in the commutative diagram

$$
\begin{array}{ccc}
M_{\gamma} & \xrightarrow{\mathrm{SO}_0(2,n+1)} & M_{\gamma'} \\
\gamma \big\uparrow & & \big\uparrow \gamma' \\
\widetilde{M} & \xrightarrow{\mathrm{SO}_0(2,n+1)} & \widetilde{M} \\
\pi \big\downarrow & & \big\downarrow \pi \\
M & \xrightarrow{\mathfrak{C}} & M
\end{array}
$$

where $\pi : \widetilde{M} \to M$ denotes the projection that sends the elements $\pm X$ of the null cone $\mathcal{K}$ into the corresponding unoriented null line.

6.3 The Cotangent Bundle to Minkowski Space

In the previous section we have defined the compactification M of Minkowski space M_0, as the manifold of the null unoriented lines through the origin of $\mathbb{R}^{2,n+1}$.

There is, however, another equivalent definition, which has some advantages. Noticing that M is a homogeneous space under the action of the conformal group, we can define it as the quotient of $\mathfrak{C}$ by the isotropy subgroup of any of its points. Remembering the expressions (6.1.1) of the conformal local diffeomorphisms $x^{\mu} \mapsto \overline{x}^{\mu}$ on M_0, one sees that the subgroup $\mathfrak{P}_0$ of $\mathfrak{C}$, defined as the closed connected subgroup having Lie algebra $\mathfrak{p} = \mathfrak{l} \oplus \mathfrak{d} \oplus \mathfrak{k}$ (that is, Lorentz + dilation + conformal translation subalgebras) does not move the origin of M_0, and thus belongs to the isotropy subgroup. Inspecting the Lie brackets of $\mathfrak{p}$, one finds the global structure:

$$\mathfrak{P}_0 = (\mathfrak{L}_0 \times \mathfrak{D}) \times_S \mathfrak{K},$$

where $\times$ is the direct and $\times_S$ the semidirect product. But $\mathfrak{P}_0$ does not exhaust the full isotropy subgroup. To see this, consider the linear action of $\mathrm{SO}_0(2, n+1)$ on $\mathbb{R}^{2,n+1}$. Let $\overline{X}$ be the vector that represents the null line corresponding to the origin of M_0. Split $\overline{X}$ as in (6.2.1). It is obvious that all transformations composed of a rotation by an angle π in the linear two-space $\mathbb{R}^2$, followed by another rotation by the same angle in a linear two-subspace of $\mathbb{R}^{n+1}$ including $\overline{X}_{(+)}$, only change the sign of the vector $\overline{X}$, and, consequently, do not move the corresponding unoriented line. All these transformations belong to the isotropy subgroup, but not to $\mathfrak{P}_0$. Pick one of these transformations, say ρ : then any other may be written as a product of ρ by an element of $\mathfrak{P}_0$ (more exactly a rotation in the subspace $\mathbb{R}^n$ orthogonal to $\overline{X}_{(+)}$). Define $\mathfrak{R}$ as the $\mathbb{Z}_2$ subgroup of $SO_0(2, n+1)$ composed of the two elements $1, \rho$. As one easily verifies, $\mathfrak{R}\mathfrak{P}_0\mathfrak{R}^{-1} = \mathfrak{P}_0$, so we can define the semidirect product

$$\mathfrak{P} = \mathfrak{R} \times_S \mathfrak{P}_0, \tag{6.3.1}$$

which is in fact the full isotropy subgroup that we were looking for. Hence

$$M \simeq \frac{SO_0(2, n+1)}{\mathfrak{P}}, \quad \widetilde{M} \simeq \frac{SO_0(2, n+1)}{\mathfrak{P}_0}.$$

The group $\mathfrak{P}$ is a so-called *parabolic* subgroup[(4)], and has dimension $\frac{1}{2}(n+2)(n+3) - (n+1)$.

Notice that, for n odd, the element -1 (which belongs to $SO_0(2, n+1)$ but not to $\mathfrak{P}_0$) can be decomposed into the product of ρ (which is defined independently from n) by an element of $\mathfrak{P}_0$.

The (transitive) action of $SO_0(2, n+1)$ on M induces a symplectic action on the cotangent bundle T^*M. This action, however, is not transitive, and it is natural to ask for its homogeneous components. These homogeneous components can in turn be identified with (in general covering spaces of) coadjoint orbits of the group, as stated by the celebrated Kirillov-Kostant-Souriau Theorem. To see this, it is convenient to use the following theorem, due essentially to Wolf (1975).

THEOREM 6.5 *Let $\mathfrak{H}$ be a closed subgroup of the Lie group $\mathfrak{G}$ with Lie algebras $\mathfrak{h}$ and $\mathfrak{g}$, respectively; let $f \in \mathfrak{g}^*$ (the dual space of the Lie algebra $\mathfrak{g}$) and $\mathfrak{G}_f$ be the isotropy subgroup of f, that is, $\mathbf{Ad}^*_{\mathfrak{G}_f} f = f$, where $\mathbf{Ad}^*$ is the coadjoint action; let $\mathfrak{g}_f$ be the Lie algebra of $\mathfrak{G}_f$. If the following three assumptions hold*

(A1) $\dim \mathfrak{h} = \frac{1}{2}(\dim \mathfrak{g} + \dim \mathfrak{g}_f)$,

(A2) $\langle f, \mathfrak{h} \rangle = 0$,

(A3) $\mathfrak{G}_f \subset \mathfrak{H}$,

then the coadjoint orbit $\mathcal{O}_f = \mathbf{Ad}^*_{\mathfrak{G}} f$, endowed with the Kirillov-Kostant-Souriau form induced by the Poisson structure

$$\Omega^{\sharp}_{(f)}(u, v) = \langle f, [u, v] \rangle, \quad u, v \in T^*_f \mathcal{O}_f \subset \mathfrak{g},$$

(4) See, e.g., Barut & Raczka (1980).

is symplectomorphic to an open orbit of $\mathfrak{G}$ *in* $T^*(\mathfrak{G}/\mathfrak{H})$, *endowed with the canonical form.*

Proof. It is enough to prove that the two manifolds are equivariantly diffeomorphic under the action of $\mathfrak{G}$, since the "infinitesimal" version of the equivariance property states precisely that the push-forward of the first form coincides with the latter: see Subsection D.3.1, in particular Equation (D.3.4) on page 400.

Firstly, we prove that $\mathbf{Ad}^*_{\mathfrak{H}} f$ is an open set in the linear subspace $\mathfrak{h}^\perp$ of $\mathfrak{g}^*$ annihilating $\mathfrak{h}$. Let $h \in \mathfrak{h}$, then $\langle \mathbf{Ad}^*_{\mathfrak{H}} f, h\rangle = \left\langle f, \mathbf{Ad}^{-1}_{\mathfrak{H}} h\right\rangle$, but $\mathbf{Ad}_{\mathfrak{H}} h \in \mathfrak{h}$, $\forall h$ since $\mathfrak{H}$ is a subgroup, hence, by assumption (A2),

$$\langle \mathbf{Ad}^*_{\mathfrak{H}} f, h\rangle = 0 \quad \forall h.$$

Thus $\mathbf{Ad}^*_{\mathfrak{H}} f \subset \mathfrak{h}^\perp$. Assumption (A3) implies $\mathfrak{g}_f \subset \mathfrak{h}$, hence

$$\dim \mathbf{Ad}^*_{\mathfrak{H}} f = \dim \mathfrak{h} - \dim \mathfrak{g}_f.$$

Moreover,

$$\dim \mathfrak{h}^\perp = \dim \mathfrak{g} - \dim \mathfrak{h} \overset{(A1)}{=} \dim \mathfrak{h} - \dim \mathfrak{g}_f.$$

Thus

$$\dim \mathbf{Ad}^*_{\mathfrak{H}} f = \dim \mathfrak{h}^\perp$$

and $\mathbf{Ad}^*_{\mathfrak{H}} f$ is an open set in the linear subspace $\mathfrak{h}^\perp$.

This allows us the following construction. Since $\mathfrak{H}$ is a closed subgroup, $\mathfrak{G}/\mathfrak{H}$ is a manifold (see Section B.3) which has tangent space $\mathfrak{g}/\mathfrak{h}$ and cotangent space $\mathfrak{h}^\perp$. Let us consider the cotangent bundle $T^*(\mathfrak{G}/\mathfrak{H})$. It is the bundle $(\mathfrak{G} \times \mathfrak{h}^\perp)/\mathfrak{H}$, consisting of all equivalence classes

$$[[g, v]] = \left\{gp^{-1}, \mathbf{Ad}^*_p v : p \in \mathfrak{H}\right\} \subset \mathfrak{G} \times \mathfrak{h}^\perp,$$

with left-action of $g' \in \mathfrak{G}$ given by

$$g'[[g, v]] = [[g'g, v]],$$

and with quotient differentiable structure from $\mathfrak{G} \times \mathfrak{h}^\perp$. Define the $\mathfrak{G}$-orbit in $T^*(\mathfrak{G}/\mathfrak{H})$

$$O_f = \mathfrak{G}[[1, f]] = \{[[g, f]] \in (\mathfrak{G} \times \mathfrak{h}^\perp)/\mathfrak{H} : g \in \mathfrak{G}\}.$$

A dimensional count gives

$$\begin{aligned}\dim O_f &= \dim(\mathfrak{G}/\mathfrak{H}) + \dim \mathbf{Ad}^*_{\mathfrak{H}} f = \dim \mathfrak{g} - \dim \mathfrak{h} + \dim \mathfrak{h}^\perp \\ &= \dim(\mathfrak{G} \times \mathfrak{h}^\perp)/\mathfrak{H} = \dim T^*(\mathfrak{G}/\mathfrak{H}),\end{aligned}$$

so that O_f is open in $T^*(\mathfrak{G}/\mathfrak{H})$.

For a fixed f, let us map $\mathcal{O}_f$ to O_f by

$$\mathbf{Ad}^*_g f \mapsto [[g,f]]\,.$$

It is visibly $\mathfrak{G}$-equivariant with image O_f. Moreover, it is well defined and 1-1, as we prove now.

(i) If $\mathbf{Ad}^*_g f = \mathbf{Ad}^*_{g'} f$, then $[[g,f]] = [[g',f]]$. In fact this hypothesis implies that $g' = g g_f$, $g_f \in \mathfrak{G}_f$, hence

$$[[g',f]] = \left[\left[g g_f, f\right]\right] \overset{(A3)}{=} \left[\left[g, \mathbf{Ad}^*_{g_f} f\right]\right] = [[g,f]]\,.$$

(ii) If $[[g,f]] = [[g',f]]$, then $\mathbf{Ad}^*_g f = \mathbf{Ad}^*_{g'} f$. In fact this hypothesis implies that $[[g^{-1}g',f]] = [[1,f]]$, from which $g^{-1}g' \in \mathfrak{G}_f$.

This proves the theorem. QED

REMARK 6.2 Notice that if we take the less restrictive assumption

(A3') $\mathfrak{g}_f \subset \mathfrak{h}$

the map $O_f \to \mathcal{O}_f$ will be in general a covering map: see point (i) of the above proof.

We will soon apply this theorem, choosing $SO_0(2,n+1)$ as $\mathfrak{G}$ and $\mathfrak{P}$ as $\mathfrak{H}$, hence M as $SO_0(2,n+1)/\mathfrak{P}$. We need to know therefore the orbits in T^*M under the action of the conformal group. By definition, the action of the conformal group transforms $\|y\|^2$, (the square of the norm of a covector), into $\omega^2(x^\mu)\,\|y\|^2$, $\omega \neq 0$. This allows us to find directly the five homogeneous components of $T^+M \overset{\text{def}}{=} T^*M - \{$zero section$\}$ under the conformal action, namely:

(i) the manifold of time-like, future-pointing covectors ($\eta^{\mu\nu} y_\mu y_\nu < 0$ with $y_0 > 0$),

(ii) the manifold of time-like, past-pointing covectors ($\eta^{\mu\nu} y_\mu y_\nu < 0$ with $y_0 < 0$),

(iii) the manifold of space-like covectors ($\eta^{\mu\nu} y_\mu y_\nu > 0$),

(iv) the manifold of null, future-pointing covectors ($\eta^{\mu\nu} y_\mu y_\nu = 0$ with $y_0 > 0$),

(v) the manifold of null, past-pointing covectors ($\eta^{\mu\nu} y_\mu y_\nu = 0$ with $y_0 < 0$).

To identify these homogeneous components with coadjoint orbits of the conformal group, we will make use of Proposition 6.6 below. But we need some definitions before.

A *bivector* of a (pseudo)-Riemannian manifold is, by definition, a skew symmetric contravariant tensor of order two. The metric tensor allows one to define the scalar product and thus the norm of a bivector. A *simple bivector* $Y \wedge X$ is obtained by taking the skew symmetric part of the tensor product $Y \otimes X$ of two vectors. In coordinates: $(Y \wedge X)_{AB} = Y_A X_B - Y_B X_A$. Let us consider the manifold $\mathcal{S}$ of the *null simple bivectors* G of $\mathbb{R}^{2,n+1}$, that is,

$$\mathcal{S} = \{G = Y \wedge X,\ 0 \neq X, Y \in \mathbb{R}^{2,n+1} \text{ and } \eta(G,G) = 0\}.$$

The nullity condition $\eta(G,G) = 0$ gives

$$X^2 Y^2 - (X \cdot Y)^2 = 0 \tag{6.3.2}$$

where, for brevity,

$$X^2 = \eta(X,X), \quad Y^2 = \eta(Y,Y), \quad X \cdot Y = \eta(X,Y).$$

The map $X, Y \mapsto X \wedge Y$ is not injective, since two other vectors X', Y' such that

$$\begin{pmatrix} X' \\ Y' \end{pmatrix} = \begin{pmatrix} a & b \\ c & d \end{pmatrix} \begin{pmatrix} X \\ Y \end{pmatrix}, \quad a, b, c, d \in \mathbb{R}, \quad ad - bc = 1,$$

give the same bivector

$$Y' \wedge X' = Y \wedge X.$$

In other words $\mathcal{S} \simeq \{\text{pairs of vectors satisfying (6.3.2)}\}/SL(2,\mathbb{R})$. Thus

$$\dim \mathcal{S} = 2(n+3) - 1 - 3 = 2(n+1).$$

Now we may state the announced

PROPOSITION 6.6 *The manifold $\mathcal{S}$ of the null simple bivectors of $\mathbb{R}^{2,n+1}$ is homeomorphic to $T^+\left(\frac{S^1 \times S^n}{\mathbb{Z}_2}\right)$, thus to T^+M.*

Proof. Every matrix may be decomposed into the product of an orthogonal and a triangular one, hence

$$\begin{pmatrix} a & b \\ c & d \end{pmatrix} = \begin{pmatrix} \lambda & 0 \\ \mu & \frac{1}{\lambda} \end{pmatrix} \begin{pmatrix} \cos\nu & -\sin\nu \\ \sin\nu & \cos\nu \end{pmatrix}, \quad \lambda, \mu \in \mathbb{R}, \ \nu \in S^1. \tag{6.3.3}$$

Choose two vectors X and Y satisfying the nullity condition (6.3.2) and, in general, let us suppose $X^2, Y^2 \neq 0$; if, instead, one of the two vectors (say X) or both has a null norm, we can continue directly with (6.3.5) below. X^2 and Y^2 may be positive or negative, despite the notation, but the nullity condition ensures that they will always have the same sign, so that $\sqrt{X^2Y^2}$ and $\sqrt{\frac{X^2}{Y^2}}$ will be real positive. Let us write (6.3.2) as $\sqrt{X^2Y^2} = \pm X \cdot Y$, where the sign is fixed by requiring the positiveness of the right-hand side. These two vectors are determined modulo the above linear transformation, and the values of λ, μ, ν can be chosen arbitrarily. Firstly, choosing $\lambda = 1$ and $\mu = 0$ yields the transformed vectors

$$X'^2 = Y^2 \left(\sqrt{\frac{X^2}{Y^2}} \cos\nu \mp \sin\nu \right)^2 ,$$

$$X' \cdot Y' = \frac{1}{2}\left(X^2 - Y^2\right) \sin 2\nu + X \cdot Y \cos 2\nu,$$

where the upper sign holds if the couple X^2, Y^2 has the same sign of $X \cdot Y$, the lower sign holds in the opposite case. Choose an angle ν such that

$$\tan\nu = \pm\sqrt{\frac{X^2}{Y^2}}, \tag{6.3.4}$$

from which

$$\sin 2\nu = \pm\frac{2\sqrt{\frac{X^2}{Y^2}}}{1+\frac{X^2}{Y^2}}, \quad \cos 2\nu = \frac{1-\frac{X^2}{Y^2}}{1+\frac{X^2}{Y^2}},$$

then the two transformed vectors will satisfy, dropping out the primes,

$$X^2 = 0, \quad X \cdot Y = 0. \tag{6.3.5}$$

Let us now consider a nontrivial first factor in the decomposition (6.3.3). This means that

$$X \sim X' = \lambda X, \quad Y \sim Y' = \frac{Y}{\lambda} + \mu X.$$

For every pair X, Y satisfying (6.3.5), we may pick another representative pair (in the same equivalence class) such that

(i) with a convenient choice of the dilation λ we can achieve

$$-\eta\left(X'_{(-)}, X'_{(-)}\right) = \eta\left(X'_{(+)}, X'_{(+)}\right) = \text{ some fixed constant,}$$

(ii) choosing μ appropriately, we can have also

$$\eta\left(X'_{(-)}, Y'_{(-)}\right) = 0 = \eta\left(X'_{(+)}, Y'_{(+)}\right).$$

That is, $X'_{(-)} \in S^1$, $X'_{(+)} \in S^n$, and $Y'_{(-)}, Y'_{(+)}$ belong to the respective (co)tangent spaces. The case $Y = 0$ is not allowed, since the definition of $\mathcal{S}$ excludes identically zero bivectors.

Once the pair of vectors X and Y satisfying the nullity condition are fixed, Equation (6.3.4) has a well defined sign. Whatever is the sign, we get *two* values of ν (differing by π), satisfying (6.3.4), since $\tan\nu = \tan(\nu+\pi)$. Moreover, also λ, hence μ, are uniquely determined up to the sign. Therefore, to every $\mathrm{SL}(2,\mathbb{R})$-equivalence class of pairs of vectors satisfying the nullity condition, i.e., to every element of $\mathcal{S}$, correspond two diametrically opposed elements of $T^+(S^1 \times S^n)$, which represent the same point of $T^+\left(\frac{S^1 \times S^n}{\mathbb{Z}_2}\right)$.

Vice versa, every element of $T^+(S^1 \times S^n)$ is made up of two vectors X, Y such that $X^2 = 0$ and $X \cdot Y = 0$, thus satisfying the nullity condition (6.3.2); to each such pair is associated a unique element of $\mathcal{S}$. QED

COROLLARY 6.7 *The submanifold of the null simple bivectors $Y \wedge X$ with $X^2 = Y^2 = X \cdot Y = 0$ is homeomorphic to (two disjoint copies of) T^+S^n.*

Proof. Since $X_{(-)}$ and $Y_{(-)}$ are 2-dimensional vectors, choosing appropriately ν we can always transform $X_{(-)}$ into a vector of prearranged direction in the $\mathbb{R}^2$ subspace. The transformed X and Y satisfy still the assumption, hence (6.3.5), then we choose λ and μ in such a way as to have the same final situation of the previous proposition. But now (since by hypothesis $Y^2 = 0$) it holds that $\eta\left(Y_{(-)}, Y_{(-)}\right) = -\eta\left(Y_{(+)}, Y_{(+)}\right)$, so the two vectors $X_{(-)}$ and $Y_{(-)}$ are frozen, apart from the orientation of the latter. This leaves us with two copies of T^+S^n, each one labelled by the orientation of $Y_{(-)}$; they are disjoint since $Y = 0$ is excluded. QED

The above proposition is useful in constructing explicitly an Ad^*-equivariant moment map

$$J : T^+M \to \mathfrak{c}^*,$$

which is a map making the diagram

$$\begin{array}{ccc} T^+M & \xrightarrow{\;\;\mathfrak{C}\;\;} & T^+M \\ {\scriptstyle J}\downarrow & & \downarrow{\scriptstyle J} \\ \mathfrak{c}^* & \xrightarrow{\;\mathbf{Ad}^*_{\mathfrak{C}}\;} & \mathfrak{c}^* \end{array}$$

commutative[5]. If $\{\mathcal{G}^*_{AB} = -\mathcal{G}^*_{BA}\}$ is the dual basis of the Lie algebra of the conformal group, the moment map is given by

$$J = \sum_{A<B} (Y_A X_B - Y_B X_A)\mathcal{G}^*_{AB}, \qquad (6.3.6)$$

where Y and X are any two vectors that satisfy the nullity condition (6.3.2). Indeed, for generic vectors $X, Y \in \mathbb{R}^{2,n+1}$ the map J is, obviously, an Ad^*-equivariant moment map: see Example D.2 on page 401; then, the nullity condition is preserved by the conformal action; finally, Proposition 6.6 ensures that the domain of J is just T^+M.

An explicit expression of the dual basis $\mathcal{G}^*_{AB}$ will be needed. The conformal group is a simple group, and the Cartan-Killing form $\mathcal{F} : \mathfrak{c} \otimes \mathfrak{c} \to \mathbb{R}$ endows its Lie algebra with a regular metric. The basis (6.1.4) is pseudo-orthonormal,

$$\mathcal{F}(\mathcal{G}_{AB}, \mathcal{G}_{CD}) = \eta_{AC}\eta_{BD} - \eta_{BC}\eta_{AD},$$

from which we infer, in particular,

$$\mathcal{F}(\mathcal{P}_\mu, \mathcal{K}_\nu) = 2\eta_{\mu\nu}, \quad \mathcal{F}(\mathcal{P}_\mu, \mathcal{P}_\nu) = 0 = \mathcal{F}(\mathcal{K}_\mu, \mathcal{K}_\nu).$$

The Cartan-Killing form allows us to identify $\mathfrak{c}$ and $\mathfrak{c}^*$. So, for $m \in \mathfrak{c}$, we define $m^* \in \mathfrak{c}^*$ by the condition

$$\langle m^*, n\rangle = \mathcal{F}(m, n), \ \forall n \in \mathfrak{c}.$$

For the complete skew symmetry of the structure constants, the isotropy algebra of m^* is given by

$$\mathfrak{c}_{m^*} = \{n \in \mathfrak{c} : [m, n] = 0\}.$$

The basis and its dual basis are

$$\begin{bmatrix} \mathcal{G}_{-1,0} & \mathcal{G}_{\alpha\beta} & \mathcal{G}_{-1,\alpha} & \mathcal{G}_{0,\alpha} \\ \mathcal{G}^*_{-1,0} & \mathcal{G}^*_{\alpha\beta} & -\mathcal{G}^*_{-1,\alpha} & -\mathcal{G}^*_{0,\alpha} \end{bmatrix},$$

or, equivalently,

$$\begin{bmatrix} \mathcal{G}_{hk} & \mathcal{G}_{0k} & \mathcal{D} & \mathcal{P}_0 & \mathcal{P}_k & \mathcal{K}_0 & \mathcal{K}_h \\ \mathcal{G}^*_{hk} & -\mathcal{G}^*_{0k} & -\mathcal{D}^* & -\frac{1}{2}\mathcal{K}^*_0 & \frac{1}{2}\mathcal{K}^*_k & -\frac{1}{2}\mathcal{P}^*_0 & \frac{1}{2}\mathcal{P}^*_h \end{bmatrix}, \qquad (6.3.7)$$

as one easily verifies: indeed, the pairing $\langle\cdot,\cdot\rangle$ between an element of the first row and one of the second row is 1 if the two elements belong to the same column, and zero otherwise.

We are ready, at last, to identify the five homogeneous components in T^+M with some coadjoint orbits in $\mathrm{so}^*(2, n+1)$.

---

[5] Since $\mathrm{SO}_0(2, n+1)$ is a simple group, this definition of moment map is equivalent to the usual more general one, where the components of a moment map are defined as the Hamiltonians of the vector fields generated by the "infinitesimal" symplectic action of a group over a manifold.

THEOREM 6.8 *Let* $f_{(-)} = -\frac{1}{2}\mathcal{K}_0^*$, $f_{(+)} = \frac{1}{2}\mathcal{K}_1^*$. *Then* $\mathcal{O}_{f_{(-)}}, \mathcal{O}_{-f_{(-)}}, \mathcal{O}_{f_{(+)}}$ *are symplectomorphic to one of the following submanifolds of* T^+M *:*

(i) *the manifold of time-like future-pointing covectors,*

(ii) *the manifold of time-like past-pointing covectors,*

(iii) *the manifold of space-like covectors,*

respectively.

Proof. Let us apply Theorem 6.5, choosing $SO_0(2, n+1)$ as $\mathfrak{G}$ and $\mathfrak{P}$ as $\mathfrak{H}$. One easily sees that the isotropy subalgebras of $f_{(-)}$ and $f_{(+)}$ have basis $\{\mathcal{K}_\mu, \mathcal{G}_{hk}\}$ and $\{\mathcal{K}_\mu, \mathcal{G}_{a0}, \mathcal{G}_{ab}\}$ respectively, and the conditions of the theorem are satisfied. The coadjoint orbits $\mathcal{O}_{f_{(-)}}, \mathcal{O}_{-f_{(-)}}, \mathcal{O}_{f_{(+)}}$ are therefore symplectomorphic to submanifolds of T^+M. Let us investigate them in some detail. To this end, define

$$X = \begin{pmatrix} \frac{1}{2} \\ 0 \\ \hline \vdots \\ 0 \\ \frac{1}{2} \end{pmatrix}, \quad Y' = \begin{pmatrix} 0 \\ 1 \\ \hline 0 \\ \vdots \\ 0 \end{pmatrix}, \quad Y'' = \begin{pmatrix} 0 \\ 0 \\ \hline 1 \\ 0 \\ \vdots \end{pmatrix},$$

which satisfy the properties (i) and (ii) in the proof of Proposition 6.6. Applying (6.3.6) and bearing in mind the expression (6.3.7) of the dual basis, one sees that

$$\begin{aligned} f_{(-)} &\text{ is generated by } -Y' \wedge X, \\ -f_{(-)} &\text{ is generated by } \quad Y' \wedge X, \\ f_{(+)} &\text{ is generated by } \quad Y'' \wedge X, \end{aligned}$$

i.e., $f_{(-)} = -\sum_{A<B}(Y'_A X_B - Y'_B X_A)\mathcal{G}^*_{AB}$, etc. Recalling the proof of Proposition 6.6, the thesis follows, since

$$\begin{aligned} \eta\left(Y'_{(-)}, Y'_{(-)}\right) &= -1, \quad \eta\left(Y'_{(+)}, Y'_{(+)}\right) = 0, \\ \eta\left(Y''_{(-)}, Y''_{(-)}\right) &= 0, \quad \eta\left(Y''_{(+)}, Y''_{(+)}\right) = 1, \end{aligned}$$

showing that Y' gives rise to a time-like covector and Y'' to a space-like covector $\in T^+_X(S^1 \times S^n)$. QED

These three orbits are $2(n+1)$-dimensional. We now come to the other two $2n$-dimensional orbits, each of which is called a *Kepler manifold*, since they will become the phase space of the regularized Kepler Problem with negative energy.

THEOREM 6.9 *Let $f_{0+} = f_{(+)} + f_{(-)}$ and $f_{0-} = f_{(+)} - f_{(-)}$; then $\mathcal{O}_{f_{0+}}$ and $\mathcal{O}_{f_{0-}}$ are symplectomorphic to the submanifolds of T^+M given by*

(iv) the manifold of null future-pointing covectors,

(v) the manifold of null past-pointing covectors

respectively, which, in turn, are both symplectomorphic to T^+S^n.

Proof. Let us apply Theorem 6.5 with the following identifications:

- $\mathfrak{G}$ is $\mathrm{SO}_0(1, n+1)$, the Lorentz subgroup of $\mathrm{SO}_0(2, n+1)$ acting linearly on $\mathbb{R}^{1,n+1}$ (i.e., the subspace of $\mathbb{R}^{2,n+1}$ obtained by deleting the X^0 coordinate) whose Lie algebra has basis $\{\mathcal{G}_{hk}, \mathcal{P}_h, \mathcal{K}_h, \mathcal{D})$;
- $\mathfrak{H}$ is the connected parabolic subgroup of $\mathrm{SO}_0(1, n+1)$ whose Lie algebra has basis $\{\mathcal{G}_{hk}, \mathcal{K}_h, \mathcal{D})$;
- f is $f_{(+)} = \frac{1}{2}\mathcal{K}_1^*$ (viewed as belonging to $\mathrm{so}^*(1, n+1)$), whose isotropy subalgebra $\mathrm{so}(1, n+1)_{f_{(+)}}$ has basis $\{\mathcal{G}_{ab}, \mathcal{K}_h\}$.

Arguing as in the previous case one sees that $\mathfrak{G}/\mathfrak{H} \simeq S^n$ and $\mathcal{O}_{f_{(+)}}$ (which is viewed as a coadjoint orbit of the subgroup $\mathrm{SO}_0(1, n+1)$ only) is symplectomorphic to T^+S^n. Let us consider the coadjoint orbits $\mathcal{O}_{f_{0+}}$ and $\mathcal{O}_{f_{0-}}$, where now the full $\mathrm{SO}_0(2, n+1)$ group is acting. Their isotropy subalgebra has basis

$$\mathrm{so}(2, n+1)_{f_{0\pm}} = \{\mathcal{G}_{ab}, \mathcal{K}_\mu, \mathcal{P}_0 \pm \mathcal{P}_1, \mathcal{G}_{0a}, \mathcal{G}_{1,0} \mp \mathcal{D}\}.$$

One easily verifies that, as vector spaces,

$$\mathrm{so}(2, n+1)_{f_{0\pm}} = \mathrm{so}(1, n+1)_{f_{(+)}} \oplus \{\mathcal{G}_{0A}\}.$$

Therefore the orbits $\mathcal{O}_{f_{0\pm}} \subset \mathrm{so}^*(2, n+1)$ have the same cotangent space as $T^+S^n \subset \mathrm{so}^*(1, n+1)$, and are hence symplectomorphic to it. Moreover, the elements $f_{0\pm}$ are generated by $(Y'' \mp Y') \wedge X$, and this completes the proof. QED

Recalling Corollary 6.7, we can then conclude that

$$T^+S^n \simeq \mathcal{O}_{f_{0\pm}} \simeq \frac{T_0^+M}{S^1},$$

where T_0^+M is the restriction of the cotangent bundle of the compactified Minkowski space to the null (but not identically zero) covectors.

6.4 Regularization of the Kepler Problem

The arguments exposed in the preceding sections of this chapter are preliminary to the present one, where we come eventually to the Kepler Problem. They are intrinsic, coordinate independent arguments; to treat the Kepler Problem, however, we must make explicit the preceding constructions.

We will need to construct some moment maps $J : T^+M_\gamma \to \mathfrak{c}^*$ for different sections γ, i.e., we must give $\frac{1}{2}(n+2)(n+3)$ functions $G_{AB}(x^\mu, y_\nu)$ with $A < B$, such that

$$J(x^\mu, y_\nu) = \sum_{A<B} G_{AB}(x^\mu, y_\nu)\mathcal{G}^*_{AB},$$

where x^μ, y_ν are canonical coordinates on T^+M_γ. The infinitesimal version of the commutativity property of the defining diagram on page 125 is

$$\{G_{AB}(x^\mu, y_\nu), G_{AC}(x^\mu, y_\nu)\} = \eta_{AA} G_{BC}(x^\mu, y_\nu),$$
$$\text{or } = 0 \text{ if all indices are different,}$$

i.e., this Poisson algebra is isomorphic to the Lie algebra (6.1.5) of the conformal group. Recalling (6.3.6), we take in general $G_{AB} = Y_A X_B - Y_B X_A$, and specify X and Y to be some functions of x^μ, y_ν. The section γ is represented by the functions $X^A = \Gamma^A(x^\mu)$ satisfying $\eta(\Gamma, \Gamma) = 0$, and the metric tensor induced on M_γ by the "ambient" metric η is

$$(g_\gamma)_{\mu\nu} = \frac{\partial \Gamma^A}{\partial x^\mu} \eta_{AB} \frac{\partial \Gamma^B}{\partial x^\nu}.$$

We must find the corresponding functions $Y_B = \Pi_B(x^\mu, y_\nu)$.

Since the action of the conformal group on T^*M is the canonical extension of the action on the base manifold M, the corresponding infinitesimal vector fields and the components of the moment map will be linear in ∂_μ and y_μ, respectively. The vector fields $\mathcal{G}_{AB} = X_B\partial_A - X_A\partial_B$, generating the infinitesimal action of $\mathrm{SO}_0(2, n+1)$ on $\mathbb{R}^{2,n+1}$, are tangent to the null cone $\mathcal{K}$, but not to the section γ in general. Actually, the group acts on the rays of the null cone, and the vectors of the infinitesimal action can be safely projected onto the tangent space to the section, which is, by hypothesis, transverse to the null rays. Let V^A be the components of a tangent vector to the null cone and V^μ the components of the γ-related vector on the section (see Definition (A.2.2) on page 330). The relation between the two vectors is by definition

$$V^A(\Gamma(x)) = \frac{\partial \Gamma^A}{\partial x^\mu} V^\mu(x) \quad \text{or equivalently } V^A \frac{\partial}{\partial X^A} = V^\mu \frac{\partial}{\partial x^\mu}. \tag{6.4.1}$$

We say that V^μ is the *projection* of V^A. As shown in Proposition A.3 on page 330, projection and Lie bracket are commuting operations, so that the projections of the vectors $\mathcal{G}_{AB}$ close up under Lie brackets, yielding a Lie algebra homomorphic to $\mathrm{so}(2, n+1)$. Really, the homomorphism is an isomorphism: the action of the conformal group on the null lines of the cone is effective and the section γ is transversal to them (this excludes that the homomorphism is trivial); moreover the conformal group is simple, hence does not admit an invariant Lie subgroup (this ensures that the kernel of the homomorphism is trivial).

Contracting the first of (6.4.1) with $\eta_{AB}\frac{\partial \Gamma^B}{\partial x^\nu}$, we find the explicit expression of the projection:

$$V^\nu = (g_\gamma)^{\mu\nu}\frac{\partial \Gamma^A}{\partial x^\mu}\eta_{AB}V^B.$$

Notice that, as expected, the vectors tangent to the rays of the null cone, i.e., of the type $V^B = \lambda(x^\mu)\Gamma^B$ with $\lambda(x^\mu) \neq 0$, have null projection, since

$$d\Gamma^A\eta_{AB}\lambda(x^\mu)\Gamma^B = \frac{1}{2}\lambda(x^\mu)\, d\left(\eta_{AB}\Gamma^A\Gamma^B\right) = 0.$$

For the cotangent map we take

$$\Pi_B(x^\mu, y_\nu) = (g_\gamma)^{\mu\nu}\frac{\partial \Gamma^A}{\partial x^\mu}\eta_{AB}y_\nu,$$

which ensures that $V^\mu y_\mu = V^A\Pi_A$, equivalent to the latter of (6.4.1).

We have thus constructed, for every section γ of the null cone $\mathcal{K}$, an associated moment map $J : T^+M_\gamma \to \mathfrak{c}^*$, whose components are

$$G_{AB}(x^\mu, y_\nu) = \Pi_A\Gamma_B - \Pi_B\Gamma_A.$$

EXAMPLE 6.3 Let us consider the section

$$\begin{aligned}
\Gamma^{-1} &= \frac{1}{2}\left(1 + \eta_{\mu\nu}x^\mu x^\nu\right),\\
\Gamma^0 &= x^0,\\
\Gamma^k &= x^k,\\
\Gamma^{n+1} &= \frac{1}{2}\left(1 - \eta_{\mu\nu}x^\mu x^\nu\right),
\end{aligned} \tag{6.4.2}$$

obtained by intersecting the null cone $\mathcal{K}$ with the hyperplane $X^{-1} + X^{n+1} = 1$. Thus

$$(g_{\mu\nu}) = \begin{pmatrix} -1 & 0 \\ 0 & \mathbb{1}_n \end{pmatrix},$$

so that the section at hand is just the Minkowski space M_0. Then

$$\begin{aligned}\Pi_{-1} &= -x^\mu y_\mu,\\ \Pi_0 &= y_0,\\ \Pi_k &= y_k,\\ \Pi_{n+1} &= -x^\mu y_\mu,\end{aligned} \tag{6.4.3}$$

and the moment map is given by

$$\begin{aligned}G_{\mu\nu} &= x_\nu y_\mu - x_\mu y_\nu,\\ P_\mu &= y_\mu,\\ K_\mu &= 2x_\mu x^\nu y_\nu - x^\nu x_\nu y_\mu,\\ D &= x^\mu y_\mu,\end{aligned} \tag{6.4.4}$$

that we could obtain from the basis $\{\mathcal{G}_{AB}\}$ of the Lie algebra $\mathfrak{c}$ by replacing $y_\mu \mapsto \partial_\mu$: compare Example D.2 on page 401.

EXERCISE 6.4 An exercise, which provides a better insight into the preceding constructions, is to verify that the Poisson structure on the coadjoint orbits of Theorems 6.8 and 6.9 coincides with the canonical one of the cotangent bundle. We only verify this for $\mathcal{O}_{f(-)}$, the other cases being entirely similar. Fix a point p in T^+M_0 with coordinates $x^\mu = 0$, $y_0 = 1$, $y_k = 0$, i.e., the origin and a unit time-like covector. To this point correspond the vector X and the covector Y' defined in the proof of Theorem 6.8, see (6.4.2) and (6.4.3). Evaluating the moment map (6.4.4) at this point, we find that all the components are zero except $P_0 = 1$, and this means that the image $J(p)$ of the moment map is $f_{(-)} = -\frac{1}{2}\mathcal{K}_0^*$, as is obvious from (6.3.7). Consider the cotangent space $T_p^*(T^+M_0)$ in the point p : it admits a natural splitting into two $(n+1)$-dimensional subspaces, the first being spanned by $\{\delta x^\mu\}$, the latter by $\{\delta y_\nu\}$. Differentiating (6.4.4) at point p shows that $T^*_{f_{(-)}}\mathcal{O}_{f_{(-)}}$ is spanned by $\{\mathcal{D}, \mathcal{G}_{0k}\} = \mathfrak{p}/\mathfrak{c}_{f_{(-)}}$ and $\{\mathcal{P}_\mu\} = \mathfrak{c}/\mathfrak{p}$, respectively. The Poisson bivector, evaluated on two elements of $\mathfrak{p}/\mathfrak{c}_{f_{(-)}}$, is zero, since $\mathfrak{p}$ is a subalgebra annihilated by $f_{(-)}$, and evaluated on two elements of $\mathfrak{c}/\mathfrak{p}$ is zero, since $\mathfrak{c}/\mathfrak{p}$ is an abelian subalgebra. Lastly, the Poisson bivector, evaluated on an element of $\mathfrak{p}/\mathfrak{c}_{f_{(-)}}$ and another of $\mathfrak{c}/\mathfrak{p}$, gives

$$\begin{aligned}\Omega^\sharp_{f_{(-)}}(\mathcal{P}_h, \mathcal{G}_{0k}) &= \left\langle -\frac{1}{2}\mathcal{K}_0^*, [\mathcal{P}_h, \mathcal{G}_{0k}]\right\rangle = -\frac{1}{2}F(\mathcal{K}_0, \delta_{hk}\mathcal{P}_0) = \delta_{hk},\\ \Omega^\sharp_{f_{(-)}}(\mathcal{P}_0, \mathcal{D}) &= \left\langle -\frac{1}{2}\mathcal{K}_0^*, [\mathcal{P}_0, \mathcal{D}]\right\rangle = -\frac{1}{2}F(\mathcal{K}_0, \mathcal{P}_0) = 1,\\ \Omega^\sharp_{f_{(-)}}(\mathcal{P}_k, \mathcal{D}) &= \left\langle -\frac{1}{2}\mathcal{K}_0^*, [\mathcal{P}_k, \mathcal{D}]\right\rangle = -\frac{1}{2}F(\mathcal{K}_0, \mathcal{P}_k) = 0,\\ \Omega^\sharp_{f_{(-)}}(\mathcal{P}_0, \mathcal{G}_{0k}) &= \left\langle -\frac{1}{2}\mathcal{K}_0^*, [\mathcal{P}_0, \mathcal{G}_{0k}]\right\rangle = -\frac{1}{2}F(\mathcal{K}_0, \mathcal{P}_k) = 0,\end{aligned}$$

as expected.

We now define three cross-sections that will be relevant for the three signs of the energy of the Kepler Problem, respectively. Next we construct the relative moment maps. Finally we reduce them, as intimated in Theorems 6.8 and 6.9, to find a moment map whose domain is the Kepler manifold T^+S^n.

i) Cross-section $M_{(-)}$

$$\begin{aligned}\Gamma^{-1} &= \cos x^0,\\ \Gamma^0 &= \sin x^0,\\ \Gamma^k &= \frac{2x^k}{x^2+1},\\ \Gamma^{n+1} &= \frac{x^2-1}{x^2+1},\end{aligned}$$

that is, the intersection of the null cone $\mathcal{K}$ with the sphere

$$\sum_A (X^A)^2 = 2.$$

The functions Γ^α are obtained by stereographic projection of $S^n \subset \mathbb{R}^{n+1}$ onto the subspace $\mathbb{R}^n$ (see Figure 4.1 on page 82). The cotangent map and the metric are given by

$$\begin{aligned}\Pi_{-1} &= -y_0 \sin x^0,\\ \Pi_0 &= y_0 \cos x^0,\\ \Pi_k &= \frac{1}{2}(x^2+1)y_k - (x,y)x_k,\\ \Pi_{n+1} &= (x,y),\end{aligned}$$

and

$$(g_{(-)})^{\mu\nu}\, y_\mu y_\nu = -y_0^2 + \left[\frac{1}{2}y\left(x^2+1\right)\right]^2,$$

respectively. This map, restricted to the coordinates $1,\ldots,n+1$, is just the inverse of the Moser map $T^*\sigma$. Notice that the term in the square brackets is the natural metric on the sphere S^n, expressed in stereographic coordinates.

ii) Cross-section $M_{(+)}$

$$\begin{aligned}\Gamma^{-1} &= \frac{x^2+1}{x^2-1},\\ \Gamma^0 &= \sinh x^0,\\ \Gamma^k &= \frac{2x^k}{x^2-1},\\ \Gamma^{n+1} &= \cosh x^0,\end{aligned}$$

that is, the intersection of the null cone $\mathcal{K}$ with the hyperboloid

$$(X^{-1})^2 - (X^0)^2 - (X^1)^2 - \cdots + (X^{n+1})^2 = 2.$$

The functions Γ^{-1}, Γ^k are obtained by projecting stereographically one sheet of the hyperboloid H^n, immersed in $\mathbb{R}^{1,n}$, into the $\mathbb{R}^n$ subspace of the latter (in fact, onto the n-dimensional Poincaré disk), followed by an inversion $x^k \mapsto \frac{x^k}{x^2}$ (assuming that $x^2 > 1$ in this case). In a similar way, for $x^2 < 1$, we project stereographically the other sheet of the hyperboloid. The cotangent map and the metric are given by

$$\begin{aligned}\Pi_{-1} &= (x,y),\\ \Pi_0 &= y_0 \cosh x^0,\\ \Pi_k &= \frac{1}{2}(x^2-1)y_k - (x,y)x_k,\\ \Pi_{n+1} &= -y_0 \sinh x^0,\end{aligned}$$

and

$$(g_{(+)})^{\mu\nu}\, y_\mu y_\nu = -y_0^2 + \left[\frac{1}{2}y\left(x^2-1\right)\right]^2.$$

Notice that the term in the square brackets is the natural metric on the hyperboloid H^n expressed in (inverted) stereographic coordinates.

iii) Cross-section $M_{(0)}$

$$\begin{aligned}\Gamma^{-1} &= 1 + \frac{1}{x^2} - \frac{(x^0)^2}{4},\\ \Gamma^0 &= x^0,\\ \Gamma^k &= \frac{2x^k}{x^2},\\ \Gamma^{n+1} &= 1 - \frac{1}{x^2} + \frac{(x^0)^2}{4},\end{aligned}$$

that is, the paraboloid given by the intersection of the null cone $\mathcal{K}$ with the hyperplane

$$\Gamma^{-1} + \Gamma^{n+1} = 2.$$

The mapping is obtained by projecting the paraboloid onto $\mathbb{R}^{1,n}$, followed by an inversion $x^k \mapsto \frac{2x^k}{x^2}$. The cotangent map and the metric are given by

$$\begin{aligned}
\Pi_{-1} &= \frac{1}{2}\left[(x,y) - x^0 y_0\right], \\
\Pi_0 &= y_0, \\
\Pi_k &= \frac{1}{2}x^2 y_k - (x,y)x_k, \\
\Pi_{n+1} &= \frac{1}{2}\left[(x,y) - x^0 y_0\right],
\end{aligned}$$

and

$$(g_{(0)})^{\mu\nu}\, y_\mu y_\nu = -y_0^2 + \left[\frac{1}{2}yx^2\right]^2,$$

respectively. Notice that the term in the square brackets is the Euclidean metric on $\mathbb{R}^n$ expressed in the inverted coordinates.

REMARK 6.5 We stress the fact that, owing to the stereographic projection in i) and the inversion in ii) and iii), one point is missing from the above parametrization of S^n, H^n or $\mathbb{R}^n$, and precisely that corresponding to $x^k \to \infty$. Restoring this point corresponds exactly to the regularization of the Kepler Problem.

With these three cross-sections we can construct the three moment maps

$$T^+M_\gamma \to \mathfrak{c}^*, \quad \gamma = (-), (+), (0),$$

given explicitly by (6.3.6), i.e., $J(x^\mu, y_\nu) = \sum_{A<B} G_{AB}(x^\mu, y_\nu) \mathcal{G}^*_{AB}$ with $G_{AB}(x^\mu, y_\nu) = \Pi_A \Gamma_B - \Pi_B \Gamma_A$. The action of the conformal group on the $2(n+1)$-dimensional manifolds T^+M_γ is not transitive, but divides each one of them into five homogeneous components: see Theorems 6.8 and 6.9. We are interested in the two $(2n+1)$-dimensional submanifolds $T_0^+M_\gamma$ characterized by the nullity condition

$$(g_\gamma)^{\mu\nu} y_\mu y_\nu = 0, \quad \text{or } y_0 = \pm K_\epsilon, \quad K_\epsilon = \frac{1}{2}y(x^2 - \epsilon),$$

where $\epsilon = -1, 1, 0$ for $\gamma = (-), (+), (0)$ respectively. They are presymplectic manifolds, i.e., they are endowed with a Poincaré-Cartan one-form

$$\theta = \pm K_\epsilon dx^0 + y_k dx^k,$$

which is the restriction (or pull-back) of the canonical one-form $y_\mu dx^\mu$ to the submanifolds $T_0^+ M_y$. The exterior derivative $\omega = d\theta$ has rank $2n$, and this implies that ω has a one-dimensional kernel. The tangent lines to this kernel are the so-called *characteristic curves* of θ and ω. Since

$$\omega = \pm\frac{\partial K_\epsilon}{\partial x^k} dx^k \wedge dx^0 \pm \frac{\partial K_\epsilon}{\partial y_k} dy_k \wedge dx^0 + dy_k \wedge dx^k,$$

the skew symmetric matrix of the two-form ω is of the type

$$\begin{pmatrix} 0 & \mp K_x & \mp K_y \\ \pm K_x & 0 & -\mathbf{1}_n \\ \pm K_y & \mathbf{1}_n & 0 \end{pmatrix}, \quad K_x = \left(\frac{\partial K_\epsilon}{\partial x^k}\right), \quad K_y = \left(\frac{\partial K_\epsilon}{\partial y_k}\right)$$

and it is immediate to check that the characteristic curves satisfy the Hamilton Equations

$$\frac{dx^k}{dx^0} = \mp\frac{\partial K_\epsilon}{\partial y_k}, \quad \frac{dy_k}{dx^0} = \pm\frac{\partial K_\epsilon}{\partial x^k}.$$

Factoring out this kernel from $T_0^+ M_y$ yields $T^+S^n, T^+H^n, T^+\mathbb{R}^n$, respectively, depending on y. The restriction to null covectors, i.e., the substitution $y_0 = \pm K_\epsilon(x^k, y_h)$, provides us with a family of equivalent ($2n$-dimensional) moment maps labelled by x^0. The conformal group acts indeed (in a way that preserves the Poincaré-Cartan one-form) on the characteristic lines, rather than merely on the points of $T_0^+ M_y$. It only moves the canonical coordinates x^k, y_h, and commutes with the translations of x^0. Hence, we may fix any value of x^0, the most suitable being $x^0 = 0$.

Let us then reduce the three moment maps, by putting

$$y_0 = \pm\frac{1}{2} y(x^2 - \epsilon), \quad x^0 = 0,$$

to obtain, in all three cases, the two equivalent moment maps

$$\begin{aligned}
G_{-1,0} &= \pm\frac{1}{2} y(x^2 + 1), \\
G_{hk} &= y_h x_k - y_k x_h, \\
G_{0,k} &= \pm y x_k, \\
G_{-1,k} &= \frac{1}{2}(x^2 + 1) y_k - (x, y) x_k, \\
G_{k,n+1} &= \frac{1}{2}(x^2 - 1) y_k - (x, y) x_k, \\
G_{-1,n+1} &= (x, y), \\
G_{0,n+1} &= \pm\frac{1}{2} y(x^2 - 1).
\end{aligned} \tag{6.4.5}$$

The two signs are related to the two orientations of $Y_{(-)}$ in Corollary 6.7. *Hereafter we will choose the sign ensuring that* $y_0 < 0$, so that

$$y_0 = \begin{cases} -\frac{1}{2}y(x^2+1), & \epsilon = -1, \quad \forall x, \\ -\frac{1}{2}yx^2, & \epsilon = 0, \quad x^2 \neq 0 \\ -\frac{1}{2}y(x^2-1), & \epsilon = 1, \quad x^2 > 1, \\ +\frac{1}{2}y(x^2-1), & \epsilon = 1, \quad x^2 < 1. \end{cases} \tag{6.4.6}$$

All that will be said in the sequel could be repeated for $y_0 > 0$ (after the required sign changes). (6.4.5) will be called the *Main Moment Map.*

To this moment map can be given a mechanical interpretation, namely by considering $G_{AB}(x_h, y_k)$ as dynamical variables on $T^+\mathbb{R}^n$, and choosing one of them as Hamiltonian. Three choices will be relevant, one for each cross-section $M_{(-)}, M_{(+)}$ and $M_{(0)}$.

i) Cross-section $M_{(-)}$

Choose as Hamiltonian the function $K = \frac{1}{2}y(x^2+1)$, so that

$$\frac{dx^k}{dx^0} = \frac{\partial K}{\partial y_k}, \quad \frac{dy_k}{dx^0} = -\frac{\partial K}{\partial x^k},$$

and x^0 appears as the "time" coordinate. Since $K = -G_{-1,0}$, from the commutation rules of the Lie algebra $\mathrm{so}(2, n+1)$ we infer at once

$$\{K, G_{hk}\} = 0, \quad \{K, A_i\} = 0 \quad \text{where } A_i \stackrel{\text{def}}{=} G_{i,n+1},$$

i.e., G_{hk} and A_i are first integrals of the motion; their Poisson algebra is isomorphic to the Lie algebra of $\mathrm{SO}(n+1)$. K and $G_{\alpha\beta}$ are the generators of the maximal compact subgroup $\mathrm{SO}(2) \times \mathrm{SO}(n+1)$. The generators of the noncompact part can be collected in two $(n+1)$-dimensional vectors

$$U_\alpha = G_{-1,\alpha}, \quad V_\beta = G_{0,\beta},$$

so that the algebraic relations of the generators G_{AB} may be written as

$$\begin{gathered} \textstyle\sum_\beta U_\beta U_\beta = \sum_\beta V_\beta V_\beta = K^2, \quad \sum_\beta U_\beta V_\beta = 0, \\ V_\alpha U_\beta - V_\beta U_\alpha = K G_{\alpha\beta}. \end{gathered} \tag{6.4.7}$$

A point of the coadjoint orbit T^+S^n can thus be parametrized by the two orthogonal vectors U_α and V_β of equal norm. Unlike the free (but local) coordinates x^h, y_k, these two vectors are globally defined. Their dynamical evolution is given by

$$\frac{dU_\alpha}{dx^0} = \{K, U_\alpha\} = V_\alpha, \quad \frac{dV_\alpha}{dx^0} = \{K, V_\alpha\} = -U_\alpha.$$

These equations are easily integrated: the two vectors rotate uniformly on their initial plane. After a canonical transformation which we present later, U_α and V_α will be identified with the Fock parameters.

ii) Cross-section $M_{(+)}$

Choose the function $K = \frac{1}{2}y(x^2 - 1)$ as Hamiltonian. From the commutation rules of the Lie algebra so$(2, n+1)$ we immediately obtain

$$\{K, G_{hk}\} = 0, \quad \{K, A_i\} = 0 \quad \text{where } A_i \stackrel{\text{def}}{=} G_{-1,i},$$

i.e., G_{hk} and A_i are first integrals of the motion; their Poisson algebra is isomorphic to the Lie algebra of SO$(1, n)$.

iii) Cross-section $M_{(0)}$

Choose $K = \frac{1}{2}yx^2$ as Hamiltonian. From the commutation rules of the Lie algebra so$(2, n+1)$ we deduce that

$$\{K, G_{hk}\} = 0, \quad \{K, A_i\} = 0 \quad \text{where } A_i \stackrel{\text{def}}{=} P_i,$$

i.e., G_{hk} and A_i are first integrals; their Poisson algebra is isomorphic to the Lie algebra of the Euclidean group SO$(n) \times_S \mathbb{R}^n$.

These last results are not very surprising in themselves: the three Hamiltonians K are, in fact, the square roots of the Hamiltonian of free motion on a sphere S^n, on a hyperboloid H^n and on a plane $\mathbb{R}^n$, respectively, so it is obvious that the invariance groups are SO$(n+1)$, SO$(1, n)$ and SO$(n) \times_S \mathbb{R}^n$. The interesting thing is the way we have obtained them, i.e., basically by restricting ourselves to the submanifold of the null covectors in T^+M_y and dividing out the kernel of the restricted canonical two-form. This procedure is intrinsic and can be carried out in any canonical coordinate system of T^+M_y. Passing to other canonical coordinates, before carrying out the procedure above, yields different looking systems that admit, nevertheless, identical dynamical properties (as invariance groups). Notice, and this is a fundamental point, that T^+M_y appears as the enlarged phase space of the reduced system, providing us with the possibility of performing an enlarged symplectic coordinate transformation, involving time and energy also.

To obtain explicitly the Kepler Problem we choose the canonical transformation

$$(q_k, p_h, t, p_t) = \mathscr{C}(x_k, y_h, x^0, y_0)$$

given by

$$q_k = -y_0 y_k, \tag{6.4.8a}$$

$$p_k = \frac{x_k}{y_0}, \tag{6.4.8b}$$

$$t = \frac{y_0^3}{\epsilon}\left[x^0 - \frac{(x,y)}{y_0}\right], \tag{6.4.8c}$$

$$p_t = \frac{\epsilon}{-2y_0^2}, \tag{6.4.8d}$$

with inverse transformation

$$y_k = q_k\sqrt{\frac{-2p_t}{\epsilon}},$$

$$x_k = -p_k\sqrt{\frac{\epsilon}{-2p_t}},$$

$$x^0 = -\epsilon t\left[\sqrt{\frac{-2p_t}{\epsilon}}\right]^3 - \sqrt{\frac{-2p_t}{\epsilon}}(q,p),$$

$$y_0 = -\sqrt{\frac{\epsilon}{-2p_t}},$$

where the signs are imposed by our choice $y_0 < 0$. To get a well defined transformation, one must plainly consider negative values for p_t when $\epsilon > 0$, and positive values when $\epsilon < 0$. The case $\epsilon = 0$ presents some problems, which we shall overcome by letting p_t go to zero while keeping $\frac{-2p_t}{\epsilon}$ finite and subsequently eliminating it[(6)]. $\mathscr{k}$ may be viewed as the composition of three canonical transformations:

(i) exchanging coordinates and momenta;

(ii) followed by (6.4.8a) and (6.4.8b), equivalent to an "energy rescaling";

(iii) concluding by (6.4.8d).

Notice that (6.4.8c) is enforced by the requirement of canonicity.

Let us substitute $\mathscr{k}^{-1}$in the constraint (that is, the nullity condition)

$$y_0 \mp \frac{1}{2}y(x^2 - \epsilon) = 0,$$

[(6)] We acknowledge that this transformation is somewhat awkward, particularly if compared with the preceding geometrical arguments. But this, in our opinion, is deeply symbolic: if the geometrical, intrinsic structure we have described is a sort of *Platonic idea* of the Kepler system, this clumsy transformation represents one of its possible concrete realizations, and suffers the inevitable imperfections of the material manifestation.

where, taking into account (6.4.6), the double sign is to ensure that $y_0 < 0$. We obtain

$$-\sqrt{\frac{\epsilon}{-2p_t}} \mp \frac{1}{2} q \sqrt{\frac{-2p_t}{\epsilon}} \left(p^2 \frac{\epsilon}{-2p_t} - \epsilon \right)$$
$$= \sqrt{\frac{\epsilon}{-2p_t}} \left[-1 \mp \frac{1}{2} q \frac{-2p_t}{\epsilon} \frac{\epsilon(p^2 + 2p_t)}{-2p_t} \right] = 0,$$

hence

$$p_t + \frac{1}{2} p^2 \pm \frac{1}{q} = 0,$$

i.e., $-p_t$ is equal to the attractive Kepler Hamiltonian $H(q_h, p_k)$ in the first three cases of (6.4.6), corresponding to minus sign in (6.4.5), and to the repulsive Hamiltonian $H_+(q_h, p_k)$ in the last case. Hereafter, we will write simply $H(q_h, p_k)$, leaving understood the subscript +. For the canonicity of $\mathscr{K}$, the pull-back, via $\mathscr{K}^{-1}$, of the Poincaré–Cartan one-form of the $(2n+1)$-dimensional presymplectic manifold $T_0^+ M_y$ is

$$\theta = -H(q_h, p_k)\, dt + p_k dq_k,$$

whose characteristic lines are solutions of the Hamilton Equations

$$\frac{dq_k}{dt} = \frac{\partial H}{\partial p_k}, \quad \frac{dp_k}{dt} = -\frac{\partial H}{\partial q_k}.$$

Arguing as before, putting $t = 0$ we factor out the kernel of the two-form $\omega = d\theta$.

From (6.4.8d) we see that $H(q_h, p_k)$ is a function of $K_\epsilon = \frac{1}{2} y(x^2 - \epsilon)$, hence has the same first integrals, transformed by $\mathscr{K}$ into

$$G_{hk} = p_h q_k - p_k q_h,$$
$$R_h = \sqrt{\frac{\epsilon}{-2p_t}} \left[p^2 q_h \mp \frac{q_h}{q} - (q, p) p_h \right].$$

The first term is the double skew symmetric tensor that generalizes, in the n-dimensional case, the angular momentum vector. The second term generalizes (for $\epsilon \neq 0$) the Runge–Lenz–Laplace vector. To have a uniform treatment for every ϵ, we define the n-dimensional generalization of the *eccentricity vector*

$$E_h = p^2 q_h \mp \frac{q_h}{q} - (q, p) p_h.$$

From the commutation rules of the Lie algebra $\mathrm{so}(2, n+1)$, we infer that

$$\{R_h, R_k\} = \left\{ \sqrt{\frac{\epsilon}{-2p_t}} E_h, \sqrt{\frac{\epsilon}{-2p_t}} E_k \right\} = -\epsilon G_{hk},$$

from which we deduce that

$$\{E_h, E_k\} = -2H\, G_{hk},$$

which is valid for every value of the energy.

Let us now verify that x^0 is the eccentric anomaly $\mathfrak{s}$. To this end, we consider Equation (6.4.8c) and interpret it as defining t as a function of x^0, x_k, y_h. For, e.g., the attractive case

$$\begin{aligned}\frac{dt}{dx^0} &= \{K_\epsilon, t\} + \frac{\partial t}{\partial x^0} = -\frac{y_0^2}{\epsilon}\{K_\epsilon, (x, y)\} + \frac{y_0^3}{\epsilon}\\ &= -\frac{y_0^2}{\epsilon}\left[-\frac{1}{2}y(x^2+\epsilon)\right] + \frac{y_0^3}{\epsilon}\\ &= \frac{y_0^2}{\epsilon}\left[\frac{1}{2}yx^2 + \frac{1}{2}y\epsilon - \frac{1}{2}yx^2 + \frac{1}{2}y\epsilon\right] = y_0^2 y = \sqrt{\frac{\epsilon}{-2p_t}}\,q.\end{aligned}$$

Comparing with (2.2.2) on page 23 and the analogous expressions for $H \geq 0$, we find, putting $\sqrt{\frac{\epsilon}{-2p_t}} = G$ for $H = 0$,

$$x^0 = \mathfrak{s}.$$

Let us restrict to the case $H < 0$. We have already seen that the generators of the maximal compact subgroup $SO(2) \times SO(n+1)$ are

$$G_{-1,0} = -\frac{1}{\sqrt{-2H(q_k, p_h)}} = -\mathcal{L}, \quad G_{hk}, \quad G_{i,n+1} = R_i.$$

The generators of the noncompact part may be collected into the two $(n+1)$-dimensional vectors

$$\begin{aligned}Z_\alpha &= G_{-1,\alpha} = -y_0 \sin x^0 \Gamma_\alpha + \Pi_\alpha \cos x^0,\\ W_\beta &= G_{0,\beta} = y_0 \cos x^0 \Gamma_\beta + \Pi_\beta \sin x^0,\end{aligned}$$

then transform them by $\mathscr{C}$, restrict to the null covectors, and put finally $t = 0$. This last condition becomes, by (6.4.8c),

$$x^0 = \frac{(x, y)}{y_0} = (q, p)\sqrt{-2H},$$

moreover

$$\Pi_\alpha = \begin{pmatrix} \frac{1}{\sqrt{-2H}}\left[-(q,p)p_k + \frac{q_k}{q}\right] \\ -(q,p) \end{pmatrix}, \quad y_0\Gamma_\alpha = \begin{pmatrix} q p_k \\ -\frac{1+2Hq}{\sqrt{-2H}} \end{pmatrix}.$$

In this way we recover the definitions and dynamical properties of the Fock and Bacry–Györgyi parameters: see Subsections 5.2.1 and 5.2.2 on pages 104–105. In fact

$$\begin{pmatrix} U_\alpha \\ V_\beta \end{pmatrix} = \begin{pmatrix} \Pi_\alpha \\ y_0\Gamma_\alpha \end{pmatrix},$$

$$\begin{pmatrix} Z_\alpha \\ W_\beta \end{pmatrix} = \begin{pmatrix} \cos\Delta & -\sin\Delta \\ \sin\Delta & \cos\Delta \end{pmatrix}\begin{pmatrix} U_\alpha \\ V_\beta \end{pmatrix}, \quad \Delta = (q,p)\sqrt{-2H}.$$

In particular, this clarifies the relation between the two sets of parameters; both are the components, pertinent to the noncompact part, of the moment map $T^+S^n \to \mathfrak{c}^*$, obtained by sectioning $T_0^+M_{(-)}$ with the hypersurfaces $x^0 = 0$ and $t = 0$, respectively. It is now clear why the Fock parameters rotate uniformly with respect to the "false time" $x^0 = \mathfrak{s}$, while the Bacry–Györgyi parameters rotate uniformly with respect to the "true time" t.

Finally, we add some comments on this chapter, which will help us to understand better the geometrical meaning of the canonical transformation $\mathscr{K}$. T_0^+M (i.e., the restriction to null nonidentically zero covectors of T^+M) is globally (pre)symplectomorphic to the product $(T^+S^n) \times S^1$. This latter will be identified, replacing the factor S^1 with its simply connected covering space $\mathbb{R}$, with the evolution space[(7)] of the geodesic flow of the n-dimensional sphere. The factor T^+S^n is the space of motions. A fixed $x^0 = \overline{x}^0$ section of the evolution space is the phase space at "time" $\overline{x}^0$, and is *globally* symplectomorphic to the space of motions for *every* $\overline{x}^0$; this property is just the one that defines a *regular problem.*

To obtain the Kepler Problem we proceed in the following way. Firstly we "deregularize" the system by removing some points from the phase space, which will no longer be homeomorphic to the space of motions; in our case, this is obtained by stereographic projection. Then, some of the trajectories (namely collision orbits) will have no intersection with the phase space. Secondly, we change the "look" of the system applying the canonical transformation $\mathscr{K}$, which, when restricted to the evolution space[(8)], sends the 1-form $y_k dx^k - K\,dx^0$ into $p_k dq^k - H\,dt$. We may interpret this transformation in the *active* sense: to every point of the evolution space is attached some *fixed* coordinate, and $\mathscr{K}$ takes the points from their initial position to another. The characteristic lines of the 1-form are thus transformed. The phase space of the system is identified with the hyperplane $x^0 = 0$ before, and with that of $t = 0$ after performing the transformation, respectively. But we may interpret $\mathscr{K}$ in the *passive* sense also: the points in the evolution space are fixed, and $\mathscr{K}$ simply changes their labels. The characteristic lines of the 1-form

---

(7) For the definition of *evolution space* and *space of motion* see the beginning of the previous chapter.

(8) To lighten the notation, we denote with the same symbol the canonical transformation and its restriction to the constraint.

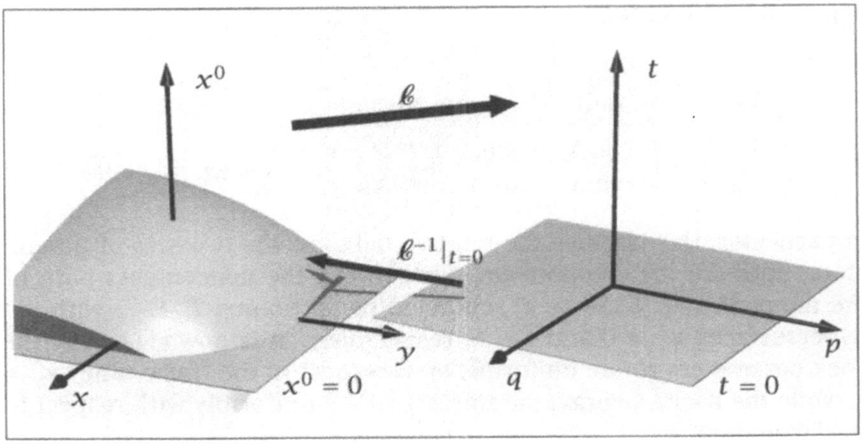

Figure 6.2: Geometric interpretation of the canonical transformation $\mathscr{C}$.

are thus fixed and $\mathscr{C}$ changes nothing in the dynamics. What is changed is the choice of the phase space: the hyperplane $x^0 = 0$ (the phase space of the motion on the sphere) becomes, under $\mathscr{C}$, the hypersurface $\mathscr{C}^{-1}|_{t=0}$, which is the preimage of the hyperplane $t = 0$ (i.e., the phase space of the Kepler Problem). Thus the *same* dynamical trajectories in the evolution space are projected onto two *different* phase spaces. The projection of the same trajectory appears different in the two cases, and this vividly shows that the geodesic motion on the sphere and the Kepler Problem for negative energy are, intrinsically, the same dynamical systems, differing only in the way they are looked upon.

In all that, nothing is peculiar to the explicit form of $\mathscr{C}$, which we might even change, obtaining some other system with the same dynamical symmetries as for the Kepler Problem. Examples will be given in Chapter 10.

CHAPTER 7

Spinorial Regularization

The results of the previous chapter do not depend on the physical dimensions of the Kepler Problem. In this chapter we will reformulate the regularization construction, but with the group $SO(2, n+1)$ replaced by its double covering, the so-called $\mathrm{Spin}(2, n+1)$ group[1].

Nothing essentially new is obtained in this way and so one might ask why we discuss spinorial regularization at all. The point is that the well known *Levi-Civita* and *Kustaanheimo-Stiefel* transformations (discovered *before* the Souriau-Moser regularization) are examples of the spinorial regularization (for $n = 2, 3$ respectively), as has been clarified by Kummer (1982).

In this chapter we will first deal with the physical case $n = 3$, where $\mathrm{Spin}(2,4) = SU(2,2)$ and the spinors are called *twistors*. Afterwards we generalize to arbitrary n, as in Cordani & Reina (1987) and Cordani (1989b).

7.1 The Homomorphism $SU(2,2) \to SO(2,4)$

It is well known that $SU(2)$ is the double covering of the $SO(3)$ group. This means that the homomorphism $SU(2) \to SO(3)$ is a local isomor-

[1]The name *Spin* of the double covering comes from the fact that $SU(2)$, the double covering of $SO(3)$, is the relevant group for the description of the elementary particles with half-integer spin.

phism and maps SU(2) onto SO(3) in an *essentially* (2-1) fashion. The term "essentially" here refers to the fact that SU(2) is connected.

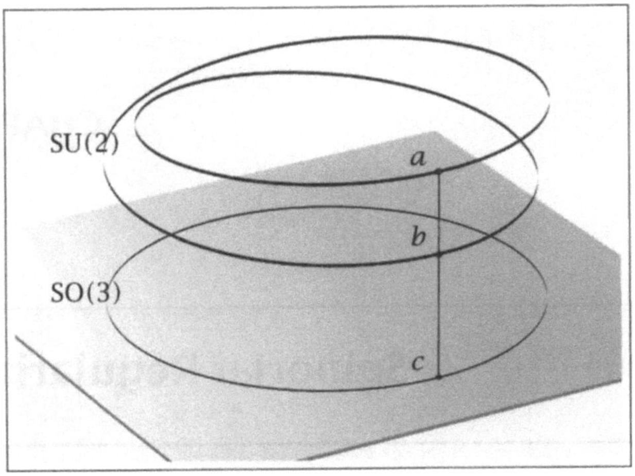

Figure 7.1: The double covering (for example of SU(2) over SO(3)).

If, as in Figure 7.1, a and b are the two elements of SU(2) that are mapped to some given element c of SO(3) (actually $b = -a$), then a may be connected to b by some curve in SU(2). The image of this curve in SO(3) is a closed curve (topologically equivalent to a continuous rotation through 2π). Neither a nor b can be preferred as *the* SU(2) image of c. For as we pass from c back to c along the curve in SO(3), the inverse image in SU(2) must pass continuously from a to b, or else back from b to a. The ambiguity between a and b is therefore essential.

There is a higher-dimensional analogue, namely the homomorphism $SU(2,2) \to SO(2,4)$, which similarly maps SU(2,2) onto the identity-connected component of SO(2,4) in an essentially (2-1) fashion, as we will now show: see Guillemin & Sternberg (1990). Let $\mathbb{C}^{2,2}$ denote a complex four-dimensional space endowed with a pseudo-Hermitian form (,) of signature $(++--)$. The group U(2,2) denotes the group of the complex linear transformations that preserves the form and SU(2,2) the subgroup of those elements of U(2,2) with determinant 1. A choice of a normalized basis q of the one-dimensional space $\bigwedge^4 \mathbb{C}^{2,2}$ determines a $*$ operator that is an antilinear map $* : \bigwedge^k \mathbb{C}^{2,2} \to \bigwedge^{4-k} \mathbb{C}^{2,2}$ determined by $u \wedge *v = (u, v)q$. Here we have continued to denote by (,) the induced scalar product on $\bigwedge^k \mathbb{C}^{2,2}$. In particular $*$ maps $\bigwedge^2 \mathbb{C}^{2,2}$ into itself, and $*^2 = 1$. We can thus think of $*$ as defining a complex conjugation on the six-dimensional complex vector space $\bigwedge^2 \mathbb{C}^{2,2}$. We let $\left(\bigwedge^2 \mathbb{C}^{2,2}\right)_{\mathbb{R}}$

denote the real points under this complex conjugation, i.e., those v that satisfy $*v = v$. Since $*$ is invariantly defined under SU(2,2), we see that SU(2,2) acts on $\left(\bigwedge^2 \mathbb{C}^{2,2}\right)_{\mathbb{R}}$ as real linear transformations. Let Q denote the complex bilinear form on $\bigwedge^2 \mathbb{C}^{2,2}$ determined by $Q(v,w)q = v \wedge w$. Notice that Q is invariant under SU(2,2). For $w \in \left(\bigwedge^2 \mathbb{C}^{2,2}\right)_{\mathbb{R}}$ we have

$$Q(w,w)q = w \wedge w = w \wedge *w = (w,w)q,$$

so that

$$Q(w,w) = (w,w).$$

Thus Q restricts to a real quadratic form on $\left(\bigwedge^2 \mathbb{C}^{2,2}\right)_{\mathbb{R}}$ which is invariant under SU(2,2). If e_1, e_2, e_3, e_4 is an orthogonal basis of $\mathbb{C}^{2,2}$ with

$$\|e_1\| = \|e_2\| = 1, \qquad \|e_3\| = \|e_4\| = -1,$$

then $\{e_i \wedge e_j\}$ is an orthogonal basis for $\bigwedge^2 \mathbb{C}^{2,2}$ under the pseudo-Hermitian form (,). Furthermore

$$\|e_1 \wedge e_2\| = \|e_3 \wedge e_4\| = 1, \quad \text{and} \left\|e_i \wedge e_j\right\| = -1$$

for the other choices. Thus (,) is a pseudo-Hermitian form of signature $(++----)$ on $\bigwedge^2 \mathbb{C}^{2,2}$. The real scalar product on $\left(\bigwedge^2 \mathbb{C}^{2,2}\right)_{\mathbb{R}}$ obtained by restriction will thus have the same signature. We may thus identify $\left(\bigwedge^2 \mathbb{C}^{2,2}\right)_{\mathbb{R}}$ with $\mathbb{R}^{2,4}$. In this way we have defined a homomorphism SU(2,2) → SO(2,4) whose kernel is ± 1.

7.1.1 Two Bases for su(2,2)

It is useful to find the explicit realization of the isomorphism between the Lie algebras su(2,2) and so(2,4). Choose a basis in su(2,2) such that the pseudo-Hermitian form (,) becomes

$$(\varphi, \psi) = \varphi_1^* \psi_1 + \varphi_2^* \psi_2 - \varphi_3^* \psi_3 - \varphi_4^* \psi_4.$$

The matrix associated to this quadratic form is

$$\mathbf{H} = \begin{pmatrix} 1 & & & \\ & 1 & & \\ & & -1 & \\ & & & -1 \end{pmatrix}.$$

Let $\mathbf{s}_k$ be the three Pauli matrices

$$\mathbf{s}_1 = \begin{pmatrix} 0 & 1 \\ 1 & 0 \end{pmatrix}, \quad \mathbf{s}_2 = \begin{pmatrix} 0 & i \\ -i & 0 \end{pmatrix}, \quad \mathbf{s}_3 = \begin{pmatrix} -1 & 0 \\ 0 & 1 \end{pmatrix}.$$

These latter satisfy

$$\mathbf{s}_h\mathbf{s}_k + \mathbf{s}_k\mathbf{s}_h = 2\delta_{hk}\mathbf{1}, \quad \mathbf{s}_h^\dagger = \mathbf{s}_h, \quad \mathrm{Tr}\,\mathbf{s}_h = 0. \tag{7.1.1}$$

A pseudo-orthogonal basis in su(2, 2) is given by

$$\begin{gathered}
\mathbf{G}_{hk} = \frac{1}{2}\begin{pmatrix} -\mathbf{s}_h\mathbf{s}_k & 0 \\ 0 & -\mathbf{s}_h\mathbf{s}_k \end{pmatrix}, \quad \mathbf{G}_{k,4} = \frac{1}{2}\begin{pmatrix} i\mathbf{s}_k & 0 \\ 0 & -i\mathbf{s}_k \end{pmatrix}, \\
\mathbf{G}_{-1,0} = \frac{1}{2}\begin{pmatrix} i\mathbf{1} & 0 \\ 0 & -i\mathbf{1} \end{pmatrix}, \\
\mathbf{G}_{0h} = \frac{1}{2}\begin{pmatrix} 0 & -\mathbf{s}_h \\ -\mathbf{s}_h & 0 \end{pmatrix}, \quad \mathbf{G}_{0,4} = \frac{1}{2}\begin{pmatrix} 0 & -i\mathbf{1} \\ i\mathbf{1} & 0 \end{pmatrix}, \\
\mathbf{G}_{-1,h} = \frac{1}{2}\begin{pmatrix} 0 & -i\mathbf{s}_h \\ i\mathbf{s}_h & 0 \end{pmatrix}, \quad \mathbf{G}_{-1,4} = \frac{1}{2}\begin{pmatrix} 0 & \mathbf{1} \\ \mathbf{1} & 0 \end{pmatrix}.
\end{gathered} \tag{7.1.2}$$

In fact these matrices have vanishing trace and satisfy

$$\mathbf{G}_{AB}^\dagger\mathbf{H} + \mathbf{H}\mathbf{G}_{AB} = 0,$$

i.e., the diagonal matrices (the generators of the compact subgroup $SO(2)\times SO(4)$) are skew Hermitian and the antidiagonal matrices Hermitian. This ensures that $\mathbf{G}_{AB} \in su(2,2)$. Moreover, as a direct calculation shows,

$$[\mathbf{G}_{AB}, \mathbf{G}_{AC}] = \eta_{AA}\mathbf{G}_{BC}, \quad \text{or} = 0 \text{ if all indices are different.}$$

Remembering Equations (6.1.5), we see that the algebra of the matrices $\mathbf{G}_{AB}$ under commutation is isomorphic to the algebra of the linear differential operators $\mathcal{G}_{AB}$ under Lie bracket. The Cartan-Killing form is

$$F\,(\mathbf{G}_{AB}, \mathbf{G}_{CD}) = -\mathrm{Tr}\,(\mathbf{G}_{AB}\mathbf{G}_{CD});$$

$\{G_{AB}\}$ is a pseudo-orthonormal basis, the generators of the compact part having square norm $+1$, those of the non compact part -1.

Besides this "natural" basis, another one will be relevant for us. Consider the linear transformation

$$\mathbf{T} = \frac{1}{\sqrt{2}}\begin{pmatrix} \mathbf{1} & \mathbf{1} \\ \mathbf{1} & -\mathbf{1} \end{pmatrix},$$

acting on $\mathbb{C}^{2,2}$, where $\mathbf{1}$ is the 2×2 unit matrix. It induces the transformations

$$\mathbf{H} \mapsto \mathbf{h} = (\mathbf{T}^\dagger)^{-1}\mathbf{H}\mathbf{T}^{-1} = \begin{pmatrix} 0 & \mathbf{1} \\ \mathbf{1} & 0 \end{pmatrix},$$

$$\mathbf{G}_{AB} \mapsto \mathbf{g}_{AB} = \mathbf{T}\,\mathbf{G}_{AB}\mathbf{T}^{-1}.$$

The explicit expression of the elements of this basis is

$$\begin{gathered}
\mathbf{g}_{hk} = \frac{1}{2}\begin{pmatrix} -\mathbf{s}_h\mathbf{s}_k & 0 \\ 0 & -\mathbf{s}_h\mathbf{s}_k \end{pmatrix}, \quad \mathbf{g}_{0k} = \frac{1}{2}\begin{pmatrix} -\mathbf{s}_k & 0 \\ 0 & \mathbf{s}_k \end{pmatrix}, \\
\mathbf{g}_{-1,4} = \frac{1}{2}\begin{pmatrix} \mathbf{1} & 0 \\ 0 & -\mathbf{1} \end{pmatrix}, \\
\mathbf{g}_{0,4} = \frac{1}{2}\begin{pmatrix} 0 & i\mathbf{1} \\ -i\mathbf{1} & 0 \end{pmatrix}, \quad \mathbf{g}_{k4} = \frac{1}{2}\begin{pmatrix} 0 & i\mathbf{s}_k \\ i\mathbf{s}_k & 0 \end{pmatrix}, \\
\mathbf{g}_{-1.0} = \frac{1}{2}\begin{pmatrix} 0 & i\mathbf{1} \\ i\mathbf{1} & 0 \end{pmatrix}, \quad \mathbf{g}_{-1,k} = \frac{1}{2}\begin{pmatrix} 0 & i\mathbf{s}_k \\ -i\mathbf{s}_k & 0 \end{pmatrix}.
\end{gathered} \tag{7.1.3}$$

We define

$$\mathbf{g}_{\mu,4} = \frac{1}{2}\left(\mathbf{P}_\mu + \mathbf{K}_\mu\right), \quad \mathbf{g}_{-1,\mu} = \frac{1}{2}\left(\mathbf{P}_\mu - \mathbf{K}_\mu\right), \quad \mathbf{g}_{4,-1} = \mathbf{D},$$

so that

$$\begin{gathered}
\mathbf{P}_0 = \begin{pmatrix} 0 & i\mathbf{1} \\ 0 & 0 \end{pmatrix}, \quad \mathbf{P}_k = \begin{pmatrix} 0 & i\mathbf{s}_k \\ 0 & 0 \end{pmatrix}, \\
\mathbf{K}_0 = \begin{pmatrix} 0 & 0 \\ -i\mathbf{1} & 0 \end{pmatrix}, \quad \mathbf{K}_k = \begin{pmatrix} 0 & 0 \\ i\mathbf{s}_k & 0 \end{pmatrix}.
\end{gathered}$$

Obviously the algebra of the matrices $\mathbf{g}_{\mu\nu}$,$\mathbf{P}_\mu$,$\mathbf{K}_\nu$,$\mathbf{D}$ under commutation is isomorphic to the algebra of the linear differential operators $\mathcal{G}_{\mu\nu}, \mathcal{P}_\mu, \mathcal{K}_\nu, \mathcal{D}$ under the Lie bracket.

7.1.2 SU(2,2) and Compactified Minkowski Space

As seen in the previous chapter, there exists an effective[(2)] action of the conformal group $\mathfrak{C} = \mathrm{SO}(2,4)/\{\pm\mathbf{1}\}$ on the compactified Minkowski space $M \simeq (S^1 \times S^3)/\mathbb{Z}_2$. Thus there must exist an effective action of $\mathrm{SU}(2,2)/\{\pm\mathbf{1}, \pm i\mathbf{1}\}$ on the same manifold. We will now exhibit this action explicitly.

Let $\mathrm{Mat}(\mathbb{C}, n)$ be the space of $n \times n$ matrices on the complex field. Define

$$\phi = \begin{pmatrix} z_1 \\ z_2 \end{pmatrix}, \quad z_1, z_2 \in \mathrm{Mat}(\mathbb{C}, 2),$$

such that

$$\phi^\dagger \mathbf{H} \phi = z_1^\dagger z_1 - z_2^\dagger z_2 = -\mathbf{1}.$$

[(2)]We recall that the action of a group is effective if the only group element whose action is trivial is the identity.

The matrix $z_1^\dagger z_1$ has positive or null eigenvalues, hence $z_2^\dagger z_2$ has strictly positive eigenvalues, so is an invertible matrix. Put $z = z_1 z_2^{-1}$, from which

$$z^\dagger z = (z_2^{-1})^\dagger z_1^\dagger z_1 z_2^{-1} = (z_2^{-1})^\dagger (z_2^\dagger z_2 - \mathbf{1}) z_2^{-1} = \mathbf{1} - (z_2 z_2^\dagger)^{-1}.$$

Thus, if we agree to write $\mathbf{M} > 0$ for a positive definite matrix $\mathbf{M}$, we have

$$\mathbf{1} - z^\dagger z = (z_2 z_2^\dagger)^{-1} > 0. \tag{7.1.4}$$

The matrices z satisfying (7.1.4) are said to belong to the *generalized Poincaré disk*. The natural linear action of $A = \begin{pmatrix} \alpha & \beta \\ \gamma & \delta \end{pmatrix} \in \mathrm{SU}(2,2)$ on ϕ, given by

$$\phi \mapsto \phi' = A\phi,$$

induces a linear fractional action on z,

$$z \mapsto z' = (\alpha z + \beta)\,(\gamma z + \delta)^{-1}, \tag{7.1.5}$$

which obviously preserves the condition (7.1.4). Similarly, for $z \neq 0$ the matrices $w = z^{-1}$ satisfy the relation

$$w^\dagger w - \mathbf{1} > 0,$$

which is preserved by the action (7.1.5). Therefore the set of unitary matrices $\mathbf{u}$, i.e., such that $\mathbf{u}^\dagger \mathbf{u} = \mathbf{1}$, is closed under the linear fractional action of $\mathrm{SU}(2,2)/\{\pm\mathbf{1}, \pm i\mathbf{1}\}$. This action is effective, since we have divided out the center of SU(2,2), i.e., $\{\pm\mathbf{1}, \pm i\mathbf{1}\}$, whose linear fractional action is the only trivial one. The set of unitary matrices $\mathbf{u}$ is called the *Bergmann-Šilov boundary*.

Under the basis change induced by $\mathbf{T}$ the unitary matrix $\mathbf{u}$ is transformed into the skew Hermitian matrix $i\mathbf{m}$, with $\mathbf{m}^\dagger = \mathbf{m}$. In fact, if

$$i\mathbf{m} = (\mathbf{u} + \mathbf{1})\,(\mathbf{u} - \mathbf{1})^{-1}, \quad \det(\mathbf{u} - \mathbf{1}) \neq 0, \tag{7.1.6}$$

since

$$\mathbf{u}^\dagger + \mathbf{1} = (\mathbf{u} - \mathbf{u}^\dagger)(\mathbf{u} - \mathbf{1})^{-1},$$
$$(\mathbf{u}^\dagger - \mathbf{1})^{-1} = (\mathbf{u} + \mathbf{1})(\mathbf{u}^\dagger - \mathbf{u})^{-1},$$

one obtains

$$(i\mathbf{m})^\dagger = (\mathbf{u}^\dagger - \mathbf{1})^{-1}(\mathbf{u}^\dagger + \mathbf{1}) = -(\mathbf{u} + \mathbf{1})(\mathbf{u} - \mathbf{1})^{-1} = -i\mathbf{m}.$$

Thus we can write

$$\mathbf{m} = x^0\mathbf{1} + x^k \mathbf{s}_k = \begin{pmatrix} x^0 - x^3 & x^1 + ix^2 \\ x^1 - ix^2 & x^0 + x^3 \end{pmatrix}, \tag{7.1.7}$$

with $\mathbf{m} \simeq M_0$, the four-dimensional Minkowski space. (7.1.6) is the *Cayley transform* and has the inverse

$$\mathbf{u} = (i\mathbf{m} - \mathbf{1})^{-1}(i\mathbf{m} + \mathbf{1}). \tag{7.1.8}$$

This transform applies all the unitary matrices, which satisfy the condition $\det(\mathbf{u} - \mathbf{1}) \neq 0$, onto M_0. To view the link with the statements of the preceding chapter, write $\mathbf{u}$ as the exponential of a skew Hermitian matrix, thus of the type

$$\mathbf{u} = \frac{1}{\Gamma^{-1} + i\Gamma^0}\left(\Gamma^4\mathbf{1} + i\mathbf{s}_k\Gamma^k\right), \quad \Gamma^{-1}, \ldots, \Gamma^4 \in \mathbb{R} \tag{7.1.9}$$

then impose $\mathbf{u}$ to be unitary

$$\mathbf{u}\mathbf{u}^\dagger = \frac{1}{(\Gamma^{-1})^2 + (\Gamma^0)^2}\left[(\Gamma^4)^2\mathbf{1} + \Gamma^k\Gamma_k\mathbf{1}\right] = \mathbf{1}.$$

This shows that the point Γ belongs to the null cone of $\mathbb{R}^{2,4}$. From (7.1.9) one sees that Γ and $\lambda\Gamma$, $\lambda \neq 0$, determine the same $\mathbf{u}$, hence $\{\mathbf{u}\}$ is homeomorphic to the space M of unoriented lines of the null cone, i.e., to the compactified Minkowski space.

EXERCISE 7.1 Substitute (7.1.7) in (7.1.8) and compare the result with (7.1.9), obtaining

$$\begin{aligned}\Gamma^{-1} &= \frac{1}{2}\left(1 + x^\mu x_\mu\right),\\ \Gamma^0 &= x^0,\\ \Gamma^k &= x^k,\\ \Gamma^4 &= \frac{1}{2}\left(1 - x^\mu x_\mu\right),\end{aligned}$$

which is just the section (6.4.2) on page 130 of the null cone $\mathcal{K}$.

It is now straightforward to verify that the linear fractional action of SU(2, 2) on $i\mathbf{m}$, i.e.,

$$i\mathbf{m} \mapsto i\overline{\mathbf{m}} = (\alpha\, i\mathbf{m} + \beta)\,(\gamma\, i\mathbf{m} + \delta)^{-1},$$

where

$$\begin{pmatrix} \alpha^\dagger & \gamma^\dagger \\ \beta^\dagger & \delta^\dagger \end{pmatrix}\begin{pmatrix} 0 & \mathbf{1} \\ \mathbf{1} & 0 \end{pmatrix}\begin{pmatrix} \alpha & \beta \\ \gamma & \delta \end{pmatrix} = \begin{pmatrix} 0 & \mathbf{1} \\ \mathbf{1} & 0 \end{pmatrix},$$

is conformal. It is enough to check that the infinitesimal linear fractional action on M_0 generates the conformal vector field (6.1.1). Bearing in mind the expression of $\mathbf{P}_\mu$, take first

$$\begin{pmatrix} \alpha & \beta \\ \gamma & \delta \end{pmatrix} = \begin{pmatrix} \mathbf{1} & i\mathbf{t} \\ 0 & \mathbf{1} \end{pmatrix},$$

where $\mathbf{t} = t^0\mathbf{1} + t^k\mathbf{s}_k$ is an infinitesimal Hermitian matrix. The linear fractional action gives

$$\mathbf{m} \mapsto \overline{\mathbf{m}} = \mathbf{m} + \mathbf{t}, \quad \text{i.e., } \overline{x}^\mu = x^\mu + t^\mu.$$

Then take, remembering the expression of $\mathbf{K}_\mu$,

$$\begin{pmatrix} \alpha & \beta \\ \gamma & \delta \end{pmatrix} = \begin{pmatrix} \mathbf{1} & 0 \\ i\mathbf{k} & \mathbf{1} \end{pmatrix},$$

where $\mathbf{k} = -k^0\mathbf{1} + k^h\mathbf{s}_h$ is an infinitesimal Hermitian matrix. The linear fractional action gives to first order

$$\mathbf{m} \mapsto \overline{\mathbf{m}} = \mathbf{m} + \mathbf{mkm}, \quad \text{i.e., } \overline{x}^\mu = x^\mu + 2x^\nu k_\nu x^\mu - x_\nu x^\nu k^\mu.$$

Analogously $\mathbf{D}$ gives the vector field generating dilations and $\mathbf{g}_{\mu\nu}$ gives those of the Lorentz group.

EXERCISE 7.2 It is very simple to find, in this basis, the isotropy subgroup of a point, for example that of the origin, of M_0. This latter group is the double covering of the parabolic group $\mathfrak{P}$, defined in (6.3.1) on page 120. We may take as origin of M_0 the point $\mathbf{m} = 0$, so that its full isotropy subgroup, under the linear fractional action, is given by matrices of the type

$$\begin{pmatrix} \alpha & 0 \\ i\gamma\alpha & (\alpha^\dagger)^{-1} \end{pmatrix}, \quad \text{with } \det\alpha \in \mathbb{R}, \quad \gamma^\dagger = \gamma.$$

It is obvious that the set of these matrices contains elements that, as

$$\rho = \begin{pmatrix} 1 & & & \\ & -1 & & \\ & & 1 & \\ & & & -1 \end{pmatrix},$$

do not belong to the double covering of $\mathfrak{P}_0$ (the identity connected component of $\mathfrak{P}$), which is characterized by $\det\alpha > 0$. We may assume $\pm\rho$ and $\pm\mathbf{1}_4$ as the four elements of a group homomorphic, with kernel $\mathbb{Z}_2$, to the group $\mathfrak{R}$ in (6.3.1).

7.2 Return to the Kustaanheimo–Stiefel Map

As seen in the previous chapter, the phase space of the regularized 3-dimensional Kepler Problem with negative energy is symplectomorphic to T^+S^3, the cotangent bundle (with the zero section deleted) to the three-dimensional sphere, which in turn is a coadjoint orbit in $so^*(2,4)$, the dual Lie algebra of the pseudo-orthogonal group $SO(2,4)$. Since $so(2,4) = su(2,2)$, T^+S^3 can be identified with a coadjoint orbit in the dual space of the pseudo-unitary algebra. To get this identification it is sufficient to find one point of the orbit.

Using the Cartan-Killing form, the dual basis of the basis (7.1.3) is simply found by merely changing the sign to the generators of the non-compact part:

$$\mathbf{g}^*_{-1,0} = \mathbf{g}_{-1,0}, \quad \mathbf{g}^*_{\alpha\beta} = \mathbf{g}_{\alpha\beta}, \quad \mathbf{g}^*_{-1,\alpha} = -\mathbf{g}_{-1,\alpha}, \quad \mathbf{g}^*_{0.\alpha} = -\mathbf{g}_{0.\alpha}.$$

The moment map $J : T^+\mathbb{R}^3 \to su^*(2,2)$ is given by

$$J = \sum_{A<B} G_{AB}(x_h, y_k)\mathbf{g}^*_{AB},$$

where the components $G_{AB}(x_h, y_k)$ are the expressions in (6.4.5) on page 135, with the minus sign for the attractive potential. We pick, as base point $\mathbf{e} \in su^*(2,2)$ of the orbit, the one with coordinates $x_k = 0, y_1 = y_2 = 0, y_3 = -1$. This choice is obviously only one of the possibilities, which we make for computational convenience. We obtain

$$\mathbf{e} = -\frac{1}{2}\mathbf{g}^*_{-1,0} - \frac{1}{2}\mathbf{g}^*_{-1,3} + \frac{1}{2}\mathbf{g}^*_{3,4} + \frac{1}{2}\mathbf{g}^*_{0,4} = \begin{pmatrix} 0 & 0 & -i & 0 \\ 0 & 0 & 0 & 0 \\ 0 & 0 & 0 & 0 \\ 0 & 0 & 0 & 0 \end{pmatrix}.$$

The fundamental point is that $\mathbf{e}$ is a matrix of rank 1, and this allows us to find a sort of square root of the coadjoint action.

To show this, we first need a definition. An element $\psi \in \mathbb{C}^{2,2}$ such that

$$(\psi, \psi) = 0, \quad \psi \neq 0$$

is called *null twistor* (Penrose 1974), (Penrose 1967). Let T_0 be the space of null twistors, and assume the equivalence relation $\psi \sim \psi e^{i\beta}$, with β a generic angle. The set $T_0/\sim$ of null twistors modulo a phase transformation is a 6-dimensional manifold, which may be equipped with the natural symplectic structure (induced by) $d\Theta$, where

$$\Theta = \mathrm{Im}(\psi, d\psi). \tag{7.2.1}$$

Since the natural linear action of $SU(2,2)$ on $T_0/\sim$

$$\psi \mapsto \psi' = g\psi, \quad g \in SU(2,2)$$

is manifestly transitive and symplectic, $T_0/\sim$ is symplectomorphic to (a covering of) a coadjoint orbit $\mathcal{O}$ of $SU(2,2)$. The moment map

$$j : T_0/\sim \to \mathcal{O} \subset su^*(2,2)$$

which connects the two manifolds is

$$j(\psi) = -i\psi\psi^{\dagger}\mathbf{h} \quad \text{with } \psi^{\dagger}\mathbf{h}\,\psi = 0, \tag{7.2.2}$$

(the choice $j(\psi) = i\psi\psi^{\dagger}\mathbf{h}$ would correspond to the upper sign in (6.4.5)). We verify that (7.2.2) is just an injective Ad^*-equivariant moment map $T_0/\sim\, \to \mathrm{su}^*(2,2)$. Indeed

(i) $j(\psi) \in \mathrm{su}^*(2,2)$ since

$$j^{\dagger}(\psi)\mathbf{h} + \mathbf{h}\,j(\psi) = 0, \quad \mathrm{Tr}\,j(\psi) = -i\psi^{\dagger}\mathbf{h}\,\psi = 0;$$

(ii) taking $T_0/\sim$ as domain, we make j injective since ψ and ψ' have the same image if and only if $\psi' = \psi e^{i\beta}$.

(iii) the diagram

$$\begin{array}{ccc} T_0/\sim & \xrightarrow{\ \mathrm{SU}(2,2)\ } & T_0/\sim \\ {\scriptstyle j}\big\downarrow & & \big\downarrow{\scriptstyle j} \\ \mathrm{su}^*(2,2) & \xrightarrow{\ \mathbf{Ad}^*\ } & \mathrm{su}^*(2,2) \end{array}$$

is commutative, since, if $g \in \mathrm{SU}(2,2)$, by definition $g^{\dagger}\mathbf{h}\,g = \mathbf{h}$, so that $g^{-1} = \mathbf{h}^{-1}g^{\dagger}\mathbf{h}$, and the coadjoint action is given by

$$j \mapsto j' = g\,j\,g^{-1} = -ig\psi\psi^{\dagger}\mathbf{h}\mathbf{h}^{-1}g^{\dagger}\mathbf{h} = -i\,(g\psi)\,(g\psi)^{\dagger}\,\mathbf{h}.$$

The orbit $\mathcal{O}$ is identified with $\mathcal{O}_{\mathbf{e}}$, the orbit through $\mathbf{e}$, choosing

$$\psi_e = \begin{pmatrix} 1 \\ 0 \\ 0 \\ 0 \end{pmatrix},$$

and checking that

$$\mathbf{e} = -i\psi_e\psi_e^{\dagger}\mathbf{h}.$$

This shows that $\mathcal{O} = \mathcal{O}_{\mathbf{e}}$, and this last is, by construction, the coadjoint orbit of the regularized Kepler Problem, thus symplectomorphic to $T^{+}S^{3}$.

We now proceed to the choice of a *local* parametrization of $\mathcal{O}$ that corresponds to the canonical extension of the stereographic projection. Define the two Hermitian matrices

$$\begin{aligned} \mathbf{x} &= \vec{x} \cdot \vec{\mathbf{s}} = \begin{pmatrix} -x_3 & x_1 + ix_2 \\ x_1 - ix_2 & x_3 \end{pmatrix}, \\ \mathbf{y} &= \frac{1}{2} y \mathbf{1} + \frac{1}{2} \vec{y} \cdot \vec{\mathbf{s}} = \frac{1}{2} \begin{pmatrix} y - y_3 & y_1 + iy_2 \\ y_1 - iy_2 & y + y_3 \end{pmatrix}, \end{aligned} \tag{7.2.3}$$

where x_h and y_k are canonical coordinates on $T^+\mathbb{R}^3$. Since $\det \mathbf{y} = 0$, there exists an element of $(\mathbb{C}^2 - \{0\}) / \sim$ which is a sort of "square root", $\sqrt{\mathbf{y}}$ say, such that

$$\mathbf{y} = \sqrt{\mathbf{y}} \sqrt{\mathbf{y}}^\dagger .$$

Define the map $\Psi : T^+\mathbb{R}^3 \to T_0/\sim$ by

$$\Psi(x_h, y_k) = \begin{pmatrix} \sqrt{\mathbf{y}} \\ i\mathbf{x}\sqrt{\mathbf{y}} \end{pmatrix} . \tag{7.2.4}$$

This map is injective but can not be surjective, as is obvious by the different topology of domain and image. In fact, the null twistors of the type $\psi = \begin{pmatrix} 0 \\ iw \end{pmatrix}$, $w \neq 0$ do not belong to the image of Ψ. Thus the space of the null twistors may be decomposed as

$$T_0 = T_0^0 \cup T_0^\infty, \quad T_0^0 \cap T_0^\infty = 0,$$

$$T_0^0 = \left\{ \begin{pmatrix} z \\ iw \end{pmatrix} \in T_0,\ z \neq 0 \right\}, \quad T_0^\infty = \left\{ \begin{pmatrix} 0 \\ iw \end{pmatrix},\ w \neq 0 \right\}, \tag{7.2.5}$$

and the orbit as

$$\mathcal{O} = (T_0^0 / \sim) \cup (T_0^\infty / \sim).$$

Note that the following homeomorphisms hold:

$$\begin{aligned} (T_0^0 / \sim) &\simeq \mathbb{R}^3 \times (\mathbb{R}^3 - \{0\}) \simeq T^+(S^3 - \{\mathrm{N}\}), \\ (T_0^\infty / \sim) &\simeq T_{\mathrm{N}}^+ S^3 \simeq \mathbb{R}^3 - \{0\}. \end{aligned}$$

As we shall see in the sequel, Ψ will prove to be (up to an exchange between the coordinates and momenta) the inverse of the Kustaanheimo-Stiefel transformation. It is a symplectic map: indeed, the pull-back of Θ is

$$\begin{aligned} \Psi^*\Theta &= \mathrm{Im} \begin{pmatrix} \sqrt{\mathbf{y}}^\dagger & -i\sqrt{\mathbf{y}}^\dagger \mathbf{x} \end{pmatrix} \begin{pmatrix} 0 & \mathbf{1} \\ \mathbf{1} & 0 \end{pmatrix} \begin{pmatrix} d\sqrt{\mathbf{y}} \\ id\mathbf{x}\sqrt{\mathbf{y}} + i\mathbf{x}\, d\sqrt{\mathbf{y}} \end{pmatrix} \\ &= \sqrt{\mathbf{y}}^\dagger d\mathbf{x} \sqrt{\mathbf{y}} = \mathrm{Tr}\, \mathbf{y}\, d\mathbf{x} = \vec{y} \cdot d\vec{x}, \end{aligned}$$

i.e., the canonical 1-form of $T^*\mathbb{R}^3$.

Write the generic element of T_0^0 as

$$\psi = \begin{pmatrix} z \\ iw \end{pmatrix}, \quad z \in \mathbb{C}^2 - \{0\}, \quad w \in \mathbb{C}^2,$$

then

$$j = \begin{pmatrix} zw^\dagger & izz^\dagger \\ iww^\dagger & -wz^\dagger \end{pmatrix}. \tag{7.2.6}$$

To find the components $g_{AB}(z, w)$ of the moment map j, we must project it on the basis. Write

$$j = \sum_{A<B} g_{AB}(z, w)\mathbf{g}_{AB}^*$$

and remember that, from the mere definition of dual basis,

$$-\mathrm{Tr}\left(\mathbf{g}_{AB}\mathbf{g}_{CD}^*\right) = \begin{cases} 1 & \text{if } AB = CD \\ 0 & \text{othewise.} \end{cases}$$

Multiply the relation

$$\sum_{A<B} g_{AB}(z, w)\mathbf{g}_{AB}^* = -i\psi\psi^\dagger \mathbf{h}$$

to the right by $\mathbf{g}_{CD}$ and take the trace. We obtain

$$g_{AB}(z, w) = i\,\mathrm{Tr}\,\psi\psi^\dagger \mathbf{h}\mathbf{g}_{AB} = i\psi^\dagger \mathbf{h}\mathbf{g}_{AB}\psi,$$

or explicitly

$$\begin{aligned} g_{-1,0} &= -\frac{1}{2}\left(z^\dagger z + w^\dagger w\right), & g_{k4} &= -\frac{1}{2}\left(z^\dagger \mathbf{s}_k z + w^\dagger \mathbf{s}_k w\right), \\ g_{hk} &= \mathrm{Re}\left(z^\dagger \mathbf{s}_h \mathbf{s}_k w\right), & g_{-1,4} &= \mathrm{Re}\left(z^\dagger w\right), \\ g_{0k} &= -\mathrm{Re}\left(z^\dagger \mathbf{s}_k w\right), & g_{0,4} &= \frac{1}{2}\left(z^\dagger z - w^\dagger w\right). \\ g_{-1,k} &= \frac{1}{2}\left(z^\dagger \mathbf{s}_k z - w^\dagger \mathbf{s}_k w\right), \end{aligned} \tag{7.2.7}$$

Comparing $g_{AB}(z, w)$ with $G_{AB}(x_h, y_k)$, given by (6.4.5) on page 135, with the minus sign for the attractive potential, we can invert (7.2.4). In fact

$$-G_{k4} + G_{-1,k} = y_k, \text{ while } -g_{k4} + g_{-1,k} = z^\dagger \mathbf{s}_k z,$$

hence

$$y_k = z^\dagger \mathbf{s}_k z, \quad y = z^\dagger z; \tag{7.2.8a}$$

moreover, comparing $G_{0k} = -y x_k$ with g_{0k}, we find

$$x_k = \frac{\mathrm{Re}\,(z^\dagger \mathbf{s}_k w)}{z^\dagger z}. \tag{7.2.8b}$$

Put

$$z = \begin{pmatrix} z_1 + i z_2 \\ z_3 + i z_4 \end{pmatrix}, \quad w = \begin{pmatrix} w_1 + i w_2 \\ w_3 + i w_4 \end{pmatrix}, \quad z_1, \dots, w_4 \in \mathbb{R},$$

Equations (7.2.8a)-(7.2.8b) give, after a canonical exchange between coordinates and momenta

$$\rightleftarrows \ : x_k \to -y_k, \ y_h \to x_h,$$

the KS map of Definition 5.1 on page 106. The constraint (5.3.1) is a rephrasing of the twistor nullity condition and the 1-dimensional kernel arises from the equivalence relation $\psi \sim \psi e^{i\beta}$. The 1-form (7.2.1) becomes

$$\Theta = z_\alpha \, dw_\alpha - w_\alpha \, dz_\alpha = 2 z_\alpha \, dw_\alpha - d\,(z_\alpha w_\alpha), \quad \alpha = 1, \dots, 4,$$

showing that $\sqrt{2} z_\alpha$ and $\sqrt{2} w_\beta$ are canonical coordinates on $(\mathbb{R}^4 - \{0\}) \times \mathbb{R}^4$. Notice that the Hamiltonian $K = -g_{-1,0}$ becomes that of a 4-dimensional harmonic oscillator, up to a numerical factor.

Most of the work done in this and in the previous chapter may be encapsulated in the commutative diagram

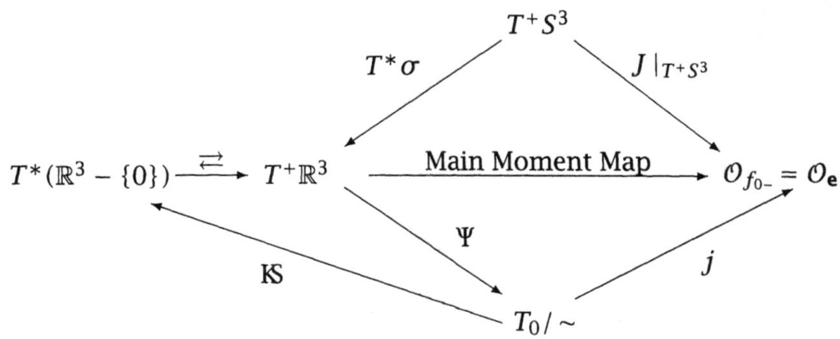

We recall that

- the Main Moment Map is given by (6.4.5) (with the minus sign) on page 135;

- T^+S^3 and $T_0/\sim$ are symplectomorphic;
- $T^*\sigma$ is the Moser symplectic extension (5.1.10) on page 101 of the stereographic projection and is a local symplectomorphism, since the North pole is missing;
- $J\,|_{T^+S^3}$ is the moment map (6.3.6) restricted to

$$\eta(X,X) = \eta(Y,Y) = \eta(X,Y) = 0,$$

(see Corollary 6.7 on page 125) whose image $\mathcal{O}_{f_{0-}}$ is a coadjoint orbit contained in so*(2, 4) (see Theorem 6.9 on page 128);
- j is the moment map (7.2.2) on page 152 whose image is $\mathcal{O}_{\mathbf{e}}$, a coadjoint orbit contained in su*(2, 2); this orbit and that of the previous item can be identified owing to the local isomorphism between SO(2, 4) and SU(2, 2) (the point f_{0-} however does *not* correspond to $\mathbf{e}$);
- the map Ψ is given by (7.2.4) and its inverse by (7.2.8a)-(7.2.8b); it is a local symplectomorphism;
- finally, the Kustaanheimo-Stiefel map is given by Ψ^{-1} composed with the canonical exchange between coordinates and momenta.

What this diagram does not show is the basic fact that T^+S^3 is embedded into $T^+\,(S^1 \times S^3)$ as the phase space of a mechanical system is embedded into the enlarged phase space, i.e., into the phase space plus time and energy as canonically conjugate variables. As we have discussed in the previous chapter, this allows us to send, with an enlarged canonical transformation, the geodesic motion on the 3-dimensional sphere, on the plane and on the hyperboloid into the motion of the Kepler Problem, for energy $\lesseqqgtr 0$, respectively.

7.3 Generalized Kustaanheimo-Stiefel Map

We will now try to extend the definition of KS map to any dimension n. Inspecting the arguments of the preceding sections one easily convinces oneself that the properties of the basis (7.1.2) and (7.1.3), relevant for what follows, come only from the fact that the three Pauli matrices satisfy relations (7.1.1). We must thus first of all construct n matrices $\mathbf{S}_k$, $k = 1, \ldots, n$ such that

$$\mathbf{S}_h\mathbf{S}_k + \mathbf{S}_k\mathbf{S}_h = 2\delta_{hk}\mathbf{1}, \quad \mathbf{S}_h^\dagger = \mathbf{S}_h, \quad \operatorname{Tr}\mathbf{S}_h = 0,$$

whose dimensions will be obviously greater than two by two. Luckily, this task is easily performed with the following recursive argument. Let us suppose we have found the n requested matrices $\mathbf{S}_k$ for n odd, then it is straightforward to verify that the matrices

$$\mathbf{S}'_k = \begin{pmatrix} 0 & i\mathbf{S}_k \\ -i\mathbf{S}_k & 0 \end{pmatrix}, \quad \mathbf{S}'_{k+1} = \begin{pmatrix} -\mathbf{1} & 0 \\ 0 & \mathbf{1} \end{pmatrix}, \quad \mathbf{S}'_{k+2} = \begin{pmatrix} 0 & \mathbf{1} \\ \mathbf{1} & 0 \end{pmatrix}$$

satisfy all the above properties for $n+2$. In this way, we can construct the matrices for a generic n starting from those of $n = 3$. The dimensions of a matrix $\mathbf{S}$ are $d = 2^{[\frac{n}{2}]}$, where $[a]$ is the integer part of the real number a. The basis (7.1.2) and (7.1.3) are generalized in the obvious way, replacing $\mathbf{s}_h$ with $\mathbf{S}_h$.

We have so constructed the Lie algebra spin$(2, n + 1)$ of the double covering Spin$(2, n + 1)$ of SO$(2, n + 1)$. As one verifies by direct inspection on the dimensions of the first few cases, spin$(2, n + 1)$ is a proper subalgebra of the Lie algebra su(d, d), with the only exception $n = 3$, when the two algebras coincide. We will continue to use the symbols $\mathbf{h}$ and $\mathbf{H}$ to denote the matrices of the two SU(d, d)-Hermitian forms; they are defined by simply replacing the 2×2 identity matrix with the $d \times d$ one.

Recalling the definition of $\mathbf{e} \in$ su*$(2, 2)$, we put

$$\mathbf{E} = \begin{pmatrix} 0 & 0 & -i\mathbf{1} & 0 \\ 0 & 0 & 0 & 0 \\ 0 & 0 & 0 & 0 \\ 0 & 0 & 0 & 0 \end{pmatrix} \in \text{spin}^*(2, n+1) \subseteq \text{su}^*(d, d),$$

where each entry is a $\frac{d}{2} \times \frac{d}{2}$ matrix. The coadjoint orbit $\mathcal{O}_{\mathbf{E}}$ under the action of Spin$(2, n + 1)$ is obviously symplectomorphic to the $2n$-dimensional manifold T^+S^n, since spin*$(2, n + 1) =$ so*$(2, n + 1)$. The $2d \times \frac{d}{2}$ matrix

$$\psi_E = \begin{pmatrix} \mathbf{1} \\ 0 \\ 0 \\ 0 \end{pmatrix}$$

is the "square root" of

$$\mathbf{E} = -i\psi_E\psi_E^\dagger \mathbf{h}.$$

Clearly any other $\psi'_E = \psi_E u$, for $u \in \mathrm{U}\left(\frac{d}{2}\right)$ (the $\frac{d}{2}$-dimensional unitary group), will give the same $\mathbf{E}$. One takes care of such an ambiguity by taking the quotient under this right action of $\mathrm{U}\left(\frac{d}{2}\right)$. The coadjoint action

$$\mathbf{E} \mapsto g\,\mathbf{E}g^{-1}, \quad g \in \text{Spin}(2, n+1)$$

is induced, as previously, by the left action of $\mathrm{Spin}(2, n+1)$ on ψ_E :

$$\psi_E \mapsto \psi = g\psi_E.$$

The orbit

$$\mathcal{J} = \left\{\psi \mid \psi = g\psi_E \bmod \mathrm{U}\left(\frac{d}{2}\right)\right\}, \quad \forall g \in \mathrm{Spin}(2, n+1), \tag{7.3.1}$$

is the generalization of $T_0/\sim$ to the n-dimensional case. It is contained in the symplectic manifold V of the complex $2d \times \frac{d}{2}$ matrices equipped with the natural symplectic form $d\Theta$, where

$$\Theta = \mathrm{Im}\,\mathrm{Tr}\left(\psi^\dagger \mathbf{h}\, d\psi\right).$$

The map $j : \mathcal{J} \to \mathrm{spin}^*(2, n+1)$ defined by

$$j = -i\psi\psi^\dagger \mathbf{h}$$

is a moment map. Let $\psi = \begin{pmatrix} z \\ iw \end{pmatrix}$ where z and w are $d \times \frac{d}{2}$ complex matrices. The moment map j appears just as in Equation (7.2.6).

The key difference between the $n = 3$ and the generic n case is the following. Consider the orbit $\mathcal{V} \subset V$ defined still by (7.3.1) but with $g \in \mathrm{SU}(d, d)$. Since

$$\psi^\dagger \mathbf{h}\, \psi = \psi_E^\dagger \mathbf{h}\, \psi_E = 0, \tag{7.3.2}$$

we notice that $\mathcal{V}$ can be obtained from V by symplectic reduction. Consider in fact the symplectic action of $\mathrm{U}\left(\frac{d}{2}\right)$ on V given by *right* multiplication and the corresponding moment map:

$$j' = i\psi^\dagger \mathbf{h}\, \psi, \quad 0 \neq \psi \in V.$$

The symplectic reduction gives the manifold

$$\mathcal{V} = j'^{-1}(0)/\mathrm{U}\left(\frac{d}{2}\right), \quad \text{with } \mathcal{J} \sqsubseteq \mathcal{V} \subset V. \tag{7.3.3}$$

In general $\mathcal{J}$ is strictly contained in $\mathcal{V}$ since $\mathrm{Spin}(2, n+1)$ is a proper subgroup of $\mathrm{SU}(d, d)$. Therefore this symplectic reduction is not sufficient to give $\mathcal{J}$, as it happens, on the contrary, when, for $n = 3$, the two manifolds $\mathcal{J}$ and $\mathcal{V}$ coincide since $\mathrm{Spin}(2, 4) = \mathrm{SU}(2, 2)$. In other words, for n generic the nullity condition (7.3.2) is no longer enough to characterize the orbit defined by (7.3.1).

There are, however, two other known exceptions to this last statement., besides $n = 3$. In fact, for $n = 2$ and $n = 5$, although the corresponding spinorial groups are strictly contained in the respective special pseudo-unitary groups, it is possible to restrict the definition of

$\mathcal{V}$ in such a way that the nullity condition becomes sufficient for the characterization of the orbit. Let us view the details.

For $n = 2$ let us choose the basis $\mathbf{S}_1 = \mathbf{s}_1$ and $\mathbf{S}_2 = \mathbf{s}_3$, i.e., the one with real entries. The relevant pseudo-unitary group is still SU(2, 2), but now Spin(2, 3) is a proper subgroup. One easily verifies that, restricting z and w to be real, the image of the moment map $j(\psi)$ given by (7.2.6) is just contained in spin$^*(2,3)$. The orbit $\mathcal{V}$ is the same 6-dimensional as for $n = 3$, whereas the orbit $\mathcal{T} \subset \mathcal{V}$ is 4-dimensional. The first step of the symplectic reduction (7.3.3) is empty, since the nullity condition is identically satisfied, but we must divide out the $\mathbb{Z}_2$ subgroup of U(1). Parametrize z and w as

$$z = \begin{pmatrix} z_1 \\ z_3 \end{pmatrix}, \quad w = \begin{pmatrix} w_1 \\ w_3 \end{pmatrix}, \quad z_1, z_3, w_1, w_3 \in \mathbb{R},$$

and from (7.2.8a)-(7.2.8b) one finds the Levi-Civita transformation.

For $n = 5$ we choose the basis

$$\vec{\mathbf{S}} = \begin{pmatrix} 0 & i\vec{\mathbf{s}} \\ -i\vec{\mathbf{s}} & 0 \end{pmatrix}, \quad \mathbf{S}_4 = \begin{pmatrix} -\mathbf{1} & 0 \\ 0 & \mathbf{1} \end{pmatrix}, \quad \mathbf{S}_5 = \begin{pmatrix} 0 & \mathbf{1} \\ \mathbf{1} & 0 \end{pmatrix}. \tag{7.3.4}$$

The relevant pseudo-unitary group is now SU(4, 4) and Spin(2, 6) is a proper subgroup. Let z and w be still 2×1 matrices, but with quaternionic entries, i.e.,

$$z = \begin{pmatrix} z_1 + i\sum_{k=2}^{4} z_k\mathbf{s}_k \\ z_5 + i\sum_{k=6}^{8} z_k\mathbf{s}_k \end{pmatrix}, \quad w = \begin{pmatrix} w_1 + i\sum_{k=2}^{4} w_k\mathbf{s}_k \\ w_5 + i\sum_{k=6}^{8} w_k\mathbf{s}_k \end{pmatrix},$$
$$z_1, \ldots, w_8 \in \mathbb{R}.$$

One verifies that $j(\psi)$ is contained in spin$^*(2,6)$. Let us apply the symplectic reduction and verify the dimensions: ψ, with z and w generic, describes a 16-dimensional manifold; $\psi^\dagger \mathbf{h}\, \psi$ being an imaginary quaternion, the nullity condition gives a 3-dimensional constraint; finally, we must divide out an imaginary quaternion. Thus the orbit $\mathcal{T}$ is 10-dimensional, as expected. The 5-dimensional KS is given always by (7.2.8a)-(7.2.8b), replacing $\mathbf{s}$ with $\mathbf{S}$.

Summing up, in all three cases we put $\psi = \begin{pmatrix} z \\ iw \end{pmatrix}$, where z and w are 2×1 matrices defined on $\mathbb{R}, \mathbb{C}, \mathbb{H}$, respectively. Enforcing the constraint $\psi^\dagger \mathbf{h}\, \psi = 0$ and dividing out $\mathbb{Z}_2$, U(1) and SU(2), we obtain that $j(\psi)$ describes T^+S^n as a coadjoint orbit of Spin$(2, n+1)$ for $n = 2, 3, 5$ respectively. The reason why all that works is the following. Let us consider the linear transformation: $\psi \mapsto \psi' = \mathbf{T}\psi$, with $\mathbf{T} = \begin{pmatrix} \mathbf{1} & 0 \\ 0 & -i\mathbf{1} \end{pmatrix}$, then $\mathbf{h} \mapsto i\mathbf{E}$, $\mathbf{E} = \begin{pmatrix} 0 & \mathbf{1} \\ -\mathbf{1} & 0 \end{pmatrix}$. In the first case, $\psi' \in \mathbb{R}^4$

and Spin(2, 3) = Sp(4, $\mathbb{R}$), the symplectic 4-dimensional group over the real numbers, while, in the third case, $\psi' \in \mathbb{H}^4$ and Spin(2, 6) = Sp(4, $\mathbb{H}$), the symplectic 4-dimensional group over the quaternions. This last is defined as the group acting linearly on $\mathbb{H}^4$ and leaving invariant the quadratic form $\varphi^\dagger \mathbf{E} \varphi$, $\varphi \in \mathbb{H}^4$.

REMARK 7.3 The Levi-Civita transformation, whose natural environment thus appears to be the real one, can also be formulated (as in the original version) in terms of complex variables:

$$y_3 + iy_1 = (z_3 + iz_1)^2.$$

Let us also mention, for curiosity, that, similarly, KS transformation, which arises in complex environment, can be recast invoking quaternionic variables. See Vivarelli (1983).

CHAPTER 8

Return to Separation of Variables

In Chapter 3 we have seen that the Hamilton–Jacobi Equation as well as the Schrödinger Equation of the 3-dimensional Kepler Problem are both separable in four different orthogonal coordinate systems. This has, however, only been verified by a direct calculation, working explicitly with those coordinates. In this chapter we show that general theorems about separable orthogonal systems, combined with the knowledge of the invariance group $SO(2) \times SO(4)$ of the Kepler Problem, enable us to *derive* these four coordinate systems, and to prove that they are in fact the only ones consistent with the separability.

In this chapter some care is required to distinguish whether repeated indices are summed or not. For this reason we momentarily ignore the Einstein convention and always explicitly indicate the summation.

The references on the separation of the Hamilton–Jacobi Equation are countless. We only quote some of the more representative recent references, in which one may find an exhaustive treatment of the problem, and an extensive bibliography: Pars (1965), Kalnins (1986), Benenti (1991), Benenti (1993), Benenti (1997).

8.1 Separable Orthogonal Systems

In this section we shall recall the main theorems on the subject, due to Stäckel, Eisenhart and Robertson. We state in advance that no attempt will be made to achieve either completeness or generality, in particular for the Eisenhart Theorem. We shall rather state the theorems in such a way that their application to the problem at hand becomes most direct and effective.

8.1.1 Stäckel Theorem

It is natural to ask if a given Hamiltonian is separable or not. The complete answer to this question is provided by a theorem of Levi-Civita, reported in Kalnins (1986). Restricting ourselves to *natural orthogonal systems*, the answer is contained in Stäckel's Theorem; see for example Pars (1965). By definition, natural systems are those whose Hamiltonian is a sum of kinetic terms quadratic in the momenta and of a potential energy which only depends on the position. Orthogonality requires that the Hamiltonian is of the type

$$H = \frac{1}{2}\sum_{k=1}^{n} g_k(q_1,\ldots,q_n)\, p_k^2 + V(q_1,\ldots,q_n),$$

i.e., does not contain product terms in the momenta.

THEOREM 8.1 *An n-dimensional system is separable if and only if there exists an invertible $n\times n$ matrix $(U_{hk}(q_h))$, called a* Stäckel matrix, *whose h-th row is a function of q_h only, and a column matrix $(w_h(q_h))$, referred to as a* Stäckel vector, *whose h-th element is a function of q_h only such that*

$$\sum_{h=1}^{n} g_h U_{hk} = \delta_{1k}, \quad \sum_{h=1}^{n} g_h w_h = V. \tag{8.1.1}$$

Proof. We show first that the conditions (8.1.1) are necessary for separability. Let us assume that the system is separable, i.e., that the Hamilton–Jacobi Equation

$$\frac{1}{2}\sum_{h=1}^{n} g_h \left(\frac{\partial S}{\partial q_h}\right)^2 + V = \alpha_1$$

admits a complete integral of the form

$$S = S_1 + S_2 + \cdots + S_n, \quad S_k = S_k(q_k, \alpha_1, \ldots, \alpha_n). \tag{8.1.2}$$

Let us substitute this complete integral into the Hamilton–Jacobi Equation, which is then identically satisfied. Partial differentiation with respect to each α yields in turn

$$\sum_{k=1}^{n} g_k \frac{\partial S}{\partial q_k} \frac{\partial^2 S}{\partial \alpha_1 \partial q_k} = 1,$$
$$\sum_{k=1}^{n} g_k \frac{\partial S}{\partial q_k} \frac{\partial^2 S}{\partial \alpha_h \partial q_k} = 0, \quad h = 2, \ldots, n. \tag{8.1.3}$$

Assuming separability, in each of these equations the coefficient of g_k is a function of q_k only. As one easily calculates, the determinant of the coefficients is

$$\Delta = \frac{\partial S}{\partial q_1} \frac{\partial S}{\partial q_2} \cdots \frac{\partial S}{\partial q_n} \det\left(\frac{\partial^2 S}{\partial \alpha_h \partial q_k}\right),$$

and does not vanish, since S is a complete integral. The Equations (8.1.3) take the form

$$\sum_{k=1}^{n} g_k U_{k1} = 1,$$
$$\sum_{k=1}^{n} g_k U_{kh} = 0, \quad h = 2, \ldots, n,$$

which have the form of the first of (8.1.1), with $\det(U_{hk}) \neq 0$. Further

$$V = \alpha_1 - \frac{1}{2} \sum_{k=1}^{n} g_k \left(\frac{\partial S}{\partial q_k}\right)^2 = \sum_{k=1}^{n} g_k \left[\alpha_1 U_{k1} - \frac{1}{2}\left(\frac{\partial S}{\partial q_k}\right)^2\right],$$

which is of the form $\sum_{k=1}^{n} g_k w_k$. This completes the proof that the conditions stated are necessary for separability.

Next we prove that the conditions are also sufficient. Assuming that the conditions are satisfied, we find a complete integral of the form (8.1.2). It is convenient to use a matrix formulation. Put

$$\mathbf{U} = (U_{hk}), \quad \mathbf{V} = \mathbf{U}^{-1}, \quad \underline{p}^2 = \begin{pmatrix} p_1^2 \\ \vdots \\ p_n^2 \end{pmatrix}, \quad \underline{w} = \begin{pmatrix} w_1 \\ \vdots \\ w_n \end{pmatrix}, \quad \underline{\alpha} = \begin{pmatrix} \alpha_1 \\ \vdots \\ \alpha_n \end{pmatrix}$$

and define the n functions

$$\underline{K}(q_h, p_k) = \frac{1}{2} \mathbf{V} \underline{p}^2 + \mathbf{V} \underline{w},$$

the first of which, owing to (8.1.1), equals the Hamiltonian. The Hamilton–Jacobi Equation can be viewed as the first of the system of n equations $\underline{K}(q_h, p_k) = \underline{\alpha}$. Inverting this last relation yields

$$\frac{1}{2} \underline{p}^2 + \underline{w} = \mathbf{U} \underline{\alpha},$$

which implicitly defines p_k as a function of $q_k, \alpha_1, \dots, \alpha_n$. Putting $\frac{\partial S_k}{\partial q_k} = p_k$, we see that $S = \sum_k S_k$ is a complete integral. Indeed, S is by construction a solution of the Hamilton–Jacobi Equation, and

$$\det\left(\frac{\partial^2 S}{\partial \alpha_h \partial q_k}\right) = \frac{\det \mathbf{U}}{\frac{\partial S}{\partial q_1} \cdots \frac{\partial S}{\partial q_n}} \neq 0.$$

Since we have assumed that **U** and $\underline{w}$ are of the Stäckel type, the Hamilton–Jacobi Equation is separable. QED

The n functions $\underline{K}(q_h, p_k)$ are therefore first integrals in involution, quadratic in the momenta and, in the coordinate system allowing separation, they do not contain product of the momenta. But our problem is just to find such coordinate systems. The Eisenhart Theorem gives us a constructive method to finding them.

REMARK 8.1 It is obvious that, if the Hamiltonian H is separable, the corresponding geodesic Hamiltonian (i.e., that with $V = 0$) must be separable in the same coordinate system also. The separation of the geodesic case is thus fundamental: this is the property that we have exploited in Chapter 3 to select, among the eleven coordinate systems admitting separation in 3-dimensional Euclidean space, those separating also in the presence of the Kepler potential.

8.1.2 Eisenhart Theorem

Before we state the Eisenhart Theorem, we need some definitions and lemmas.

DEFINITION 8.2 *To every first integral quadratic in the momenta of a natural motion*

$$K(q_h, p_k) = \frac{1}{2}\sum_{hk} K_{hk}(q_1, \dots, q_n) p_h p_k + U(q_1, \dots, q_n),$$

we associate a symmetric matrix **K**, *with* $(\mathbf{K})_{hk} = K_{hk}$, *called the* Killing matrix. *Given two first integrals* $K^{(1)}$ *and* $K^{(2)}$, *they are said to be in involution if their Poisson bracket is zero, while we will speak of* algebraic commutativity *if*

$$\left[\mathbf{K}^{(1)}, \mathbf{K}^{(2)}\right] \stackrel{def}{=} \mathbf{K}^{(1)}\mathbf{K}^{(2)} - \mathbf{K}^{(2)}\mathbf{K}^{(1)} = 0,$$

where $[\cdot, \cdot]$ *is the commutator of the two matrices.*

Involutivity and algebraic commutativity are obviously independent properties. The following lemma shows that both the conditions are necessary to construct a Stäckel matrix: remember in fact that two symmetric matrices can be simultaneously diagonalized if and only if they commute.

LEMMA 8.3 *If n independent Killing matrices are simultaneously diagonalized in some coordinate system and the corresponding first integrals $K^{(a)} = \frac{1}{2}\sum_h K^{(a)}_{hh} p_h^2 + U_a$ are in involution, then the $n \times n$ matrix*

$$(\mathbf{V})_{ah} = K^{(a)}_{hh}$$

is the inverse of a Stäckel matrix; putting $V_{ah} = (\mathbf{V})_{ah}$, then U_a equals $\sum_i V_{ai} w_i$ where $(w_1, \dots, w_n)$ is a Stäckel vector.

Proof. If $K^{(a)}$ and $K^{(b)}$ are in involution, then

$$\left\{K^{(a)}, K^{(b)}\right\} = \frac{1}{2}\sum_{ih} \left(V_{ai}\partial_i V_{bh} - V_{bi}\partial_i V_{ah}\right) p_i p_h^2 + \sum_i \left(V_{ai}\partial_i U_b - V_{bi}\partial_i U_a\right) p_i = 0.$$

Since the matrix $\mathbf{V}$ only depends on the coordinates, not on the momenta, the above relation is satisfied if and only if *every* coefficient of $p_i p_h^2$ and of p_i vanishes, hence

$$V_{ai}\partial_i V_{bh} - V_{bi}\partial_i V_{ah} = 0, \quad i \text{ not summed,} \tag{8.1.4a}$$

$$V_{ai}\partial_i U_b - V_{bi}\partial_i U_a = 0, \quad i \text{ not summed.} \tag{8.1.4b}$$

Let us consider the first equation. Put $\check{V}_{ah} = \left(\mathbf{V}^{-1}\right)_{ah}$. Multiply by $\check{V}_{hc}\check{V}_{ja}\check{V}_{kb}$ and sum over h, a, b;

$$\sum_{hab} \check{V}_{hc}\check{V}_{ja}\check{V}_{kb} \left(V_{ai}\partial_i V_{bh} - V_{bi}\partial_i V_{ah}\right) = \delta_{ki}\partial_i \check{V}_{jc} - \delta_{ji}\partial_i \check{V}_{kc} = 0, \quad i \text{ not summed.}$$

If $i = j = k$ or $i \neq k, j$, this relation is identically satisfied for all c, while if $i = j \neq k$ it is equivalent to $\partial_i \check{V}_{kc} = 0$, which is just the defining property of a Stäckel matrix. Analogously if $i = k \neq j$.

Let us consider the second equation. Define a vector w_h such that

$$U_a = \sum_h V_{ah} w_h.$$

Thus it must be that

$$V_{ai}\partial_i \sum_h V_{bh} w_h - V_{bi}\partial_i \sum_h V_{ah} w_h = 0,$$

which, taking into account Equation (8.1.4a), can be written

$$V_{ai} \sum_h V_{bh}\partial_i w_h - V_{bi} \sum_h V_{ah}\partial_i w_h = 0.$$

Multiply by $\check{V}_{aj}\check{V}_{bk}$ and sum over a, b. Equation (8.1.4b) is therefore equivalent to

$$\delta_{ij}\partial_i w_k - \delta_{ik}\partial_i w_j = 0, \quad i \text{ not summed},$$

which, arguing as in the previous case, shows that w_k is a Stäckel vector.
QED

Let us consider a generic Riemannian manifold, with metric tensor g_{hk} not necessarily diagonal. Let $X_1, \dots, X_n$ be the orthonormal eigenvectors of the metric tensor, i.e., n vectors satisfying the two equivalent conditions

$$X_a \cdot X_b \stackrel{\text{def}}{=} \sum_{hk} g_{hk} X_a^h X_b^k = \delta_{ab} \quad \text{or} \quad \sum_a X_a^h X_a^k = g^{hk},$$

where $\sum_k g_{hk} g^{ki} = \delta_h^i$. These n orthonormal vectors span in (almost) every point of the manifold a (in general nonholonomic) basis of the tangent space. The Hamiltonian T (the kinetic energy) of the geodesic motion and one of the dynamical variables K can be written

$$T = \frac{1}{2}\sum_a \pi_a^2, \quad K = \frac{1}{2}\sum_a k_a \pi_a^2,$$

where $\pi_a = \sum_h X_a^h p_h$, and $k_1, \dots, k_n$ are the eigenvalues of $\mathbf{K}$. The variables π_a are obviously generalized, noncanonical momenta, since the basis is nonholonomic, and the Poisson brackets are

$$\{\pi_a, \pi_b\} = \sum_c \Omega_{abc}\pi_c, \quad \Omega_{abc} = [X_a, X_b] \cdot X_c.$$

DEFINITION 8.4 *In a manifold, n pointwise independent vector fields (called a* frame*) are said to be* hypersurface forming *if all the $(n-1)$-dimensional distributions complementary to each one of vector fields are integrable. In this case the n tangent lines to the vector fields provide us with a coordinate system.*

The following two propositions, first pointed out by Kalnins & Miller (1980), weaken some of the conditions required by the original version of the Eisenhart Theorem. We report here the neat proofs of Benenti, Chanu & Rastelli (2001).

PROPOSITION 8.5 *In a manifold, n pointwise independent vector fields are hypersurface forming if and only if for each pair of indices $a \neq b$ the distribution Δ_{ab}, spanned by the two vectors X_a, X_b, is completely integrable.*

Proof. Assuming that the frame is hypersurface forming, let us consider the integral foliations $S_{\hat{a}}$ of $\Delta_{\hat{a}}$, the distributions made of all the vectors except X_a. These foliations are made of submanifolds of codimension 1. Submanifolds belonging to different foliations are transversal. For any choice of $a \neq b$, the foliation given by the intersection of all $S_{\hat{c}}$ with $c \neq a, b$ is made of submanifolds of dimension 2 that are tangent to X_a, X_b. Hence, Δ_{ab} is integrable.

Conversely, suppose that all Δ_{ab} are integrable and let us consider the corresponding foliations S_{ab} made of submanifolds of dimension 2. For any fixed a, let us consider the foliation $S_{\hat{a}}$ given by the union of the foliations S_{bc} with $b, c \neq a$. This is a foliation of submanifolds of codimension 1 tangent to $\Delta_{\hat{a}}$. Then this distribution is integrable and by definition the frame is hypersurface forming. QED

PROPOSITION 8.6 *Suppose that in a Riemannian manifold n Killing matrices are (i) pointwise independent (ii) simultaneously diagonalized in a frame of orthogonal vectors $X_1, \ldots, X_n$ and that (iii) the associated functions $K^{(1)}, \ldots, K^{(n)}$ (quadratic in the momenta) are in involution. Then the distributions Δ_{ab} spanned by pairs of vectors X_a, X_b are completely integrable.*

Proof. By assumption (ii), $K^{(a)} = \frac{1}{2} \sum_b k_b^{(a)} \pi_b^2$ holds and assumption (i) is equivalent to

$$\det\left(k_b^{(a)}\right) \neq 0.$$

The Poisson bracket $\{K^{(a)}, K^{(b)}\}$ is a homogeneous function of third degree in the generalized momenta, and by assumption (iii) all the coefficients vanish. Let us consider the coefficients of the terms $\pi_c \pi_d \pi_e$ with c, d, e all different. In particular, the coefficients of $\pi_1 \pi_2 \pi_3$ (as well as for all possible choices of three distinct indices) give rise to the equation

$$\begin{aligned} k_1^{(a)} k_2^{(b)} \Omega_{123} + k_1^{(a)} k_3^{(b)} \Omega_{132} + k_2^{(a)} k_3^{(b)} \Omega_{231} \\ + k_2^{(a)} k_1^{(b)} \Omega_{213} + k_3^{(a)} k_1^{(b)} \Omega_{312} + k_3^{(a)} k_2^{(b)} \Omega_{321} = 0. \end{aligned}$$

Using the skew symmetry of Ω this equation can be recast into the form

$$\det\begin{pmatrix} \Omega_{231} & \Omega_{312} & \Omega_{123} \\ k_1^{(a)} & k_2^{(a)} & k_3^{(a)} \\ k_1^{(b)} & k_2^{(b)} & k_3^{(b)} \end{pmatrix} = 0.$$

This means that the three vectors of $\mathbb{R}^3$

$$\begin{aligned} \vec{\Omega} &= (\Omega_{231}, \Omega_{312}, \Omega_{123}) \\ \vec{k}_a &= (k_1^{(a)}, k_2^{(a)}, k_3^{(a)}) \\ \vec{k}_b &= (k_1^{(b)}, k_2^{(b)}, k_3^{(b)}) \end{aligned}$$

are linearly dependent. If $\vec{\Omega} \neq 0$, then all the vectors $\vec{k}_a$, $a = 1, \dots n$ belong to the same 3-plane containing $\vec{\Omega}$, i.e.,

$$\det\begin{pmatrix} k_1^{(a)} & k_2^{(a)} & k_3^{(a)} \\ k_1^{(b)} & k_2^{(b)} & k_3^{(b)} \\ k_1^{(c)} & k_2^{(c)} & k_3^{(c)} \end{pmatrix} = 0.$$

As a consequence, by applying the Laplace rule to the lines $(1, 2, 3)$ of the matrix $(k_b^{(a)})$, we infer that its determinant vanishes, which contradicts assumption (i). This means that $\Omega_{123} = \Omega_{231} = \Omega_{312} = 0$ and, in general, $\Omega_{abc} = 0$ for any choice of distinct indices, thus by definition $[X_a, X_b]$ is orthogonal to all the vectors X_c with $c \neq a, b$. The vectors being all orthogonal, this means that $[X_a, X_b]$ belongs to the 2-space spanned by X_a, X_b. The statement follows using the Frobenius Theorem. QED

Now we can state and prove the Eisenhart Theorem.

THEOREM 8.7 (EISENHART) *The Hamilton-Jacobi Equation relative to the natural Hamiltonian on an n-dimensional Riemannian manifold is separable in some coordinate system if*

(i) there exist n independent first integrals $K^{(a)}$ in involution and quadratic in the momenta, with $K^{(1)} = H$;

(ii) the associated Killing matrices commute algebraically.

Proof. For the last two propositions, the orthogonal eigenvectors of the Killing matrices are hypersurface forming, thus determine a coordinate system. In this system $K^{(a)} = \frac{1}{2}\sum_i V_{ai} p_i^2 + U_a$, where, for Lemma 8.3, $(\mathbf{V})_{ah} = K_{hh}^{(a)}$ is the inverse of a Stäckel matrix, while U_a is generated by a Stäckel vector. Thus, by the Stäckel Theorem, the natural Hamiltonian system is separable. QED

COROLLARY 8.8 *In a two-dimensional Riemannian manifold, if the two eigenvalues of a Killing matrix are functionally independent, they provide separable coordinates.*

Proof. Calculating the Poisson bracket between T and K defined above one finds

$$\{T,K\} = \sum_{abc} \frac{1}{2}\left(\sum_k X_a^k \partial_k \lambda_b\right) \pi_a \pi_b^2 + \lambda_b \Omega_{abc} \pi_a \pi_b \pi_c = 0,$$

i.e., a linear combination of symmetric triple products of the generalized momenta. All the coefficients must be zero. In particular, for $a = b = c$ and taking into account the skew symmetry in the first two indices of Ω_{abc}, one finds

$$\sum_k X_a^k \partial_k \lambda_a = 0 \quad \forall a,$$

i.e., every eigenvector X_a is tangent to the $(n-1)$-dimensional hypersurface λ_a =constant. For $n = 2$ the statement follows. QED

At this point the strategy that one must follow to seek the orthogonal coordinate systems separating a problem should be clear. Firstly, find n first integrals in involution which are quadratic in the momenta; secondly, verify that the associated Killing matrices commute algebraically; lastly, find their n common eigenvectors and the coordinate lines tangent to them.

8.1.3 Robertson Theorem

A direct calculation has shown, in Chapter 4, that the four coordinate systems separating additively the Hamilton–Jacobi Equation of the Kepler Problem do separate multiplicatively also the corresponding Schrödinger Equation. This is not a mere coincidence, but follows from the Robertson Theorem, which we state in the Eisenhart form.

THEOREM 8.9 *If the Hamilton–Jacobi Equation of a natural problem of a Riemannian manifold is separable in a coordinate system, then the corresponding Schrödinger Equation is separable in the same system if and only if the Ricci tensor R_{ij} is diagonal.*

For the original proof, see Robertson (1927) and Eisenhart (1949), while a more recent reference is Kalnins (1986). For the Kepler Problem the condition is trivially satisfied since the metric is Euclidean and the Ricci tensor is identically zero.

REMARK 8.2 Benenti *et al.* (2001) have pointed out that, while the requirements of the above theorem are sufficient for the separability, the proof, given by Robertson and accepted by Eisenhart, that they are also necessary is in fact not satisfactory. This is basically due to the fact that these authors do not consider a "complete" solution of the Schrödinger Equation, which is one containing $2n$ integration parameters. In that reference the entire argument is accurately revised.

8.2 Finding Coordinate Systems Separating the Kepler Problem

In this section we want to apply the above strategy to find all the coordinate systems which separate the Kepler Problem. We must thus find all the possible triplets of first integrals in involution which are quadratic in the momenta, then verify if the algebraic commutativity is satisfied. More exactly, we want to find all the possible equivalence classes of triplets of first integrals, where the equivalence relation is stated in the following definition.

DEFINITION 8.10 *Two triplets of first integrals are equivalent if*

(i) the first integrals of a triplet are a linear combination of those of the other;

(ii) the first integrals of a triplet are obtained from those of the other with a rigid rotation.

The first condition is due to the fact that taking a linear combination of commuting matrices does not change their common eigendirections, leaving thus unaltered (if it does exist) the coordinate system generated by them. The second condition is due to the fact that the action of the rotation group on the first integrals generates a rotation of the relative coordinate system: obviously, we consider the various rigidly rotated coordinate systems as being the "same".

Now we state and prove the central theorem of the chapter.

THEOREM 8.11 *For the Kepler Problem, there exist four inequivalent triplets of first integrals that are quadratic in the momenta and are in involution. As representative elements for each one of the four triplets one may just choose those already considered in Chapter 3, that is* *(i)* H, G^2, G_3^2 *(ii)* H, E_3, G_3^2 *(iii)* $H, G^2 - dE_3, G_3^2$ *(iv)* $H, G^2, b^2G_1^2 - a^2G_3^2$, *where* d, a *and* b *are non zero arbitrary real parameters.*

Proof. For the Kepler Problem all first integrals are known: they are the Hamiltonian H, the three components of the angular momentum $\vec{G}$ and the three components of the eccentricity vector $\vec{E}$. Since the Hamiltonian H is in involution with $\vec{G}$ and $\vec{E}$, our task is reduced to "extract", from these two vectors, two independent functions, Γ_1 and Γ_2 say, which are quadratic in the momenta and are in involution. Since $\vec{E}$ is quadratic in the momenta while $\vec{G}$ is linear, the most general functions Γ_1 and Γ_2 will be of the type

$$\Gamma_1 = \sum_h a_h E_h + \sum_{hk} a_{hk} G_h G_k,$$
$$\Gamma_2 = \sum_h b_h E_h + \sum_{hk} b_{hk} G_h G_k.$$

We must find all real parameters a_h, a_{hk}, b_h, b_{hk} for which $\{\Gamma_1, \Gamma_2\} = 0$. Firstly, we investigate if it is possible to satisfy this requirement by taking $a_h = b_k = 0$, $\forall h, k$. It is clear that the condition

$$\left\{ \sum_{hk} a_{hk} G_h G_k, \sum_{hk} b_{hk} G_h G_k \right\} = 0$$

is satisfied if and only if one of the two terms is a multiple of the Casimir invariant G^2 of $SO(3)$. Choose $a_{hk} = \alpha\delta_{hk}$. With a rotation we can always diagonalize b_{ij}. Thus

$$\Gamma_1 = \alpha G^2,$$
$$\Gamma_2 = \beta_1 G_1^2 + \beta G_2^2 + \beta G_3^2,$$

where we may assume $\beta_1 \geq \beta_2 \geq \beta_3$, the case $\beta_1 = \beta_2 = \beta_3$ being trivial. With a linear combination of Γ_1 and Γ_2 we obtain the case (iv) of the theorem if the β's are all different, and the case (i) if two are equal.

Let us now consider the general case. Since $\{E_h, E_k\}$ is equal to a product of H and G_j, it is the only Poisson bracket that will contain terms of the type $p^3 q$, thus

$$\left\{ \sum_h a_h E_h, \sum_h b_h E_h \right\} = 0$$

must hold. This imposes $a_h = \lambda b_h$. Taking a convenient linear combination of Γ_1 and Γ_2 we can make one of the two terms to vanish, and transform the other into μE_3 by a rotation. Thus

$$\Gamma_1 = \mu E_3 + \sum_{hk} a_{hk} G_h G_k,$$
$$\Gamma_2 = \sum_{hk} b_{hk} G_h G_k.$$

(These a_{hk} and b_{hk} are obviously different from the preceding expressions). Again, we must have

$$\left\{E_3, \sum_{hk} b_{hk} G_h G_k\right\} = 0,$$

$$\left\{\sum_{hk} a_{hk} G_h G_k, \sum_{hk} b_{hk} G_h G_k\right\} = 0.$$

Bearing in mind that now we can no longer perform an arbitrary rotation, having already exploited this possibility, we deduce, from the first of these equations, that

$$\sum_{hk} b_{hk} G_h G_k = \nu G_3^2,$$

must hold. Substituting into the latter we obtain

$$\left\{\sum_{hk} a_{hk} G_h G_k, G_3^2\right\} = 0,$$

which is equivalent either to

$$\sum_{hk} a_{hk} G_h G_k = \gamma G_3^2 \quad \text{or} \quad \sum_{hk} a_{hk} G_h G_k = \gamma G^2.$$

Therefore

$$\Gamma_1 = \mu E_3 + \gamma G_3^2, \quad \Gamma_2 = \nu G_3^2,$$

or

$$\Gamma_1 = \mu E_3 + \gamma G^2, \quad \Gamma_2 = \nu G_3^2,$$

from which, with two suitable linear combinations, we obtain case (ii) and (iii) of the theorem, respectively. QED

REMARK 8.3 In cases (iii) and (iv) an overall scale factor can be fixed arbitrarily to comply with the choices of Chapter 3. Hereafter we put $d = 2$ and $a^2 + b^2 = 1$.

Having found all the triplets of first integrals in involution, we must verify their algebraic commutativity. To each first integral is associated its Killing matrix, and to the Hamiltonian $H = \frac{1}{2}p^2 - \frac{1}{q}$ is plainly associated the unit matrix, which commutes with everything. Examining the other integrals, we note that it is enough to check that

$$\left[\mathbf{G}^2, \mathbf{G}_1^2\right] = 0, \quad \left[\mathbf{G}^2, \mathbf{G}_3^2\right] = 0, \quad \left[\mathbf{E}_3, \mathbf{G}_3^2\right] = 0, \qquad (8.2.1)$$

where, with an obvious notation, $\mathbf{G}^2$ is the Killing matrix associated to G^2, and so forth. The explicit expression of these matrices is

$$\mathbf{G}_1^2 = \begin{pmatrix} 0 & 0 & 0 \\ 0 & q_3^2 & -q_2q_3 \\ 0 & -q_2q_3 & q_2^2 \end{pmatrix}, \qquad \mathbf{G}_3^2 = \begin{pmatrix} q_2^2 & -q_1q_2 & 0 \\ -q_1q_2 & q_1^2 & 0 \\ 0 & 0 & 0 \end{pmatrix},$$

$$\mathbf{G}^2 = \begin{pmatrix} q_2^2+q_3^2 & -q_1q_2 & -q_1q_3 \\ -q_1q_2 & q_1^2+q_3^2 & -q_2q_3 \\ -q_1q_3 & -q_2q_3 & q_1^2+q_2^2 \end{pmatrix}, \quad \mathbf{E}_3 = \begin{pmatrix} q_3 & 0 & -\frac{1}{2}q_1 \\ 0 & q_3 & -\frac{1}{2}q_2 \\ -\frac{1}{2}q_1 & -\frac{1}{2}q_2 & 0 \end{pmatrix}$$

and it is a simple matter of calculation to verify that (8.2.1) holds.

For the third and last step, i.e., to finding the coordinate lines, we will be able, in all four cases, to reduce the problem to a two-dimensional one, and hence to apply Corollary 8.8.

Let us now examine separately the four cases, bearing in mind that the Killing matrix associated to the Hamiltonian is the unit matrix, and does not give information in the search for common eigendirections.

8.2.1 Spherical Coordinates

The spherical coordinates are related to the case (i) of Theorem 8.11. One of the first integrals is G_3, the third component of angular momentum, and this means that the coordinate system we seek must be rotationally symmetric with respect to the third axis. We can thus restrict ourselves to a vertical plane, for example that of q_2q_3. In this plane $\mathbf{G}^2$ is reduced to $\mathbf{G}_1^2$, whose eigenvalues are $\lambda_1 = 0$ and $\lambda_2 = q_2^2 + q_3^2$, with eigenvectors $v_1 = \begin{pmatrix} q_2 \\ q_3 \end{pmatrix}$ and $v_2 = \begin{pmatrix} q_3 \\ -q_2 \end{pmatrix}$ respectively.

It is obvious that the associated coordinate system is that of the spherical coordinates. In Figure 8.1 are plotted the eigendirections of $\mathbf{G}_1^2$ and the associated spherical coordinates.

The Stäckel matrix and the Stäckel vector are

$$\mathbf{U} = \begin{pmatrix} 1 & -\frac{1}{q^2} & 0 \\ 0 & 1 & -\frac{1}{\sin^2\vartheta} \\ 0 & 0 & 1 \end{pmatrix}, \quad \underline{w} = \begin{pmatrix} -\frac{1}{q} \\ 0 \\ 0 \end{pmatrix}.$$

8.2.2 Parabolic Coordinates

The parabolic coordinates are related to case (ii) of Theorem 8.11. As in the preceding case, we can restrict ourselves to the plane q_2q_3. In this

Figure 8.1: Case (i): the eigendirections of $\mathbf{G}_1^2$ and the associated spherical coordinates.

plane the eigenvalues of $\mathbf{E}_3$ are

$$\lambda_1 = \frac{1}{2}(q_3 + r), \quad \lambda_2 = \frac{1}{2}(q_3 - r), \quad r = \sqrt{q_2^2 + q_3^2}$$

with eigenvectors

$$v_1 = \begin{pmatrix} q_2 \\ q_3 + r \end{pmatrix}, \quad v_2 = \begin{pmatrix} q_2 \\ q_3 - r \end{pmatrix}.$$

The eigenvalues are independent functions of q_2 and q_3; for Corollary 8.8, some of their functions can therefore be taken as separable coordinates; we take $\xi^2 = 2\lambda_1$, $\eta^2 = -2\lambda_2$.

In Figure 8.2 are plotted the eigendirections of $\mathbf{E}_3$ and the associated parabolic coordinates.

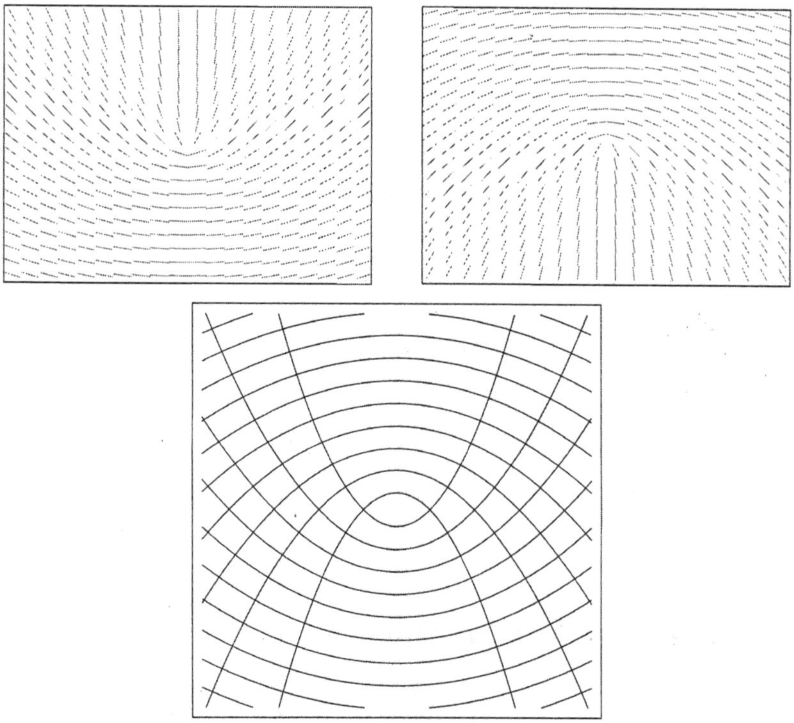

Figure 8.2: Case (ii): the eigendirections of $\mathbf{E}_3$ and the associated parabolic coordinates.

The Stäckel matrix and the Stäckel vector are

$$\mathbf{U} = \begin{pmatrix} \xi^2 & -1 & -\frac{1}{\xi^2} \\ \eta^2 & 1 & -\frac{1}{\eta^2} \\ 0 & 0 & 1 \end{pmatrix}, \quad \underline{w} = \begin{pmatrix} -1 \\ -1 \\ 0 \end{pmatrix}.$$

8.2.3 Elliptic Coordinates

The elliptic coordinates are related to case (iii) of Theorem 8.11. As in the preceding case, we can restrict ourselves to the plane q_2q_3. In this plane the eigenvalues of $\mathbf{G}^2 - 2\mathbf{E}_3$ are

$$\lambda_1 = \frac{1}{4}(r + r')^2 - 1, \quad r' = \sqrt{q_2^2 + (q_3 - 2)^2},$$

$$\lambda_2 = \frac{1}{4}(r - r')^2 - 1,$$

with eigenvectors

$$v_1 = \begin{pmatrix} q_2 (q_3 - 1) \\ (q_3 - 1)^2 - \frac{1}{4}(r + r') \end{pmatrix}, \quad v_2 = \begin{pmatrix} q_2 (q_3 - 1) \\ (q_3 - 1)^2 - \frac{1}{4}(r - r') \end{pmatrix}.$$

The eigenvalues are independent functions of q_2 and q_3; for Corollary 8.8, some of their functions can therefore be taken as separable coordinates:

$$\xi = \sqrt{\lambda_1 + 1}, \quad \eta = \sqrt{\lambda_2 + 1}.$$

In Figure 8.3 are plotted the eigendirections of $\mathbf{G}^2 - 2\mathbf{E}_3$ and the associated elliptic coordinates.

The Stäckel matrix and the Stäckel vector are

$$\mathbf{U} = \begin{pmatrix} 1 & -\frac{1}{\xi^2-1} & -\frac{1}{(\xi^2-1)^2} \\ 1 & \frac{1}{1-\eta^2} & -\frac{1}{(1-\eta^2)^2} \\ 0 & 0 & 1 \end{pmatrix}, \quad \underline{w} = \begin{pmatrix} -\frac{\xi}{\xi^2-1} \\ \frac{\eta}{1-\eta^2} \\ 0 \end{pmatrix}.$$

8.2.4 Spheroconical Coordinates

The spheroconical coordinates are related to case (iv) of Theorem 8.11. In this case, the coordinate system is no longer rotationally symmetric, and we must consider both first integrals G^2 and $b^2G_1^2 - a^2G_3^2$. The matrix $\mathbf{G}^2$ has rank 2, thus admits a one-dimensional kernel that is given by the radial directions: $\mathbf{G}^2 \begin{pmatrix} q_1 \\ q_2 \\ q_3 \end{pmatrix} = 0$. The image of the linear operator $\mathbf{G}^2$ is therefore an integrable 2-dimensional distribution orthogonal in every point to the ray vector: the integral surfaces are plainly spheres with the origin as their common center.

The problem in $\mathbb{R}^3$ is reduced to a problem on a sphere, i.e., to a 2-dimensional one. The matrix $b^2\mathbf{G}_1^2 - a^2\mathbf{G}_3^2$ has rank 2. Compute its eigenvalues λ_1 and λ_2 : they are functions of q_1, q_2, q_3, that, when restricted to the sphere, provide us with separable coordinates by Corollary 8.8. Together with the radial coordinate they form the triplet we were searching for. A somewhat lengthy but straightforward calculation shows that λ_1 and λ_2 satisfy the two identities

$$\frac{q_1^2}{b^2 + \lambda_1(q_1, q_2, q_3)} + \frac{q_2^2}{\lambda_1(q_1, q_2, q_3)} - \frac{q_3^2}{a^2 - \lambda_1(q_1, q_2, q_3)} \equiv 0,$$

$$-\frac{q_1^2}{b^2 - \lambda_2(q_1, q_2, q_3)} + \frac{q_2^2}{\lambda_2(q_1, q_2, q_3)} + \frac{q_3^2}{a^2 + \lambda_2(q_1, q_2, q_3)} \equiv 0,$$

Figure 8.3: Case (iii): the eigendirections of $\mathbf{G}^2 - 2\mathbf{E}_3$ and the associated elliptic coordinates.

and can therefore be identified with the spheroconical coordinates $\xi^2 = \lambda_1$ and $\eta^2 = \lambda_2$: see (3.1.2) on page 42 and Figure 8.4.

The Stäckel matrix and the Stäckel vector are

$$\mathbf{U} = \begin{pmatrix} 1 & -\frac{1}{q^2} & 0 \\ 0 & \frac{\xi^2}{(b^2+\xi^2)(a^2-\xi^2)} & \frac{1}{(b^2+\xi^2)(a^2-\xi^2)} \\ 0 & \frac{\eta^2}{(b^2-\eta^2)(a^2+\eta^2)} & -\frac{1}{(b^2-\eta^2)(a^2+\eta^2)} \end{pmatrix}, \quad \underline{w} = \begin{pmatrix} -\frac{1}{q} \\ 0 \\ 0 \end{pmatrix}.$$

8.3 Integrable Perturbations

While deferring to Part III the study of generic perturbations, in this section, however, we investigate those perturbations of the Kepler Problem that, even though they destroy the complete degeneration (thus giving

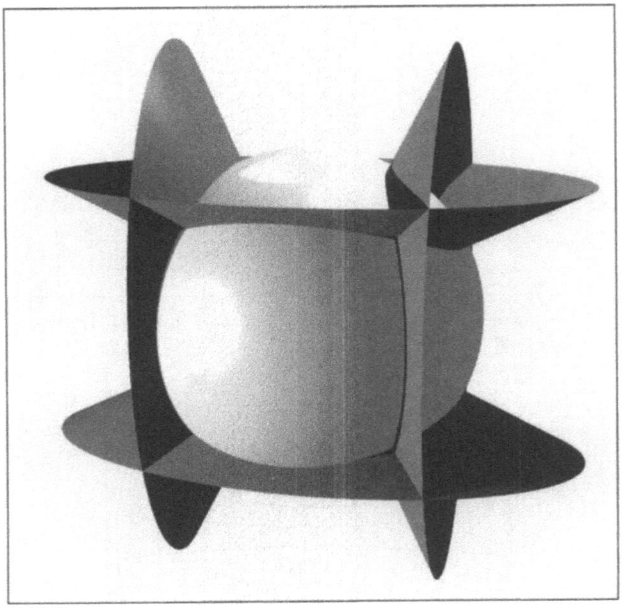

Figure 8.4: Case (iv): the coordinate surfaces of the spheroconical coordinates are concentric spheres and elliptic cones with horizontal and vertical axes.

rise to nonperiodic orbits), they nevertheless maintain the integrability.

These integrable perturbations arise in the following way. In one of the above four coordinate systems for which the Kepler Problem is separable, substitute, in the three first integrals,

$$p_k \mapsto p_k + A_k(q_k), \quad w_k(q_k) \mapsto w_k(q_k) + w'_k(q_k),$$

where $A_k(q_k)$ and $w'_k(q_k)$ are arbitrary functions of the coordinate q_k only. This amounts to adding to the Hamiltonian of the Kepler Problem a vector potential A_k and a scalar potential $\sum_{k=1}^{3} V_{1k} w'_k$, respectively. The resulting *finitely*[(1)] perturbed Hamiltonian remains obviously separable, and hence integrable. Choosing suitably the functions A_k and w'_k, one may obtain interesting physical systems, for which it is certain that the calculation of the motion is reduced to quadratures.

As an example, we want to study to some extent the Euler problem, i.e., the motion of a material particle subjected to the gravitational attraction of two fixed points of arbitrary masses. The quadratures involve elliptical integrals, but one can reach a good understanding of the

[(1)]In the sense that the perturbations are not supposed to be small, as in the generic, nonintegrable, case.

properties of the various orbits with a qualitative analysis, as shown in Deprit (1962). Anyway, the program EULER, written in MAPLE, is supplied, which calculates and displays the orbits for generic initial conditions.

Another interesting example is the Stark problem (mostly considered in atomic physics) i.e., the Kepler Problem with a constant force field, which can be viewed as the limiting case of the Euler Problem when, while keeping fixed one of the two masses, the other goes to infinity.

8.3.1 Euler Problem

To simplify the plot of the orbits, we limit ourselves to the plane case, nothing essentially new being added in the 3-dimensional one. Consider the elliptic coordinates

$$x = \sqrt{(\xi^2 - 1)(1 - \eta^2)}, \quad y = \xi\eta + 1,$$
$$\xi \geq 1, \quad -1 \leq \eta \leq 1.$$

The lines ξ = constant are ellipses with foci at $F_1 \equiv (0,0)$ and $F_2 \equiv (0,2)$, and the lines η = constant are confocal hyperbolas. See Figure 8.5.

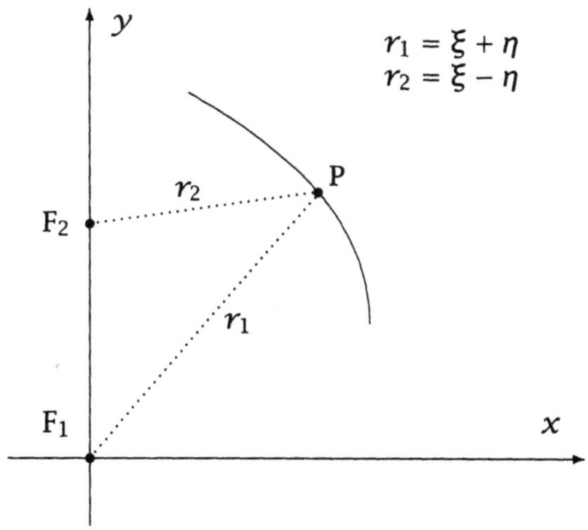

Figure 8.5: The Euler Problem.

Taking

$$\underline{w}' = \begin{pmatrix} -m\frac{\xi}{\xi^2-1} \\ -m\frac{\eta}{\eta^2-1} \end{pmatrix}, \quad 0 \leq m \leq 1 \quad \text{and } A_k = 0,$$

the Hamiltonian of the Kepler Problem is transformed into that of the Euler Problem

$$H = \frac{1}{2}\left(\frac{\xi^2 - 1}{\xi^2 - \eta^2} p_\xi^2 + \frac{1 - \eta^2}{\xi^2 - \eta^2} p_\eta^2\right) - \frac{1}{\xi + \eta} - \frac{m}{\xi - \eta},$$

the parameter m playing the role of a mass placed at F_2.

To separate the variables, let us multiply by $\xi^2 - \eta^2$ and, after a reordering of the terms, we find the second first integral

$$H\xi^2 - \frac{1}{2}(\xi^2 - 1)p_\xi^2 + (1 + m)\xi = -\gamma,$$
$$H\eta^2 + \frac{1}{2}(1 - \eta^2)p_\eta^2 + (1 - m)\eta = -\gamma.$$

A direct calculation shows that

$$\gamma = E_y - \frac{1}{2}G^2 + m\frac{y - 2}{r_2} - H,$$

which, apart from the inessential term H, differs by the corresponding first integral of the Kepler Problem for the term $m\frac{y-2}{r_2}$. We seek the complete integral of the Hamilton-Jacobi Equation in the form

$$S(\xi, \eta, H, \gamma, t) = S_\xi(\xi, H, \gamma) + S_\eta(\eta, H, \gamma) - Ht,$$

from which

$$\left(\frac{\partial S_\xi}{\partial \xi}\right)^2 = 2\frac{H\xi^2 + (1 + m)\xi + \gamma}{\xi^2 - 1},$$
$$\left(\frac{\partial S_\eta}{\partial \eta}\right)^2 = 2\frac{H\eta^2 + (1 - m)\eta + \gamma}{\eta^2 - 1}$$

and the problem is reduced to the quadratures.

The function S generates the canonical transformation

$$\begin{pmatrix} p_\xi & p_\eta \\ \xi & \eta \end{pmatrix} \mapsto \begin{pmatrix} H & \gamma \\ \beta_1 & \beta_2 \end{pmatrix}.$$

Since the transformed Hamiltonian vanishes identically, also the new coordinates β_1, β_2 are first integrals. In particular, from $\frac{\partial S}{\partial \gamma}$ = constant we find

$$\int \frac{d\xi}{\sqrt{(\xi^2 - 1)\,[H\xi^2 + (1 + m)\xi + \gamma]}}$$
$$+ \int \frac{d\eta}{\sqrt{(\eta^2 - 1)\,[H\eta^2 + (1 - m)\eta + \gamma]}} = \text{constant},$$

from which

$$d\tau = \frac{d\xi}{\sqrt{(\xi^2 - 1)\,[H\xi^2 + (1+m)\xi + \gamma]}},$$
$$d\tau = -\frac{d\eta}{\sqrt{(\eta^2 - 1)\,[H\eta^2 + (1-m)\eta + \gamma]}}, \tag{8.3.1}$$

τ being a parameter. If we succeed in integrating the two right-hand sides, we find, after inversion, the parametric expressions $\xi(\tau)$ and $\eta(\tau)$ of the orbit.

We are concerned with integrals of the type $\int \frac{dw}{\sqrt{P(w)}}$, where $P(w)$ is a quartic polynomial. The point is to find a transformation $w \mapsto z(w)$ such that the transformed integral acquires the Legendre canonical form of an elliptic integral of first kind:

$$\int \frac{dz}{\sqrt{(1-z^2)(1-k^2z^2)}}, \quad 0 < k < 1.$$

This last integral is solved invoking the inverse of Jacobi elliptic function $z = \mathrm{sn}(\tau + c)$, where c is an integration constant. See the appendix to this chapter for a very short account on the Jacobi elliptic functions $\mathrm{sn}(\tau), \mathrm{cn}(\tau), \mathrm{dn}(\tau)$.

The transformations reducing the right-hand side of (8.3.1) to canonical form depend, obviously, on the roots of the quartic polynomial. In the generic case of distinct roots, the program EULER, written in the MAPLE language, performs analytically the two integrations, implementing the general method described in Whittaker & Watson (1952, Section 22.7 of Chapter XXII) and Bowman (1961, Chapter IX). Given the initial position and velocity, the program displays the resulting orbit; see Codegoni (2000) for details.

The properties of the orbits can be studied with an accurate qualitative analysis of the position of the roots of the polynomials $H\xi^2 + (1 + m) + \gamma$ and $H\eta^2 + (1 - m) + \gamma$ with respect to the other two roots -1 and $+1$. Elementary considerations allow one to find the position of the two roots x_1 and x_2 of the second degree algebraic equation

$$ax^2 + bx + c = 0, \quad b \geq 0,$$

when varying the coefficients a, b, c; see Figure 8.6. Put

$$M(\xi) = H\xi^2 + (1+m)\xi + \gamma, \quad Z(\eta) = H\eta^2 + (1-m)\eta + \gamma.$$

By definition $\xi \geq 1$ and $-1 \leq \eta \leq 1$, so that

$$M(\xi) \geq 0, \quad Z(\eta) \leq 0, \tag{8.3.2}$$

must hold.

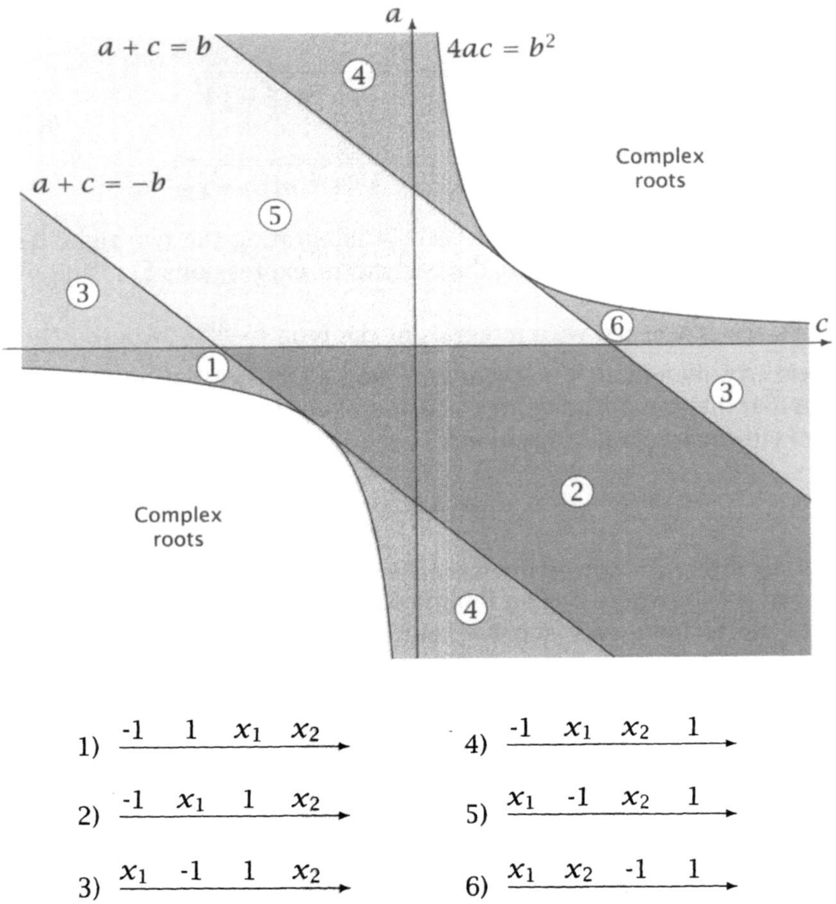

Figure 8.6: Root position for the equation $ax^2 + bx + c = 0$.

PROPOSITION 8.12 *The two roots of* $M(\xi) = 0$ *are always real.*

Proof. Let us suppose that the two roots are complex. Hence

$$(1 - m)^2 < (1 + m)^2 < 4\gamma H,$$

and also $Z(\eta) = 0$ would have complex roots. This implies that $M(\xi)$ and $Z(\eta)$ would have the same sign (namely that of H), contradicting (8.3.2). Obviously, $Z(\eta)$ may instead have complex roots. QED

Let us examine the position of the roots ξ_1, ξ_2 of $M(\xi)$ and η_1, η_2 of $Z(\eta)$, discussing separately the two cases of negative or positive energy, which correspond to the bound and unbound orbits, respectively. We

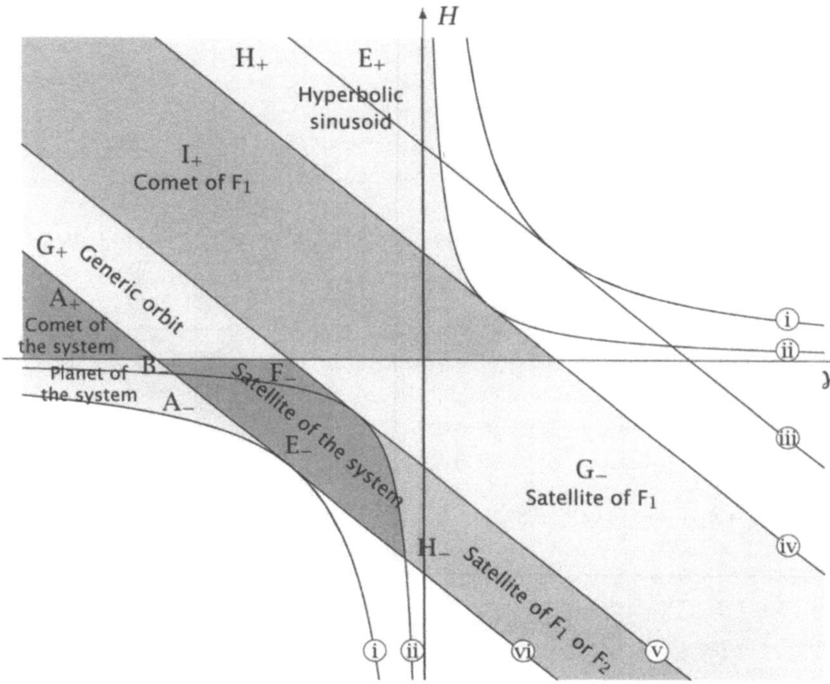

(i) $4\gamma H = (1+m)^2$ (iii) $\gamma + H = 1 + m$ (v) $\gamma + H = -(1-m)$

(ii) $4\gamma H = (1-m)^2$ (iv) $\gamma + H = 1 - m$ (vi) $\gamma + H = -(1+m)$

Figure 8.7: Orbit types for the Euler Problem.

first consider the generic case, i.e., we suppose that the quartic polynomials do not have double roots.

Negative Energy

For $H < 0$, to satisfy (8.3.2) we must consider the values of ξ belonging to the closed interval $[\xi_1, \xi_2]$ and the values of η external to the open interval (η_1, η_2). Moreover, the restrictions imposed by elliptic coordinates must be taken into account. All, in principle, possible combinations are presented in the table below, where the numbers refer to Figure 8.6. The various pairs are labeled by A_-,B_-,C_-, etc.

	A_-	B_-	C_-	D_-	E_-	F_-	G_-	H_-	I_-	L_-	M_-	N_-
$M(\xi)$	1	1	1	1	2	2	2	2	3	3	3	3
$Z(\eta)$	c.r.	1	2	4	c.r.	1	2	4	c.r.	1	2	4

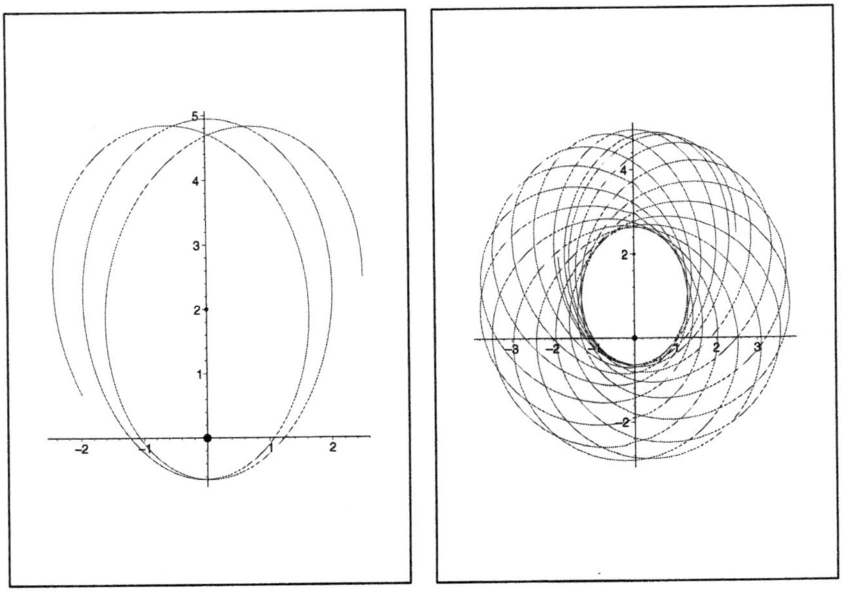

Figure 8.8: Planet of the system.

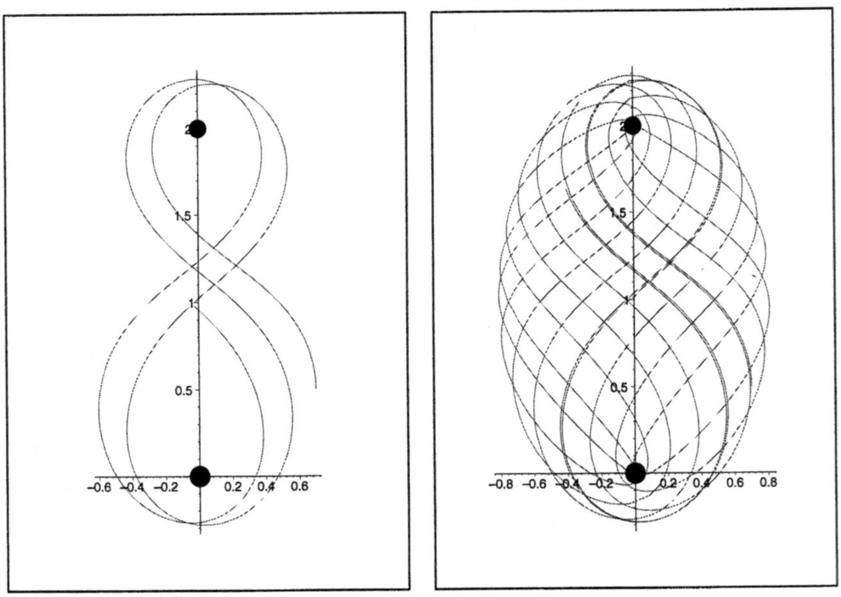

Figure 8.9: Satellite of the system.

Then we investigate when the two areas (one relative to $M(\xi)$, the other to $Z(\eta)$) of a pair have a nonempty intersection. The final result is displayed in Figure 8.7. It turns out that four types of bound orbits exist.

(i) In the areas A_- and B_- we have $\xi_1 \le \xi \le \xi_2$, with $\xi_1, \xi_2 > 1$, while $-1 \le \eta \le 1$. The point P turns around F_1 and F_2, touching alternatively the two ellipses $\xi = \xi_1$ and $\xi = \xi_2$ and filling an "elliptic channel", as shown in Figure 8.8, where the same orbit for two different time intervals is depicted. We say that P is a planet of the system.

(ii) In the areas E_- and F_- we have $1 \le \xi \le \xi_2$, where ξ_2 is the only root > 1, while $-1 \le \eta \le 1$. The point P moves inside the ellipse $\xi = \xi_2$, passing across the y-axis also between F_1 and F_2. See Figure 8.9. We say that P is a satellite of the system.

(iii) In the area G_- we have $1 \le \xi \le \xi_2$, where ξ_2 is the only root > 1 of $M(\xi)$, while $-1 \le \eta \le \eta_1$, where η_1 is the only root of $Z(\eta)$ strictly contained in $(-1, 1)$. The point P turns around F_1 filling the area delimited by the ellipse $\xi = \xi_2$ and the hyperbola $\eta = \eta_1$. See Figure 8.10. We say that P is a satellite of F_1.

(iv) In the area H_- we have $1 \le \xi \le \xi_2$, where ξ_2 is the only root > 1 of $M(\xi)$, while $-1 \le \eta \le \eta_1$ or $\eta_2 \le \eta \le 1$, where $\eta_1 < \eta_2$ are the two roots of $Z(\eta)$ strictly contained in $(-1, 1)$. The point P moves inside the area, delimited by the ellipse $\xi = \xi_2$ and the hyperbola $\eta = \eta_1$, which contains F_1. Alternatively, it moves inside the area, delimited by the ellipse $\xi = \xi_2$ and the hyperbola $\eta = \eta_2$, which contains F_2. The two cases differ by the initial positions. See Figure 8.11. We say that P is a satellite of F_1 or F_2.

Positive Energy

For $H > 0$, to satisfy (8.3.2) we must consider the values of ξ external to the open interval (ξ_1, ξ_2) and the values of η belonging to the closed interval $[\eta_1, \eta_2]$. The relative table is now the following.

	A_+	B_+	C_+	D_+	E_+	F_+	G_+	H_+	I_+	L_+	M_+	N_+
$M(\xi)$	3	3	3	4	4	4	5	5	5	6	6	6
$Z(\eta)$	3	4	5	3	4	5	3	4	5	3	4	5

Then we investigate when the two areas (one relative to $M(\xi)$, the other to $Z(\eta)$) of each label A_+,B_+,$C_+ \dots$ have a nonempty intersection.

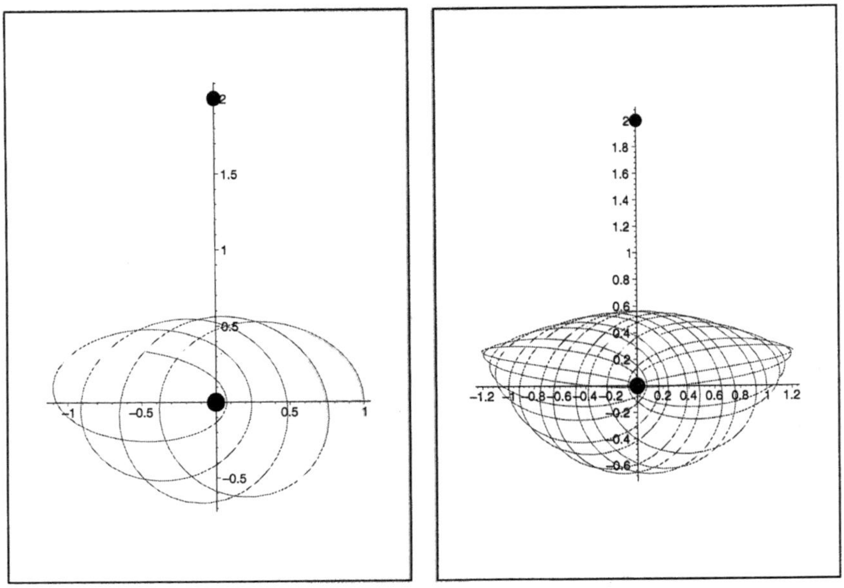

Figure 8.10: Satellite of the primary mass.

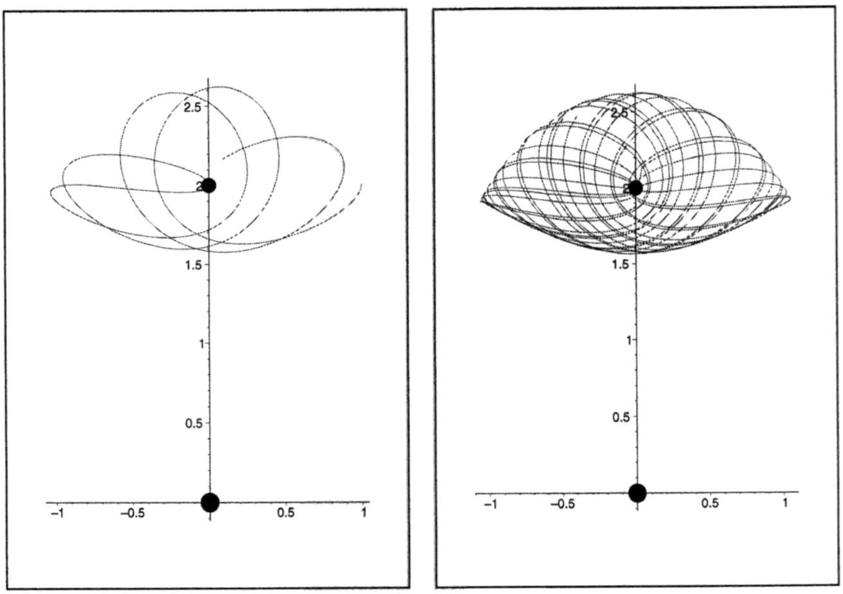

Figure 8.11: Satellite of the secondary mass.

The final result is displayed in Figure 8.7. It turns out that four types of unbound orbits exist.

(i) In the area A_+ we have $\xi_2 \le \xi < \infty$, where ξ_2 is the only root >1 of $M(\xi)$, while $-1 \le \eta \le 1$. The point P comes from infinity, turns around the ellipse $\xi = \xi_2$ and returns to infinity. See Figure 8.12a. We say that P is a comet of the system.

(ii) In the area G_+ we have $1 \le \xi < \infty$ and $-1 \le \eta \le 1$. These orbits have no particular characterization. See Figure 8.12b.

(iii) In the areas E_+ and H_+ we have $1 \le \xi < \infty$, while $\eta_1 \le \eta \le \eta_2$, where $\eta_1 < \eta_2$ are the two roots of $Z(\eta)$ contained in the open interval $(-1, 1)$. The point P travels from infinity to infinity, staying in a channel bound below by the hyperbola $\eta = \eta_1$ and above by the hyperbola $\eta = \eta_2$. See Figure 8.13a. We say that the orbit is a hyperbolic sinusoid.

(iv) In the area I_+ we have $1 \le \xi < \infty$, while $-1 \le \eta \le \eta_2$, where η_2 is the only root of $Z(\eta)$ in $(-1, 1)$. The point P comes from infinity, turns around F_1 without cutting the hyperbola $\eta = \eta_2$ and returns to infinity. See Figure 8.13b. We say that the P is a comet of F_1.

In the above qualitative analysis[(2)] we have supposed $m < 1$ (we recall that the mass at F_1 is equal to 1). If $m = 1$, the cases for which the mass at F_1 has a distinguished role (satellite and comet of F_1 only) disappear, the two straight lines (iv) and (v) of Figure 8.7 collapse into the one passing through the origin, and the hyperbola (ii) collapses into the axes.

If the two quartic polynomials have double roots, the two corresponding integrals can be expressed in terms of elementary functions; the program EULER is not able to treat these particular cases. The corresponding orbits may be divided into *asymptotic* and *singular* ones.

(i) The asymptotic orbits are those which separate two different types of motions. They correspond, on the plane yH of Figure 8.7, to the points of those lines separating two areas: the part of the straight line (vi) above the contact point with hyperbola (i); the part of the straight line (v) above the contact point with the hyperbola (ii); the part of the hyperbola (ii) separating E_- and H_-; the part of the straight line (iv) above the contact point with the hyperbola (ii). Notice that the part of the straight line (v) below the contact point with (ii) belongs rather to the following group.

(2) But we stress that the pictures of the orbits are obtained with *exact* analytical integration.

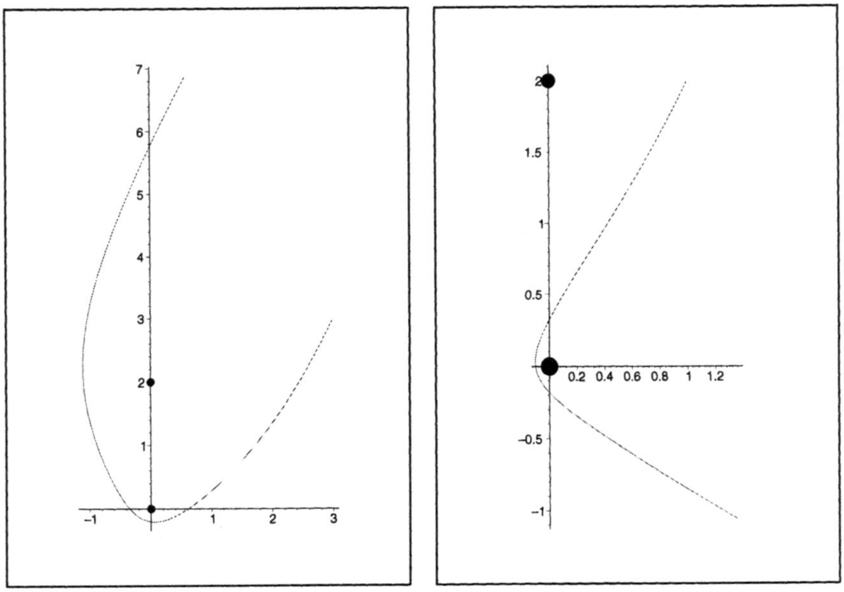

Figure 8.12: a) Comet of the system; b) Generic orbit.

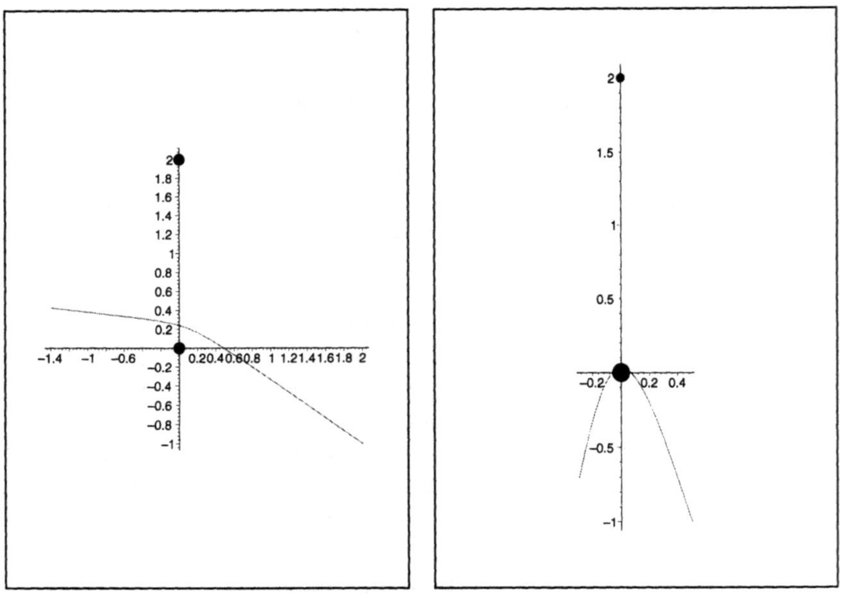

Figure 8.13: a) Hyperbolic sinusoid; b) Comet of the primary mass.

(ii) The singular orbits are those which correspond to a *constant* value of ξ or η, and consist obviously of (arcs of) ellipses or hyperbolas or, in the limit cases, of intervals of the straight lines $x = 0$ or $y = 1$. We may classify them in two groups.

(a) The representative point on the plane γH lies on the border of the admissible values: the arc of (i) above the contact point with (vi); the part of (vi) below the contact point with (i); the arc of (ii) above the contact point with (iv); the part of (iv) below the contact point with (ii).

(b) The representative point on the plane γH belongs, as for the asymptotic orbits, to one of the lines separating two areas of different types of motion but, moreover, the initial position satisfies some particular conditions, for example, lies on the y-axis.

8.3.2 Stark Problem

In the parabolic plane coordinates

$$x = \xi\eta, \quad y = \frac{1}{2}(\xi^2 - \eta^2)$$

choose $\underline{w}' = \frac{1}{2}\mathcal{E}\begin{pmatrix} \xi^4 \\ -\eta^4 \end{pmatrix}$ and $A_k = 0$, obtaining

$$H = \frac{1}{2}\frac{p_\xi^2 + p_\eta^2}{\xi^2 + \eta^2} - \frac{2}{\xi^2 + \eta^2} + \mathcal{E}\frac{\xi^2 - \eta^2}{2},$$

where $\mathcal{E}$ is a parameter fixing the value of the constant force.

To separate the variables, let us multiply by $\xi^2 + \eta^2$ and, after a reordering of the terms, we find the second first integral

$$H\xi^2 - \frac{1}{2}p_\xi^2 + 1 - \frac{1}{2}\mathcal{E}\xi^4 = \gamma,$$
$$H\eta^2 - \frac{1}{2}p_\eta^2 + 1 - \frac{1}{2}\mathcal{E}\eta^4 = -\gamma.$$

A direct calculation shows that

$$\gamma = E_y - \frac{1}{2}\mathcal{E}x^2,$$

which differs from the corresponding first integral of the Kepler Problem by the term $\frac{1}{2}\mathcal{E}x^2$.

APPENDIX

8.A Jacobian Elliptic Functions

The elementary trigonometric function $z = \sin\tau$ may be implicitly defined by

$$\tau = \int_0^z \frac{dz'}{\sqrt{1-z'^2}},$$

and all its properties (periodicity, range, composition rules, etc.) can be deduced from this definition. More generally, consider the *elliptic integral*

$$\tau = \int_0^z \frac{dz'}{\sqrt{(1-z'^2)(1-k^2z'^2)}}, \quad 0 < k < 1, \tag{8.A.1}$$

and define $z = \mathrm{sn}(\tau, k)$; k is called the *modulus*. Moreover, define

$$\mathrm{cn}^2(\tau,k) = 1 - \mathrm{sn}^2(\tau,k), \quad \mathrm{dn}^2(\tau,k) = 1 - k^2\mathrm{sn}^2(\tau,k).$$

sn, cn and dn are the Jacobian elliptic functions, a sort of generalized trigonometric functions. Put

$$K(k) = \int_0^1 \frac{dz'}{\sqrt{(1-z'^2)(1-k^2z'^2)}},$$

thus $\mathrm{sn}(\tau,k)$ and $\mathrm{cn}(\tau,k)$ have period $4K$, while $\mathrm{dn}(\tau,k)$ has period $2K$. Indeed, Definition (8.A.1) is equivalent to

$$\left(\frac{dz}{d\tau}\right)^2 = (1-z^2)(1-k^2z^2), \quad z(0,k) = 0,$$

from which we deduce $-1 \le z(\tau,k) \le 1$ (if we want $z \in \mathbb{R}$) and that an integral curve translated along the τ-axis remains still an integral curve, since the right-hand side does not depend explicitly on τ; the statement on the periodicity thus follows from the symmetries of the function to be integrated.

Let us consider the limiting values of the modulus. For $k = 0$, we have $K = \frac{\pi}{2}$ and $\mathrm{sn}(\tau,0) = \sin\tau$, $\mathrm{cn}(\tau,0) = \cos\tau$, $\mathrm{dn}(\tau,0) = 1$, while $K \to \infty$ for $k \to 1$ and $\mathrm{sn}(\tau,1) = \tanh\tau$, $\mathrm{cn}(\tau,1) = \mathrm{dn}(\tau,1) = \dfrac{1}{\cosh\tau}$. See Figure 8.14.

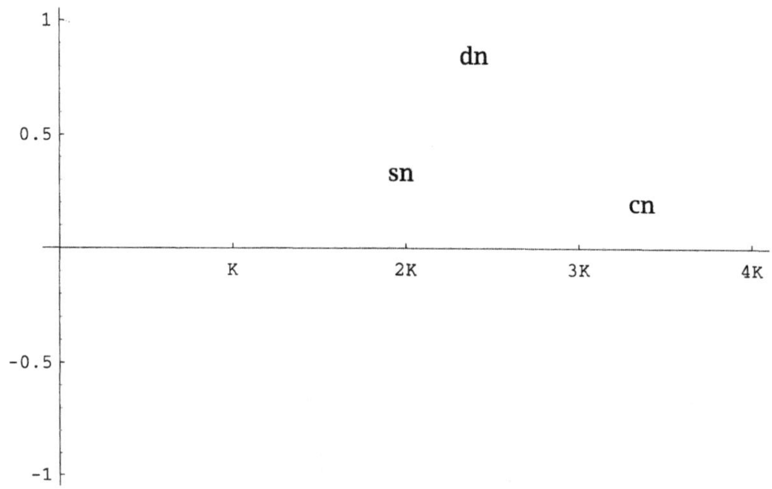

Figure 8.14: The Jacobian elliptic functions for $k = 0.9$.

This very short account is restricted to the real domain; for a deeper understanding one must consider the extension to the complex field, where the elliptic functions appear as meromorphic functions defined on a complex torus. See for example Chandrasekharan (1985) for a modern introduction to the argument. See also Dubrovin (1981) for applications of the (hyper)elliptic functions to integrable systems.

CHAPTER 9

Geometric Quantization

In Chapter 4 we have quantized the Kepler Problem using the Schrödinger method, i.e., we have found the eigenvalues and the eigenfunctions of the Hermitian operator

$$\hat{H} = \frac{1}{2}\hat{p}^2 - \frac{1}{q}, \quad \text{where} \quad \hat{p}_k = -i\frac{\partial}{\partial q^k}.$$

Subsequently, we have quantized the same problem using the Fock method, i.e., we have found the eigenvalues and the eigenfunctions of the Hermitian operator

$$\hat{\mathcal{L}}^{-1}\Phi(X) = \frac{\Gamma\left(\frac{n-1}{2}\right)}{2\pi^{\frac{n+1}{2}}}\int_{S^n}\frac{\Phi(X')}{\|X - X'\|^{n-1}}d\Omega', \tag{9.0.1}$$

where $\mathcal{L} = \frac{1}{\sqrt{-2H}}$ is the classical action of the Kepler Problem. We pointed out that the Moser regularization, described in Chapter 5, is, in some sense, the classical analogue of the quantum Fock method.

In this chapter we want to make precise this relation, exploiting the ideas of the so-called *geometric quantization* (Souriau 1997), (Kostant 1970), (Woodhouse 1980). The starting point of geometric quantization is the identification of the phase space of a Hamiltonian system with a coadjoint orbit of a Lie group, called a *dynamical group*, and of the relevant dynamical quantities (typically the energy and the first integrals) with the components of a moment map. In this way one can give a rigorous and mathematically well founded definition of the term "quantization": a unitary irreducible representation of the dynamical group on the Hilbert space L_2 of the square-integrable functions defined over the base space of a polarization of the coadjoint orbit[(1)], so that the corresponding Lie algebra acts by means of Hermitian operators. If one is able to exhibit this representation, to every relevant dynamical quantity, i.e., to every *observable*, is associated a Hermitian operator, and the algebra of the observables under Poisson brackets is isomorphic to the algebra of the Hermitian operators under commutation. In this setting the canonical quantization, that is the prescription $p_k \mapsto -i\partial_k$, turns out to be, a posteriori, a mere recipe, which sometimes, but not always, gives the right result.

With regard to the Kepler Problem, the first task of the above program has already been carried out, since in Chapter 6 we have obtained the Main Moment Map (6.4.5) (see also the commutative diagram on page 155). It is a moment map $T^+\mathbb{R}^n \to \mathfrak{c}^*$, or, after a canonical exchange between coordinates and momenta, $T^*(\mathbb{R}^n - \{0\}) \to \mathfrak{c}^*$. All the first integrals of the Kepler Problem (energy, angular momentum, Runge-Lenz-Laplace vector) appear as components of this moment map. In this chapter our task is to "quantize" it by constructing, for every observable G_{AB} (i.e., for every component of the moment map), a Hermitian operator $\hat{G}_{AB}$, acting on $L_2(\mathbb{R}^n)$, in such a way that

$$\left[\hat{G}_{AB}, \hat{G}_{AC}\right] = -i\eta_{AA}\hat{G}_{BC}$$

holds, which is the quantum version of the relation (6.1.5) on page 115.

The realization of this program is not straightforward, and this is due basically to the fact that the natural (local) action of $SO_0(2, n+1)$ is on $\mathbb{R}^{1,n}$, not on $\mathbb{R}^n$. This implies that the Main Moment Map is not linear in the momenta (indeed it contains the norm y of the momentum vector) and this is obviously an obstruction, if one would apply the recipe of the canonical quantization. Actually, one could notice that the Main Moment Map becomes linear and quadratic in the momenta after a canonical exchange between x_k and y_k, making it possible to apply the canonical quantization recipe once solved the problem of the ordering of the operators. In particular the square of the momentum vector is sent into the Laplacian operator. This is the route followed in Barut & Bornzin (1971), where the quantization of the Main Moment Map is achieved only for $n = 3$ in a purely computational way (i.e., with a "trial and error" method) which is not systematic and not entirely satisfying.

In order to quantize the Main Moment Map in a rational way, we shall use the theory of the representations of the semisimple noncompact Lie

[(1)]We stress however that this definition does not touch the physically fundamental point of the *interpretation* of quantum mechanics.

groups, which goes back to Wigner (1939) and to Bargmann (1946), reviewed in the first section. The surprising point is that the Fock operator $\hat{\mathfrak{L}}^{-1}$ arises quite naturally in this construction as an intertwining operator, which is the standard tool to make unitary a representation of the supplementary series (these terms are defined below). Moreover, the reader will notice that to achieve our goal we will follow very closely the construction of Chapter 6, confirming that this is the "right" way to treat the Kepler Problem.

The exposition of this chapter is taken from Cordani (1988) and Cordani (1989).

9.1 Multiplier Representations

Let us consider a Lie group $\mathfrak{G}$ acting on a manifold[(2)] M and the Hilbert space $L_2(M)$ of the square-integrable functions over M, with inner product $\langle f, g\rangle$, $f, g \in L_2(M)$. Let $a, b, \ldots$ be elements of $\mathfrak{G}$ and x^k local coordinates on M. With $\overline{x} = ax$ we mean that the action of $\mathfrak{G}$ sends the point with coordinates $x^1, x^2, \ldots$ into the point with coordinates $\overline{x}^1, \overline{x}^2, \ldots$. Obviously it must be that $a(bx) = (ab)x$, where ab is the group multiplication. A *representation* of $\mathfrak{G}$ is the assignment, to every element $a \in \mathfrak{G}$, of a linear operator $\hat{T}_a$, acting on $L_2(M)$, in such a way that

$$\hat{T}_a \hat{T}_b f = \hat{T}_{ab} f \quad \forall f \in L_2(M). \tag{9.1.1}$$

If $\left\langle \hat{T}_a f, \hat{T}_a g\right\rangle = \langle f, g\rangle$, i.e., if $\hat{T}_a$ is a unitary operator $\forall a \in \mathfrak{G}$, the representation preserves the inner product of $L_2(M)$, and will be said to be *unitary*. An element $f \in L_2(M)$ is said to be *smooth* if the map $f \mapsto \hat{T}_a f$ is smooth. The collection of smooth vectors is a $\mathfrak{G}$-invariant subspace which is usually not closed. The infinitesimal action of $\mathfrak{G}$ on these subspaces defines a Lie algebra representation of $\mathfrak{g}$, whose operators will be denoted by $\hat{A}$. Let us remark that if $\hat{A}$ is the operator corresponding to an element of $\mathfrak{g}$ generating the one-parameter subgroup $a(t) \in \mathfrak{G}$, then the action may be expressed in the form $\hat{T}_{a(t)} f = \exp(-it\hat{A}) f$. See for example Vogan (1987) for more details. If the representation is unitary, the operator $\hat{A}$ is selfadjoint, i.e.,

$$\left\langle f, \hat{A} g\right\rangle = \left\langle \hat{A} f, g\right\rangle.$$

Since every one-parameter subgroup is generated by an element of the algebra $\mathfrak{g}$ of the group $\mathfrak{G}$, there is a correspondence between the operators $\hat{A}$ and the elements of the Lie algebra $\mathfrak{g}$. This correspondence will

[(2)]For a moment, M is a generic manifold. Later on, it will be identified with the compactified Minkowski space and $\mathfrak{G}$ with $SO(2, n+1)$.

be referred to as *infinitesimal representation*, or simply *representation* when there is no risk of confusion.

The most natural example of a representation is the standard one, (also called *left-regular*) given by

$$\hat{T}_a f(x) = f(a^{-1}x),$$

with the corresponding infinitesimal representation

$$\hat{v}_a = -i v_a^k \partial_k,$$

$v_a^k \partial_k$ being the vector field induced by the infinitesimal action of $a(t) \in \mathfrak{G}$ on M. If the action of $\mathfrak{G}$ on M preserves some Riemannian metric, this representation is unitary with respect to the inner product

$$\langle f, g \rangle = \int_M f^* g \, dV, \quad f^* \text{ complex conjugate of } f,$$

which involves the volume form induced by the metric.

A generalization is the so-called *multiplier representation*, i.e.,

$$\hat{T}_a f(x) = \mu(a, a^{-1}x) f(a^{-1}x), \tag{9.1.2}$$

where μ is some function $\mathfrak{G} \times M \to \mathbb{R}^+$, or equivalently

$$\hat{T}_a f(x) = \mu(a, \overline{x}) f(\overline{x}), \quad \overline{x} = a^{-1}x.$$

PROPOSITION 9.1 *The function μ (called* multiplier*) must satisfy the two relations (e is the identity of $\mathfrak{G}$)*

$$\mu(e, x) = 1, \quad \mu(ab, x) = \mu(a, bx)\, \mu(b, x). \tag{9.1.3}$$

Proof. The two relations follow obviously from the composition property (9.1.1). The first relation follows from $\hat{T}_e f = f$. To obtain the second, set

$$f_1 = \hat{T}_b f, \quad f_2 = \hat{T}_a f_1$$

for an arbitrary function $f(x)$. From (9.1.2) we find

$$\begin{aligned} f_1(x) &= \mu(b, b^{-1}x)\, f(b^{-1}x), \\ f_2(y) &= \mu(a, a^{-1}y)\, f_1(a^{-1}y) \\ &= \mu(a, a^{-1}y)\, \mu(b, b^{-1}a^{-1}y)\, f(b^{-1}a^{-1}y). \end{aligned}$$

The function $f_2(y) = \hat{T}_a \hat{T}_b f$ coincides with

$$f_2(y) = \hat{T}_{ab} f = \mu(ab, b^{-1}a^{-1}y)\, f(b^{-1}a^{-1}y)$$

if and only if the second relation holds, as is seen by inserting $x = b^{-1}a^{-1}y$ and $bx = a^{-1}y$. QED

The following properties of the multipliers are straightforward. Any complex power of a multiplier is again a multiplier and, in particular, the inverse of a multiplier is a multiplier. The product of two multipliers is a multiplier. This shows that the multipliers themselves form a group with respect to the multiplication. Finally, on setting $b = a^{-1}$ in (9.1.3), we find

$$\mu(a,\overline{x})\,\mu(a^{-1},x) = 1. \tag{9.1.4}$$

Let us put

$$\hat{T}_a f(x) = \mu^h(a,\overline{x})f(\overline{x}), \quad h \in \mathbb{C}, \tag{9.1.5}$$

(h has nothing to do with the Planck constant). To find the infinitesimal representation, let us consider an element a very close to the group identity. Put

$$\tau_a(x) = \left.\frac{\partial\mu(a,x)}{\partial a}\right|_{a=e};$$

τ is called *infinitesimal multiplier.* Thus, up to second order terms,

$$\hat{T}_a = 1 - it\hat{A} = 1 + t\left(h\tau_a - v_a^k\partial_k\right),$$

from which

$$\hat{A} = \hat{v}_a + ih\tau_a. \tag{9.1.6}$$

Denote by $\hat{G}_k$ the infinitesimal operator corresponding to the k-th element of a basis of $\mathfrak{g}$. Then

$$[\hat{G}_h, \hat{G}_k] = -ic^j_{hk}\hat{G}_j,$$

where c^j_{hk} are the structure constants of the Lie algebra $\mathfrak{g}$.

REMARK 9.1 In the recent literature the multiplier representations are known as *induced representations.* There is in fact a one-to-one correspondence between them, as shown, for example, in Barut & Raczka (1980, page 484). Moreover it can be shown that every unitary representation of a real semisimple group can be obtained as some induced representation.

Let us now examine when a multiplier representation is unitary. The fundamental points (Gel'fand & Shilov 1964), (Barut & Raczka 1980), (Knapp 1986) are the following. For a multiplier representation of the type (9.1.5), the exponent can be written as $h = N + \lambda$, with N a real integer, depending on the dimensions of the group[3], and λ a complex number such that the representation is unitary if $\lambda^* = -\lambda$. In this case, i.e., if λ is purely imaginary, the representation is said to belong to the *principal series*, otherwise, i.e., if λ is real, to the *supplementary series*. This last can be made unitary, if one is able to define an *intertwining operator* $\hat{I}_\lambda$ such that

$$\hat{I}_\lambda \hat{R}_\lambda = \hat{R}_{-\lambda} \hat{I}_\lambda ,$$

where $\hat{R}_\lambda$ stands for each of the operators (9.1.5). In fact, one proves that the representation $\hat{R}_\lambda$ becomes unitary with respect to the new inner product

$$\langle f, g \rangle_\lambda \stackrel{\text{def}}{=} \left\langle f, \hat{I}_\lambda g \right\rangle .$$

9.2 Quantization of the Geodesic Flow on the Sphere

Following the analysis, pursued at classical level, of the Chapter 6, we firstly quantize the geodesic flow on the sphere S^n, this last being just the base space of the natural polarization of the coadjoint orbit $T^+S^n \subset \mathrm{so}^*(2, n+1)$. In other words, we construct an (infinitesimal) irreducible representation of $\mathrm{SO}_0(2, n+1)$, by means of operators acting on the Hilbert space $L_2(S^n)$. This representation, however, turns out to be unitary only when restricted to the maximal compact subgroup $\mathrm{SO}(2) \times \mathrm{SO}(n+1)$. To construct it, we can not apply directly (9.1.2) or (9.1.5), since the natural action of $\mathrm{SO}_0(2, n+1)$ is on $S^1 \times S^n$. Rather, we apply (9.1.5) but considering this natural action, obtaining a multiplier representation of the supplementary series with operators acting on $L_2(S^1 \times S^n)$. For a judicious choice of the parameter λ, this representation turns out to be reducible and two (isomorphic to each other) of the irreducible components are just the representation we are looking for.

We start from those functions $F : \mathcal{K} \to \mathbb{C}$, where $\mathcal{K}$ is the null cone

$$\mathcal{K} = \{X \in \mathbb{R}^{2,n+1} : \eta(X, X) = 0\}, \quad \eta = \mathrm{diag}(- - + \cdots +),$$

which can be restricted on $\widetilde{M}$, the manifold of null oriented lines diffeomorphic to $S^1 \times S^n$. The null cone $\mathcal{K}$ can be parametrized as

$$\begin{aligned} X^{-1} &= r \cos x^0, \\ X^0 &= r \sin x^0, \\ X^\alpha &= r \Gamma^\alpha(x^k), \end{aligned}$$

[3]We are vague about this and the following point, but when we examine the concrete case of our interest, the integer N, with the successively defined *intertwining operator*, will acquire a well defined geometrical meaning.

with $\Gamma^\alpha \Gamma_\alpha = 1$ and $r > 0$. Specifying r to be some function $r_\gamma(x^\mu)$, we fix a cross-section γ of the null cone. We may identify $\widetilde{M}$ with the section $r \equiv 1$. The functions F are homogeneous functions of generic degree $-h \in \mathbb{C}$ (the minus sign has been inserted to simplify some of the following formulae) in the variables X^A :

$$F(X) = r^{-h} f(x), \quad x = x^0, \ldots, x^n.$$

To obtain a multiplier representation $\hat{T}_a$ acting on $L_2(\widetilde{M})$, we start from the standard representation $\hat{T}_a^0$, acting on $L_2(\mathcal{K})$,

$$\hat{T}_a^0 F(X) = F(a^{-1}X) = (a^{-1}r)^{-h} f(a^{-1}x).$$

Since the action of $SO_0(2, n+1)$ on $\mathbb{R}^{2,n+1}$ is linear and transforms each null line into another null line, we have the map

$$r \mapsto a^{-1}r = \mu(a^{-1}, x)\, r, \tag{9.2.1}$$

where μ is some function. The representation $\hat{T}_a$ that we seek is, by definition, that making the diagram

$$\begin{array}{ccc} F(X) & \xrightarrow{\hat{T}_a^0} & F(a^{-1}X) \\ {\scriptstyle r=1}\big\downarrow & & \big\downarrow{\scriptstyle r=1} \\ f(x) & \xrightarrow{\hat{T}_a} & \mu^{-h}(a^{-1},x) f(a^{-1}x) \end{array}$$

commutative, from which it is clear that μ is a multiplier. For (9.1.4) we may write

$$\hat{T}_a f(x) = \mu^h(a, \overline{x}) f(\overline{x}),$$

which is a multiplier representation of type (9.1.5).

Each point $P \in \widetilde{M}$ is transformed by the action of a^{-1} into the point $P_\gamma \in M_\gamma$ (as shown in Figure 9.1), to which corresponds the point $\overline{P} \in \widetilde{M}$. The coordinates in $\mathbb{R}^{2,n+1}$ of these points are

$$\begin{aligned} P \equiv \{X^A\} &= \{\cos x^0, \sin x^0, \Gamma^\alpha(x^k)\}, \\ P_\gamma \equiv \{X_\gamma^A\} &= \{r_\gamma(a,\overline{x}) \cos \overline{x}^0, r_\gamma(a,\overline{x}) \sin \overline{x}^0, r_\gamma(a,\overline{x}) \Gamma^\alpha(\overline{x}^k)\}, \\ \overline{P} \equiv \{\overline{X}^A\} &= \{\cos \overline{x}^0, \sin \overline{x}^0, \Gamma^\alpha(\overline{x}^k)\}. \end{aligned}$$

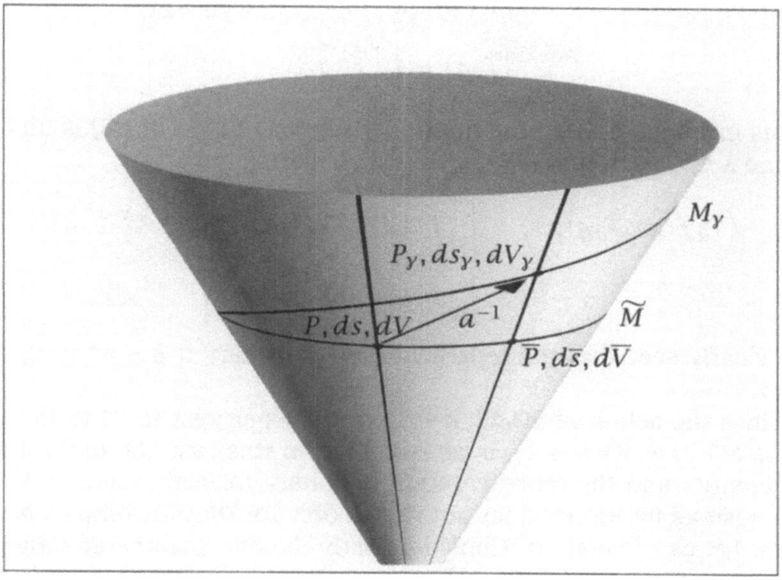

Figure 9.1: The conformal action on the null cone.

The transformation $P \xrightarrow{a^{-1}} \overline{P}$ is a conformal transformation, as described in Chapter 6. Anyway, this is easily verified, bearing in mind that the action of $a^{-1} \in SO_0(2, n+1)$ is an isometry of $\mathbb{R}^{2,n+1}$; thus

$$ds^2 = \eta(dX, dX) = ds_\gamma^2 = \eta(dX_\gamma, dX_\gamma) = r_\gamma^2(a, \overline{x})\, d\overline{s}^2.$$

Since $\dim \widetilde{M} = n + 1$, the above expression shows that the relations among the volume forms are

$$dV = dV_\gamma = r_\gamma^{n+1}(a, \overline{x})\, d\overline{V}.$$

From (9.2.1) and (9.1.4), since $\widetilde{M}$ is the section of the null cone characterized by $r \equiv 1$ we can write

$$r_\gamma(a, \overline{x}) = \frac{1}{\mu(a, \overline{x})},$$

hence

$$d\overline{V} = \mu^{n+1}(a, \overline{x})\, dV.$$

This shows how the volume form of $\widetilde{M}$ changes under the conformal action.

The natural inner product of the Hilbert space $L_2(\widetilde{M})$ is

$$\langle f, g \rangle = \int_{\widetilde{M}} f^* g \, dV.$$

Let us investigate when the multiplier representation (9.1.5) is unitary. Define $\lambda \in \mathbb{C}$ such that $h = \frac{n+1}{2} + \lambda$. Calculating

$$\begin{aligned}\left\langle \hat{T}_a f(x), \hat{T}_a g(x) \right\rangle &= \int_{\widetilde{M}} f^*(\overline{x})\, g(\overline{x})\, [\mu(a, \overline{x})]^{n+1+\lambda+\lambda^*} \, dV \\ &= \int_{\widetilde{M}} f^*(\overline{x})\, g(\overline{x})\, [\mu(a, \overline{x})]^{\lambda+\lambda^*} \, d\overline{V},\end{aligned}$$

one clearly sees that the representation is unitary if $\lambda + \lambda^* = 0$, or if $\mu \equiv 1$.

Since the action of $SO_0(2, n+1)$ on $\widetilde{M}$, when restricted to the subgroup $SO(2) \times SO(n+1)$, is an isometry, in this case the multiplier is the identity and the representation is unitary for each value of λ. The infinitesimal multiplier does not vanish only for the noncompact generators. Let us calculate it. Consider firstly the one-parameter subgroup generated by $G_{-1,1}$: for the other generators $G_{-1,\alpha}$, $\alpha = 2, \dots, n+1$ all runs very similarly. The explicit action of this subgroup on the null cone is given by

$$\begin{pmatrix} \cosh s & 0 & \sinh s & & & \\ 0 & 1 & 0 & & & \\ \sinh s & & \cosh s & & & \\ & & & 1 & & \\ & & & & \ddots & \\ & & & & & 1 \end{pmatrix} \begin{pmatrix} r\cos x^0 \\ r \sin x^0 \\ r\Gamma^1 \\ r\Gamma^2 \\ \vdots \\ r\Gamma^{n+1} \end{pmatrix} = \begin{pmatrix} r\cos x^0 \cosh s + r\Gamma^1 \sinh s \\ r\sin x^0 \\ r \cos x^0 \sinh s + r\Gamma^1 \cosh s \\ r\Gamma^2 \\ \vdots \\ r\Gamma^{n+1} \end{pmatrix},$$

from which

$$\begin{aligned} r_\gamma^2 &= (\cos x^0 \cosh s + \Gamma^1 \sinh s)^2 + \sin^2 x^0 \\ &= (\cos x^0 \sinh s + \Gamma^1 \cosh s)^2 + (\Gamma^2)^2 + \cdots + (\Gamma^{n+1})^2. \end{aligned}$$

The multiplier is the square root of the right-hand side, so that the infinitesimal multipliers turn out to be

$$\tau_{-1,1} = \left.\frac{\partial \mu}{\partial s}\right|_{s=0} = \cos x^0 \Gamma^1,$$

while, for the other generators,

$$\tau_{-1,\alpha} = \cos x^0 \Gamma^\alpha.$$

Arguing in a similar way, one obtains

$$\tau_{0,\alpha} = \sin x^0 \Gamma^\alpha.$$

As said before, the multipliers relative to $SO(2) \times SO(n+1)$ are null

$$\tau_{-1,0} = 0 = \tau_{\alpha\beta}.$$

We are now ready to find the multiplier representation of $SO_0(2, n+1)$ acting on $L_2(S^1 \times S^n)$. Remembering that the cross-section $M_{(-)}$ on page 132 is just $\widetilde{M}$, we put

$$\begin{aligned}
\Gamma^{-1} &= \cos x^0, \\
\Gamma^0 &= \sin x^0, \\
\Gamma^k &= \frac{2x^k}{x^2+1}, \\
\Gamma^{n+1} &= \frac{x^2-1}{x^2+1},
\end{aligned}$$

and, taking into account the expressions of the corresponding Π,

$$\begin{aligned}
\hat{\Pi}_{-1} &= -\sin x^0\, \hat{y}_0, \\
\hat{\Pi}_0 &= \cos x^0\, \hat{y}_0, \\
\hat{\Pi}_k &= \frac{1}{2}(x^2+1)\hat{y}_k - x_k x^h \hat{y}_h, \\
\hat{\Pi}_{n+1} &= x^h \hat{y}_h,
\end{aligned}$$

where $\hat{y}_\mu = -i\frac{\partial}{\partial x^\mu}$. From that we can write

$$\hat{v}_{AB} = \Gamma_B \hat{\Pi}_A - \Gamma_A \hat{\Pi}_B,$$

and for (9.1.6) the multiplier representation is given by

$$\hat{G}_{AB} = \hat{v}_{AB} + i\left(\frac{n+1}{2} + \lambda\right)\tau_{AB}.$$

The previous work ensures that the commutators of the operators $\hat{G}_{AB}$ close up, yielding the Lie algebra of the conformal group. We are mainly interested in the noncompact ones, which is useful, in what follows, to write as $\hat{G}^{\pm}_{\alpha} = \hat{G}_{-1,\alpha} \pm i\hat{G}_{0,\alpha}$. The corresponding vector fields and infinitesimal multipliers are

$$\hat{v}^{\pm}_{\alpha} = e^{\pm ix^0}\left(\hat{\Pi}_\alpha \pm i\Gamma_\alpha \hat{y}_o\right), \quad \tau^{\pm}_{\alpha} = e^{\pm ix^0}\Gamma_\alpha.$$

As a basis for the representation space we assume the functions $\Phi : S^1 \times S^n \to \mathbb{C}$ of the type

$$\Phi_{\ell m} = h_\ell(\Gamma) e^{imx^0}, \quad \ell = 0, 1, 2, \ldots,$$
$$m = \begin{cases} \ldots -\frac{3}{2}, -\frac{1}{2}, \frac{1}{2}, \frac{3}{2}, \ldots & (n \text{ even}) \\ \ldots -2, -1, 0, 1, 2, \ldots & (n \text{ odd}) \end{cases} \tag{9.2.2}$$

where h_ℓ is a harmonic homogeneous polynomial of degree ℓ (see Section 4.A.2 on page 89). We recall that the harmonic homogeneous polynomials restricted to the sphere are the spherical harmonics, a complete orthogonal set of functions.

It is easy to verify that

$$\hat{\Pi}_\alpha \Gamma_\beta = -i(\delta_{\alpha\beta} - \Gamma_\alpha \Gamma_\beta),$$

so that, because of the homogeneity of h_ℓ,

$$\hat{\Pi}_\alpha h_\ell = \frac{\partial h_\ell}{\partial \Gamma_\beta} \hat{\Pi}_\alpha \Gamma_\beta = -i\left(\frac{\partial}{\partial \Gamma_\alpha} - \ell \Gamma_\alpha\right) h_\ell.$$

When applied to an element $\Phi_{\ell m}$ of the basis of the representation space, the operators $\hat{G}^\pm_\alpha$ can be written as

$$\hat{G}^\pm_\alpha = -ie^{\pm ix^0}\left[\frac{\partial}{\partial \Gamma_\alpha} - \left(\pm m + \ell + \frac{n+1}{2} + \lambda\right)\Gamma_\alpha\right]. \tag{9.2.3}$$

Define $\mathcal{F}_{\ell m}$ as the space of the functions of the type (9.2.2). For each fixed pair (ℓ, m), $\mathcal{F}_{\ell m}$ is an invariant subspace under the action of the maximal compact subgroup. Define $\mathcal{F} = \bigoplus_{\ell m} \mathcal{F}_{\ell m}$. The crucial point for the construction of an irreducible representation acting over $L_2(S^n)$ is contained in the following theorem.

THEOREM 9.2 *For $\lambda = -1$ the two subspaces of $\mathcal{F}$ determined by the pairs $(\ell, m^\spadesuit)$ and $(\ell, m^\clubsuit)$, with*

$$m^\spadesuit = \ell + \frac{n-1}{2}, \quad m^\clubsuit = -m^\spadesuit,$$

are invariant under the action of the operators (9.2.3), thus of the whole representation, and are manifestly isomorphic to the space of the functions on S^n.

Proof. Let us consider the case $(\ell, m^\spadesuit)$, the proof for the other being similar. Examine the action of the generators

$$\hat{G}^-_\alpha = -ie^{-ix^0}\frac{\partial}{\partial \Gamma_\alpha}$$

over $\mathcal{F}_{\ell m^\spadesuit}$. Since $\frac{\partial}{\partial \Gamma_\alpha}$ commutes with the Laplacian $\frac{\partial}{\partial \Gamma_\beta}\frac{\partial}{\partial \Gamma^\beta}$, we have that $\frac{\partial h_\ell}{\partial \Gamma_\alpha}$ is a harmonic homogeneous polynomial of degree $(\ell - 1)$. Therefore

$$\hat{G}_\alpha^- \mathcal{F}_{\ell m^\spadesuit} \subset \mathcal{F}_{\ell-1, m^\spadesuit-1}, \quad \forall \ell.$$

Consider now the action of $\hat{G}_\alpha^+$. Since $\Gamma_\beta \Gamma^\beta = 1$, we may write

$$\hat{G}_\alpha^+ = -ie^{ix^0}\left[\Gamma_\beta \Gamma^\beta \frac{\partial}{\partial \Gamma_\alpha} - 2\left(\ell + \frac{n-1}{2}\right)\Gamma_\alpha\right].$$

It is clear that the operator in the square bracket, when applied to a harmonic homogeneous polynomial of degree ℓ, gives a homogeneous polynomial of degree $\ell + 1$. Let us verify that this last is also harmonic. Recalling that

$$\Delta(fg) = g\Delta f + f\Delta g + 2\nabla f \cdot \nabla g,$$

we find

$$\begin{aligned}
&\frac{\partial}{\partial \Gamma^\gamma}\frac{\partial}{\partial \Gamma_\gamma}\left[\Gamma_\beta \Gamma^\beta \frac{\partial h_\ell}{\partial \Gamma_\alpha} - 2\left(\ell + \frac{n-1}{2}\right)\Gamma_\alpha h_\ell\right] \\
&\qquad = 2(n+1)\frac{\partial h_\ell}{\partial \Gamma_\alpha} + 4\Gamma^\beta \frac{\partial}{\partial \Gamma_\beta}\frac{\partial h_\ell}{\partial \Gamma_\alpha} - 4\left(\ell + \frac{n-1}{2}\right)\frac{\partial h_\ell}{\partial \Gamma_\alpha} \\
&\qquad = \left[2(n+1) + 4(\ell - 1) - 4\ell - 2(n-1)\right]\frac{\partial h_\ell}{\partial \Gamma_\alpha} = 0.
\end{aligned}$$

This proves that

$$\hat{G}_\alpha^+ \mathcal{F}_{\ell m^\spadesuit} \subset \mathcal{F}_{\ell+1, m^\spadesuit+1} \quad \forall \ell.$$

The noncompact part of the representation allows one to jump from $\mathcal{F}_{\ell m^\spadesuit}$ to $\mathcal{F}_{\ell-1, m^\spadesuit-1}$ or to $\mathcal{F}_{\ell+1, m^\spadesuit+1}$, while the compact part acts transitively on every $\mathcal{F}_{\ell m^\spadesuit}$. This proves the theorem. QED

The content of the theorem is graphically displayed in the following picture

⋮	♣	♡	◇	♡	◇	⋯	◇	♡	◇	♡	♠
3	♡	♣	♡	◇	♡	⋯	♡	◇	♡	♠	♡
2	◇	♡	♣	♡	◇	⋯	◇	♡	♠	♡	◇
1	♡	◇	♡	♣	♡	⋯	♡	♠	♡	◇	♡
0	◇	♡	◇	♡	♣	⋯	♠	♡	◇	♡	◇
⇑ ℓ			⋯	$\frac{1-n}{2}-1$	$\frac{1-n}{2}$	⋯	$\frac{n-1}{2}$	$\frac{n-1}{2}+1$	⋯		

$m \Rightarrow$

where each suit represents one of the subspaces $\mathcal{F}_{\ell m}$. The sets of the spades and of the clubs represent the two invariant subspaces of Theorem 9.2. The other suits, i.e., the red ones, may in turn be split in two groups. Start in fact from an arbitrary pair (ℓ, m), with $m \neq \pm m^{\spadesuit}$. In this case, the operators $\hat{G}^{\pm}_{\alpha}$, when applied to $h_\ell(\Gamma)$, produce terms of the type $\Gamma_\alpha\, h_\ell(\Gamma)$, which are homogeneous (but not harmonic) polynomials of degree $\ell + 1$. They are decomposable into homogeneous harmonic polynomials of degree $\ell + 1, \ell - 1, \ell - 3, \ldots$. The result is displayed in the following picture

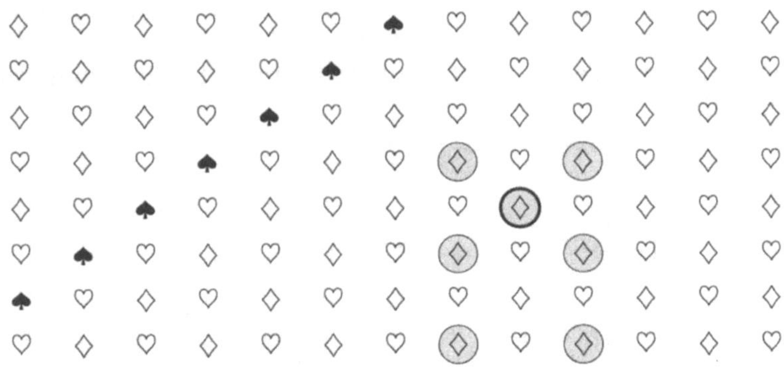

If the starting point is the shaded suit enclosed in the fat circle, the action of one of the operators $\hat{G}^{\pm}_{\alpha}$ allows us to reach the other shaded suits. Further applications extend their number, but obviously, if one starts from a diamonds (hearts) suit only diamonds (hearts) suits can be reached. The representation is thus reducible, splitting into two disjoint invariant subspaces. Each one may contain an invariant subspace (i.e., one from Theorem 9.2), whose complementary subspace however is not invariant: starting from a diamonds suit one may reach a spades suit. Notice that if one starts, as in the above picture, from a diamonds suit placed in the right-below zone with respect to the spades line, this last becomes an insurmountable frontier, giving rise to another invariant subspace. Thus the representation is a so-called *indecomposable* one, being reducible but not fully reducible.

We claim to have obtained an almost unitary[4] representation of the conformal group by means of pseudo-differential operators acting on $L_2(S^n)$. In fact, consider, for example, the operator $\hat{G}_{-1,0} = \hat{y}_0$, corresponding to the classical action $\mathcal{L}$. The eigenvalues of $\hat{G}_{-1,0}$, acting on $\mathcal{F}_{\ell m^{\spadesuit}}$, are $\left(\ell + \frac{n-1}{2}\right)$ with multiplicity $\frac{(2\ell+n-1)(\ell+n-2)!}{\ell!(n-1)!}$ and equal those of $\sqrt{-\Delta_{S^n} + \left(\frac{n-1}{2}\right)^2}$, acting on $L_2(S^n)$. The use of pseudo-differential

[4] In the sense that only the representation of the subgroup $\mathrm{SO}(2)\times\mathrm{SO}(n+1)$ is unitary.

operators is obviously due to the fact that our dynamical group does not act effectively on S^n. Theorem 9.2 is the representation-theoretic analogue of the reduction $T^+M \to T^+S^n$ described in Chapter 6. In fact, the two subspaces of Theorem 9.2 are annihilated by the operators

$$-i\frac{\partial}{\partial x^0} \pm \sqrt{-\Delta_{S^n} + \left(\frac{n-1}{2}\right)^2}$$

and this is the "quantum" analogue of the "classical" conditions

$$y_0 \pm \frac{1}{2}y(x^2+1) = 0,$$

giving the restriction to null covectors.

REMARK 9.2 There are several papers concerning the computation of the spectrum and multiplicity of the energy operator of the geodesic motion on a sphere. We refer to Simms (1974) for the 3-dimensional case and Mladenov & Tsanov (1985) for the n-dimensional one. However in these works only the so-called "pre-quantization" is carried out. Results similar to those exposed in this section have been found by Akyildiz (1980) and Malkin & Man'ko (1963), but in a different framework. The philosophy of our approach is instead reminiscent of that of Onofri (1976); owing to the difficulty to quantize the Kepler manifold T^+S^n, in both cases one starts from another manifold, that can be easily quantized. There is a crucial technical difference: we start from a reducible representation of the supplementary series (which quantizes T^+M), Onofri starts instead from the discrete series, which quantizes the Kaehler manifold $\frac{SO(2,n+1)}{SO(2)\times SO(n+1)}$. To obtain, with a limiting process, the supplementary series from the discrete one is not an easy task, which seems difficult to be realized in a rigorous way; in fact a choice, basically equivalent to our $\lambda = -1$, is made by Onofri in a heuristic way.

9.3 Quantization of the Kepler Problem

At this point it is natural to ask: what is the precise relation between the Fock and the classical procedure? The Fock procedure seems to be very closely related to the canonical transformation $\mathscr{k}$ on page 138, but, for example, why must we redefine the wave function as in (4.3.4)?

In this section we implement the quantization of $\mathscr{k}$. The difficult task concerns the interchange between coordinates and momenta. To be more precise, we want to find how the representation on S^n of the previous section is transformed when in the Main Moment Map (6.4.5) we interchange coordinates and momenta.

It is well known that to the interchange between canonical coordinates and momenta corresponds a Fourier transform at quantum level. But we cannot perform directly this transform, since the wave function is defined on S^n, rather than on $\mathbb{R}^n$. It would be possible to obtain a representation on $\mathbb{R}^n$ following a reasoning analogous to that of the previous section, but in this way we meet a difficulty: the generators that turn out to be Hermitian would be now the generators of the isometry subgroup of $\mathbb{R}^n$, i.e., of the Euclidean subgroup only. The Hamiltonian $\hat{G}_{-1,0}$ of the Kepler Problem for negative energy would not be Hermitian and this is obviously a disaster. The only way to escape this difficulty is to make *fully* unitary the representation. At first sight, it seems we fall into another problem: a general method to obtain this does exists (for what concerns us) for the Lorentz subgroup $\mathrm{SO}_0(1, n+1)$ only. Luckily, as we shall see, this will prove to be sufficient.

Let us view how to find and make fully unitary a representation of the Lorentz group belonging to the supplementary series. We may repeat almost all we have said on $\mathrm{SO}_0(2, n+1)$, with only some minor adjustments to take into account the change of dimensions. The Lorentz group acts linearly on $\mathbb{R}^{1,n+1}$ and preserves the null cone $\eta(X, X) = 0$, where $\eta = (-, +, \ldots, +)$. It acts thus conformally on the manifold of the null rays through the origin, which is homeomorphic to S^n, the compactification of $\mathbb{R}^n$. If $a \in \mathrm{SO}_0(1, n+1)$ and $x \in S^n$, the multiplier representation $\hat{R}_\rho$ is given by

$$\hat{T}_a \psi(x) = \nu^{\frac{n}{2}+\rho} (a, \overline{x})\, \psi(\overline{x}),$$

where $\nu(a, x)$ is the multiplier of the representation and ρ an arbitrary complex number. The natural scalar product is obviously given by

$$\langle \psi, \phi \rangle = \int_{S^n} \psi^* \phi \, d\Omega,$$

where $d\Omega$ is the Riemannian volume element of the unit sphere S^n. Thus $d\overline{\Omega} = \nu^n (a, \overline{x})\, d\Omega$ and $\hat{R}_\rho$ is unitary if ρ is purely imaginary:

$$\langle \hat{R}_\rho \psi, \hat{R}_\rho \phi \rangle = \int_{S^n} \psi^* (\overline{x}) \phi(\overline{x})\, \nu^{\rho+\rho^*} d\overline{\Omega} = \langle \psi, \phi \rangle.$$

Consider now the case ρ real. A representation of the supplementary series $\hat{R}_\rho$ can be made unitary by means of an intertwining operator $\hat{I}_\rho$. This latter is defined by the property

$$\hat{I}_\rho \hat{R}_\rho = \hat{R}_{-\rho} \hat{I}_\rho.$$

Finding such an operator is equivalent to defining a new scalar product in the Hilbert space of the representation:

$$\langle \psi, \phi \rangle_\rho \stackrel{\text{def}}{=} \langle \psi, \hat{I}_\rho \phi \rangle, \tag{9.3.1}$$

so that $\hat{R}_\rho$ becomes fully unitary, as is easily verified:

$$\begin{aligned}\langle \hat{R}_\rho \psi, \hat{R}_\rho \phi \rangle_\rho &= \langle \hat{R}_\rho \psi, \hat{I}_\rho \hat{R}_\rho \phi \rangle = \langle \hat{R}_\rho \psi, \hat{R}_{-\rho} \hat{I}_\rho \phi \rangle \\ &= \int_{S^n} (\hat{R}_\rho \psi)^* \hat{R}_{-\rho} \hat{I}_\rho \phi \, d\Omega \\ &= \int_{S^n} \psi^*(\overline{x}) \hat{I}_\rho \phi(\overline{x}) \, \nu^{\rho-\rho} d\overline{\Omega} = \langle \psi, \phi \rangle_\rho .\end{aligned}$$

With the following proposition we can explicitly calculate the intertwining operator. Consider the null cone in $\mathbb{R}^{1,n+1}$ and let Γ and Γ' be two points belonging to the intersection with the hyperplane $X^0 = 1$, thus to the unitary sphere S^n.

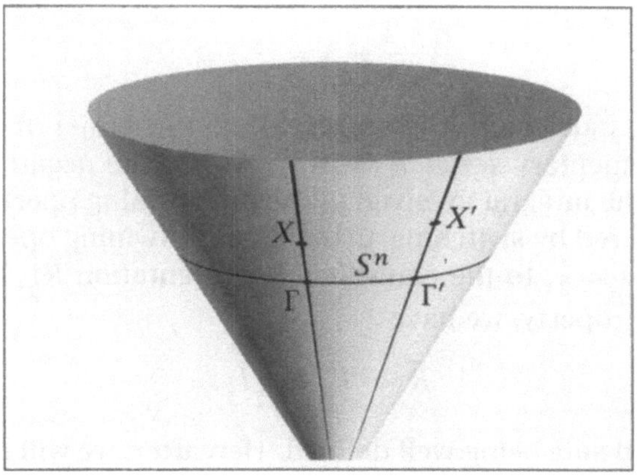

Figure 9.2: The null cone in $\mathbb{R}^{1,n+1}$.

PROPOSITION 9.3 *For the Lorentz group* $SO_0(1, n+1)$ *the intertwining operator is given by*

$$\hat{I}_\rho \phi(\Gamma) = \int_{S^n} \frac{\phi(\Gamma')}{\|\Gamma - \Gamma'\|^{n-2\rho}} d\Omega',$$

where $\|\cdot\|$ *is the Euclidean norm in* $\mathbb{R}^{n+1}$ *(indeed,* $\Gamma - \Gamma'$ *belongs to the Euclidean subspace of* $\mathbb{R}^{1,n+1}$*).*

Proof. See Figure 9.2. Let X and X' be two points on the null lines through Γ and Γ', respectively. Put $X = r\Gamma$ and $X' = r'\Gamma'$. Define the inner product

$$\langle \psi, \phi \rangle_\rho = \int_{S^n} \psi^*(\Gamma) \, d\Omega \int_{S^n} \left[-2 \frac{\eta(X, X')}{r r'} \right]^{\rho - \frac{n}{2}} \phi(\Gamma') \, d\Omega'.$$

Taking into account the invariance of $\eta(X, X')$ under the group action and that $\overline{r} = \frac{r}{\nu(a,\overline{x})}$, one comes to the conclusion that the representation $\hat{R}_\rho$ is unitary with respect to this inner product. But

$$-2\frac{\eta(X,X')}{r\,r'} = -2\eta(\Gamma,\Gamma') = \eta(\Gamma - \Gamma', \Gamma - \Gamma') = \|\Gamma - \Gamma'\|^2,$$

and, by (9.3.1), we obtain the statement. QED

Let us consider the irreducible representation of $SO_0(2, n+1)$, found in the previous section, and restrict to the Lorentz subgroup $SO_0(1, n+1)$. This restriction is obtained by excluding the $n+2$ operators $\hat{G}_{-1,0}$ and $\hat{G}_{0,\alpha}$, i.e., those realized in pseudo-differential form. Since the relation

$$\frac{n+1}{2} + \lambda = \frac{n}{2} + \rho,$$

with $\lambda = -1$, must hold, the representation of $SO_0(1, n+1)$ belongs to the supplementary series $\hat{R}_\rho$ with $\rho = -\frac{1}{2}$. The negative value of ρ implies that the integral involved in the intertwining operator diverges. This can be cured by switching, using the intertwining operator relative to the value $\rho = \frac{1}{2}$, to the equivalent representation $\hat{R}_{\frac{1}{2}}$. Indeed, from the defining property, we have

$$\hat{R}_{\frac{1}{2}} = \hat{I}_{\frac{1}{2}}^{-1}\hat{R}_{-\frac{1}{2}}\hat{I}_{\frac{1}{2}},$$

the right-hand side being well defined. Hereafter, we will only consider this positive value.

At this point, a sort of miracle arises. Since

$$\hat{I}_{\frac{1}{2}}\Phi(X) = \int_{S^n} \frac{\Phi(X')}{\|X - X'\|^{n-1}} d\Omega',$$

we see that $\hat{I}_{\frac{1}{2}}$ *is equal*, up to a numerical factor, *to the inverse of the action operator* (9.0.1) of the Kepler Problem. This is a fundamental point: $\hat{I}_{\frac{1}{2}}$ is an operator that is Hermitian with respect to the inner product $\langle\cdot,\cdot\rangle$ and therefore also with respect to the inner product $\langle\cdot,\cdot\rangle_{\frac{1}{2}}$. This allows us to conclude that the excluded pseudo-differential operators also are Hermitian, making the whole representation of $SO_0(2, n+1)$ fully unitary with respect to $\langle\cdot,\cdot\rangle_{\frac{1}{2}}$. In fact $\hat{G}_{-1,0}$ is, up a numerical factor, the inverse of $\hat{I}_{\frac{1}{2}}$, and is therefore Hermitian. Moreover, the operators $\hat{G}_{0,\alpha}$ will be also Hermitian, since they can be obtained as commutators of Hermitian operators.

Having made the representation fully unitary, our aim is now to carry over the wave function of the representation from S^n to $\mathbb{R}^n$, in order to be able to perform a Fourier transform. Recall that the carrier space of the multiplier representation $\hat{R}_\rho$ of $SO_0(1, n+1)$ is the space of the homogeneous functions of degree $-\left(\frac{n}{2}+\rho\right)$ on the null cone in $\mathbb{R}^{1,n+1}$, i.e., the space of the functions defined over the null lines. As the manifold on which the wave function is defined, we may choose a section of the null cone. Intersecting this last with the hyperplane $X^0 = 1$, we obtain S^n. Intersecting with the hyperplane $X^0 - X^{n+1} = 1$, we obtain a paraboloid that can be identified with $\mathbb{R}^n$, since its induced metric is flat. Explicitly, we have the two sections

- for S^n : $X^0 = 1, \quad X^k = \frac{2x^k}{x^2+1}, \quad X^{n+1} = \frac{x^2-1}{x^2+1}$;
- for $\mathbb{R}^n$: $X^0 = \frac{x^2+1}{2}, \quad X^k = x^k, \quad X^{n+1} = \frac{x^2-1}{2}$.

As one sees by inspection of the above formulas, each X of $\mathbb{R}^n$ may be obtained from the corresponding X of S^n by multiplying this latter by $r = \frac{x^2+1}{2}$, so that to every n-tuple $(x^1 \dots x^n)$ correspond, on the two sections, two points that lie on the same null line. To a function of the representation R_ρ defined on S^n there corresponds a homogeneous function of degree $-\left(\frac{n}{2}+\rho\right)$ on the null cone, thus a function on $\mathbb{R}^n$; given a function on S^n, the corresponding function on $\mathbb{R}^n$ is obtained through multiplication by $\left(\frac{2}{x^2+1}\right)^{\frac{n}{2}+\rho}$.

The representation at hand is characterized by the value $\rho = \frac{1}{2}$, and we can conclude that, to implement the interchange between the canonical coordinates and momenta, we must

(i) multiply the wave function on S^n by $\left(\frac{2}{x^2+1}\right)^{\frac{n+1}{2}}$, so that it becomes defined on $\mathbb{R}^n$, and

(ii) perform a Fourier transform.

We will now find the implementation of the representation from S^n to $\mathbb{R}^n$, beginning with the action operator $\hat{G}_{-1,0}$. Choose $m = m^{\clubsuit}$ and start from the eigenvalue equation $-G_{-1,0}\Phi = \eta\Phi$ which we can write in the form

$$\frac{\Gamma\left(\frac{n-1}{2}\right)}{2\pi^{\frac{n+1}{2}}} \int_{S^n} \frac{\Phi(X')}{\|X - X'\|^{n-1}} d\Omega' = \frac{1}{\eta}\Phi(X).$$

Performing (i) and (ii) above is equivalent to following the Fock procedure backward, but without energy rescaling. This is exactly the procedure followed to get Equation (5.1.9) on page 101, which we may write as

$$\frac{1}{2}x(\hat{y}^2 + 1)\phi(x) = \eta\phi(x). \tag{9.3.2}$$

The implementation of the whole representation on $\mathbb{R}^n$ is

$$
\begin{aligned}
\hat{G}_{-1,0} &= -\frac{1}{2}x(\hat{y}^2+1),\\
\hat{G}_{hk} &= x_k\hat{y}_h - x_h\hat{y}_k,\\
\hat{G}_{0,k} &= -x\hat{y}_k,\\
\hat{G}_{-1,k} &= \frac{1}{2}x_k(\hat{y}^2+1) - \left[\langle x,\hat{y}\rangle - i\frac{n-1}{2}\right]\hat{y}_k,\\
\hat{G}_{k,n+1} &= \frac{1}{2}x_k(\hat{y}^2-1) - \left[\langle x,\hat{y}\rangle - i\frac{n-1}{2}\right]\hat{y}_k,\\
\hat{G}_{-1,n+1} &= \langle x,\hat{y}\rangle - i\frac{n-1}{2},\\
\hat{G}_{0,n+1} &= -\frac{1}{2}x(\hat{y}^2-1).
\end{aligned}
\tag{9.3.3}
$$

The simplest way to obtain this result is the following. The form of $\hat{G}_{-1,0}$ comes directly from what was said above. That of $\hat{G}_{hk}$ and $\hat{G}_{-1,k} - \hat{G}_{k,n+1}$ is obvious. The form of the other generators can be determined through brackets of these $2n+1$ operators, and the results of this chapter ensure that $\hat{G}_{-1,0}, \hat{G}_{hk}$ and $\hat{G}_{-1,k} - \hat{G}_{k,n+1}$ must generate the whole conformal algebra.

$\hat{I}_{\frac{1}{2}}$ is a convolution operator and its Fourier transform will be a multiplicative operator: recalling the Fourier transform (4.3.1) on page 82, it follows that the representation (9.3.3) is unitary with respect to the inner product

$$(\psi,\phi) = \int_{\mathbb{R}^n} \psi^*\phi\frac{1}{x}\,d^nx.$$

The infinitesimal representation (9.3.3) is clearly the quantization of the Main Moment Map (6.4.5).

Lastly, to obtain the Schrödinger Equation for the Kepler Problem we must still implement the "energy rescaling", but this is an easy task. Let us consider the eigenvalue Equation (9.3.2) and perform the rescaling

$$x_k = \frac{q_k}{\eta}, \quad \hat{y}_h = \eta\hat{p}_h.$$

We obtain

$$\left[\frac{1}{2}\hat{p}^2 - \frac{1}{q}\right]\phi = \frac{1}{-2\eta^2}\phi,$$

i.e., the eigenvalue equation of the Kepler Problem for negative energy (for positive energy we start instead with $-\hat{G}_{0,n+1}\phi = \eta\phi$). It is clear that (i) and (ii), together with the above rescaling and the change in the eigenvalue parameter: $\eta \mapsto \frac{1}{-2\eta^2}$, are the quantum implementation of the canonical transformation $\mathscr{C}$ on page 138.

CHAPTER 10

Kepler Problem with Magnetic Monopole

The Kepler Problem, described in the previous chapters, is a highly symmetric one, and it would seem, at first sight, impossible to change anything in the above constructions without destroying the beautiful machinery. It is thus somewhat surprising that adding the vector potential $\vec{\alpha}$ of a Dirac magnetic monopole, such that $\nabla \times \vec{\alpha} = \frac{\vec{q}}{q^3}$, plus a repulsive scalar potential decreasing as $\frac{1}{q^2}$, does nevertheless preserve the $SU(2,2)$, or $SO(2,4)$, symmetry.

As we show below, this symmetry, found by McIntosh & Cisneros (1970) and Zwanziger (1968), is recovered simply by considering, (unlike as in Chapter 7), twistors with a nonvanishing norm, followed by a canonical transformation. Modifying this latter, we find, furthermore, another interesting system, which arises namely in the study of geodesic motion in Euclidean Taub-NUT space. This latter problem approximates in fact the scattering of self dual monopoles (Atiyah & Hitchin 1988), (Gibbons & Manton 1986). The introduction of the magnetic monopole is, ultimately, the reason why one investigates the spinor regularization.

This construction is based on the Hopf fibration $S^3 \to S^2$, and is therefore dimension dependent. It can be extended to the case $n = 5$, with analogous Hopf fibration $S^7 \to S^4$, considering twistors over the quaternions. In this case, the monopole is called the *BPST instanton* (Belavin, Polyakov, Schwarz & Tyupkin 1975).

In this chapter we will follow Cordani (1986b) and Cordani, Fehér & Horváthy (1990).

10.1 Nonnull Twistors and Magnetic Monopoles

As in Chapter 7, let us consider a twistor $\psi = \begin{pmatrix} z \\ iw \end{pmatrix}$, viewed as an element of $\mathbb{C}^{2,2}$. Now we no longer require the nullity condition, but put rather

$$\psi^\dagger \mathbf{h} \psi = 2\mu, \ \mu \in \mathbb{R}^+. \tag{10.1.1}$$

Let T_μ be the space of twistors satisfying (10.1.1), with the equivalence relation $\psi \sim \psi e^{i\beta}$, and β a generic angle. The set $T_\mu / \sim$ is a 6-dimensional manifold, which may be equipped with the natural symplectic structure (induced by) $d\Theta$, where Θ is the 1-form

$$\Theta = \operatorname{Im} \psi^\dagger \mathbf{h}\, d\psi.$$

PROPOSITION 10.1 *$T_\mu / \sim$ is diffeomorphic to $(\mathbb{R}^3 - \{0\}) \times \mathbb{R}^3$, and carries a symplectic structure that differs from the canonical one in a magnetic term.*

Proof. Let $\vec{\chi}, \vec{\pi} \in (\mathbb{R}^3 - \{0\}) \times \mathbb{R}^3$. Define the two matrices

$$\boldsymbol{\pi} = i\frac{\mu}{\chi}\mathbf{1} + \vec{\pi} \cdot \vec{\mathbf{s}} = \begin{pmatrix} i\frac{\mu}{\chi} - \pi_3 & \pi_1 + i\pi_2 \\ \pi_1 - i\pi_2 & i\frac{\mu}{\chi} + \pi_3 \end{pmatrix},$$

$$\boldsymbol{\chi} = \frac{1}{2}\chi\mathbf{1} + \frac{1}{2}\vec{\chi} \cdot \vec{\mathbf{s}} = \frac{1}{2}\begin{pmatrix} \chi - \chi_3 & \chi_1 + i\chi_2 \\ \chi_1 - i\chi_2 & \chi + \chi_3 \end{pmatrix},$$

where we denote, as usual, the norm of the vectors $\vec{\chi}$ and $\vec{\pi}$ by χ and π. Since $\det\boldsymbol{\chi} = 0$, there exists an element, denoted by $\sqrt{\boldsymbol{\chi}}$, of $(\mathbb{C}^2 - \{0\}) / \sim$ such that

$$\boldsymbol{\chi} = \sqrt{\boldsymbol{\chi}}\sqrt{\boldsymbol{\chi}}^\dagger.$$

Define the map $\Psi^{(\mu)} : (\mathbb{R}^3 - \{0\}) \times \mathbb{R}^3 \to T_\mu / \sim$ by

$$\Psi^{(\mu)}(\chi_h, \pi_k) = \begin{pmatrix} \sqrt{\boldsymbol{\chi}} \\ -i\boldsymbol{\pi}\sqrt{\boldsymbol{\chi}} \end{pmatrix}.$$

This parametrization differs from that of (7.2.3) and (7.2.4) on page 153 owing to the presence of the μ term, and also to the canonical exchange $\vec{\chi} = \vec{y}, \vec{\pi} = -\vec{x}$. Remembering the discussion on the decomposition (7.2.5) of the null twistor space T_0, it is clear that now the map $\Psi^{(\mu)}$ is not only injective but also surjective; in fact, twistors of the type T_0^∞, being null twistors, are excluded from T_μ.

As to the symplectic structure, one calculates

$$\left(\Psi^{(\mu)}\right)^* \Theta = \operatorname{Im}\begin{pmatrix} \sqrt{\boldsymbol{X}}^\dagger & i\sqrt{\boldsymbol{X}}^\dagger \boldsymbol{\pi}^\dagger \end{pmatrix} \begin{pmatrix} 0 & 1 \\ 1 & 0 \end{pmatrix} \begin{pmatrix} d\sqrt{\boldsymbol{X}} \\ -i\, d\boldsymbol{\pi}\sqrt{\boldsymbol{X}} - i\boldsymbol{\pi}\, d\sqrt{\boldsymbol{X}} \end{pmatrix}$$
$$= -\frac{1}{2}\sqrt{\boldsymbol{X}}^\dagger d(\boldsymbol{\pi}+\boldsymbol{\pi}^\dagger)\sqrt{\boldsymbol{X}} - i\frac{\mu}{\chi}\left(\sqrt{\boldsymbol{X}}^\dagger d\sqrt{\boldsymbol{X}} - d\sqrt{\boldsymbol{X}}^\dagger \sqrt{\boldsymbol{X}}\right).$$

$\left(\Psi^{(\mu)}\right)^* \Theta$ is the sum of two terms. The first term gives, up to an exact 1-form, the canonical form of $T^*\mathbb{R}^3$. To calculate the latter, put

$$\sqrt{\boldsymbol{X}} = \begin{pmatrix} \sqrt{\chi}\sin\frac{\vartheta}{2} e^{i\frac{\varphi+\beta}{2}} \\ \sqrt{\chi}\cos\frac{\vartheta}{2} e^{i\frac{-\varphi+\beta}{2}} \end{pmatrix}, \tag{10.1.2}$$

that is equivalent to introducing spherical coordinates in $\mathbb{R}^3$:

$$\chi_1 = \chi\sin\vartheta\cos\varphi, \quad \chi_2 = \chi\sin\vartheta\sin\varphi, \quad \chi_3 = \chi\cos\vartheta.$$

Since

$$\sqrt{\boldsymbol{X}}^\dagger d\sqrt{\boldsymbol{X}} = \frac{1}{2}d\chi + \frac{i}{2}\chi\,(-\cos\vartheta\, d\varphi + d\beta),$$

the final result is

$$\left(\Psi^{(\mu)}\right)^* \Theta = \vec{\pi}\cdot d\vec{\chi} - \mu\cos\vartheta\, d\varphi,$$

up to the exact 1-form $\mu\, d\beta$. But

$$d(-\cos\vartheta\, d\varphi) = d\vartheta \wedge \sin\vartheta\, d\varphi$$
$$= \frac{\chi_1}{\chi^3}d\chi_2 \wedge d\chi_3 + \frac{\chi_2}{\chi^3}d\chi_3 \wedge d\chi_1 + \frac{\chi_3}{\chi^3}d\chi_1 \wedge d\chi_2,$$

which is the surface form on the sphere S^2. The Poisson structure on $T_\mu/\sim$ results therefore in

$$\{\chi_i, \chi_k\} = 0, \quad \{\pi_i, \chi_k\} = \delta_{ik}, \quad \{\pi_i, \pi_h\} = -\mu\,\varepsilon_{ihk}\frac{\chi_k}{\chi^3}, \tag{10.1.3}$$

where the right-hand side of the third term is just the magnetic field of a Dirac monopole. QED

We may therefore view $T_\mu/\sim$ as the symplectic reduction of $\mathbb{C}^{2,2}$ under the U(1)-action

$$\psi \mapsto \psi' = \psi g, \quad \psi \in \mathbb{C}^{2,2}, \quad g \in \mathrm{U}(1).$$

Since the natural linear action of U(2,2) on $T_\mu/\sim$

$$\psi \mapsto \psi' = g\psi, \quad g \in \mathrm{U}(2,2) \tag{10.1.4}$$

is manifestly symplectic, we can construct a moment map

$$j^{(\mu)} : T_\mu/\sim \to \mathrm{u}^*(2,2)$$

by putting

$$j^{(\mu)}(\psi) = -i\psi\psi^\dagger \mathbf{h} \quad \text{with } \psi^\dagger \mathbf{h}\, \psi = 2\mu.$$

One can verify, just we did for (7.2.2), that $j^{(\mu)}$ is a moment map. It can obviously be restricted to a moment map of $\mathrm{su}^*(2,2)$. Since the action of SU(2, 2) on $T_\mu/\sim$ is manifestly transitive, the image of $j^{(\mu)}$ is a coadjoint orbit, $\mathcal{O}_\mu$ say. The explicit expression of $j^{(\mu)}$ is obtained by applying (7.2.7) on page 154, where

$$z = \sqrt{\mathbf{X}}, \quad w = -\boldsymbol{\pi}\sqrt{\mathbf{X}}.$$

Using 3-dimensional notation, the map $j^{(\mu)} \circ \Psi^{(\mu)}$ is given explicitly by

$$\begin{aligned}
K_{(-)} &= \frac{1}{2}\chi\left(\pi^2+1\right) + \frac{1}{2}\frac{\mu^2}{\chi} \Rightarrow -G_{-1,0},\\
\vec{G} &= \vec{\chi}\times\vec{\pi} - \mu\frac{\vec{\chi}}{\chi} \Rightarrow -\epsilon_{ihk}G_{hk},\\
\vec{L} &= \chi\vec{\pi} \Rightarrow G_{0k},\\
\vec{A}_{(+)} &= \vec{\pi}\times\vec{G} - \frac{\vec{\chi}}{\chi}K_{(+)} \Rightarrow G_{-1,k},\\
\vec{A}_{(-)} &= \vec{\pi}\times\vec{G} - \frac{\vec{\chi}}{\chi}K_{(-)} \Rightarrow G_{k,4},\\
D &= \vec{\chi}\cdot\vec{\pi} \Rightarrow -G_{-1,4},\\
K_{(+)} &= \frac{1}{2}\chi\left(\pi^2-1\right) + \frac{1}{2}\frac{\mu^2}{\chi} \Rightarrow -G_{0,4},
\end{aligned} \tag{10.1.5}$$

while μ appears as the generator of the U(1)-action that, being the center of U(2, 2), commutes with the action of SU(2, 2).

We may also obtain the moment map (10.1.5) in a slightly different way. Let us in fact consider (10.1.4) as describing a U(2, 2)-action on $(\mathbb{C}^2 - \{0\}) \times \mathbb{C}^2 = T^*(\mathbb{R}^4 - \{0\})$, endowed with the 1-form

$$\Theta = z_\alpha dw_\alpha - w_\alpha dz_\alpha, \quad \alpha = 1,\ldots,4,$$

instead of on $T_\mu/\sim$. Take as coordinate system the canonical extension of (10.1.2). We define $p_\chi, p_\vartheta, p_\varphi, p_\beta$ as the conjugate momenta to $\chi, \vartheta, \varphi, \beta$, then

$$\Theta = p_\chi d\chi + p_\vartheta d\vartheta + p_\varphi d\varphi + p_\beta d\beta.$$

The action of the center U(1) of U(2,2) on $\mathbb{C}^2 - \{0\}$, given by $z \mapsto z e^{i\frac{\beta}{2}}$, commutes with that of SU(2,2), and thus its Hamiltonian p_β is a constant we call μ. It is easy to verify that, composing the canonical extension of (10.1.2) with $j^{(\mu)}$, we obtain just (10.1.5), where the components of $\vec{\pi}$ are $(p_\chi, p_\vartheta, p_\varphi + \mu \cos\vartheta)$: in other words, the symplectic structure of $T^*(\mathbb{R}^3 - \{0\})$ is the canonical one, but the potential $\vec{A} \equiv (0, 0, \mu\cos\vartheta)$ of the Dirac monopole appears. For example

$$K_{(-)} = \frac{1}{2}\left[p_\chi^2 + \frac{1}{\chi^2}\left(p_\vartheta^2 + \frac{(p_\varphi + \mu\cos\vartheta)^2}{\sin^2\vartheta}\right) + 1\right] + \frac{\mu^2}{2\chi}.$$

The above symplectic reduction $(\mathbb{C}^2 - \{0\}) \times \mathbb{C}^2 \to (\mathbb{R}^3 - \{0\}) \times \mathbb{R}^3$ is based on the Hopf fibration $S^3/S^1 = S^2$, as one sees by writing

$$\frac{\mathbb{C}^2 - \{0\}}{\mathrm{U}(1)} = \frac{\mathbb{R}^+ \times S^3}{S^1} = \mathbb{R}^+ \times S^2 = \mathbb{R}^3 - \{0\}. \qquad (10.1.6)$$

When $\mu = 0$, the moment map (10.1.5) reduces to (6.4.5) on page 135, with the lower signs, after the canonical exchange $\vec{\chi} = \vec{y}$ and $\vec{\pi} = -\vec{x}$. (6.4.5) is an injective, but nonsurjective map $T^+\mathbb{R}^3 \to \mathcal{O}_{\mathbf{e}} \subset \mathrm{su}^*(2,2)$, where $\mathcal{O}_{\mathbf{e}}$ is a coadjoint orbit, symplectomorphic to T^+S^3. The lack of surjectivity is easily understood by remembering that the map is based on the (inverse of) the stereographic projection. But what about (10.1.5)? The answer is given by the following two propositions.

PROPOSITION 10.2 *Removing a sphere S^2 from the image $\mathcal{O}_\mu$ of $j^{(\mu)}$, one obtains a manifold homeomorphic to T^+S^3.*

Proof. Recalling the definition of the Fock parameters, let us consider the two 4-vectors $U_\alpha = G_{-1,\alpha}$ and $V_\beta = G_{0,\beta}$:

$$U = \begin{pmatrix} \vec{U} \\ U_4 \end{pmatrix} = \begin{pmatrix} \vec{A}_{(+)} \\ -D \end{pmatrix}, \quad V = \begin{pmatrix} \vec{V} \\ V_4 \end{pmatrix} = \begin{pmatrix} \vec{L} \\ -K_{(+)} \end{pmatrix}.$$

A direct calculation shows that

$$\|U\|^2 = \|V\|^2 = K_{(-)}^2 - \mu^2, \quad U \cdot V = 0, \qquad (10.1.7)$$

where norm and scalar product are Euclidean. These relations are similar to the first two of (6.4.7) on page 136, with the difference that now the norm can also be null, since $K_{(-)}$ reaches its absolute minimum value μ just when $\pi = 0$ and $\chi = \mu$. The vectors U and V (if they are different from zero) span therefore a 6-dimensional manifold, homeomorphic to T^+S^3. That the whole T^+S^3 is filled is ensured by the coadjoint action of SU(2, 2). Another calculation shows that

$$\begin{aligned}\vec{G} &= \frac{K_{(-)}}{K_{(-)}^2 - \mu^2}\vec{U}\times\vec{V} + \frac{\mu}{K_{(-)}^2 - \mu^2}(U_4\vec{V} - V_4\vec{U}),\\ \vec{A}_{(-)} &= \frac{K_{(-)}}{K_{(-)}^2 - \mu^2}(U_4\vec{V} - V_4\vec{U}) + \frac{\mu}{K_{(-)}^2 - \mu^2}\vec{U}\times\vec{V},\end{aligned} \tag{10.1.8}$$

which generalize the last relations (6.4.7). For the first of (10.1.7), $\vec{G}$ and $\vec{A}_{(-)}$ only depend on the direction of U and V, rather than on the two vectors themselves, and are therefore well defined also when U and V go to zero. In fact, since $\vec{U}\times\vec{V}$ and $U_4\vec{V} - V_4\vec{U}$ are orthogonal,

$$\left\|\vec{U}\times\vec{V} + U_4\vec{V} - V_4\vec{U}\right\|^2 = \left\|\vec{U}\times\vec{V}\right\|^2 + \left\|U_4\vec{V} - V_4\vec{U}\right\|^2$$

holds and the right-hand side is equal to the norm of the 4-dimensional bivector $U\wedge V$:

$$\|U\wedge V\|^2 = \|U\|^2\,\|V\|^2 - (U\cdot V)^2 = (K_{(-)}^2 - \mu^2)^2.$$

For $U, V \to 0$, then $K_{(-)} \to \mu$ and

$$\vec{G} \text{ and } \vec{A}_{(-)} \to \mu\frac{\vec{U}\times\vec{V} + U_4\vec{V} - V_4\vec{U}}{K_{(-)}^2 - \mu^2} = \mu\,\vec{e},$$

where $\vec{e}$ is a unit arbitrary vector. The two vectors $\vec{G}$ and $\vec{A}_{(-)}$ span therefore in the limit a sphere of radius μ. The same conclusion is reached by putting $\pi = 0$ and $\chi = \mu$ in (10.1.5). Cutting out this sphere amounts to requiring that U, V are both different from zero, leaving us, at last, with a manifold homeomorphic to T^+S^3. QED

PROPOSITION 10.3 *The map* $j^{(\mu)}\circ\Psi^{(\mu)} : (\mathbb{R}^3 - \{0\})\times\mathbb{R}^3 \to \mathcal{O}_\mu$ *is injective and surjective.*

Proof. We must prove that, for every point of $\mathcal{O}_\mu$, its pre-image in $(\mathbb{R}^3 - \{0\})\times\mathbb{R}^3$ exists and is unique. Let us distinguish the two cases.

(i) $\|U\| = \|V\| \neq 0$. We exhibit explicitly the inverse map $(j^{(\mu)}\circ\Psi^{(\mu)})^{-1}$. Owing to (10.1.7) and (10.1.8), a point of $\mathcal{O}_\mu$ is uniquely determined by two orthogonal 4-dimensional vectors U and V, of equal (arbitrary) norm. By (10.1.5) we can write

$$\begin{aligned}\vec{\chi} &= \vec{A}_{(+)} - \vec{A}_{(-)}\\ &= \vec{U} - \frac{K_{(-)}}{K_{(-)}^2 - \mu^2}(U_4\vec{V} - V_4\vec{U}) - \frac{\mu}{K_{(-)}^2 - \mu^2}\vec{U}\times\vec{V},\end{aligned} \tag{10.1.9}$$

$$\vec{\pi} = \frac{\vec{V}}{\chi},$$

where it is understood that $K_{(-)}$, which is always positive, is expressed as a function of U or V :

$$K_{(-)} = \sqrt{\|U\|^2 + \mu^2} = \sqrt{\|V\|^2 + \mu^2}.$$

Expressions (10.1.9) represent the sought inverse map. We must only check that χ never vanishes. To this end, notice that χ could vanish only if $\vec{A}_{(-)}$ is proportional to $\vec{A}_{(+)} = \vec{U}$, thus, by the second of (10.1.8), if $\vec{V} = 0$ or $\vec{V}$ proportional to $\vec{U}$. In both cases $\vec{U} \times \vec{V} = 0$, so that

$$\left\| U_4 \vec{V} - V_4 \vec{U} \right\| = \|U \wedge V\| = K_{(-)}^2 - \mu^2.$$

Thus

$$\vec{\chi} = \vec{U} - K_{(-)} \frac{\vec{U}}{\left\|\vec{U}\right\|}, \quad \text{with } \left\|\vec{U}\right\|^2 \le \|U\|^2 = K_{(-)}^2 - \mu^2,$$

from which it is obvious that, for $\mu \neq 0$, $\vec{\chi}$ never vanishes.

(ii) $U = V = 0$, i.e., $K_{(-)} = \mu$, and this is achieved only when $\pi = 0$ and $\chi = \mu$. The choice of $\vec{G} = \vec{A}_{(-)} = -\vec{\chi}$ obviously determines the vector $\vec{\chi}$ in a unique way.

This completes the proof. QED

REMARK 10.1 At first sight, these two propositions above may appear surprising, since they establish a homeomorphism

$$(\mathbb{R}^3 - \{0\}) \times \mathbb{R}^3 - S^2 \simeq T^+ S^3.$$

But this is indeed what happens. In fact, the (co)tangent bundle of S^3 is trivial, because S^3 is a (Lie) group. We can write therefore

$$\begin{aligned} T^+ S^3 &\simeq S^3 \times (\mathbb{R}^3 - \{0\}) \simeq S^3 \times S^2 \times \mathbb{R}^+ \\ &\simeq (\mathbb{R}^4 - \{0\}) \times S^2 \simeq \mathbb{R}^4 \times S^2 - S^2. \end{aligned}$$

But

$$\mathbb{R}^4 \times S^2 \simeq \mathbb{R}^3 \times \mathbb{R} \times S^2 \simeq \mathbb{R}^3 \times \mathbb{R}^+ \times S^2 \simeq \mathbb{R}^3 \times (\mathbb{R}^3 - \{0\}).$$

In particular, the homeomorphism $\mathbb{R}^+ \simeq \mathbb{R}$ is realized by the map

$$K_{(+)} \big|_{\pi=0} = -\frac{1}{2}\chi + \frac{1}{2}\frac{\mu}{\chi}.$$

For a more intuitive understanding, one may think of the same situation but in a lower dimension, with SU(2, 2) replaced by SU(1, 1). T^+S^3 corresponds to one sheet of the null semicone with its origin removed (which has therefore the topology of T^+S^1) in su* (1, 1) $\simeq \mathbb{R}^{2,1}$, while to $\mathcal{O}_\mu$ corresponds one sheet of the hyperboloid (which has the topology of $\mathbb{R} \times \mathbb{R}$) tending asymptotically to the null cone. Cutting out S^2 from $\mathcal{O}_\mu$ corresponds, in lower dimension, to puncturing the vertex of the hyperboloid. Considering the canonical extension of the stereographic projection corresponds in turn to cutting the null semicone along a generatrix, providing us with a manifold homeomorphic to $\mathbb{R} \times \mathbb{R}$.

In full analogy with the Kepler Problem without magnetic term, we consider $K_{(-)}$ or $K_{(+)}$ as Hamiltonians of a dynamical system: they generate the bound or unbound motions, respectively[(1)]. Moreover, other Hamiltonian systems with the same symmetries can be generated. Indeed, let us define an enlarged phase space by adding to $\vec{\pi}$, $\vec{\chi}$ the canonical variables π_α, α. This space is endowed with the symplectic 2-form

$$\Omega = d\pi_\alpha \wedge d\alpha + d\Theta.$$

Let us now consider the two 7-dimensional "energy surfaces"

$$\pi_\alpha \mp K_{(\epsilon)} = 0, \quad \epsilon = +, -. \tag{10.1.10}$$

Let the map $\mathscr{K}^{-1}_{\mathrm{GEN}}(t, p_t, \vec{q}, \vec{p}) = (\alpha, \pi_\alpha, \vec{\chi}, \vec{\pi})$ be defined by

$$\begin{aligned} \vec{\chi} &= \sigma(p_t)\vec{q}, \\ \vec{\pi} &= \frac{\vec{p}}{\sigma(p_t)}, \\ \pi_\alpha &= \kappa(p_t), \\ \alpha &= \frac{1}{\kappa'}\left[t + \frac{\sigma'}{\sigma}\vec{q} \cdot \vec{p}\right], \end{aligned} \tag{10.1.11}$$

where the *scaling factor* σ and the *energy transformation* κ are differentiable functions, the prime denoting differentiation. As one easily verifies, (10.1.11) is a canonical transformation for every μ, and obviously generalizes (up to an exchange between coordinates and momenta) the transformation $\mathscr{K}$ of the Kepler Problem. Enforcing constraint (10.1.10), after performing this canonical transformation, one obtains, as in Chapter 6, transformed Hamiltonians with the same symmetries of $K_{(-)}$ or $K_{(+)}$.

Exploiting the group and geometric properties of (10.1.5), let us now study the motion generated by $K_{(-)}$ and $K_{(+)}$.

---

(1) We inherit the suffix "minus" and "plus" from the Kepler Problem, where it is related to the sign of the Hamiltonian for the bound and unbound motions respectively, although in the present case $K_{(-)} > 0$ and $-\infty < K_{(+)} < \infty$.

10.1.1 Bound Motions

We consider the Hamiltonian system with phase space $(\mathbb{R}^3 - \{0\}) \times \mathbb{R}^3$, Poisson structure (10.1.3) and Hamiltonian $K_{(-)}$. The commutation rules of su(2,2) imply at once that $\vec{G}$ and $\vec{A}_{(-)}$ are first integrals, and that they generate the invariance subgroup SO(4). These seven first integrals satisfy two identities that we want to find here. To this end, let us define the two vectors

$$\overrightarrow{G^0} = \frac{\vec{U} \times \vec{V}}{\sqrt{K_{(-)}^2 - \mu^2}}, \quad \overrightarrow{A^0}_{(-)} = \frac{U_4 \vec{V} - V_4 \vec{U}}{\sqrt{K_{(-)}^2 - \mu^2}},$$

which satisfy the identities

$$\left\|\overrightarrow{G^0}\right\|^2 + \left\|\overrightarrow{A^0}_{(-)}\right\|^2 = K_{(-)}^2 - \mu^2, \quad \overrightarrow{G^0} \cdot \overrightarrow{A^0}_{(-)} = 0. \tag{10.1.12}$$

Since

$$\begin{pmatrix} \vec{G} \\ \vec{A}_{(-)} \end{pmatrix} = \begin{pmatrix} \frac{K_{(-)}}{\sqrt{K_{(-)}^2 - \mu^2}} & \frac{\mu}{\sqrt{K_{(-)}^2 - \mu^2}} \\ \frac{\mu}{\sqrt{K_{(-)}^2 - \mu^2}} & \frac{K_{(-)}}{\sqrt{K_{(-)}^2 - \mu^2}} \end{pmatrix} \begin{pmatrix} \overrightarrow{G^0} \\ \overrightarrow{A^0}_{(-)} \end{pmatrix},$$

we easily find the identities we are looking for, namely:

$$\left\|\vec{G}\right\|^2 + \left\|\vec{A}_{(-)}\right\|^2 = K_{(-)}^2 + \mu^2, \quad \vec{G} \cdot \vec{A}_{(-)} = \mu K_{(-)}.$$

The equations of motion,

$$\frac{dU}{d\alpha} = \{K_{(-)}, U\} = V, \quad \frac{dV}{d\alpha} = \{K_{(-)}, V\} = -U,$$

are easily integrated as

$$\begin{aligned} U &= \overline{U} \cos\alpha + \overline{V} \sin\alpha, \\ V &= -\overline{U} \sin\alpha + \overline{V} \cos\alpha, \end{aligned}$$

where $\overline{U}$ and $\overline{V}$ are the initial values. Since $\vec{\chi} = \vec{U} - \vec{A}_{(-)}$, the "time" evolution of the position vector is

$$\vec{\chi}(\alpha) = \vec{\overline{U}} \cos\alpha + \vec{\overline{V}} \sin\alpha - \vec{A}_{(-)},$$

showing that the orbit is an ellipse, whose center is $-\vec{A}_{(-)}$. This ellipse can also be seen as a conic section. In fact

$$\vec{\chi} \cdot \vec{U} \times \vec{V} = -\vec{A}_{(-)} \cdot \vec{U} \times \vec{V},$$

and, since $\vec{A}_{(-)}$ and $\vec{U} \times \vec{V}$ are first integrals, it follows that $\vec{\chi}$ moves on a fixed plane, orthogonal to $\vec{U} \times \vec{V}$ and passing through $-\vec{A}_{(-)}$. Moreover, remembering the expression of $\vec{G}$ given by (10.1.5), we get

$$\frac{\vec{\chi}}{\chi} \cdot \vec{G} = -\mu,$$

i.e., the position vector $\vec{\chi}$ moves on a fixed cone with axis described by $\vec{G}$ and opening angle $\zeta = \arccos \frac{\mu}{G}$. The angle ζ is well defined, since the minimum value of the norm of $\vec{G}$ is just μ. See Figure 10.1.

The two vectors $\vec{G}$ and $\vec{A}_{(-)}$ are related to the cone, while the two other vectors $\overrightarrow{G^0}$ and $\overrightarrow{A^0}_{(-)}$ are related to the shape of the elliptic orbit. To see this, let us choose the initial vector $\vec{\overline{U}}$ such that it coincides with the semimajor axis of the ellipse. Then $\overline{U}_4 = 0$, and $\vec{\overline{V}} \cdot \vec{\overline{U}} = 0$, i.e., $\vec{\overline{V}}$ coincides with the semiminor axis. Moreover, the semimajor axis has length

$$\left\|\vec{\overline{U}}\right\|^2 = \|U\|^2 = K_{(-)}^2 - \mu^2.$$

As $\vec{\overline{U}}$ and $\vec{\overline{V}}$ are orthogonal, we have

$$\left\|\vec{\overline{U}} \times \vec{\overline{V}}\right\|^2 = \left\|\vec{\overline{U}}\right\|^2 \left\|\vec{\overline{V}}\right\|^2 = \left(K_{(-)}^2 - \mu^2\right) \left\|\vec{\overline{V}}\right\|^2.$$

From the definition of $\overrightarrow{G^0}$ it follows that

$$\left\|\vec{\overline{U}} \times \vec{\overline{V}}\right\|^2 = \left(K_{(-)}^2 - \mu^2\right) \left\|\overrightarrow{G^0}\right\|^2,$$

hence $\left\|\vec{\overline{V}}\right\| = \left\|\overrightarrow{G^0}\right\|$. Therefore, the norm of $\overrightarrow{G^0}$ is equal to the semiminor axis length, and the first of the two identities (10.1.12) shows that $\overrightarrow{A^0}_{(-)}$ coincides in fact with the vector joining the focus to the center of the ellipse.

The most relevant difference with respect to the case $\mu = 0$ is that no collision orbit exists now. There are still radial motions, and also a sphere of equilibrium points. In fact, when $\vec{\chi}$ and $\vec{\pi}$ become parallel, $\vec{G}$ and $\vec{A}_{(-)}$ take also the same direction, as is obvious from their definition in (10.1.5). Moreover, $\left\|\vec{G}\right\|$ reaches its absolute minimum μ, so that the cone degenerates into a half-line. The particle moves along radial orbits that never reach the origin, since, in this case, the Hamiltonian would diverge. For each fixed value of $K_{(-)}$, the radial motion has two

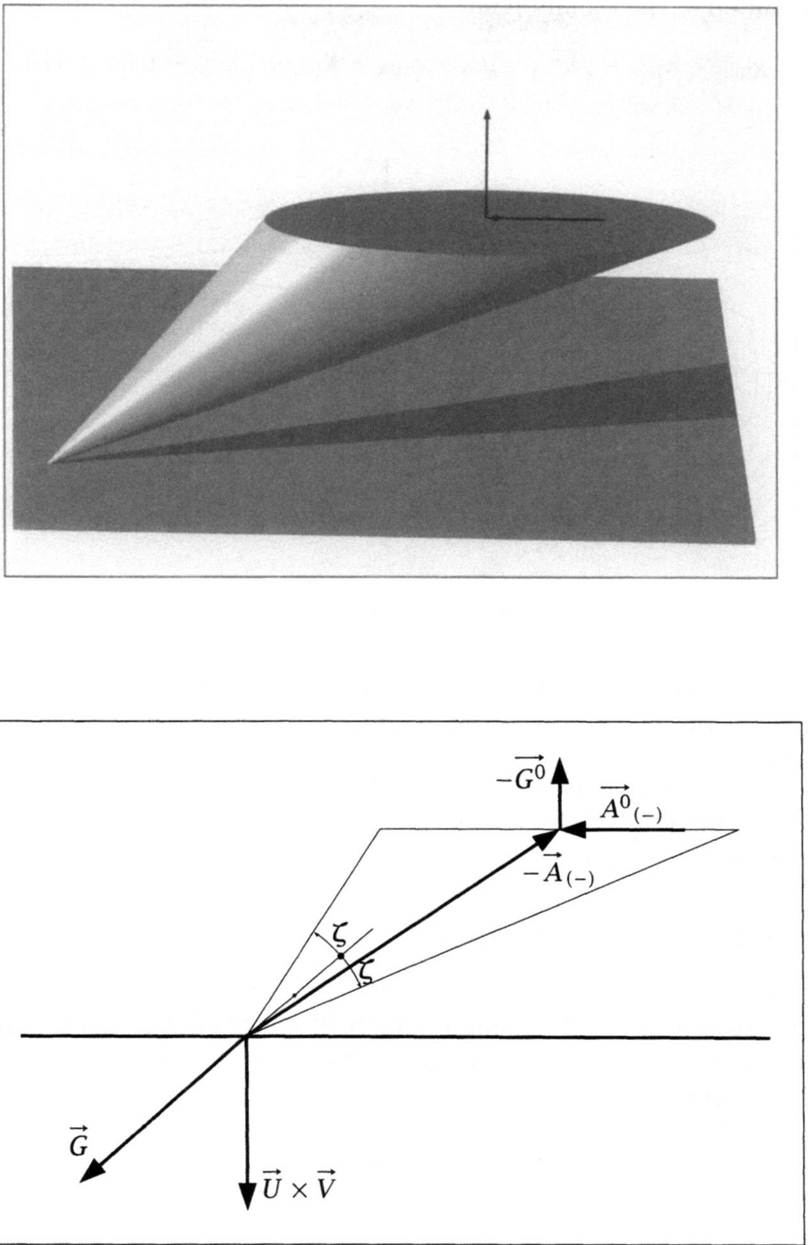

Figure 10.1: Geometric description of the orbits of the Hamiltonian $K_{(-)}$.

turning points, $\chi_{\min}$ and $\chi_{\max}$, which one computes putting $\pi = 0$ in the definition of the Hamiltonian:

$$\chi_{\min} = K_{(-)} - \sqrt{K_{(-)}^2 - \mu^2}, \quad \chi_{\max} = K_{(-)} + \sqrt{K_{(-)}^2 - \mu^2}. \tag{10.1.13}$$

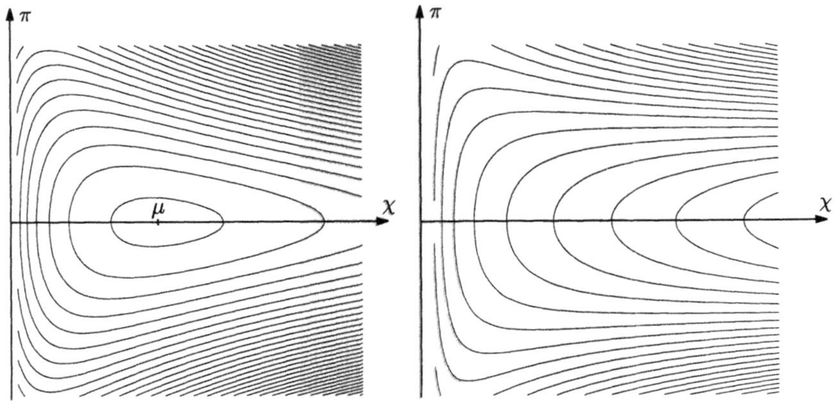

Figure 10.2: Level curves of $K_{(-)}$ (left) and $K_{(+)}$ (right).

The level curves of $K_{(-)}$ are plotted on Figure 10.2. All points of the sphere with center at the origin and radius μ are, obviously, equilibrium points.

Physically, the disappearance of the singularity is due to the repulsive potential $\frac{\mu^2}{2\chi}$, while topologically it is due to the fact that the moment map (10.1.5) is surjective, i.e., no point of the coadjoint orbit is missing.

10.1.2 Unbound Motions

Let us assume that the Hamiltonian $K_{(+)}$ describes, just as for $\mu = 0$, the unbound motions. The commutation relations of su(2, 2) imply at once that $\vec{G}$ and $\vec{A}_{(+)}$ are first integrals, and that they generate the invariance subgroup SO(1, 3).

The Fock parameters are

$$U = \begin{pmatrix} D \\ \vec{A}_{(-)} \end{pmatrix}, \quad V = \begin{pmatrix} K_{(-)} \\ -\vec{L} \end{pmatrix}.$$

These vectors satisfy the equations of motion

$$\frac{dU}{d\alpha} = \{K_{(+)}, U\} = V, \quad \frac{dV}{d\alpha} = \{K_{(+)}, V\} = U,$$

which are easily integrated to yield

$$U = \overline{U}\cosh\alpha + \overline{V}\sinh\alpha,$$
$$V = \overline{U}\sinh\alpha + \overline{V}\cosh\alpha.$$

Since $\vec{\chi} = \vec{A}_{(+)} - \vec{U}$, the "time" evolution of the position vector is

$$\vec{\chi} = \vec{A}_{(+)} - \vec{\overline{U}}\cosh\alpha - \vec{\overline{V}}\sinh\alpha.$$

The orbit is again a conic section, namely a hyperbola, whose center is $\vec{A}_{(+)}$. In fact,

$$\vec{\chi}\cdot\vec{U}\times\vec{V} = \vec{A}_{(+)}\cdot\vec{U}\times\vec{V},$$

and, $\vec{A}_{(+)}$ and $\vec{U}\times\vec{V}$ being first integrals, it follows that $\vec{\chi}$ moves on a fixed plane, orthogonal to $\vec{U}\times\vec{V}$, and passing through $\vec{A}_{(+)}$. Moreover,

$$\frac{\vec{\chi}}{\chi}\cdot\vec{G} = -\mu,$$

i.e., the position vector $\vec{\chi}$ moves on a fixed cone, with axis described by $\vec{G}$ and opening angle $\zeta = \arccos\frac{\mu}{G}$.

When $\vec{\chi}$ and $\vec{\pi}$ become parallel, the cone degenerates into a half-line and the motion becomes radial. The level curves of the Hamiltonian $K_{(+)}$ are plotted on Figure 10.2. There is only one turning point, namely

$$\chi_{\min} = -K_{(+)} + \sqrt{K_{(+)}^2 + \mu^2}, \tag{10.1.14}$$

since the particle plainly escapes always to infinity.

10.1.3 Quantization

We now proceed to quantize the Hamiltonian $K_{(-)}$, following two different methods.

The first, due to Mladenov & Tsanov (1987), is a direct generalization of the work of Simms (1974) on the quantization of the hydrogen atom. This last is a mathematically refined version of the Pauli quantization. The two vectors

$$\vec{\gamma} = \frac{1}{2}(\vec{G} + \vec{A}_{(-)}),\quad \vec{\alpha} = \frac{1}{2}(\vec{G} - \vec{A}_{(-)})$$

have the norm

$$\gamma^2 = \frac{1}{4}(G^2 + A_{(-)}^2 + 2\vec{G}\cdot\vec{A}_{(-)}) = \frac{1}{4}(K_{(-)} + \mu)^2,$$
$$\alpha^2 = \frac{1}{4}(G^2 + A_{(-)}^2 - 2\vec{G}\cdot\vec{A}_{(-)}) = \frac{1}{4}(K_{(-)} - \mu)^2.$$

Therefore, the space of the orbits for a fixed energy $K_{(-)}$ is again $S^2 \times S^2$, but with two different radii. The quantization process requires that

$$K_{(-)} + \mu = N_1 + 1, \quad K_{(-)} - \mu = N_2 + 1, \quad \mathbb{Z} \ni N_1, N_2 \geq 0,$$

from which

$$K_{(-)} = \frac{1}{2}(N_1 + N_2 + 2), \quad \mu = \frac{1}{2}(N_1 - N_2).$$

The monopole strength is quantized in half-integers: $\mu = 0, \frac{1}{2}, 1, \frac{3}{2}, \ldots$, while

$$K_{(-)} \equiv K_{(-)} - \mu + \mu = N + \mu, \quad N = 1, 2, \ldots. \tag{10.1.15}$$

The second method, see Mladenov (1989), is even more interesting: it uses the *Dirac quantization of the constraints.* Dirac's idea is to embed first the system which we want to quantize into another, bigger one, whose quantization is, however, simpler, then to perform a reduction at the quantum level to obtain the Hilbert space. Note that our quantization procedure in Chapter 9 follows basically the same philosophy. In the present case, the constraint is given by the condition (10.1.1) on the norm of the twistors.

To apply the method, it is convenient to pass to another basis, namely the first one discussed in Chapter 7. Denoting by Z_1, Z_2, Z_3, Z_4 the components of the twistors in this basis, it is straightforward to find

$$\frac{1}{2}(Z_1\overline{Z}_1 + Z_2\overline{Z}_2 + Z_3\overline{Z}_3 + Z_4\overline{Z}_4) = K_{(-)},$$
$$\frac{1}{2}(Z_1\overline{Z}_1 + Z_2\overline{Z}_2 - Z_3\overline{Z}_3 - Z_4\overline{Z}_4) = \mu.$$

To these we could add all the other generators that, being quadratic, are quantizable by the Groenewold-Van Hove Theorem: see Mack & Todorov (1969). The quantization is achieved by solving the two eigenvalue equations

$$\left(Z_1\frac{\partial}{\partial Z_1} + Z_2\frac{\partial}{\partial Z_2}\right)\Psi = (K_{(-)} + \mu)\Psi,$$
$$\left(Z_3\frac{\partial}{\partial Z_3} + Z_4\frac{\partial}{\partial Z_4}\right)\Psi = (K_{(-)} - \mu)\Psi,$$
$$\Psi = \psi_1(Z_1, Z_2)\psi_2(Z_3, Z_4)\sqrt{dZ_1 \wedge dZ_2 \wedge dZ_3 \wedge dZ_4},$$

where ψ_1 and ψ_2 are homogeneous polynomials of degree N_1 and N_2, respectively, with $N_1, N_2 \geq 0$. The presence of the *half-form* $(dZ_1 \wedge \ldots \wedge dZ_4)^{\frac{1}{2}}$ is a subtle and important point; without entering into details, we note that it is necessary in order to include into the quantization scheme a phenomenon well known in the geometric optics limit: the phase of the wave function jumps discontinuously at the caustics. This phenomenon also leads to the Maslov correction to the Bohr–Sommerfeld quantization conditions, see Maslov (1972). For a thorough study of this fascinating argument, see Guillemin & Sternberg (1977).

The two eigenvalue equations above are easily solved. Recalling (A.3.3), we find, for every k,

$$Z_k \frac{\partial}{\partial Z_k}(dZ_1 \wedge dZ_2 \wedge dZ_3 \wedge dZ_4) = dZ_1 \wedge dZ_2 \wedge dZ_3 \wedge dZ_4, \quad k \text{ not summed},$$

so that

$$K_{(-)} + \mu = N_1 + 1, \quad K_{(-)} - \mu = N_2 + 1,$$

which confirms the previous result.

The multiplicity of the spectrum is easily found also. Since the wave function is the product of homogeneous polynomials in two variables of degree N_1, N_2 respectively, the dimensionality m of the Hilbert space is then

$$m = (N_1 + 1)(N_2 + 1) = K_{(-)}^2 - \mu^2 = N^2 + 2\mu, \quad N = 1, 2, \ldots.$$

As in Guillemin & Sternberg (1982), Puta (1984), Gotay (1986), the above result confirms an old conjecture going back to Guillemin and Sternberg, namely that quantization commutes with the reduction. For a very nice treatment of this last argument with an up to date bibliography, see Ginzburg, Guillemin & Karshon (2002)

10.2 The MICZ System

The McIntosh–Cisneros–Zwanziger (MICZ) system has phase space $T^*\mathbb{R}^3$ endowed with the Poisson structure

$$\{q_i, q_j\} = 0, \quad \{p_i, q_j\} = \delta_{ij}, \quad \{p_i, p_k\} = -\mu \epsilon_{ikj} \frac{q_j}{q^3},$$

and Hamiltonian

$$H_{\text{MICZ}} = \frac{1}{2}p^2 - \frac{1}{q} + \frac{\mu^2}{2q^2}.$$

This system describes the motion of a test particle in the asymptotic field of a self-dual monopole (Feher 1986). It was studied previously in Zwanziger (1968) and McIntosh & Cisneros (1970) because of its remarkable symmetries, but without the physical interpretation above.

The equations of motion can be explicitly integrated by the Souriau method (Bates 1988). In order to get an intuitive understanding of the properties of the motion we prefer, however, to follow an approach similar to that of the previous section.

A somewhat intricate calculation (Bates 1988) reveals that H_{MICZ} admits, in full analogy with the Kepler Problem, the two vectorial first integrals

$$\vec{G} = \vec{q} \times \vec{p} - \mu \frac{\vec{q}}{q},$$
$$\vec{R} = \frac{1}{\sqrt{|2H_{\text{MICZ}}|}} \left(\vec{p} \times \vec{G} - \frac{\vec{q}}{q} \right),$$

which yield a closed algebra under Poisson bracket, namely so(4) for $H_{\text{MICZ}} < 0$ and so(1,3) for $H_{\text{MICZ}} > 0$, respectively.

The trajectories are again conic sections. The relation $\vec{G} \cdot \frac{\vec{q}}{q} = -\mu$ shows in fact that they lie on the cone with opening angle $\zeta = \arccos \frac{\mu}{G}$, and symmetry axis defined by $\vec{G}$. Moreover,

$$\left(\vec{p} \times \vec{G} - \frac{\vec{q}}{q} - \frac{\vec{G}}{\mu} \right) \cdot \vec{q} = \left(\vec{q} \times \vec{p} \right) \cdot \vec{G} - q + q \qquad (10.2.1)$$
$$= \left(\vec{G} + \mu \frac{\vec{q}}{q} \right) \cdot \vec{G} = G^2 - \mu^2,$$

which shows that the motion lies in a plane. The first vector in the bracket on the left-hand side and the last scalar on the right-hand side are in fact both constant.

At this point, it should be obvious that the Hamiltonian H_{MICZ} can be deduced from the two Hamiltonians $K_{(-)}$ and $K_{(+)}$, just as that of the Kepler Problem is deduced from the analogous expressions in the Main Moment Map (6.4.5). If one defines $\mathscr{k}_{\text{MICZ}}$ choosing in (10.1.11)

$$\sigma(p_t) = \sqrt{\frac{-2p_t}{\epsilon}}, \quad \kappa(p_t) = -\frac{1}{\sigma(p_t)},$$

(the same as for the Kepler Problem), the constraint (10.1.10) of the energy surface becomes

$$p_t + \frac{1}{2}p^2 \pm \frac{1}{q} + \frac{\mu^2}{2q^2} = 0, \quad \text{that is } p_t + H_{\text{MICZ}} = 0,$$

but allowing for a repulsive potential $\frac{1}{q}$ also. Moreover, it is straightforward to verify that $\vec{G}$ and $\vec{A}_{(-)}$ or $\vec{A}_{(+)}$ are transformed into the angular momentum and the Runge-Lenz-Laplace vector, respectively. The transformation being canonical, these vectors close obviously into an so(4) or so(1,3) algebra. All properties of the motion follow from those presented in the previous section.

Recalling (10.1.15), the expression $H_{\rm MICZ} = -\frac{1}{2K_{(-)}^2}$ gives the energy spectrum for the quantized system.

REMARK 10.2 With a slightly different approach, Iwai & Uwano (1986) arrived, independently, at the same results. Using our notation, their starting point is still the Hamiltonian

$$K_{\rm IU} = \frac{1}{2}(w_\alpha w_\alpha + z_\alpha z_\alpha), \quad \Theta = z_\alpha dw_\alpha - w_\alpha dz_\alpha, \quad \alpha = 1,\dots,4,$$

i.e., the $-G_{-1,0}$ generator. Then, they define the other Hamiltonian

$$H_{\rm IU} = \frac{1}{z_\alpha z_\alpha}(K_{\rm IU} - 1) - \frac{1}{2} = \frac{1}{2}\frac{w_\alpha w_\alpha}{z_\alpha z_\alpha} - \frac{1}{z_\alpha z_\alpha},$$

called the *conformal Kepler Hamiltonian.* When restricted to the "energy" hypersurface $K_{\rm IU} = 1$ or $H_{\rm IU} = -\frac{1}{2}$ and after a change of the "time" variable similar to (2.2.2) on page 23, the two Hamiltonian flows coincide. "Pulling-back" $H_{\rm IU}$ with the map $\Psi^{(\mu)}$, one obtains $H_{\rm MICZ}$. This makes the diagram

$$\begin{array}{ccc}
 & \text{IU-map} & \\
K_{\rm IU} = \frac{1}{2}(w_\alpha w_\alpha + z_\alpha z_\alpha) & \longrightarrow & H_{\rm IU} = \frac{1}{2}\frac{w_\alpha w_\alpha}{z_\alpha z_\alpha} - \frac{1}{z_\alpha z_\alpha} \\
\left(\Psi^{(\mu)}\right)^* \Big\downarrow & & \Big\downarrow \left(\Psi^{(\mu)}\right)^* \\
 & \text{Moser-map} & \\
K_{(-)} = \frac{1}{2}\chi\,(\pi^2 + 1) + \frac{1}{2}\frac{\mu^2}{\chi} & \longrightarrow & H_{\rm MICZ} = \frac{1}{2}p^2 - \frac{1}{q} + \frac{\mu^2}{2q^2}
\end{array}$$

commute, where the Moser-map is defined by (5.1.6) on page 100. We have taken into account that, because of the restriction to the "energy" hypersurface $H = -\frac{1}{2}$, $p = \pi$ and $q = \chi$ hold. The commutative diagram on page 155 recalls that the map $\Psi^{(\mu)}$ is the inverse of the Kustaanheimo-Stiefel transformation. Note that the Moser-map, together with the change of the "time" variable, is equivalent to the transformation $\ell_{\rm MICZ}$.

10.3 The Taub-NUT System

The phase space of the Taub-NUT system is

$$T^*(\mathbb{R}^3 - \{0\}) - S^2_{\text{sing}}, \quad S^2_{\text{sing}} = \{\vec{q} : q^2 = 1\},$$

with Poisson structure as in the previous section, and Hamiltonian

$$H_{\text{TNUT}} = \frac{1}{2}\frac{p^2}{1-\frac{1}{q}} + \frac{\mu^2}{2}\left(1 - \frac{1}{q}\right).$$

This expression is singular for $q = 0$ and $q^2 = 1$, so that the origin as well as the unit sphere S^2_{sing} must be excluded.

Things work exactly as in the MICZ case discussed above. The two vectorial first integrals,

$$\vec{G} = \vec{q} \times \vec{p} - \mu\frac{\vec{q}}{q},$$
$$\vec{R} = \frac{1}{\sqrt{|\mu^2 - 2H_{\text{TNUT}}|}}\left[\vec{p} \times \vec{G} - (\mu^2 - H_{\text{TNUT}})\frac{\vec{q}}{q}\right]$$

close under Poisson bracket, as before, into so(4) or so(1, 3) algebras, depending on $H_{\text{TNUT}} < \frac{\mu^2}{2}$ or $H_{\text{TNUT}} > \frac{\mu^2}{2}$, respectively. The trajectories are conic sections; in fact, to (10.2.1) corresponds

$$\left[\vec{p} \times \vec{G} - (\mu^2 - H_{\text{TNUT}})\left(\frac{\vec{q}}{q} + \frac{\vec{G}}{\mu}\right)\right] \cdot \vec{q} = G^2 - \mu^2.$$

The Hamiltonian H_{TNUT} is obviously another first integral. The velocity is

$$\vec{v} = \frac{\partial H_{\text{TNUT}}}{\partial \vec{p}} = \frac{\vec{p}}{1-\frac{1}{q}},$$

and the Hamiltonian can be written as

$$H_{\text{TNUT}} = \frac{1}{2}\left(1 - \frac{1}{q}\right)\left(v^2 + \mu^2\right). \tag{10.3.1}$$

Clearly, the particle can only escape to infinity if $H_{\text{TNUT}} \geq \frac{\mu^2}{2}$. If $H_{\text{TNUT}} < \frac{\mu^2}{2}$ the motion is bounded, but in two different ways, as we now show. Via (10.3.1), the sign of the energy depends on q being smaller or larger than 1. The excluded points form the singular sphere S^2_{sing}, which divides the space in two regions. Energy conservation implies that a particle cannot go from one region to the other (although it can hit the boundary S^2_{sing}, see below). If a finite-energy motion approaches S^2_{sing}, its velocity v goes to infinity as $\frac{1}{\sqrt{1-\frac{1}{q}}}$, as seen from (10.3.1). Its momentum p goes hence to zero as $\sqrt{1-\frac{1}{q}}$. The motions in the interior of S^2_{sing} have negative energy: they are the *tightly bound motions.* Those in the exterior and having energy $0 < H_{\text{TNUT}} < \frac{\mu^2}{2}$ are the *lightly bound motions.*

Let us consider the lightly bound motions. In the generic case, the orbital angular momentum $\vec{q} \times \vec{p}$ is nonzero, and the cone has opening angle $\zeta \neq 0$. Such motions avoid S^2_{sing}. Indeed, for nonvanishing orbital angular momentum, $\vec{G}$ cannot be radial, while, when hitting the singular sphere, the orbital part necessarily vanishes, since $p = 0$, requiring $\vec{G}$ to be radial.

Consider now the radial motions. Fixing a direction, we work with q, p. If the initial velocity is inward, the particle reaches the singularity in finite time and leaves the "physical" space. If the velocity is outward (but sufficiently low so the motion remains bounded), we have

$$v = \sqrt{\frac{(2H_{\text{TNUT}} - \mu^2)q + \mu^2}{q - 1}}$$

which shows that there is a unique turning point $q_{\max}$ where v vanishes, namely at

$$q_{\max} = \frac{\mu^2}{\mu^2 - 2H_{\text{TNUT}}} > 1.$$

After reaching $q_{\max}$, the particle returns, and falls inward until it disappears in the singular sphere. At this very moment, another radial motion leaves the singularity and follows the same phase space trajectory backwards. In order to regularize the problem, such motions should (and will) be identified. Then all motions become periodic.

The situation is basically the same for the tightly bound motions, i.e., for $H_{\text{TNUT}} < 0$. The nonradial motions are ellipses that avoid the origin as well the singular sphere. A radial motion has an internal turning point

$$q_{\min} = \frac{\mu^2}{\mu^2 - 2H_{\text{TNUT}}} < 1.$$

All radial motions fall into S^2_{sing} in finite time from the inside, with infinite velocity and zero momentum. Such a motion should be identified with the motion that leaves the singular sphere in the opposite direction along the same trajectory.

The close similarities between the properties of this system and that of the first section hint that there may exist a mapping $\mathscr{C}_{\text{TNUT}}$ that con-

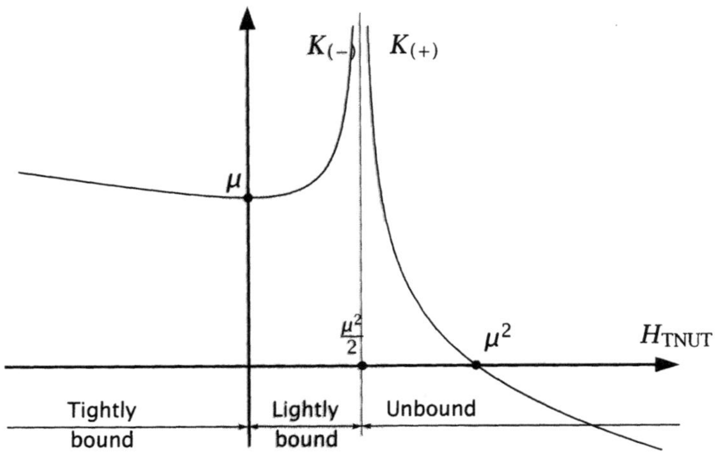

Figure 10.3: Energy transformation for the Taub-NUT system.

nects the two systems. We show now that this is indeed so. Choose, in (10.1.11),

$$\sigma(p_t) = \sqrt{-\frac{2p_t + \mu^2}{\epsilon}}, \quad \kappa(p_t) = -\frac{p_t + \mu^2}{\sigma(p_t)}, \quad \epsilon = \mathrm{sgn}(2H_{\mathrm{TNUT}} - \mu^2).$$

Then the usual mechanism gives

$$p_t + \frac{1}{2}\frac{p^2}{1 \pm \frac{1}{q}} + \frac{\mu^2}{2}\left(1 \pm \frac{1}{q}\right) = 0,$$

i.e., a slight generalization of the Taub-NUT Hamiltonian. The positive sign here is much less interesting. It gives rise always to hyperbolic motions, not requiring any regularization; it is the analogue of the repulsive potential in the Kepler case. In what follows we will only consider the negative sign, thus the positive sign in (10.1.10). The Runge-Lenz-Laplace type vector $\vec{R}$ is the transform of $\vec{A}_{(-)}$ or $\vec{A}_{(+)}$, depending on the two values of ϵ.

A closer analysis of the transformation $\mathscr{C}_{\mathrm{TNUT}}$ (the choice of σ and κ determines in effect its inverse) explains the dynamical properties of the Taub-NUT system in a direct way. As obvious from Figure 10.3, κ is not injective, and inverting $\mathscr{C}_{\mathrm{TNUT}}^{-1}$ requires therefore some care. We now proceed to the inversion, studying separately the three intervals in which κ is monotone.

The level curves of H_{TNUT} are plotted on Figure 10.4 .

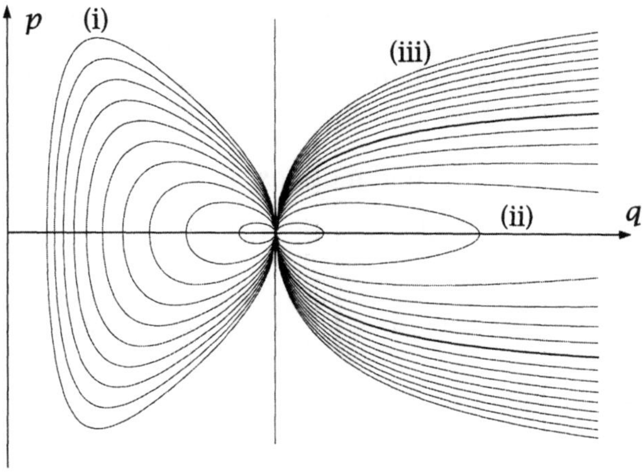

Figure 10.4: Level curves of H_{TNUT}.

(i) $H_{\text{TNUT}} < 0$, i.e., the tightly bound motions.

Inverting $K_{(-)} = \frac{\mu^2 - H_{\text{TNUT}}}{\sqrt{\mu^2 - 2H_{\text{TNUT}}}}$ one finds

$$H_{\text{TNUT}} = -\sqrt{K_{(-)}^2 - \mu^2}\left(K_{(-)} + \sqrt{K_{(-)}^2 - \mu^2}\right),$$

which, recalling (10.1.15), gives the energy spectrum in the quantum case. The scaling factor is

$$\sigma = \sqrt{\mu^2 - 2H_{\text{TNUT}}} = K_{(-)} + \sqrt{K_{(-)}^2 - \mu^2},$$

which equals χ_{max} in (10.1.13). The first equation of (10.1.11) shows that all the motions are mapped into trajectories *internal* to S^2_{sing}. The singular sphere is only hit by radial motions, since χ_{max} is reached only if $\pi = 0 = p$. See Figure 10.4.

(ii) $0 < H_{\text{TNUT}} < \frac{\mu^2}{2}$, i.e., the lightly bound motions.

$K_{(-)}$ is as before; inverting it yields rather

$$H_{\text{TNUT}} = \sqrt{K_{(-)}^2 - \mu^2}\left(K_{(-)} - \sqrt{K_{(-)}^2 - \mu^2}\right),$$

which, as in the previous item, gives the energy spectrum in the quantum case. The scaling factor is

$$\sigma = \sqrt{\mu^2 - 2H_{\text{TNUT}}} = K_{(-)} - \sqrt{K_{(-)}^2 - \mu^2},$$

which equals $\chi_{\min}$ in (10.1.13). The first equation of (10.1.11) shows that all the motions are sent into trajectories *external* to S^2_{sing}. The singular sphere can be hit only by the radial motions, since $\chi_{\min}$ is reached only if $\pi = 0 = p$. See Figure 10.4.

(iii) $H_{\text{TNUT}} > \frac{\mu^2}{2}$, i.e., unbound motions.

Inverting $K_{(+)} = \frac{\mu^2 - H_{\text{TNUT}}}{\sqrt{2H_{\text{TNUT}} - \mu^2}}$ one finds

$$H_{\text{TNUT}} = \sqrt{\mu^2 + K^2_{(+)}}\left(\sqrt{\mu^2 + K^2_{(+)}} - K_{(+)}\right).$$

The scaling factor is

$$\sigma = \sqrt{2H_{\text{TNUT}} - \mu^2} = -K_{(+)} + \sqrt{K^2_{(+)} + \mu^2},$$

that equals the $\chi_{\min}$ of (10.1.14). Considerations similar to those above follow in a straightforward manner, the only difference being that the motions are now unbounded.

At last, note that the sphere of the equilibrium points is transformed into the singular sphere S^2_{sing}.

10.4 The BPST Instanton

The work done in the first section of this chapter can be generalized to the 5-dimensional case, replacing the Dirac monopole by the BPST instanton, and (10.1.6) by

$$\frac{\mathbb{H}^2 - \{0\}}{\text{SU}(2)} = \frac{\mathbb{R}^+ \times S^7}{S^3} = \mathbb{R}^+ \times S^4 = \mathbb{R}^5 - \{0\}.$$

Let us consider a quaternionic twistor of the type $\psi = \begin{pmatrix} z \\ iw \end{pmatrix}$ with $z, w \in \mathbb{H}^2$. A quaternion $q \in \mathbb{H}$ is defined as $q = q_0\mathbf{1} + i\vec{s}\cdot\vec{q}$, $q_0 \in \mathbb{R}$, $\vec{q} \in \mathbb{R}^3$, and conjugation as $q^\dagger = q_0\mathbf{1} - i\vec{s}\cdot\vec{q}$. Let us endow the quaternionic space with a pseudo-Hermitian form $(\psi,\psi) = \psi^\dagger \mathbf{h}\psi$. As seen in Chapter 7, the left action $\psi \mapsto g\psi$, $g \in \text{Spin}(2,6) = \text{Sp}(4,\mathbb{H})$ leaves this form invariant. Another natural action relevant for our purposes is the right action $\psi \mapsto \psi g$, $g \in \text{SU}(2) = \text{U}(1,\mathbb{H})$. Define the equivalence relation $\psi \sim \psi g$. The space $T_0/\sim$ (where T_0 is the space of null twistors), when endowed with the two-form $d\Theta$, $\Theta = \frac{1}{2}\,\text{Im}\,\text{Tr}(\psi, d\psi)$, is symplectomorphic to $T^*\mathbb{R}^5$. Moreover, $j = -i\psi\psi^\dagger\mathbf{h}$ defines a moment map $T_0/\sim \to \text{spin}^*(2,6)$.

Let us generalize this construction considering nonnull twistors. Defining

$$T_\mu = \{\psi \in \mathbb{H}^{2,2} : (\psi,\psi) = 2\vec{\mu}\cdot\vec{\mathbf{s}},\ \vec{\mu} \in \mathbb{R}^3\},$$

$T_\mu/\sim$, carrying the two-form $d\Theta$, is a 10-dimensional symplectic manifold. To investigate the form Θ, we parametrize T_μ in a way similar to the 3-dimensional case. Recall first the polar form of a quaternion

$$q = \varrho e^{i\phi},\ \varrho \in \mathbb{R},\ e^{i\phi} = \cos\phi\mathbf{1} + i\sin\phi\vec{n}\cdot\vec{\mathbf{s}},\quad \vec{n}\cdot\vec{n} = 1.$$

Plainly, $e^{i\phi}e^{-i\phi} = \mathbf{1}$; quaternions of this type are called *unitary*. Putting

$$e^{i\frac{\phi}{2}} \stackrel{\text{def}}{=} \cos\frac{\phi}{2}\mathbf{1} + i\sin\frac{\phi}{2}\vec{n}\cdot\vec{\mathbf{s}},$$

we have $e^{i\frac{\phi}{2}}e^{i\frac{\phi}{2}} = e^{i\phi}$. Moreover,

$$e^{-i\frac{\phi}{2}}(\vec{\mu}\cdot\vec{\mathbf{s}})e^{i\frac{\phi}{2}} = \vec{\mu}'\cdot\vec{\mathbf{s}},$$

the vector $\vec{\mu}'$ being the vector $\vec{\mu}$ rotated by an angle ϕ around the axis $\vec{n}$, as one sees by the decomposition $\vec{\mu} = \vec{\mu}_\parallel + \vec{\mu}_\perp$ with $\vec{\mu}_\parallel \times \vec{n} = 0$ and $\vec{\mu}_\perp\cdot\vec{n} = 0$. Recalling the Definition (7.3.4) of the matrices $\mathbf{S}_k$, put

$$\boldsymbol{X} = \frac{1}{2}X\mathbf{1}_4 + \frac{1}{2}X_k\mathbf{S}_k,\quad X = \sqrt{X_kX_k},$$

$$\boldsymbol{\pi} = \pi_h\mathbf{S}_h + \frac{i}{X}\mathbf{M},$$

where the 4×4 matrix $\mathbf{M}$ (analogous to $\mu\mathbf{1}_2$ of the 3-dimensional case) must be defined in such a way that the condition $(\psi,\psi) = 2\vec{\mu}\cdot\vec{\mathbf{s}}$ holds. To this end, put

$$\sqrt{\boldsymbol{X}} = \begin{pmatrix} \sqrt{X}\sin\frac{\vartheta}{2}e^{i\frac{\phi}{2}}e^{i\frac{\beta}{2}} \\ \sqrt{X}\cos\frac{\vartheta}{2}e^{-i\frac{\phi}{2}}e^{i\frac{\beta}{2}} \end{pmatrix},$$

from which $\sqrt{\boldsymbol{X}}\sqrt{\boldsymbol{X}}^\dagger = \boldsymbol{X}$, and

$$\mathbf{M} = \begin{pmatrix} e^{i\frac{\phi}{2}}e^{i\frac{\beta}{2}}\vec{\mu}\cdot\vec{\mathbf{s}}\,e^{-i\frac{\beta}{2}}e^{-i\frac{\phi}{2}} & 0 \\ 0 & e^{-i\frac{\phi}{2}}e^{i\frac{\beta}{2}}\vec{\mu}\cdot\vec{\mathbf{s}}\,e^{-i\frac{\beta}{2}}e^{i\frac{\phi}{2}} \end{pmatrix},$$

$$\psi = \begin{pmatrix} \sqrt{\boldsymbol{X}} \\ -i\boldsymbol{\pi}\sqrt{\boldsymbol{X}} \end{pmatrix},$$

so that the above condition is identically satisfied. We now are ready to calculate the pull-back of the 1-form

$$\begin{aligned}\Theta &= \frac{1}{2}\operatorname{Im}\operatorname{Tr}\psi^\dagger\mathbf{h}\,d\psi \\ &= \frac{1}{2}\operatorname{Tr}\left[-\frac{1}{2}\sqrt{\boldsymbol{X}}^\dagger d(\boldsymbol{\pi}+\boldsymbol{\pi}^\dagger)\sqrt{\boldsymbol{X}} - \frac{i}{X}(\sqrt{\boldsymbol{X}}^\dagger\mathbf{M}\,d\sqrt{\boldsymbol{X}} - d\sqrt{\boldsymbol{X}}^\dagger\mathbf{M}\sqrt{\boldsymbol{X}})\right].\end{aligned}$$

Just as in the 3-dimensional case, the first term in the square brackets turns out to be, up to an exact 1-form, the canonical 1-form $\pi_k\, d\chi_k$ of $T^*\mathbb{R}^5$. The second term may be rearranged in a more suitable form. Define

$$i\omega(\phi) = e^{-i\frac{\phi}{2}}\, de^{i\frac{\phi}{2}} = -de^{-i\frac{\phi}{2}} e^{i\frac{\phi}{2}},$$

i.e., the left invariant 1-form with values in su*(2), hence $\omega = \dfrac{\vec{\omega}}{2} \cdot \vec{\mathbf{s}}$. Moreover,

$$i\omega^*(\phi) = e^{i\frac{\phi}{2}}\, de^{-i\frac{\phi}{2}} = -de^{i\frac{\phi}{2}} e^{-i\frac{\phi}{2}}$$

is, by definition, the right invariant 1-form. After some calculation, we obtain

$$\begin{aligned}
&-\frac{1}{2}\frac{i}{\chi}\mathrm{Tr}\left[\sqrt{\mathbf{X}}^{\dagger}\mathbf{M}\, d\sqrt{\mathbf{X}} - d\sqrt{\mathbf{X}}^{\dagger}\mathbf{M}\sqrt{\mathbf{X}}\right] \\
&\quad = \frac{1}{2}\mathrm{Tr}\left[\begin{array}{c} e^{i\frac{\beta}{2}}\vec{\mu}\cdot\vec{\mathbf{s}}\, e^{-i\frac{\beta}{2}}\left(\sin^2\frac{\vartheta}{2}\vec{\omega}(\phi)\cdot\vec{\mathbf{s}} + \cos^2\frac{\vartheta}{2}\vec{\omega}^*(\phi)\cdot\vec{\mathbf{s}}\right) \\ +(\vec{\mu}\cdot\vec{\mathbf{s}})\,(\vec{\omega}(\beta)\cdot\vec{\mathbf{s}}) \end{array}\right] \\
&\quad = \vec{\mu}'\cdot\left[\sin^2\frac{\vartheta}{2}\vec{\omega}(\phi) + \cos^2\frac{\vartheta}{2}\vec{\omega}^*(\phi)\right] + \vec{\mu}\cdot\vec{\omega}(\beta),
\end{aligned}$$

where now $\vec{\mu}'$ is the vector $\vec{\mu}$ rotated by $e^{-i\frac{\beta}{2}}$. The first term is the instantonic connection 1-form, while the latter is the canonical 1-form on T^*S^3.

For a deeper study of the problem see Iwai (1990).

Part III

Perturbation Theory

CHAPTER 11

General Perturbation Theory

Having studied in the preceding chapters the Kepler Problem to some extent, we want now to investigate the corresponding perturbed problem. In general, given the Hamiltonian H_0 of a completely integrable system, we define the *perturbed problem* to be that with Hamiltonian[(1)]

$$H = H_0 + \varepsilon H_p,$$

where H_p is a generic function whose numerical value is of the same order of H_0, while $\varepsilon << 1$. The perturbed problem differs thus slightly from the unperturbed one, but unfortunately the same can not be said for the solution: a small perturbation can give rise to *secular* effects, i.e., to a slow but progressive wandering from the unperturbed (and known for infinite time) solution.

In this chapter we shall describe the methods of general perturbation theory, deferring to the following two chapters their application to the Kepler Problem. In the first section of the chapter only *formal* series expansion are considered, while in the second section the subtle problem of their convergence is considered.

In the exposition, we will follow some unpublished lectures of Giorgilli for the formalism of the Lie series, and Benettin, Galgani, Giorgilli & Strelcyn (1984) for the proof of the Kolmogorov Theorem.

[(1)]In this chapter the symbol H denotes a generic Hamiltonian, not necessarily that of the Kepler Problem.

11.1 Formal Expansions

The basic idea of perturbation theory, i.e., to consider the perturbation in some sense "small" and thus to proceed to series expansions, goes back to the early ages of the classical mechanics. Its actual form, however, is due to Poincaré (1892-1893-1899), and may be summarized in this way: find a canonical transformation, which differs from the identity by a quantity of order ε, such that the transformed Hamiltonian is integrable up to the second order terms. One may obviously iterate the procedure, pushing the perturbation to third order, and so on. If the process converges (but this is the key point), to higher orders one obtains better and better approximations.

In the sequel we adopt this point of view, but with a technical improvement that simplifies some calculations: as suggested in Hori (1966) and Deprit (1969), use is made of the Lie series, which avoids those inversion problems which occur typically for methods involving the generating function of a canonical transformation.

11.1.1 Lie Series and Formal Canonical Transformations

Let us consider a one-parameter family of diffeomorphisms Φ_τ, $\tau \in \mathbb{R}$, defined in some analytic manifold N. Such a family is said to be a *flow* when

(i) for $\tau = 0$ one has $\Phi_0 =$ identity,

(ii) $\Phi_{\tau'} \circ \Phi_{\tau''} = \Phi_{\tau'+\tau''}$.

If the flow is defined for every τ, it is a one-parameter group of diffeomorphisms. Otherwise, Φ_τ represents a *local* flow. Let $x = x_1, \ldots, x_n$ be local coordinates on N, and $x_\tau = \Phi_\tau x$ the coordinates of the transformed point. To every flow Φ_τ corresponds the vector field (a sort of velocity field)

$$X(x) = \frac{d}{d\tau}(\Phi_\tau x)\bigg|_{\tau=0},$$

called the *generator* of the flow.

One may reverse the argument, starting from a vector field X and defining the corresponding flow Φ_τ as that which satisfies

$$\frac{d}{d\tau}(\Phi_\tau x) = X(\Phi_\tau x),$$

i.e., as the solution of the dynamical system

$$\frac{dx}{d\tau} = X(x). \tag{11.1.1}$$

Let now $f : N \to \mathbb{R}$ be any function of the state of the system, which we shall also call a *dynamical variable*. Then the evolution of f under the flow generated by the vector field X is naturally defined as the function $f(\Phi_\tau x)$. If f is differentiable, then its derivative with respect to τ is

$$\frac{df}{d\tau} = \sum_i X^i \frac{\partial f}{\partial x^i}.$$

This suggests defining the operator

$$\mathcal{L}_X f = \sum_i X^i \frac{\partial f}{\partial x^i},$$

which maps the space of the analytic functions on N into itself. $\mathcal{L}_X f$ is called the *Lie derivative* of f along the flow generated by X. It is plainly a linear operator. Moreover, it satisfies the Leibniz rule

$$\mathcal{L}_X(fg) = f\mathcal{L}_X g + g\mathcal{L}_X f, \tag{11.1.2}$$

which, by induction, can be generalized to

$$\mathcal{L}_X^s(fg) = \sum_{j=0}^{s} \binom{s}{j} (\mathcal{L}_X^j f)(\mathcal{L}_X^{s-j} g), \tag{11.1.3}$$

where $\binom{s}{j} = \frac{s!}{(s-j)!j!}$ are the binomial coefficients. In fact

$$\begin{aligned}
\mathcal{L}_X^s(fg) &= \mathcal{L}_X \left(\mathcal{L}_X^{s-1}(fg)\right) \\
&= \mathcal{L}_X \sum_{j=0}^{s-1} \binom{s-1}{j} (\mathcal{L}_X^j f)(\mathcal{L}_X^{s-1-j} g) \\
&= \sum_{j=1}^{s} \binom{s-1}{j-1} (\mathcal{L}_X^j f)(\mathcal{L}_X^{s-j} g) + \sum_{j=0}^{s-1} \binom{s-1}{j} (\mathcal{L}_X^j f)(\mathcal{L}_X^{s-j} g) \\
&= f\mathcal{L}_X^s g + \sum_{j=1}^{s-1} \left[\binom{s-1}{j-1} + \binom{s-1}{j}\right] (\mathcal{L}_X^j f)(\mathcal{L}_X^{s-j} g) + g\mathcal{L}_X^s f,
\end{aligned}$$

and the claim follows from the property of the binomial coefficients

$$\binom{s-1}{j-1} + \binom{s-1}{j} = \binom{s}{j}.$$

Given two vector fields X and Y, the commutator $[\mathcal{L}_X, \mathcal{L}_Y]$ is again a derivative with respect to a vector field. In fact

$$[\mathcal{L}_X, \mathcal{L}_Y] = \mathcal{L}_{[X,Y]},$$

where, by definition,

$$[X,Y]^j = \sum_{h=1}^{n} \left(X^h \frac{\partial Y^j}{\partial x^h} - Y^h \frac{\partial X^j}{\partial x^h} \right) \qquad (11.1.4)$$

is the *Lie bracket*. In fact, the second derivatives cancel out.

As one can easily verify, the commutator between vector fields possesses the following properties:

(i) it is linear, i.e., $[X, \alpha Y + \beta Z] = \alpha[X,Y] + \beta[X,Z]$;

(ii) it is skew symmetric, i.e., $[X,Y] + [Y,X] = 0$;

(iii) it satisfies the Jacobi identity, i.e.,

$$[X,[Y,Z]] + [Y,[Z,X]] + [Z,[X,Y]] = 0$$

for every vector field. This shows that the vector fields form a Lie algebra with respect to the commutator.

The Lie derivative allows one to construct explicitly, at a formal level, the general solution of the dynamical system (11.1.1). In fact, let us look for a solution which satisfies the initial conditions $x(0) = x_0$, of the form

$$x(\tau) = x_0 + \tau x^{(1)} + \tau^2 x^{(2)} + \cdots ,$$

namely as a power expansion of the solution, the coefficients being

$$x^{(s)} = \frac{1}{s!} \frac{d^s}{d\tau^s} (\Phi_\tau x_0) \bigg|_{\tau=0} .$$

On the other hand, since Φ_τ is the flow generated by the vector field X, it is immediately seen that

$$\frac{d}{d\tau} (\Phi_\tau x_0) \bigg|_{\tau=0} = X|_{x=x_0}, \ldots, \frac{d^s}{d\tau^s} (\Phi_\tau x_0) \bigg|_{\tau=0} = \mathcal{L}_X^{s-1} X \Big|_{x=x_0}, \ldots,$$

where $\mathcal{L}_X X = (\mathcal{L}_X X_1, \ldots, \mathcal{L}_X X_n)$. Remarking that $\mathcal{L}_X x = X$, the solution can be given in the explicit form

$$\begin{aligned} x(\tau) &= x_0 + \tau \mathcal{L}_X x|_{x=x_0} + \frac{\tau^2}{2!} \mathcal{L}_X^2 x \Big|_{x=x_0} + \frac{\tau^3}{3!} \mathcal{L}_X^3 x \Big|_{x=x_0} + \cdots \\ &\stackrel{\text{def}}{=} \exp(\tau \mathcal{L}_X) x_0 . \end{aligned}$$

The (just now defined) exponential Lie operator $\exp(\tau\mathcal{L}_X)$ is clearly linear and provides us with a flow. In fact, for $\tau = 0$ it is the identity, and moreover

$$\exp(t\mathcal{L}_X)\circ\exp(\tau\mathcal{L}_X)=\sum_{r\geq 0}\frac{t^r}{r!}\mathcal{L}_X^r\sum_{s\geq 0}\frac{\tau^s}{s!}\mathcal{L}_X^s$$

$$=\sum_{k\geq 0}\frac{1}{k!}\mathcal{L}_X^k\sum_{j=0}^{k}\binom{k}{j}t^j\tau^{k-j}=\sum_{k\geq 0}\frac{(t+\tau)^k}{k!}\mathcal{L}_X^k=\exp\left((t+\tau)\mathcal{L}_X\right).$$

A basic feature of the exponential Lie operator is the so-called *exchange property*, which will be used in the reduction to normal form of a perturbed Hamiltonian. Put

$$x_\tau=\exp(\tau\mathcal{L}_X)x=x+\tau X+\cdots,$$

and consider it as a coordinate change: $x\mapsto x_\tau$. Given some $f(x)$, define $f_\tau(x)=f(x_\tau)$. Then

$$\left.\frac{df_\tau(x)}{d\tau}\right|_{\tau=0}=\left.\frac{df(x_\tau)}{d\tau}\right|_{\tau=0}=\left.\sum_{k=1}^{n}\frac{\partial f(x_\tau)}{\partial x_\tau^k}\frac{dx_\tau^k}{d\tau}\right|_{\tau=0}$$

$$=\sum_{k=1}^{n}X^k\frac{\partial f(x)}{\partial x^k}=\mathcal{L}_Xf(x),$$

from which

$$f(\exp(\tau\mathcal{L}_X)x)=f(x)+\tau\left.\frac{df(x_\tau)}{d\tau}\right|_{\tau=0}+\frac{1}{2!}\tau^2\left.\frac{d^2f(x_\tau)}{d\tau^2}\right|_{\tau=0}+\ldots$$

$$=f(x)+\tau\mathcal{L}_Xf(x)+\frac{1}{2!}\tau^2\mathcal{L}_X^2f(x)+\cdots=\exp(\tau\mathcal{L}_X)f(x),$$

showing that the exponential Lie operator can be moved in front of the function, so that the transformed function can be calculated without substitution of variables.

Up to now the manifold N has been supposed to be a generic one. From now on N will be a $2n$-dimensional symplectic manifold, and the vector field X_χ the symplectic gradient of some Hamiltonian $\chi : N\to\mathbb{R}$, i.e.,

$$(X_\chi)^k=\sum_{h=1}^{2n}\Omega^{kh}\frac{\partial\chi}{\partial x^h},$$

(Ω^{kh}) being the inverse of the matrix associated to a symplectic (i.e., closed and invertible) 2-form. To simplify the notation we put $\mathcal{L}_{X_\chi}=\mathcal{L}_\chi$. The Poisson bracket between two dynamical variables $F,G:N\to\mathbb{R}$ can thus be written, for the skew symmetry of the symplectic 2-form,

$$\{F,G\}\stackrel{\text{def}}{=}\sum_{h,k=1}^{2n}\frac{\partial G}{\partial x^h}\Omega^{hk}\frac{\partial F}{\partial x^k}=\mathcal{L}_FG=-\mathcal{L}_GF.$$

The closure property of the 2-form ensures the validity of the classical Jacobian formula

$$\{H, \{F, G\}\} + \{F, \{G, H\}\} + \{G, \{H, F\}\} = 0,$$

which can be written as

$$\mathcal{L}_H\{F, G\} = \{F, \mathcal{L}_H G\} + \{\mathcal{L}_H F, G\},$$

similar to the Leibniz formula (11.1.2). As for (11.1.3), it is easy then to prove, recurrently, that

$$\mathcal{L}_H^s\{F, G\} = \sum_{j=0}^{s} \binom{s}{j} \{\mathcal{L}_H^{s-j} F, \mathcal{L}_H^j G\}. \tag{11.1.5}$$

We can now prove the following basic property:

PROPOSITION 11.1 *The transformation* $x \mapsto x' = \exp \mathcal{L}_X x$ *is canonical, i.e., the Poisson bracket of two transformed functions* F, G *is equal to the transform of the Poisson bracket.*

Proof. Simply verify that

$$\begin{aligned}
\{\exp \mathcal{L}_X F, \exp \mathcal{L}_X G\} &= \sum_{m \geq 0} \sum_{j \geq 0} \frac{1}{m!} \frac{1}{j!} \{\mathcal{L}_X^m F, \mathcal{L}_X^j G\} \\
&= \sum_{m \geq 0} \sum_{j \geq 0} \frac{1}{(m+j)!} \binom{m+j}{j} \{\mathcal{L}_X^m F, \mathcal{L}_X^j G\} \\
&= \sum_{s \geq 0} \sum_{j=0}^{s} \frac{1}{s!} \binom{s}{j} \{\mathcal{L}_X^{s-j} F, \mathcal{L}_X^j G\} \\
\text{(see (11.1.5))} &= \sum_{s \geq 0} \frac{1}{s!} \mathcal{L}_X^s \{F, G\} = \exp \mathcal{L}_X \{F, G\},
\end{aligned}$$

which is the statement. QED

Lastly, we note that the Hamilton Equations may be written as

$$\frac{dx}{dt} = \mathcal{L}_H x,$$

where H is the Hamiltonian.

With these tools at hand we can pass to the study of the perturbations of Hamiltonian systems.

11.1.2 Homological Equation and its Formal Solution

Given a perturbed Hamiltonian system

$$H(x) = H_0(x) + \varepsilon H_p(x), \quad x \in N,$$

we want to find a canonical transformation

$$x \mapsto x' = \exp(\varepsilon \mathcal{L}_\chi)x = x + \varepsilon \mathcal{L}_\chi x + O(\varepsilon^2)$$

with generator χ, such that converts the Hamiltonian into a completely integrable one up to second order terms. Even this somewhat "minimal" requirement can be in general satisfied, as we shall see, only under certain conditions.

Recalling the exchange property, we can write

$$\begin{aligned} H(x') &= \exp(\varepsilon \mathcal{L}_\chi)H = H + \varepsilon \mathcal{L}_\chi H + \cdots \\ &= H_0 + \varepsilon(H_p + \mathcal{L}_\chi H_0) + O(\varepsilon^2). \end{aligned}$$

We must find the unknown generator χ in such a way that the first order term $H_p + \mathcal{L}_\chi H_0$ becomes integrable. To this end, let us assume that the local coordinates x's are action-angle variables

$$x = I_1, \ldots, I_n, \varphi_1, \ldots, \varphi_n.$$

Hence the unperturbed Hamiltonian H_0, being completely integrable by assumption, will be a function of (in general not all) the action variables only. Without loss of generality, we suppose that $H_0 = H_0(I_1, \ldots, I_d)$, $d \leq n$, and that this dependence is nontrivial, i.e.,

$$\det\left(\frac{\partial H_0}{\partial I_a \partial I_b}\right) \neq 0, \quad 1 \leq a, b \leq d.$$

In this case, H_0 is said to be $(n-d)$-fold *degenerate*. If $d = n$, the unperturbed Hamiltonian is said to be *nondegenerate*.

Let us suppose first that the unperturbed system is nondegenerate, and put

$$\omega_i(I) = \frac{\partial H_0}{\partial I_i}, \quad i = 1, \ldots, n.$$

The vector with components $\omega_i(I)$ is called the *frequency vector*. This name is due to the following fact. The Hamilton Equations of the unperturbed system

$$\dot{I}_h = 0, \quad \dot{\varphi}_h = \omega_h(I),$$

are easily integrated

$$I_h(t) = I_h^0, \quad \varphi_h(t) = \omega_h(I^0)t + \varphi_h^0,$$

with I_h^0 and φ_h^0 integration constants. Thus the unperturbed angles evolve with constant angular velocity, whence the name.

Write the first order perturbing term as

$$H_p - \mathcal{L}_{H_0}\chi = H_p - \omega \cdot \partial\chi.$$

Obviously H_p is a function of the angles also, otherwise the perturbed system would be integrable. It is tempting to try to find the unknown χ such that the disturbing term goes to zero, but this is not possible: in fact the term $\omega \cdot \partial\chi$, as a function of the angles, has vanishing mean value, while H_p is a generic function of the angles. Hereafter an overbar will denote the mean value of a function, obtained by averaging over the angles, i.e.,

$$\overline{f} = \frac{1}{(2\pi)^n} \int_{\mathbb{T}^n} f(\varphi_1, \dots, \varphi_n)\, d\varphi_1 \dots d\varphi_n,$$

where $\mathbb{T}^n = \overbrace{S^1 \times \cdots \times S^1}^{n \text{ times}}$ is the n-dimensional torus. Expanding into a Fourier series

$$\chi(I, \varphi) = \overline{\chi}(I) + \sum_{k \in \mathbb{Z}^n - \{0\}} \chi_k(I) e^{ik\cdot\varphi},$$

$$H_p(I, \varphi) = \overline{H}_p(I) + \sum_{k \in \mathbb{Z}^n - \{0\}} H_{pk}(I) e^{ik\cdot\varphi},$$

where

$$\chi_k(I) = \frac{1}{(2\pi)^n} \int_{\mathbb{T}^n} \chi(I, \varphi) e^{-ik\cdot\varphi}\, d\varphi_1 \dots d\varphi_n,$$

$$H_{pk}(I) = \frac{1}{(2\pi)^n} \int_{\mathbb{T}^n} H_p(I, \varphi) e^{-ik\cdot\varphi}\, d\varphi_1 \dots d\varphi_n,$$

and $k \in \mathbb{Z}^n - \{0\}$ is a multi-index, i.e., an n-dimensional vector with integer components $(k_1, \dots, k_n)$ not all zero. Choosing

$$\chi_k(I) = \frac{H_{pk}(I)}{ik \cdot \omega},$$

plainly annihilates the disturbing term except for its mean value part, leaving us with the transformed Hamiltonian

$$H(I, \varphi) = H_0(I) + \varepsilon \overline{H}_p(I) + O(\varepsilon^2). \tag{11.1.6}$$

We call *resonant* those frequency vectors ω for which there exists some $k \in \mathbb{Z}^n - \{0\}$ such that $k \cdot \omega = 0$. Excluding the resonant frequency vectors, we are thus able to solve the equation

$$H_p - \overline{H}_p = \omega \cdot \partial\chi \tag{11.1.7}$$

and to push the unwanted dependence on the angles up to second order terms. Equation (11.1.7) was called by Poincaré the *homological equation.*

The procedure can be obviously iterated. Applying to (11.1.6) the operator $\exp(\varepsilon^2 \mathcal{L}_{\chi_2})$ and solving a second homological equation, the unwanted dependence on the angles is pushed to third order, and so forth. To give an idea of how the iterative process works we report explicitly the first four terms, putting, for simplicity, $\mathcal{L}_{\chi_j} = \mathcal{L}_j$. Then

$$\prod_{j=4}^{1} \exp(\varepsilon^j \mathcal{L}_j) H = H_0 + \varepsilon\left(\mathcal{L}_1 H_0 + H_p\right) + \varepsilon^2\left(\mathcal{L}_2 H_0 + \frac{1}{2}\mathcal{L}_1^2 H_0 + \mathcal{L}_1 H_p\right)$$
$$+ \varepsilon^3\left(\mathcal{L}_3 H_0 + \frac{1}{2}\mathcal{L}_1^2 H_p + \mathcal{L}_2 H_p + \mathcal{L}_2\mathcal{L}_1 H_0 + \frac{1}{6}\mathcal{L}_1^3 H_0\right)$$
$$+ \varepsilon^4\left(\begin{array}{c} \mathcal{L}_4 H_0 + \frac{1}{2}\mathcal{L}_2\mathcal{L}_1^2 H_0 + \mathcal{L}_2\mathcal{L}_1 H_p + \mathcal{L}_3\mathcal{L}_1 H_0 + \\ + \mathcal{L}_3 H_p + \frac{1}{2}\mathcal{L}_2^2 H_0 + \frac{1}{24}\mathcal{L}_1^4 H_0 + \frac{1}{6}\mathcal{L}_1^3 H_p \end{array}\right) + O(\varepsilon^5).$$

As one easily sees, the term of order r is of the type

$$\varepsilon^r\left(H_r(\chi_{r-1}, \chi_{r-2}, \ldots, \chi_1, H_0, H_p) - \omega \cdot \partial\chi_r\right),$$

H_r being some function (besides H_0 and H_p) of the generators of order $\leq r-1$. Having solved the homological equations of order $\leq r-1$, these generators are all known, and the homological equation of order r

$$H_r - \overline{H}_r = \omega \cdot \partial\chi_r$$

can, at least formally, be solved, leaving us with the term $\varepsilon^r \overline{H}_r(I)$. The transformed Hamiltonian is said to be in *normal form* to order r, and the process we have described is referred to as *normalization procedure.*

We have proved that, at least for a nondegenerate H_0 and excluding the resonant frequency vectors, a perturbed problem, truncated at some finite order, can be canonically transformed into a completely integrable one. Let us now consider an $(n-d)$-fold degenerate system. In this case, only the first d components $\omega_a(I)$ of the frequency vector are different from zero, so that only the first d angles (the *fast angles*) evolve for the unperturbed Hamiltonian. If the perturbation is switched on, the other angles (called the *slow angles*) evolve also, but very slowly due to the smallness of the perturbation:

$$\dot{I}_j = -\varepsilon\frac{\partial H_p}{\partial\varphi_j}, \quad \dot{\varphi}_a = \omega_a + \varepsilon\frac{\partial H_p}{\partial I_a}, \quad \dot{\varphi}_h = \varepsilon\frac{\partial H_p}{\partial I_h},$$
$$j = 1, \ldots, n, \quad a = 1, \ldots, d, \quad h = d+1, \ldots, n.$$

The normalization process can clearly be performed only for the fast angles. These are eliminated from the perturbation up to some finite order, while the slow angles survive also in the normalized Hamiltonian; loosely speaking, there is no fast angular evolution on which to average. This means that the normalized Hamiltonian is not integrable, and one must devise a further strategy if wants to continue in an analytical way.

In every case, i.e., either degeneracy of some order or nondegeneracy, a perturbed Hamiltonian is, therefore, said to be in normal form at order r if

$$H = H_0 + \sum_{j=1}^{r} \varepsilon^j \overline{H}_j + O(\varepsilon^{r+1}), \quad \{H_0, \overline{H}_j\} = 0 \quad \forall j \leq r.$$

11.2 The Convergence Problem

In this section we consider non degenerate analytical Hamiltonians. The crucial question is: does the normalization process converge when $r \to \infty$? Till the fundamental work of Kolmogorov (1954), two, diametrically opposed answers were given. For the astronomers, interested in the computation of perturbed orbits, the answer was (more or less tacitly) affirmative[(2)]: the negative one seemed to be a disaster, making pointless the series expansions. On the contrary, for the statistical physicists, interested in the possibility of applying the ergodic theorem, the answer was negative. That of Kolmogorov was, in some sense, a Solomon's verdict: the ultimate fate of a slightly perturbed orbit depends on the initial conditions, so that for the same perturbed Hamiltonian coexist, in general, orbits that stay in the neighbourhood of an unperturbed one *for infinite time*, and others that depart indefinitely.

To make more precise this statement, let us consider the unperturbed motions. The system being completely integrable, the phase space is foliated with n-dimensional tori. If the normalization process would converge, the transformed Hamiltonian would be still integrable, and the tori, though slightly deformed, would fill continuously the phase space. But complete integrability is an exception, and even a very small perturbation will destroy it. The point is that this destruction does not happen abruptly: for small perturbations only a few tori disappear, their number growing with the intensity of the perturbation.

But what kind of tori are firstly destroyed? Since by assumption H_0 is nondegenerate, i.e., $\det\left(\frac{\partial \omega_i}{\partial I_j}\right) \neq 0$, the frequency vectors can be used to label the tori instead of the actions, and the answer may be given in terms of them. The frequency vectors can be classified (with the so-called *Diophantine theory*) on the basis of the more or less reciprocal irrationality of their components. The Kolmogorov Theorem then states

(2) With obviously some notable exceptions, for example Poincaré.

that those tori whose frequency vectors have components reciprocally close to rationality are the first, as ε grows, to be destroyed.

The Kolmogorov Theorem answers thus the above question in a precise mathematical way, which appears however unsatisfactory for the physical applications. Since both the rational and the irrational numbers are dense, in an arbitrary neighbourhood of a conserved torus there exists one which is destroyed, and vice versa. Conserved and destroyed tori form a complicated set of the Cantor type, and it appears impossible to decide whether a given orbit (whose initial conditions are never known with absolute precision) is stable or not.

Nekhoroshev observed that this disappointing situation is due to the unphysical fact that we want answers valid for *infinite* time. If one is satisfied with an answer valid for a *finite* time Δt, Nekhoroshev shows that, for arbitrary initial conditions, the variation of the actions is less than a positive power of ε for a Δt that is greater than some positive power of $e^{\frac{1}{\varepsilon}}$. In other words, for a small perturbation the variation of the actions remains small even for a very long lapse of time. If Δt is of the order, e.g., of the age of the universe, we can speak of *practical* stability.

The Kolmogorov Theorem was stated in Kolmogorov (1954) with a sketch of the proof only. A complete and rigorous proof was subsequently given in Arnold (1963), Moser (1955), Moser (1962) and Moser (1967), but following different lines. The original idea of Kolmogorov was reconsidered in Benettin *et al.* (1984) and completed. In this section we prove the Kolmogorov Theorem, following closely this last work. For the Nekhoroshev Theorem, the original works are Nekhoroshev (1977) and Nekhoroshev (1979), but see Benettin, Galgani & Giorgilli (1985) for a probably more accessible exposition. We will give only a sketch of the proof.

Before presenting the two theorems, we must yet return to the Lie series and to the solution (through Fourier series) of the homological equation, to investigate their convergence: remember, indeed, that the solution already found is a only formal one.

Let us fix some notation. For $v \in \mathbb{C}^n$, we put $\|v\| = \max_i |v_i|$. Analogously the norm of a matrix (M_{ij}) is defined as for an n^2-vector, namely by $\|\mathbf{M}\| = \max_{ij} \left|M_{ij}\right|$. Moreover we put

$$I = \{I_1, \dots, I_n\} \in \mathcal{B} \subset \mathbb{R}^n, \quad \varphi = \{\varphi_1, \dots, \varphi_n\} \in \mathbb{T}^n,$$

$\mathcal{B}$ being an open ball in $\mathbb{R}^n$. Being interested in the analytic case, we will consider complex extensions of subsets of $\mathbb{R}^{2n}$ and of real analytic functions defined there. Having fixed $I^* \in \mathcal{B}$ and a positive $\varrho < 1$ so small that the real closed ball of radius ϱ centered at I^* is contained in

$\mathcal{B}$, a central role will be played by the subset $\mathcal{D}_\varrho$ of $\mathbb{C}^{2n}$ defined by

$$\mathcal{D}_\varrho = \{(I,\varphi) \in \mathbb{C}^{2n};\ \|I - I^*\| \le \varrho,\ \|\mathrm{Im}\,\varphi\| \le \varrho\},$$

where $\mathrm{Im}\,\varphi = (\mathrm{Im}\,\varphi_1, \dots, \mathrm{Im}\,\varphi_n)$. For what concern functions, let $\mathcal{A}_\varrho$ be the set of all complex analytic functions in the interior of $\mathcal{D}_\varrho$ that are real for real values of the variables: their norm is defined by

$$\|f\|_\varrho = \sup |f(z)|\,,\ z \in \mathcal{D}_\varrho,\ f \in \mathcal{A}_\varrho.$$

In the case of vector valued functions $f = (f_1, \dots, f_n)$ with values in $\mathbb{C}^n$ we also write $f \in \mathcal{A}_\varrho$ if $f_h \in \mathcal{A}_\varrho$, $h = 1, \dots, n$ and define

$$\|f\|_\varrho = \max_h \|f_h\|_\varrho\,.$$

Analogously, if $\mathbf{D}$ is an $(n \times n)$ matrix whose elements D_{hk} belong to $\mathcal{A}_\varrho$, we set

$$\|\mathbf{D}\|_\varrho = \max_{hi} \|D_{hi}\|_\varrho\,.$$

Being interested in giving estimates, we will make use of the Cauchy inequality. Given $f \in \mathcal{A}_\varrho$, a positive $\delta < \varrho$ and nonnegative integers j_i, h_i $(i = 1, \dots, n)$, one has (see Appendix 11.B to this chapter)

$$\left\| \frac{\partial^{j_1+\cdots+j_n+h_1+\cdots+h_n}}{\partial I_1^{j_1} \dots \partial I_n^{j_n} \partial\varphi_1^{h_1} \dots \partial\varphi_n^{h_n}} f(I,\varphi) \right\|_{\varrho-\delta} \le \frac{j_1! \dots j_n! h_1! \dots h_n!}{\delta^{j_1+\cdots+j_n+h_1+\cdots+h_n}} \|f\|_\varrho\,.$$

Let us comment on this formula, which will be used to check the convergence of some series. For that, we would like the right-hand side to be small, but δ is the only parameter at our disposal: choosing it large we restrict the domain in the left-hand side, so that iterating the application of the Cauchy inequality we risk that $\varrho - \delta_1 - \delta_2 - \cdots$ will vanish. We are between the two fires of conflicting requirements, and the delicate point is to prove convergence keeping at the same time controlled, i.e., finite, the domain measure.

11.2.1 Convergence of Lie Series

Firstly we note that, given two functions $\chi, f \in \mathcal{A}_\varrho$ with

$$\chi_\varrho^* = \max\left(\left\|\frac{\partial\chi}{\partial I}\right\|_\varrho, \left\|\frac{\partial\chi}{\partial\varphi}\right\|_\varrho \right), \tag{11.2.1}$$

and any positive $\delta < \varrho$, the inequalities

$$\begin{aligned} \|\{\chi, f\}\|_{\varrho-\delta} &\le 2n\frac{\chi_\varrho^*}{\delta}\|f\|_\varrho\,, \\ \|\{\chi,\{\chi, f\}\}\|_{\varrho-\delta} &\le 4n(2n+1)\left(\frac{\chi_\varrho^*}{\delta}\right)^2 \|f\|_\varrho \end{aligned} \tag{11.2.2}$$

hold. They are deduced from the Cauchy inequality and by counting the terms in Poisson brackets.

Let χ be the generator of a (for a moment formal) canonical transformation

$$Q : J, \psi \to I, \varphi .$$

The existence of such a canonical transformation Q with associated operator $\exp \mathcal{L}_\chi$, and some relevant estimates are then given by the following

PROPOSITION 11.2 *Assume $\chi_\varrho^* < \frac{\delta}{2}$. Then, for all initial data $(J, \psi) \in \mathcal{D}_{\varrho-\delta}$, the solution of the canonical system*

$$\frac{dI}{d\tau} = -\frac{\partial \chi}{\partial \varphi}, \quad \frac{d\varphi}{d\tau} = \frac{\partial \chi}{\partial I}$$

at $\tau = 1$ exists in $\mathcal{D}_\varrho$, thus defining a canonical transformation $Q : \mathcal{D}_{\varrho-\delta} \to \mathcal{D}_\varrho$, $Q \in \mathcal{A}_{\varrho-\delta}$. Then, the operator $\exp \mathcal{L}_\chi : \mathcal{A}_\varrho \to \mathcal{A}_{\varrho-\delta}$, with $\exp \mathcal{L}_\chi f = f \circ Q$, is well defined. Moreover, one has the estimates

$$\|Q - identity\|_{\varrho-\delta} \le \chi_\varrho^* \tag{11.2.3}$$

and

$$\begin{aligned} \|\exp \mathcal{L}_\chi f\|_{\varrho-\delta} &\le \|f\|_\varrho , \\ \|\exp \mathcal{L}_\chi f - f\|_{\varrho-\delta} &\le 4n \frac{\chi_\varrho^*}{\delta} \|f\|_\varrho , \\ \|\exp \mathcal{L}_\chi f - f - \{\chi, f\}\|_{\varrho-\delta} &\le 16n(2n+1) \left(\frac{\chi_\varrho^*}{\delta}\right)^2 \|f\|_\varrho . \end{aligned} \tag{11.2.4}$$

Proof. Using the Cauchy inequality, one guarantees that in the subdomain $\mathcal{D}_{\varrho-\delta/2}$ the second member of the above canonical system has finite derivative. Moreover, as $\max\left(\left\|\frac{dI}{d\tau}\right\|, \left\|\frac{d\varphi}{d\tau}\right\|\right) \le \chi_\varrho^* < \frac{\delta}{2}$, if one takes initial data (J, ψ) in $\mathcal{D}_{\varrho-\delta}$, the standard existence and uniqueness theorem guarantees that the corresponding solutions $(I(\tau), \varphi(\tau))$ with $(I(0), \varphi(0)) = (J, \psi)$ exist in $\mathcal{D}_{\varrho-\delta/2}$ for any τ with $0 \le \tau \le 1$. Thus, for any such τ, one has a mapping

$$Q^\tau : \mathcal{D}_{\varrho-\delta} \to \mathcal{D}_{\varrho-\delta/2} \subset \mathcal{D}_\varrho ,$$

given by

$$Q^\tau(J, \psi) = (I(\tau), \varphi(\tau)) ,$$

and, in particular, the mapping Q^1 is also defined. This mapping is obviously canonical, being the "time-one" solution of a canonical system.

Let us now come to the estimates. Estimate (11.2.3) is an immediate consequence of the mean value theorem. The first of (11.2.4) is trivial. The second and the third ones follow from Taylor's formula for f of first and second order respectively, namely

$$\exp \mathcal{L}_\chi f - f = \left.\frac{df}{d\tau}\right|_{\tau'} = \{\chi, f\}|_{\tau'},$$

$$\exp \mathcal{L}_\chi f - f - \frac{df}{d\tau} = \frac{1}{2}\left.\frac{d^2 f}{d\tau^2}\right|_{\tau''} = \frac{1}{2}\{\chi, \{\chi, f\}\}|_{\tau''},$$

with $0 < \tau', \tau'' < 1$. Remembering that one has $Q(J, \psi) \in \mathcal{D}_{\varrho-\delta/2}$ if $0 < \tau \le 1$ and $(J, \psi) \in \mathcal{D}_{\varrho-\delta}$, one has then to estimate $\|\{\chi, f\}\|_{\varrho-\delta/2}$ and $\|\{\chi, \{\chi, f\}\}\|_{\varrho-\delta/2}$. The second and the third estimates (11.2.4) then follow by (11.2.2), with δ replaced by $\frac{\delta}{2}$. QED

Having verified that the Lie series expansions are not only formal, we investigate the solution of the homological equation.

11.2.2 Homological Equation and its Solution

In this subsection we are interested in functions of the angles only, so that the actions will be considered as parameters, thus disregarded.

As said, the homological equation

$$\sum_i \lambda_i \frac{\partial F(\varphi)}{\partial \varphi_i} = G(\varphi)$$

is formally solvable if $\overline{G} = 0$. Here $\lambda = (\lambda_1, \ldots, \lambda_n) \in \mathbb{R}^n$ is a known *fixed arbitrary* frequency vector that we suppose to be nonresonant. The formal solution is

$$F(\varphi) = \sum_{k \in \mathbb{Z}^n - \{0\}} \frac{g_k}{ik \cdot \lambda} e^{ik\cdot\varphi}, \qquad (11.2.5)$$

where g_k are the Fourier coefficients of G. At first sight, it seems that this sum can not converge in general, i.e., for arbitrary, even if nonresonant, frequency vectors $(\lambda_1, \ldots, \lambda_n)$. In fact, the set in $\mathbb{R}^n$ for which the denominator vanishes is dense (since the rational numbers are dense on the real line, i.e., every nonempty open set in $\mathbb{R}$ contains at least one rational number), and for hoping in the convergence we must exclude from the sum not only the resonant term, but also those for which $k \cdot \lambda$ is very small. This is the celebrated problem of the "small denominators". The situation may appear hopeless but, fortunately, a classical result in Diophantine theory (see Appendix 11.A to this chapter) guarantees that

$k \cdot \lambda$ can be bounded from below, without to be left with an empty set. In more precise terms, the inequality

$$|k \cdot \lambda| \geq \frac{\gamma}{|k|^n} \quad \forall k \in \mathbb{Z}^n - \{0\}, \quad |k| \stackrel{\text{def}}{=} \sum_i \left| k_j \right|, \tag{11.2.6}$$

for some positive γ is satisfied by a set Ω_γ of real vectors λ of large relative measure, the complement of this set having Lebesgue measure $O(\gamma)$. (11.2.6) is referred to as the *Diophantine condition.*

This is a key point. In fact, since G is by hypothesis analytic, we prove in the following two lemmas that its Fourier coefficients g_k decay exponentially with $|k|$ (and vice versa), while $\frac{1}{|k \cdot \lambda|}$ grows at most as a power owing to the Diophantine condition. This will allow us to prove the convergence of (11.2.5) in the further proposition.

It is easy to estimate the Fourier coefficients of an analytical function $G(\varphi)$, $\varphi \in S^1$. The set $|\mathrm{Im}\,\varphi| < \varrho$ can be represented as a thin circular strip surrounding the unit circle in the complex plane of the variable $z = e^{i\varphi}$. Then

$$g_k = \frac{1}{2\pi} \int_0^{2\pi} G(\varphi) e^{-ik\varphi} d\varphi = \frac{1}{2\pi} \int_0^{2\pi} G(\varphi \pm i\varrho) e^{\pm k\varrho} e^{-ik\varphi} d\varphi, \quad k \in \mathbb{Z}$$

since, for the analyticity of G, it is possible to shift the integration path from the unit to the inner or outer circle of the strip. Taking the minus sign for $k > 0$ and the plus sign for $k < 0$, we obtain

$$\left| g_k \right| \leq \|G\|_\varrho \, e^{-|k|\varrho}.$$

The argument can be extended to the n-dimensional case as follows.

LEMMA 11.3 *If* $G \in \mathcal{A}_\varrho$, $G(\varphi) = \sum_k g_k e^{ik \cdot \varphi}$, *then for every* $k \in \mathbb{Z}^n$ *one has*

$$\left| g_k \right| \leq \|G\|_\varrho \, e^{-|k|\varrho}.$$

Proof. By definition one has

$$g_k = \frac{1}{(2\pi)^n} \int_{\mathbb{T}^n} G(\varphi) e^{-ik \cdot \varphi} d\varphi_1 \ldots d\varphi_n.$$

As $G \in \mathcal{A}_\varrho$, one can shift the integration path as above and write

$$g_k = \frac{1}{(2\pi)^n} \int_{\mathbb{T}^n} G\left(\varphi_1 - i\frac{k_1}{|k_1|}\varrho, \ldots, \varphi_n - i\frac{k_n}{|k_n|}\varrho \right) \times \prod_{j=1}^n e^{-ik_j\left(\varphi_j - i\frac{k_j}{|k_j|}\varrho\right)} d\varphi_1 \ldots d\varphi_n,$$

where, by convention, we may assume $\frac{0}{|0|} = 0$. One has then

$$|g_k| \le \|G\|_\varrho \, e^{-|k|\varrho} \frac{1}{(2\pi)^n} \int_{\mathbb{T}^n} \left| e^{-ik\cdot\varphi} \right| \, d\varphi_1 \dots d\varphi_n = \|G\|_\varrho \, e^{-|k|\varrho},$$

which is the claim. QED

We will need also this other lemma that is, in some sense, the inverse of the previous.

LEMMA 11.4 *Suppose that for some positive constants C and ϱ with $\varrho \le 1$ and for each $k \in \mathbb{Z}^n$ one has $|f_k| \le Ce^{-|k|\varrho}$, and define the function $F(\varphi) = \sum_k f_k e^{ik\cdot\varphi}$. Then, for any positive $\delta < \varrho$ one has that, obviously, $F \in \mathcal{A}_{\varrho-\delta}$ and that*

$$\|F\|_{\varrho-\delta} \le C \left(\frac{4}{\delta}\right)^n .$$

Proof. Let $\|\mathrm{Im}\,\varphi\| \le \varrho - \delta$. Then

$$\begin{aligned}
\|F\|_{\varrho-\delta} &= \sup \left| \sum_{k\in\mathbb{Z}^n} f_k e^{ik\cdot\varphi} \right| \le C \sum_{k\in\mathbb{Z}^n} e^{-|k|\varrho} e^{|k|(\varrho-\delta)} \\
&= C \sum_{k\in\mathbb{Z}^n} e^{-|k|\delta} = C \left(\sum_{k\in\mathbb{Z}} e^{-|k|\delta} \right)^n \\
&< 2^n C \left(\sum_{k=0}^{\infty} e^{-\delta k} \right)^n = 2^n C \frac{1}{(1-e^{-\delta})^n} < C \left(\frac{4}{\delta}\right)^n ,
\end{aligned}$$

because $\frac{1}{1-e^{-\delta}} < \frac{2}{\delta}$ for any positive $\delta < 1$. QED

We can now prove the

PROPOSITION 11.5 *Consider the homological equation*

$$\sum_i \lambda_i \frac{\partial F(\varphi)}{\partial \varphi_i} = G(\varphi),$$

and assume $\lambda \in \Omega_\gamma$ and $G \in \mathcal{A}_\varrho$ for some positive γ and ϱ with $\overline{G} = 0$. Then, for any positive $\delta < \varrho$, the homological equation admits a unique solution $F \in \mathcal{A}_{\varrho-\delta}$ with $\overline{F} = 0$. For this solution one has the basic estimates

$$\|F\|_{\varrho-\delta} \le \frac{\sigma}{\gamma\delta^{2n}} \|G\|_\varrho , \qquad \left\| \frac{\partial F}{\partial \varphi} \right\|_{\varrho-\delta} \le \frac{\sigma}{\gamma\delta^{2n+1}} \|G\|_\varrho ,$$

where $\sigma = 2^{4n+1} \left(\frac{n+1}{e}\right)^{n+1}$.

Proof. The formal solution of the homological equation is given by (11.2.5), where now we assume that the frequency vector λ satisfies the Diophantine condition (11.2.6). By $G \in \mathcal{A}_\varrho$ and Lemma (11.3), one has

$$|f_k| \le \frac{|k|^n}{\gamma} \|G\|_\varrho e^{-|k|\varrho}. \tag{11.2.7}$$

Since for any $K, s \in \mathbb{R}^+$ one easily verifies (putting $\frac{K\delta}{s} = x$) that

$$K^s \le \left(\frac{s}{e\delta}\right)^s e^{K\delta},$$

thus, with $K = |k|$ and $s = n$,

$$|f_k| \le \frac{1}{\gamma}\left(\frac{n}{e\delta}\right)^n \|G\|_\varrho e^{-|k|(\varrho-\delta)} \stackrel{\text{def}}{=} Ce^{-|k|(\varrho-\delta)}.$$

For any $\delta < \varrho$, Lemma (11.4), with $\varrho - \delta$ in place of ϱ, ensures that

$$F(\varphi) \stackrel{\text{def}}{=} \sum_{k\neq 0} f_k e^{ik\cdot\varphi} \in \mathcal{A}_{\varrho-2\delta},$$

$$\|F\|_{\varrho-2\delta} \le \frac{1}{\gamma}\left(\frac{4n}{e\delta^2}\right)^n \|G\|_\varrho .$$

Taking $\frac{\delta}{2}$ instead of δ, the statement of the proposition and the first estimate are proved, because $\sigma > \left(\frac{16n}{e}\right)^n$.

Let us prove the other estimate. The Fourier coefficients of $\frac{\partial F}{\partial \varphi_j}$ are of the form $h_{k_j} = \frac{k_j g_k}{\lambda\cdot k}$, $j = 1, \dots, n$, so that as for (11.2.7) one gets

$$\left|h_{k_j}\right| \le \frac{|k|^{n+1}}{\gamma} \|G\|_\varrho e^{-|k|\varrho}.$$

But, with $K = |k|$ and $s = n + 1$, this gives

$$\left|h_{k_j}\right| \le \frac{1}{\gamma}\left(\frac{n+1}{e\delta}\right)^{n+1} \|G\|_\varrho e^{-|k|(\varrho-\delta)},$$

so that from Lemma (11.4), with $\varrho - \delta$ in place of ϱ, one obtains

$$\left\|\frac{\partial F}{\partial \varphi}\right\|_{\varrho-2\delta} \le \frac{4^n}{\gamma\delta^{2n+1}}\left(\frac{n+1}{e}\right)^{n+1} \|G\|_\varrho .$$

By taking again $\frac{\delta}{2}$ instead of δ, one therefore obtains the second estimate also. QED

REMARK 11.1 The above estimates are not optimal. With deeper arguments optimal estimates can be given with δ^{-2n} and $\delta^{-(2n+1)}$ replaced by δ^{-n} and $\delta^{-(n+1)}$ respectively, and σ replaced by $\sigma_R = \frac{2^{4n+1} n!}{2^n - 1}$. See Rüssmann (1975).

11.2.3 Kolmogorov Theorem

An important thing we learned before is that the goal of classical perturbation theory to eliminate completely the angles in the transformed Hamiltonian is hopeless. Indeed, even a very small perturbation seems to destroy the foliation of the phase space in invariant tori, since the n first integrals in involution of the unperturbed problem do no longer exist.

The basic idea of Kolmogorov was to focus the attention on the conservation problem of the tori, instead of on the existence problem of the first integrals in involution, and to investigate if *all* the tori are destroyed, or some survive for infinite times. To this end, he *fixes* the frequency vector, and states that, if this vector satisfies the Diophantine condition and the perturbation is sufficiently small, the corresponding torus is only slightly deformed, but not destroyed, by the perturbation.

To state and prove the Kolmogorov Theorem, we start with a rearrangement of the perturbed Hamiltonian $H(I,\varphi)$, which we suppose already expressed as a function of the action-angle variables. Having fixed the values of the action variables, which we can safely put to zero after perhaps translating I^* to the origin, by a Taylor expansion in I one can write the Hamiltonian H in the form

$$H(I,\varphi) = a + A(\varphi) + [\lambda + B(\varphi)]\cdot I + \frac{1}{2}I\cdot \mathbf{D}(\varphi)\, I + R(I,\varphi).$$

Here λ is the frequency vector of the unperturbed Hamiltonian, $a \in \mathbb{R}$ is a constant that is uniquely defined by the condition $\overline{A} = 0$, while $A, B_i, D_{ij}, R \in \mathcal{A}_\varrho$ and R is, as a function of I, of the order $\|I\|^3$. One has clearly

$$a = \overline{H}(0), \qquad A(\varphi) = H(0,\varphi) - a,$$
$$B_i = \frac{\partial H}{\partial I_i}(0,\varphi) - \lambda_i, \quad D_{ij}(\varphi) = \frac{\partial^2 H}{\partial I_i \partial I_j}(0,\varphi).$$

Let us suppose for a moment that $A(\varphi)+B(\varphi)\cdot I = 0$. Then the Hamilton Equations would be

$$\dot\varphi_i = \lambda_i + \sum_j D_{ij} I_j + \frac{\partial R}{\partial I_i},$$
$$\dot I_i = -\frac{\partial}{\partial \varphi_i}\left[\frac{1}{2} I\cdot \mathbf{D} I + R\right],$$

which admit the *particular* solution

$$I_i(t) = 0, \quad \varphi_i(t) = \lambda_i t + \varphi_i^0,$$

as one sees using the fact that R is of the order $\|I\|^3$. In other words, if one is not guaranteed to have a foliation of the whole phase space into invariant tori, however one sees by inspection the invariance of *one* torus supporting quasi-periodic motions with angular frequency λ_i.

The Kolmogorov Theorem states just that the disturbing term $A(\varphi)+B(\varphi)\cdot I$ can be removed with a canonical transformation $Q : J, \psi \to I, \varphi$. In the new canonical variables J, ψ the torus $J_i = 0$ is thus invariant for the Hamiltonian flow induced by the transformed Hamiltonian $H \circ Q$. In terms of the original variables I, φ this torus is described by the parametric equations $(I, \varphi) = Q(0, \psi)$, and is invariant for the Hamiltonian flow induced by H. This torus is a small perturbation of the torus $I_i = 0$, which is by assumption invariant for the unperturbed Hamiltonian, supporting quasi-periodic motions with the same angular frequencies λ_i.

We can now state the main theorem of this chapter.

THEOREM 11.6 (KOLMOGOROV) *Consider the Hamiltonian*

$$H(I,\varphi) = H^0(I,\varphi) + H^1(I,\varphi)$$

defined in the domain $\mathfrak{D}_\varrho$ *by*

$$H^0(I,\varphi) = a + \lambda\cdot I + \frac{1}{2} I \cdot \mathbf{D}(\varphi) I + R(I,\varphi),$$
$$H^1(I,\varphi) = A(\varphi) + B(\varphi)\cdot I$$

with $\|H\|_\varrho < 1$ *and* R *of the order* $\|I\|^3$. *Let us suppose that*

(i) $\lambda \in \Omega_\gamma$, *i.e., the unperturbed frequency vector satisfies the Diophantine condition,*

(ii) $\det\mathbf{D} \neq 0$,

then there exist positive numbers η *and* ϱ' *with* $\varrho' < \varrho$ *such that, if* H^1 *is sufficiently small, i.e.,*

(iii) $max\big(\|A\|_\varrho, \|B\|_\varrho\big) \le \eta$,

one can construct a canonical analytical transformation

$$Q : \mathfrak{D}_{\varrho'} \to \mathfrak{D}_\varrho,\ Q \in \mathcal{A}_{\varrho'},$$

which brings the Hamiltonian H *into the form*

$$H'(J,\psi) \equiv (H \circ Q)(J,\psi) = a' + \lambda\cdot J + R'(J,\psi),$$

where $R' \in \mathcal{A}_{\varrho'}$ *is, as a function of* J, *of the order* $\|J\|^2$. *This canonical transformation is near the identity, in the sense that* $\|Q - \text{identity}\|_{\varrho'} \to 0$ *as* $\|H^1\|_\varrho \to 0$.

Notice that $D_{ij} \in \mathcal{A}_\varrho$ and (ii) are equivalent to the existence of a positive m such that

(ii') $\|\mathbf{D}v\|_\varrho \le \frac{1}{m}\|v\|, \quad m\|v\| \le \|\overline{\mathbf{D}}v\|, \quad \forall v \in \mathbb{C}^n.$

The scheme of the proof of the Kolmogorov Theorem is as follows. One performs a sequence of canonical analytical transformations such that the disturbing term H^1 at step r decreases with r, its norm $\|H_r^1\|_{\varrho_r}$ being essentially of the order of $\left\|H_{r-1}^1\right\|_{\varrho_{r-1}}^2$, while the other parameters ϱ_r and m_r are kept controlled, i.e., strictly positive. The convergence of the scheme with $\|H_r^1\|_{\varrho_r} \to 0$, $\varrho_r \to \varrho_\infty > 0$ and $m_r \to m_\infty > 0$ as $r \to \infty$ is then established.

Each step of the iterative process satisfies the following lemma, which we show before proving the Kolmogorov Theorem.

LEMMA 11.7 (ITERATIVE) *For given positive number $\gamma, \varrho, m, \eta < 1$ and $0 < \varrho_* < \varrho$, $0 < m_* < m$ consider the Hamiltonian $H(I,\varphi)$ as in the Kolmogorov Theorem and satisfying (i), (ii), (iii). For any positive δ, so small that $\varrho - 3\delta > \varrho_*$, let us assume that η is so small that*

$$m - \frac{n}{\varrho_*^2}\Lambda\frac{\eta}{m^3\delta^\tau} > m_* \tag{11.2.8}$$

where

$$\Lambda = 2(4n+1)^2\frac{\sigma^2}{\gamma^2}, \qquad \tau = 4n+3,$$

with σ defined in Proposition 11.5. Then one can find an analytical canonical transformation $Q : \mathcal{D}_{\varrho-3\delta} \to \mathcal{D}_\varrho$, $Q \in \mathcal{A}_{\varrho-3\delta}$, such that the transformed Hamiltonian $H' = H \circ Q$ can be decomposed in a way analogous to H with corresponding primed quantities $a', A', B', \mathbf{D}'$ and R', but with the same λ, and satisfies analogous conditions with positive parameters $\varrho', m', \eta' < 1$ given by

$$\varrho' = \varrho - 3\delta > \varrho_*, \tag{11.2.9}$$

$$m' = m - \frac{n}{\varrho_*^2}\Lambda\frac{\eta}{m^3\delta^\tau} > m_*, \tag{11.2.10}$$

$$\eta' = \frac{\Lambda^2}{m^6\varrho_*}\frac{\eta^2}{\delta^{2\tau}}, \tag{11.2.11}$$

with $\|H'\|_{\varrho'} \le 1$. Moreover, for any $f \in \mathcal{A}_\varrho$,

$$\|f \circ Q - f\|_{\varrho'} < \frac{1}{2}\sqrt{\varrho_*\eta'}\,\|f\|_\varrho\,. \tag{11.2.12}$$

Proof. If a canonical change of variables with generating function χ is performed, one obtains in place of H the new Hamiltonian H' and one can make a decomposition $H' = H'^0 + H'^1$ in a way analogous to the decomposition $H = H^0 + H^1$. Precisely, one has

$$H'^0 = a' + \lambda \cdot J + \frac{1}{2} J \cdot \mathbf{D}'(\psi) J + R'(J, \psi),$$
$$H'^1 = A'(\psi) + B'(\psi) \cdot J$$

with $R'(J, \psi)$ of order $\|J\|^3$. In the spirit of the perturbation theory, one thinks that both H'^1 and the generating function χ are of first order and chooses χ in order to eliminate the undesired terms of the same order in the new Hamiltonian H'. To this end one first writes the identity

$$H' = H^0 + H^1 + \{\chi, H^0\} + \left[\{\chi, H^1\} + \exp \mathcal{L}_\chi H - H - \{\chi, H\} \right], \quad (11.2.13)$$

where all terms which have to be considered of the second order have been collected into the bracket [...], in agreement with the estimate given in the third of (11.2.4). Then one tries to choose χ in such a way that the first-order terms in H', namely $H^1 + \{\chi, H^0\}$, do not contribute to H'^1. This is obtained by imposing

$$H^1 + \{\chi, H^0\} = c + O\left(\|I\|^2\right),$$

where c is a constant.

Following Kolmogorov (1954), we show that this condition is met by a generating function χ of the form

$$\chi = \xi \cdot \varphi + X(\varphi) + Y(\varphi) \cdot I,$$

where the constant $\xi \in \mathbb{R}^n$ and the functions $X(\varphi), Y_i(\varphi)$ are to be suitably determined. A straightforward calculation gives

$$H^1 + \{\chi, H^0\} = -\sum_i \xi_i \lambda_i + A(\varphi) - \sum_i \lambda_i \frac{\partial X}{\partial \varphi_i}$$
$$+ \sum_h \left[B_h(\varphi) - \sum_i D_{ih}(\varphi) \left(\xi_i + \frac{\partial X}{\partial \varphi_i} \right) - \sum_i \lambda_i \frac{\partial Y_h}{\partial \varphi_i} \right] I_h + O\left(\|I\|^2\right).$$

Then one must impose

$$\sum_i \lambda_i \frac{\partial X}{\partial \varphi_i} = A(\varphi), \quad (11.2.14a)$$

$$\sum_i \lambda_i \frac{\partial Y_h}{\partial \varphi_i} = B_h(\varphi) - \sum_i D_{ih}(\varphi) \left(\xi_i + \frac{\partial X}{\partial \varphi_i} \right). \quad (11.2.14b)$$

i.e., two homological equations. By Proposition 11.5, Equation (11.2.14a) can be solved for the unknown X as $\overline{A} = 0$. Then one has to determine the unknown constants ξ_i in such a way that the mean value of the right-hand side of (11.2.14b) vanishes. This leads to a linear system of equations for ξ_i, which in matrix notation can be written as

$$\overline{\mathbf{D}}\xi = \overline{B - \mathbf{D}\frac{\partial X}{\partial \varphi}}. \tag{11.2.15}$$

This equation can be solved since, by hypothesis, $\det \overline{\mathbf{D}} \neq 0$. Equation (11.2.14b) in the unknown Y can therefore be solved also.

The existence of the wanted generating function χ is thus guaranteed, and one must now prove the existence of the corresponding canonical transformation Q.

To this end, we exhibit at first the following basic estimate involving the generating function χ : Equations (11.2.14a), (11.2.14b) and (11.2.15) in the unknown $\xi_i, X(\varphi)$ and $Y_i(\varphi)$, which define the generating function χ, can be solved with $X, Y_i \in \mathcal{A}_{\tilde{\varrho}}$, $\tilde{\varrho} = \varrho - 2\delta$ for any positive $\delta < \frac{\varrho}{2}$, and, for the quantity $\chi^*_{\tilde{\varrho}}$ defined by (11.2.1), one gets

$$\chi^*_{\tilde{\varrho}} \leq (4n+1)\frac{\sigma^2}{\gamma^2}\frac{\eta}{m^3\delta^{\tau-1}}. \tag{11.2.16}$$

This inequality, concerning a function of the solutions of the two homological Equations (11.2.14a), (11.2.14b), can obviously be verified with a twofold application of Proposition 11.5; see Benettin *et al.* (1984) for details.

Remarking now that the inequality (11.2.16) can be written, owing to (11.2.8), as

$$\frac{\chi^*_{\tilde{\varrho}}}{\delta} < \frac{1}{2(4n+1)}(m - m_*)\frac{\varrho_*^2}{n},$$

one sees that $\chi^*_{\tilde{\varrho}} < \frac{\delta}{2}$, so that Proposition 11.2 can be applied and χ generates a canonical transformation with domain $\mathcal{D}_{\varrho'}$, where $\varrho' = \tilde{\varrho} - \delta = \varrho - 3\delta$.

We arrive now to the proof of the estimates that we review briefly, referring to Benettin *et al.* (1984) for the details. The inequality $\|H'\|_{\varrho'} \leq \|H\|_{\varrho} \leq 1$ is just the first of (11.2.4). Let us then come to η'. For what concerns A', by its definition, using expression (11.2.13) for H' and recalling that, in virtue of the choice of χ, only the last term contributes to A', one gets

$$\|A'\|_{\varrho'} \leq 2\,\|\{\chi, H_1\} + H \circ Q - H - \{\chi, H\}\|_{\varrho'},$$

from which, after some calculations, one obtains $\|A'\|_{\varrho'} \le \frac{\Lambda^2\eta^2}{m^6\delta^{2\tau}}$. Analogously $\|B'\|_{\varrho'} \le \frac{1}{2}\frac{\Lambda^2\eta^2}{\varrho_* m^6\delta^{2\tau}}$, so that

$$\max\left(\|A'\|_{\varrho'}, \|B'\|_{\varrho'}\right) \le \eta',$$

with $\varrho' = \varrho - 3\delta$ and $\eta' = \frac{\Lambda^2\eta^2}{\varrho_* m^6\delta^{2\tau}}$. For what concerns the inequality (11.2.12), from the second of (11.2.4), with $\tilde{\varrho}$ in place of ϱ, one has

$$\|\exp \mathcal{L}_\chi f - f\|_{\tilde{\varrho}-\delta} \le 4n\frac{\chi^*_{\tilde{\varrho}}}{\delta}\|f\|_{\tilde{\varrho}},$$

from which (11.2.12) follows using (11.2.16) and

$$\|f\|_{\tilde{\varrho}} \le \|f\|_{\varrho}, \quad \frac{2n}{4n+1} < \frac{1}{2}.$$

Lastly, let us come to the estimate for m'. From definition of **D** and **D**$'$, from Cauchy inequality and (11.2.12) one gets

$$\|(\mathbf{D}' - \mathbf{D})v\|_{\varrho'} \le \frac{n}{\varrho_*^2}\frac{\Lambda\eta}{m^3\delta^\tau}\|v\|,$$

from which it is easy to obtain conditions (ii') with **D**$'$ in place of **D** and m', defined in (11.2.10), in place of m. QED

Having proved the iterative lemma, we have at hand all the tools for the proof of the Kolmogorov Theorem.

Proof (Kolmogorov Theorem). One has to apply repeatedly the iterative lemma in order to eliminate the perturbation $H^1(\varphi, I) = A(\varphi) + B(\varphi) \cdot I$, verifying firstly that, as the iterative order $r \to \infty$, the parameters ϱ_r and m_r do not vanish: this is the *consistency problem.* Let us choose the sequence

$$\eta_r = \frac{\eta_0}{2^{2\tau r}}, \quad \eta_0 > 0. \tag{11.2.17}$$

Since $\eta_r \to 0$ for $r \to \infty$, and

$$\max\left(\|A\|_{\varrho_r}, \|B\|_{\varrho_r}\right) \le \eta_r,$$

then $H^1_r \to 0$. Let us fix arbitrarily the initial values $\varrho_0, m_0 > 0$ and the "bounding parameters" $0 < \varrho_* < \varrho_0$, $0 < m_* < m_0$, leaving for the moment unspecified η_0, whose value will be found accordingly with the consistency problem. From (11.2.10), (11.2.11) and (11.2.9) we have

$$m_{r+1} = m_r - \frac{n}{\sqrt{\varrho_*^3}}\sqrt{\eta_{r+1}}, \tag{11.2.18}$$

$$\delta_r^{2\tau} = \frac{\Lambda^2}{m_r^6\varrho_*}\frac{\eta_r^2}{\eta_{r+1}}, \tag{11.2.19}$$

$$\varrho_{r+1} = \varrho_r - 3\delta_r. \tag{11.2.20}$$

The relation (11.2.18) gives m_r at the generic order r, (11.2.19) gives δ_r, (11.2.20) gives ϱ_r, all as some functions of η_0. Clearly, the successions $\{m_r\}$ and $\{\varrho_r\}$ are decreasing: we want to show that for every ϱ_* and m_* it is possible to find η_0 in such a way as to satisfy the consistency problem, i.e.,

$$\varrho_r > \varrho_*, \; m_r > m_* \quad \forall r.$$

From (11.2.18) we obtain

$$m_\infty = m_0 - \frac{n}{\sqrt{\varrho_*^3}}\sum_{r=0}^{\infty}\sqrt{\eta_{r+1}}.$$

Since $\sum_{r=0}^{\infty}\frac{1}{2^{\tau(r+1)}} = \frac{1}{2^\tau - 1}$, by imposing $m_\infty > m_*$ we obtain that η_0 must satisfy the inequality

$$\eta_0 < \frac{\varrho_*^3(m_0 - m_*)^2}{n^2}\left(2^\tau - 1\right)^2. \tag{11.2.21}$$

(11.2.20) gives $\varrho_\infty = \varrho_0 - \sum_{r=0}^{\infty}3\delta_r$. Taking into account that, for the chosen sequence (11.2.17),

$$\left(\frac{\eta_r^2}{\eta_{r+1}}\right)^{\frac{1}{2\tau}} = \frac{\eta_0^{\frac{1}{2\tau}}}{2^{r-1}}$$

holds, we obtain

$$\sum_{r=0}^{\infty}\left(\frac{\eta_r^2}{\eta_{r+1}}\right)^{\frac{1}{2\tau}} = 4\eta_0^{\frac{1}{2\tau}}.$$

Let us impose $\varrho_\infty > \varrho_*$. Since we have already guaranteed that $m_r > m_*$, we obtain, taking into account (11.2.19), that η_0 must satisfy the inequality

$$\eta_0 < \frac{\varrho_* m_*^6}{\Lambda^2}\left(\frac{\varrho_0 - \varrho_*}{12}\right)^{2\tau}. \tag{11.2.22}$$

Obviously one must respect the more restrictive of the two conditions (11.2.21) or (11.2.22).

Finally, we come to the ultimate task: the convergence of the sequence of canonical transformations. Starting from the initial perturbed Hamiltonian H defined in $\mathcal{D}_{\varrho_0}$ with $\|H\|_{\varrho_0} \le 1$, characterized by positive parameters $\gamma, \varrho_0, m_0 < 1$, consider the quantity η_0, satisfying the more restrictive of the two above conditions, and assume

$$\max\left(\|A\|_{\varrho_0}, \|B\|_{\varrho_0}\right) < \eta_0.$$

Then one can apply recursively the iterative lemma, defining at every step $r \ge 1$ a canonical transformation $Q_r : \mathcal{D}_{\varrho_r} \to \mathcal{D}_{\varrho_{r-1}}$, with the corresponding operator $\exp \mathcal{L}_{\chi_r} : \mathcal{A}_{\varrho_{r-1}} \to \mathcal{A}_{\varrho_r}$. Furthermore, from (11.2.12), one has the estimate

$$\left\|\exp \mathcal{L}_{\chi_r} f - f\right\|_{\varrho_r} < \sqrt{\varrho_* \eta_r}\, \|f\|_{\varrho_{r-1}}. \tag{11.2.23}$$

We can now define the composite canonical transformation $\hat{Q}_r : \mathcal{D}_{\varrho_r} \to \mathcal{D}_{\varrho_0}$ by $\hat{Q}_r = Q_1 \circ \cdots \circ Q_r$ and the corresponding composite operator $\hat{\mathcal{U}}_r : \mathcal{A}_{\varrho_0} \to \mathcal{A}_{\varrho_r}$, defined by $\hat{\mathcal{U}}_r f = f \circ \hat{Q}_r$, or equivalently by $\hat{\mathcal{U}}_r = \exp \mathcal{L}_{\chi_r} \hat{\mathcal{U}}_{r-1}$, with $\hat{\mathcal{U}}_0 =$ identity. Clearly, in order to prove the convergence of the sequence $\{\hat{Q}_r\}$ of canonical transformations restricted to $\mathcal{D}_{\varrho_\infty}$, it is sufficient to prove the convergence of the corresponding sequence $\{\hat{\mathcal{U}}_r\}$ of operators for every $f \in \mathcal{A}_{\varrho_0}$. This in turn is seen by remarking that, from (11.2.23) and the first of (11.2.4), one has

$$\begin{aligned}\left\|\left(\hat{\mathcal{U}}_{r+1} - \hat{\mathcal{U}}_r\right) f\right\|_{\varrho_{r+1}} &= \left\|\exp \mathcal{L}_{\chi_{r+1}} \left(\hat{\mathcal{U}}_r f\right) - \hat{\mathcal{U}}_r f\right\|_{\varrho_{r+1}} \\ &\le \sqrt{\varrho_* \eta_{r+1}} \left\|\hat{\mathcal{U}}_r f\right\|_{\varrho_r} \le \sqrt{\varrho_* \eta_{r+1}}\, \|f\|_{\varrho_0}\end{aligned}$$

and also, applying the triangular inequality, for any $j \ge 1$

$$\begin{aligned}&\left\|\left(\hat{\mathcal{U}}_{r+j} - \hat{\mathcal{U}}_r\right) f\right\|_{\varrho_{r+j}} \\ &= \left\|\left(\hat{\mathcal{U}}_{r+j} - \hat{\mathcal{U}}_{r+j-1}\right) f + \left(\hat{\mathcal{U}}_{r+j-1} - \hat{\mathcal{U}}_{r+j-2}\right) f + \ldots\right\|_{\varrho_{r+j}} \\ &\le \|f\|_{\varrho_0} \sum_{s=r}^{r+j-1} \sqrt{\varrho_* \eta_{s+1}}.\end{aligned}$$

Thus, as the series

$$\sum_{r=1}^{\infty} \sqrt{\varrho_* \eta_r} = \sum_{r=1}^{\infty} \frac{\sqrt{\varrho_* \eta_0}}{2^{\tau r}} = \frac{\sqrt{\varrho_* \eta_0}}{2^{\tau} - 1}$$

converges, one deduces that the sequence $\hat{\mathcal{U}}_r f$ converges uniformly for any $f \in \mathcal{A}_{\varrho_0}$. Therefore, by the Weierstrass Theorem one has then

$$\lim_{r \to \infty} \hat{\mathcal{U}}_r f = \hat{\mathcal{U}}_\infty f \in \mathcal{A}_{\varrho_\infty}.$$

In particular, for the Hamiltonian $H_\infty = \lim_{r\to\infty} \hat{\mathcal{U}}_r H$, one has $H_\infty = H_\infty^0 + H_\infty^1$, where, by construction, $H_\infty^1(I,\varphi) = A_\infty(\varphi) + B_\infty(\varphi)\cdot I = 0$. Finally, the mapping $\hat{Q}_\infty = \lim_{r\to\infty} \hat{Q}_r$ turns out to be canonical again by virtue of the Weierstrass Theorem, as a uniform limit of canonical mappings.

QED

Let us review and comment on the hypotheses of the Kolmogorov Theorem, and the role that they play in the proof.

(i) The unperturbed frequencies λ_i must satisfy the Diophantine condition

$$|k\cdot\lambda| \geq \frac{\gamma}{|k|^n}, \quad \forall k \in \mathbb{Z}^n - \{0\}.$$

This condition is necessary to ensure that, owing to Proposition 11.5, the two homological Equations (11.2.14a), (11.2.14b) are solvable, so that an iterative step can be performed. This condition means that the unperturbed torus at hand does not support periodic orbits or even too close to periodicity; it is physically intuitive that, in the opposite case, the effects of the perturbation are cumulative, and the torus will be, sooner or later, destroyed. Note, however, that the Kolmogorov Theorem does not dictate the destruction of *all* the periodic orbits on the torus, so that some of them may survive, though slightly deformed.

(ii) The unperturbed Hamiltonian must be nondegenerate. This condition is necessary to ensure the invertibility of the relation between actions and frequencies, so that one can, at each iterative step, tune the actions to keep fixed the frequencies.

(iii) The perturbation must be sufficiently small, and the threshold η_0 is given by the more restrictive of the two conditions (11.2.21), (11.2.22), usually the latter. In this case, recalling the definition of Λ, one sees that $\eta_0 = O(\gamma^4)$, which is a sensible result: lowering γ weakens the Diophantine condition, so that more tori would be conserved, and this requires a smaller perturbation. The $2n$-dimensional Lebesgue measure of the set of tori, whose existence is guaranteed by the Kolmogorov Theorem, is positive and the measure of its complement tends to zero as the size of the perturbation tends to zero. Thinking topologically, one would call the invariant tori exceptional, as the complement of an open dense set (see Appendix 11.A to this chapter). However, since they have a large Lebesgue measure, they are more the rule than the exception, and a slightly perturbed Hamiltonian system behaves practically as an integrable one.

It is now immediate to deduce from the above theorem the following corollary, more suited for the applications and that is probably known as the "true" Kolmogorov or Kolmogorov-Arnold-Moser (KAM) Theorem.

COROLLARY 11.8 *Given the total perturbed Hamiltonian*

$$H(I,\varphi) = H_0(I) + \varepsilon H_p(I,\varphi),$$

with a nondegenerate unperturbed Hamiltonian, i.e., $\det\left(\frac{\partial^2 H_0}{\partial I_h \partial I_j}\right) \neq 0$, *then for every set I_h^* of the actions such that the unperturbed frequencies $\omega_h(I^*) = \frac{\partial H_0}{\partial I_h}(I^*)$ satisfy the Diophantine condition, the tori I_h^* =constant survive, though slightly deformed, to sufficiently small perturbations.*

Proof. We may safely suppose $I^* = 0$, after a possible translation in the actions. From the mere definitions one has that

$$\lambda_i = \omega_i(0), \quad a = H_0(0) + \varepsilon \overline{H}_p(0),$$

$$A(\varphi) = \varepsilon[H_p(0,\varphi) - \overline{H}_p(0)], \quad B_h(\varphi) = \varepsilon \frac{\partial H_p}{\partial I_h}(0,\varphi),$$

$$(\mathbf{D}(0,\varphi))_{hi} = \frac{\partial^2 H_0}{\partial I_h \partial I_i}(0) + \varepsilon \frac{\partial^2 H_p}{\partial I_h \partial I_i}(0,\varphi).$$

From the expressions in the second line it is obvious that the condition on the smallness of the perturbation, i.e., of ε, implies that condition (iii) of the Kolmogorov Theorem holds, and, moreover, that ε and η have basically the same meaning. From the expression in the third line and the nondegeneracy of H_0, condition (ii) follows. Therefore, for the Kolmogorov Theorem there exists a canonical transformation which brings the perturbed Hamiltonian into a form for which the conservation of the torus $I^* = 0$ is obvious. QED

11.2.4 Nekhoroshev Theorem

Let us state the

THEOREM 11.9 (NEKHOROSHEV) *Let us consider the total perturbed Hamiltonian*

$$H(I,\varphi) = H_0(I) + \varepsilon H_p(I,\varphi)$$

defined on $\mathcal{B}_\varrho \times \mathbb{T}^n$, where $\mathcal{B}_\varrho$ is an open set in $\mathbb{R}^n$ and suppose H_0 to be convex,[(3)] *i.e., for some $m \in \mathbb{R}^+$,*

$$|\mathbf{A}(I)\underline{v} \cdot \underline{v}| \geq m\underline{v} \cdot \underline{v}, \quad \forall I \in \mathcal{B}, \quad \underline{v} \in \mathbb{R}^n, \quad (\mathbf{A})_{hi} = \frac{\partial^2 H_0}{\partial I_h \partial I_i}.$$

(3) In the original works, Nekhoroshev assumes the slightly more general condition of *steepness*.

Then there exist positive constants $\varepsilon_*, a, b, C, T, \delta$ such that, for all the initial data in $\mathcal{B}_{\varrho-\delta} \times \mathbb{T}^n$,

$$\|I(t) - I(0)\| \le C \left(\frac{\varepsilon}{\varepsilon_*}\right)^a ,$$

for all times t satisfying

$$t \le T \exp\left(\frac{\varepsilon_*}{\varepsilon}\right)^b .$$

Sketch of the Proof. We want to see what happens when the initial data do *not* satisfy the nonresonance condition, thus when the Equation $k \cdot \frac{\partial H_0}{\partial I} = 0$ admits nontrivial solutions $k \neq 0$. Let us suppose that this resonance is simple, that is the space of the solutions is one-dimensional, thus the corresponding resonant torus is $(n-1)$-dimensional (the extension to the generic case is immediate). Let $\psi = \sum k_i \varphi_i$ be the angular variable invariant for the resonance. We can partially average the perturbed Hamiltonian, obtaining

$$H = H_0(I) + \varepsilon \overline{H}_p(I, \psi) + \cdots ,$$

the averaged term $\overline{H}_p$ depending from the angle variables through the linear combination ψ only. The evolution for the action variables is given by

$$\dot{I}_i = \varepsilon \frac{\partial \overline{H}_p}{\partial \varphi_i} = \varepsilon \frac{\partial \overline{H}_p}{\partial \psi} k_i,$$

that is the variation vector of the actions has the same direction as the resonance vector k satisfying $k \cdot \omega = 0$. Under the assumption of convexity one proves that the motion of the actions is transversal to the resonance surface, thus the resonance is rapidly violated, and the evolution of the actions takes place in a brief time interval, so that its averaged speed is bounded from above by an exponentially small quantity. QED

An interesting and, in some sense, natural connection between the KAM and Nekhoroshev Theorems is established in Morbidelli & Giorgilli (1995), where it is proved that there exists a hierarchy of nested domains in phase space such that, at each more inner level, the evolution speed of the actions is exponentially small with respect to the previous one. The limit of these domains is the Cantor set of invariant KAM tori.

APPENDICES

11.A Results from Diophantine Theory

A frequency vector $\lambda \in \mathbb{R}^n$ is said to be *resonant* if there exists a $k \in \mathbb{Z}^n$, $k \neq 0$ such that $\lambda \cdot k = 0$. A frequency vector is said to be *strongly nonresonant* if one can find a positive function ψ such that

$$|k \cdot \lambda| \geq \psi(|k|) \quad \forall k \in \mathbb{Z}^n,\ k \neq 0,$$

where the notation $|k| = \sum_i |k_i|$ has been used. Given an open bounded subset $\mathcal{D} \subset \mathbb{R}^n$, the question is whether one can determine ψ in such a way that the subset of the strongly nonresonant frequencies in $\mathcal{D}$, namely the set

$$\Omega = \{\lambda \in \mathcal{D} : |k \cdot \lambda| \geq \psi(|k|)\},$$

has a large measure in $\mathcal{D}$.

A simple procedure to determine such a ψ is presented in Giorgilli (1989). Pick a nonzero $k \in \mathbb{Z}^n$, and consider

$$\tilde{\Omega}_k = \{\lambda \in \mathcal{D} : |k \cdot \lambda| < \psi(|k|)\},$$

which is in fact the set of the λ's that are close to resonance with k. Consider the plane in $\mathbb{R}^n$ through the origin orthogonal to k, i.e., the plane which contains all the resonant frequency vectors, and the set of points whose distance from the plane is less than $\frac{\psi(|k|)}{\sqrt{k_1^2+\ldots+k_n^2}} \leq \frac{\sqrt{n}\psi(|k|)}{|k|}$: this set of points contains clearly $\tilde{\Omega}_k$ and its measure $\mu\left(\tilde{\Omega}_k\right)$ is bounded by

$$\mu\left(\tilde{\Omega}_k\right) \leq \sqrt{n}C\frac{\psi(|k|)}{|k|},$$

where C is a constant that depends only on the domain $\mathcal{D}$. Then the measure of the complement of Ω in $\mathcal{D}$ cannot exceed

$$\mu\left(\bigcup_{k\neq 0}\tilde{\Omega}_k\right) \leq \sum_{k\neq 0}\mu\left(\tilde{\Omega}_k\right) \leq \sqrt{n}C\sum_{k\neq 0}\frac{\psi(|k|)}{|k|}.$$

Writing now

$$\sum_{k\neq 0}\frac{\psi(|k|)}{|k|} = \sum_{s>0}\sum_{|k|=s}\frac{\psi(s)}{s},$$

and using the fact that the number of vectors $k \in \mathbb{Z}^n$ that satisfies $|k| = s$ does not exceed $2^n s^{n-1}$, one finally gets

$$\mu\left(\bigcup_{k\neq 0} \tilde{\Omega}_k\right) \le 2^n \sqrt{n} C \sum_{s>0} s^{n-2}\psi(s).$$

Then it is enough to choose $\psi(s) = \frac{\gamma}{s^\tau}$ with suitable constants $\gamma > 0$ and $\tau \ge n$ in order to get that the complement of Ω in $\mathcal{D}$ has a measure that is small with γ. Such a result, although obtained with rough estimates, is optimal with relation to the value of τ. Indeed, for $\tau < n-1$ the set Ω is empty, while for $\tau = n-1$ the set Ω is nonempty, but has zero measure: see for example Rüssmann (1975).

The set Ω is therefore the complement of a "strange" set, which is of a small measure, though open and dense. This last resembles

$$\bigcup_{i=1}^{\infty}\left(\nu_i - \frac{\varepsilon}{2^i}, \nu_i + \frac{\varepsilon}{2^i}\right), \quad \varepsilon > 0, \quad \nu_1, \nu_2, \ldots \text{ rational numbers.}$$

The measure of this set is less than $\sum_{i=1}^{\infty} \frac{\varepsilon}{2^{i-1}} = 2\varepsilon$, that is it is arbitrarily small, although the set is open and dense.

11.B Cauchy Inequality

Consider an open disk $\Delta_\varrho(z_0)$ centered at the point z_0 of the complex plane $\mathbb{C}$. Consider a function f analytic and bounded on the disk $\Delta_\varrho(z_0)$. The *supremum norm* $\|f\|_\varrho$ of f in the domain $\Delta_\varrho(z_0)$ is defined as

$$\|f\|_\varrho = \sup_{z\in\Delta_\varrho(z_0)} |f(z)|\,.$$

The Cauchy inequality for the derivative f' at the origin states that

$$|f'(z_0)| \le \frac{1}{\varrho}\|f\|_\varrho\,.$$

More generally, for the s-th derivative $f^{(s)}$ one has the estimate

$$\left|f^{(s)}(z_0)\right| \le \frac{s!}{\varrho^s}\|f\|_\varrho\,.$$

This estimate cannot be improved in general. In fact, let $z_0 = 0$, $\varrho = 1$, and consider the function $f(z) = z^s$: it is then easy to verify that $\|f\|_1 = 1$, so that the Cauchy inequality gives $\left|f^{(s)}(0)\right| \le s!$. The proof of the inequalities above is an easy consequence of the Cauchy integral formula

$$f^{(s)}(z) = \frac{s!}{2\pi i} \oint \frac{f(\zeta)}{(\zeta - z)^{s+1}} \, d\zeta.$$

Indeed, let $z = z_0$ and let the integration path be the circle of radius ϱ with center at z_0. If $\zeta - z_0 = \varrho e^{i\theta}$, then $d\zeta = i(\zeta - z_0)\, d\theta$ and

$$\left| f^{(s)}(z_0) \right| \le \frac{s!}{2\pi} \oint \left| \frac{f(\zeta)}{(\zeta - z_0)^{s+1}} \right| \varrho \, d\theta,$$

so that the claim follows by replacing $|\zeta - z_0| = \varrho$ and $|f(\zeta)| \le \|f\|_{\varrho}$.

In the case of n variables the extension is straightforward. Let the domain $\Delta_{\varrho}(z_0)$ be the polydisk of radius ϱ centered at $z_0 \in \mathbb{C}^n$, namely

$$\Delta_{\varrho}(z_0) = \{z \in \mathbb{C}^n : |z - z_0| \le \varrho\},$$

where $|z| = \max_j \left| z_j \right|$. This is nothing but the Cartesian product of complex disks of radius ϱ in the complex plane. Define the supremum norm of an analytic function f as above. Then the Cauchy estimate reads

$$\left| \frac{\partial f}{\partial z_j}(z_0) \right| \le \frac{1}{\varrho} \|f\|_{\varrho}, \quad 1 \le j \le n.$$

The estimates above can be extended to more general domains. Indeed, let U be a subset of $\mathbb{C}^n$, and consider the domain

$$\mathcal{D}_{\varrho} = \bigcup_{z \in U} \Delta_{\varrho}(z),$$

that is: the domain $\mathcal{D}_{\varrho}$ is the union of closed polydisks of radius ϱ around every point of U.

The supremum norm of a complex function $f : \mathcal{D}_{\varrho} \to \mathbb{C}$ is defined in a natural way as

$$\|f\|_{\varrho} = \sup_{z \in \mathcal{D}_{\varrho}} |f(z)|.$$

With this setting, the Cauchy inequality is stated by the following

PROPOSITION 11.10 *Let f be analytic in the interior of $\mathcal{D}_{\varrho}$, and $\|f\|_{\varrho}$ be finite. Then for any positive $\delta \le \varrho$ one has*

$$\left\| \frac{\partial f}{\partial z_j} \right\|_{\varrho - \delta} \le \frac{1}{\delta} \|f\|_{\varrho}.$$

Proof. If $z \in \mathfrak{D}_{\varrho-\delta}$, then applying Cauchy inequality to the polydisk $\Delta_\delta(z)$ we get

$$\left|\frac{\partial f}{\partial z_j}(z)\right| \le \frac{1}{\delta} \sup_{z' \in \Delta_\delta(z)} |f(z')|.$$

Since $\Delta_\delta(z) \subset \mathfrak{D}_\varrho$, the supremum over the polydisk does not exceed the norm $\|f\|_\varrho$. Since the inequality holds for every point of $\mathfrak{D}_{\varrho-\delta}$, it holds for the supremum norm of the derivative. QED

CHAPTER 12

Perturbations of the Kepler Problem

In this chapter we study the perturbations of the Kepler problem, i.e., the motion of a particle with mass m and Hamiltonian

$$H(\vec{Q}, \vec{P}) = \frac{1}{2m}P^2 - \frac{k}{Q} + \varepsilon H_p(\vec{Q}, \vec{P}),$$

where $H_p(\vec{Q}, \vec{P})$ is the perturbation Hamiltonian and ε a dimensionless, "small" parameter. The canonical variables $\vec{Q}, \vec{P}$ differ from the previously defined $\vec{q}, \vec{p}$ and $\vec{x}, \vec{y}$ by rescaling factors; see (12.1.3) and Remark 12.2 below.

Consistently with the theory exposed in the previous chapter, for the analytical and numerical integration one usually performs a canonical transformation, passing to action–angle variables, for example to those of Delaunay, discussed in Chapter 3. In these new variables the unperturbed Hamiltonian is a function of only one action variable, and the motion generated by the unperturbed Hamiltonian is very simple: only one angle, the mean anomaly, evolves linearly with time, while the other five coordinates remain constant.

Unquestionably, the use of these coordinates presents some advantages.

- Well established methods exist to push the perturbation to higher orders, so that, truncating the Hamiltonian to some order, we have an approximate but, if another first integral is known[1], integrable problem.

- The Delaunay variables have a geometrical meaning directly related to the description of the orbits and their variations are much more significant for the astronomers than those of $\vec{Q}$ and $\vec{P}$.

- In the perturbed motion the mean value of the constants of the unperturbed motion evolves with velocity of order ε. For numerical computations this means that, integrating only this slow motion, errors do not affect the integration of the known (and predominant) part.

But, unfortunately, we have also some drawbacks.

- The explicit form of the Delaunay map is rather intricate and to write the perturbed Hamiltonian in these variables is not an easy task. The Delaunay variables are in fact transcendental functions of the physical coordinates (this is related to the Kepler Equation), whose inversion cannot be achieved in closed form.

- The Delaunay variables are not global. For the circular orbits or those lying on the plane orthogonal to the z-axis they are not unambiguously defined. Analogously for any other system of action-angle variables: this is related to the fact that the global topology of the phase space of the (regularized) Kepler Problem is that of T^+S^3 (the cotangent bundle of the 3-dimensional sphere without the null-section, as discussed in Chapter 6) while an action-angle variable system describes a manifold with the topology of $\mathbb{T}^3 \times \mathbb{R}^3$ (where $\mathbb{T}^3$ is the 3-dimensional torus).

The aim of this chapter is to present a method that avoids the drawbacks without losing the advantages. In the subsequent chapter, we will apply the method to some concrete cases. Let us briefly explain the main idea. Since, as said, the topology of the Kepler Problem is that of T^+S^3, we describe its phase space using two vectors $u, v \in \mathbb{R}^4$, such that

$$\|u\| = \|v\|, \quad u \cdot v = 0,$$

where norm and scalar products are induced by the Euclidean metric. This is analogous, for example, to the parametrization of a sphere with a vector subject to the constraint of the constant norm, rather than with two local variables, like latitude and longitude, that leave out the poles.

[1] A typical example is that of an axially symmetric perturbation, as in the subsequent chapter.

It results that the two vectors are algebraic functions of the physical coordinates position and velocity, and one can easily write the perturbed Hamiltonian as a function of u, v. The new dynamical variables u_α, v_β ($\alpha, \beta = 1, \dots, 4$) are not canonical but are (global) coordinates of a so-called 8-dimensional Poisson manifold of rank 6. After a rescaling of the time, the unperturbed Hamiltonian becomes equal to $\|u\| = \|v\|$ and the unperturbed motion is a fast uniform rotation of the two vectors on their initial plane. The bivector $u \wedge v$ describes the angular momentum and the Runge-Lenz-Laplace vector respectively, and the averaged perturbation induces mean slow variations on the bivector but preserves the constraints. The projection of u onto the subspace $Q_1Q_2Q_3$ gives the trajectory. The known methods of normalization can still be used, since the basic tools are the Poisson brackets and their properties, whatever is their explicit form. As for the numerical integration:

- the method of the variation of the constants is used, avoiding those errors which would arise integrating numerically also the unperturbed and exactly known motion;
- a stabilization method (see for example Ascher (1997)) is used to avoid the violation of the constraints: a term is added to the perturbed equations of motion, which acts like a force pulling back the point to the 6-dimensional phase space.

We try to make this and the subsequent chapter as much as possible self-contained, and reading of Chapters 4–10, though advisable, not strictly necessary. References, which concern more involved topics, are relegated to some remarks.

In this and in the subsequent chapter we use the symbols u, v and U, V in place of U, V and $\overline{U}, \overline{V}$, respectively, as in Part I and II; thus u, v are the variable 4-vectors and U, V their initial values.

The exposition of this chapter has been strongly influenced by Moser (1970) and Kummer (1983). The section on numerical integration follows Cordani (2000) and Cordani & Merlini (2001).

12.1 A More Convenient Hamiltonian

Let us consider the *total* Hamiltonian[(2)]

$$H(\vec{Q}, \vec{P}) = \frac{1}{2m} P^2 - \frac{k}{Q} + \varepsilon H_p(\vec{Q}, \vec{P}), \qquad (12.1.1)$$

[(2)] In this and the subsequent chapter, with H_0 and K_0 we denote those *unperturbed* Kepler Hamiltonians previously called H and K, keeping aside these last symbols for the corresponding *total* Hamiltonians.

sum of the unperturbed Hamiltonian $H_0(\vec{Q}, \vec{P})$ of the Kepler Problem and the perturbing one $H_p(\vec{Q}, \vec{P})$. Since the perturbation is time independent, the Hamiltonian (12.1.1) is a constant of the motion whose value, h, depends on the initial conditions $\vec{Q_0}, \vec{P_0}$,

$$H(\vec{Q}, \vec{P}) = h.$$

Let us call $K(\vec{Q}, \vec{P})$ the function obtained by solving algebraically this equation for k:

$$K(\vec{Q}, \vec{P}) = Q\left(\frac{1}{2m}P^2 - h + \varepsilon H_p\right).$$

The units can be rescaled in such a way that

$$m = 1, \quad h = -\frac{1}{2}, \quad k = 1.$$

Define, indeed, the rescaling factors

$$\lambda_m = m, \quad \lambda_Q = \frac{k}{-2h}, \quad \lambda_T = \sqrt{\frac{mk^2}{-8h^3}}, \tag{12.1.2}$$

and pass to the dimensionless variables

$$\vec{Q} \mapsto \vec{x} = \frac{\vec{Q}}{\lambda_Q}, \quad \vec{P} \mapsto \vec{y} = \frac{\vec{P}}{\lambda_P}, \quad T \mapsto t = \frac{T}{\lambda_T}, \tag{12.1.3}$$

where $\lambda_P = \lambda_m \frac{\lambda_Q}{\lambda_T}$. Replacing in $H(\vec{Q}, \vec{P})$ and $K(\vec{Q}, \vec{P})$ we obtain

$$H(\vec{x}, \vec{y}) = \frac{1}{2}y^2 - \frac{1}{x} + \varepsilon H_p(\vec{x}, \vec{y}) = -\frac{1}{2}, \tag{12.1.4}$$

$$K(\vec{x}, \vec{y}) = \frac{1}{2}x(y^2 + 1) + \varepsilon x H_p(\vec{x}, \vec{y}) = 1. \tag{12.1.5}$$

The physical meaning of the rescaling factors is the following. Consider an unperturbed circular motion of total energy h: the constant radius of the orbit is λ_Q, the constant norm of the linear momentum is λ_P and the revolution period is $2\pi\lambda_T$. Hereafter we will use the dimensionless variables $\vec{x}, \vec{y}$.

Then consider the Hamilton Equations generated by $H(\vec{x}, \vec{y})$:

$$\frac{d\vec{x}}{dt} = \vec{y} + \varepsilon\frac{\partial H_p}{\partial \vec{y}}, \quad \frac{d\vec{y}}{dt} = -\frac{\vec{x}}{x^3} - \varepsilon\frac{\partial H_p}{\partial \vec{x}},$$

and those generated by $K(\vec{x}, \vec{y})$, but with a new, for the moment unspecified, "fictitious time" s

$$\frac{d\vec{x}}{ds} = x\left(\vec{y} + \varepsilon\frac{\partial H_p}{\partial \vec{y}}\right), \quad \frac{d\vec{y}}{ds} = -x\left(K(\vec{x}, \vec{y})\frac{\vec{x}}{x^3} + \varepsilon\frac{\partial H_p}{\partial \vec{x}}\right).$$

From the very definition of $K(\vec{x},\vec{y})$, with the same rescaled initial conditions, we infer

$$K(\vec{x},\vec{y}) = 1,$$

so that, if we define

$$dt = x\,ds, \tag{12.1.6}$$

the two systems of Hamilton's Equations are identical. As should be clear (see Chapter 2), the "fictitious time" is the eccentric anomaly. The reason why we have defined the new unperturbed Hamiltonian $K_0 = \frac{1}{2}x(y^2+1)$ is that it is more suitable, for subsequent study of the perturbed problem, than the original one $H_0 = \frac{1}{2}y^2 - \frac{1}{q}$.

To show this, let us introduce a new set of dynamical variables; instead of the six independent variables $\vec{x},\vec{y}$ we define the eight variables u_α, v_α, $\alpha = 1,\dots,4$, i.e.,

$$u = \begin{pmatrix} \frac{1}{2}\vec{x}(y^2+1) - (\vec{x}\cdot\vec{y})\vec{y} \\ -(\vec{x}\cdot\vec{y}) \end{pmatrix}, \quad v = \begin{pmatrix} x\vec{y} \\ -\frac{1}{2}x(y^2-1) \end{pmatrix}, \tag{12.1.7}$$

which satisfy the two constraints

$$\|u\| = \|v\|, \quad u\cdot v = 0, \tag{12.1.8}$$

the norm and scalar product being induced by Euclidean metric of $\mathbb{R}^4$. Moreover,

$$\|u\| = \|v\| = K_0. \tag{12.1.9}$$

As one may check with a direct calculation (but see the remark below for a "rational" explanation of this and subsequent properties), the canonical Poisson brackets

$$\{y_i, y_k\} = \{x_i, x_k\} = 0, \quad \{y_i, x_k\} = \delta_{ik}$$

induce the Poisson brackets

$$\begin{aligned} \{u_\alpha, u_\beta\} = \{v_\alpha, v_\beta\} &= \frac{1}{K_0}(u_\alpha v_\beta - u_\beta v_\alpha), \\ \{u_\alpha, v_\beta\} &= -K_0\delta_{\alpha\beta}. \end{aligned} \tag{12.1.10}$$

We may thus take as phase space a symplectic leaf of the Poisson manifold $(\mathbb{R}^4 - \{0\})\times(\mathbb{R}^4 - \{0\})$ with *global* coordinates u, v. In fact, the two scalar functions

$$C_1 = \|u\|^2 - \|v\|^2, \quad C_2 = u\cdot v$$

are Casimir functions, i.e.,

$$\{C_i, u_\alpha\} = \{C_i, v_\alpha\} = 0, \quad \forall\alpha$$

on the symplectic leaf, given by $C_1 = C_2 = 0$, of the Poisson manifold $(\mathbb{R}^4 - \{0\})\times(\mathbb{R}^4 - \{0\})$. This symplectic leaf is a manifold symplectomorphic to T^+S^3.

The first integrals of the Hamiltonian K_0 are the angular momentum and the Runge-Lenz-Laplace vectors

$$\vec{G} = \vec{x} \times \vec{y}, \quad \vec{A} = \frac{1}{2}\vec{x}(y^2 - 1) - (\vec{x}\cdot\vec{y})\vec{y},$$

which, when expressed in terms of u and v, become

$$G_i = \frac{1}{K_0}\epsilon_{ihk}(u_h v_k - u_k v_h),$$
$$A_i = \frac{1}{K_0}(u_4 v_i - u_i v_4).$$

They satisfy

$$\{G_i, G_j\} = -\epsilon_{ijk}G_k, \quad \{G_i, A_j\} = -\epsilon_{ijk}A_k, \quad \{A_i, A_j\} = -\epsilon_{ijk}G_k, \tag{12.1.11}$$

which are the commutation rules of the 4-dimensional rotation group. Moreover, the following algebraic relations hold:

$$G^2 + A^2 = K_0^2, \quad \vec{G}\cdot\vec{A} = 0. \tag{12.1.12}$$

We may gather the vectors $\vec{G}$ and $\vec{A}$ in a 4-dimensional skew symmetric tensor by putting

$$G_{hk} = -\epsilon_{ihk}G_i, \quad G_{k4} = A_k,$$

from which

$$G_{\alpha\beta} = \frac{1}{K_0}(v_\alpha u_\beta - v_\beta u_\alpha), \quad \alpha, \beta = 1, \ldots, 4. \tag{12.1.13}$$

Moreover, we may gather $K_0, G_{\alpha\beta}, u_\alpha, v_\beta$ in the unique skew symmetric expression $G_{AB} = -G_{BA}$, $A, B, C \ldots = -1, 0, \ldots, 4$, with

$$-G_{-1,0} = K_0, \quad G_{-1,\alpha} = u_\alpha, \quad G_{0,\alpha} = v_\alpha,$$

so that all the Poisson brackets can be written as

$$\{G_{AB}, G_{AC}\} = \eta_{AA}G_{BC}, \quad (\eta_{AB}) = \mathrm{diag}\,(--++++),$$

and $\{G_{AB}, G_{CD}\} = 0$ if all indices are different.

The Hamilton Equations,

$$\frac{du_\alpha}{ds} = \{K_0, u_\alpha\} = v_\alpha, \quad \frac{dv_\beta}{ds} = \{K_0, v_\alpha\} = -u_\beta,$$

are easily integrated as

$$\begin{pmatrix} u_\alpha(\mathrm{s}) \\ v_\alpha(\mathrm{s}) \end{pmatrix} = \begin{pmatrix} \cos \mathrm{s} & \sin \mathrm{s} \\ -\sin \mathrm{s} & \cos \mathrm{s} \end{pmatrix} \begin{pmatrix} U_\alpha \\ V_\alpha \end{pmatrix} \tag{12.1.14}$$

where U_α, V_α are the initial values at "time" $\mathrm{s} = 0$. The motion of the two vectors is a uniform rotation on their initial plane. We may express U_α and V_α as some functions of the angular momentum and Runge-Lenz-Laplace vector. Put

$$U = \begin{pmatrix} K_0 \frac{\vec{A}}{A} \\ 0 \end{pmatrix}, \quad V = \begin{pmatrix} \vec{G} \times \frac{\vec{A}}{A} \\ A \end{pmatrix}.$$

Because of the identities (12.1.12), the two vectors U_α and V_β satisfy the constraints (12.1.8), moreover they commute with the Hamiltonian K_0, and therefore can be chosen as initial values.

Since, by definition,

$$\vec{x} = \vec{u} - \vec{A} \tag{12.1.15}$$

we can write the orbit equation

$$\vec{x}(\mathrm{s}) = K_0 \frac{\vec{A}}{A} \cos \mathrm{s} + \vec{G} \times \frac{\vec{A}}{A} \sin \mathrm{s} - \vec{A}, \tag{12.1.16}$$

from which

$$x(\mathrm{s}) = K_0 - A \cos \mathrm{s}. \tag{12.1.17}$$

Plainly, $\mathrm{s} = 0$ corresponds to the pericentre. Integrating

$$\int_0^t dt = \int_{\mathrm{s}_0}^{\mathrm{s}} x(\mathrm{s})\, d\mathrm{s},$$

we get the Kepler Equation

$$t - t_p = K_0(\mathrm{s} - E \sin \mathrm{s}), \quad E = \frac{A}{K_0}, \quad t_p = -K_0(\mathrm{s}_0 - E \sin \mathrm{s}_0),$$

where t_p is the time of the passage at the pericentre.

By definition also

$$\vec{y} = \frac{\vec{v}}{x} \tag{12.1.18}$$

and the relations (12.1.15), (12.1.18) give the physical coordinates as functions of u and v, which allows us to write explicitly the perturbations as some functions of u and v.

REMARK 12.1 The above "hand-made" results can be easily deduced in a more rational way from the work of Chapter 6. The fifteen dynamical variables $u_\alpha, v_\beta, \vec{G}, \vec{A}, K_0$ are the components of the moment map $T^+S^3 \to \mathrm{so}^*(2,4)$, which explains why their Poisson brackets close so nicely. Moreover, the action of SO(2,4) on T^+S^3 being transitive, this last may be identified with a 6-dimensional coadjoint orbit in $\mathrm{so}^*(2,4)$: this explains the existence of the nine independent algebraic relations (12.1.8), (12.1.9) and (12.1.13) between the fifteen dynamical variables. Lastly, the relation between K_0 and the unperturbed Hamiltonian H_0 of the Kepler Problem is given by the extended canonical transformation $\mathscr{K}$ on page 138 (up to the exchange between coordinates and momenta). Notice that this extended canonical transformation is a rescaling of coordinates and momenta

$$\vec{x} \mapsto \vec{x}\sqrt{-2H_0}, \quad \vec{y} \mapsto \frac{\vec{y}}{\sqrt{-2H_0}},$$

In the unperturbed case, since $H_0 = -\frac{1}{2}$, the rescaling is the identity, which implies that $K_0 = \frac{1}{\sqrt{-2H_0}}$ and $\vec{A} = \vec{R}$. This fact obviously fails for the perturbed case, where $H = -\frac{1}{2}$.

REMARK 12.2 The rescaling (12.1.3) may be factorized into the product of a trivial one(3) (independent from the initial conditions)

$$\vec{Q} \mapsto \vec{q} = \frac{\vec{Q}}{k}, \quad \vec{P} \mapsto \vec{p} = \frac{\vec{P}}{\sqrt{m}}, \quad T \mapsto t = \frac{T}{\sqrt{mk^2}},$$

which transforms the Hamiltonian (12.1.1) into

$$H(\vec{q}, \vec{p}) = \frac{1}{2}p^2 - \frac{1}{q} + \varepsilon H_p(\vec{q}, \vec{p}),$$

and of a more significant one

$$\vec{q} \mapsto \frac{\vec{q}}{\lambda^2}, \quad \vec{p} \mapsto \lambda\vec{p}, \quad t \mapsto \frac{t}{\lambda^3}, \quad \text{where } \lambda = \frac{1}{\sqrt{-2h}},$$

which implies $H \mapsto \lambda^2 H$. This last is the rescaling of Moser, Equation (5.1.8) on page 100 and Kummer (1982, Equation (5.7)), making commutative the diagram

(3) It is trivial since, choosing judiciously the physical units of measure, one can make $k = 1, m = 1$.

$$
\begin{array}{ccc}
(x_\mu, y_\nu) & \xrightarrow{\text{Covector dilation}} & \left(x_\mu, \dfrac{y_\nu}{\lambda}\right) \\
\Big\downarrow \mathscr{C} & & \Big\downarrow \mathscr{C} \\
(\vec{q}, \vec{p}, t, H) & \xrightarrow{\text{Rescaling}} & \left(\dfrac{\vec{q}}{\lambda^2}, \lambda\vec{p}, \dfrac{t}{\lambda^3}, \lambda^2 H\right)
\end{array}
$$

Here $\mathscr{C}$ is the extended canonical transformation (6.4.8a)-(6.4.8d) that intertwines the somewhat "artificial" rescaling of Moser-Kummer with the more natural covector dilations. As pointed out in Guillemin & Sternberg (1990, page 16), the Moser-Kummer rescaling is essentially a restatement of Kepler's third law.

Composing the perturbed Hamiltonian $K(\vec{x}, \vec{y})$ with (12.1.15) and (12.1.18) we obtain

$$K(u, v) = K_0 + \varepsilon K_p(u, v), \quad K_p = x H_p. \tag{12.1.19}$$

The Hamilton Equations are

$$\frac{du_\alpha}{ds} = \{K, u_\alpha\}, \quad \frac{dv_\alpha}{ds} = \{K, v_\alpha\}, \tag{12.1.20}$$

where the Poisson brackets are given by (12.1.10). We stress the fact that, due to the Poisson structure, the constraints (12.1.8) are preserved by the solutions of these Hamilton Equations, independently of the explicit form of $K(u, v)$.

As finding an *exact* analytical solution is in general out of the question, we can proceed with

(i) normalization (or averaging) method, truncated at some order, or

(ii) numerical integration.

12.2 Normalization (or Averaging) Method

Let us apply the machinery of Subsection 11.1.2. Multiply from the left the total Hamiltonian (12.1.19) with the operator $\exp(\varepsilon \mathcal{L}_{\chi_1})$ where χ_1 is unknown, and retain the terms up to second order

$$\exp(\varepsilon\mathcal{L}_{\chi_1})(K_0+\varepsilon K_p)$$
$$= K_0+\varepsilon(\mathcal{L}_{\chi_1}K_0+K_p)+\varepsilon^2\left(\mathcal{L}_{\chi_1}K_p+\frac{1}{2}\mathcal{L}^2_{\chi_1}K_0\right)+O(\varepsilon^3).$$

Since $\mathcal{L}_{\chi_1}K_0 = -\mathcal{L}_{K_0}\chi_1$, we are led to solving the equation

$$\mathcal{L}_{K_0}\chi_1 = K_p - \overline{K}_1, \tag{12.2.1}$$

where $\overline{K}_1$ is the mean value of K_p

$$\overline{K}_1 = \frac{1}{2\pi}\int_0^{2\pi}\phi_s^* K_p\, ds.$$

ϕ_s is the dynamical flow generated by K_0 and ϕ_s^* the pull-back (see Appendix A for the definition). The solution of (12.2.1) is given by

$$\chi_1 = \frac{1}{2\pi}\int_0^{2\pi} s\,\phi_s^*(K_p-\overline{K}_1)\, ds,$$

in fact

$$\begin{aligned}\mathcal{L}_{K_0}\chi_1 &= \frac{1}{2\pi}\int_0^{2\pi} s\frac{d}{ds}\phi_s^*(K_p-\overline{K}_1)\, ds\\ &= \frac{1}{2\pi}\left[s\,\phi_s^*(K_p-\overline{K}_1)\Big|_0^{2\pi} - \int_0^{2\pi}\phi_s^*(K_p-\overline{K}_1)\, ds\right]\\ &= K_p-\overline{K}_1,\end{aligned}$$

using the periodicity of $K_p\,(u(s),v(s))$.

We have normalized the perturbed Hamiltonian to first order. Iterating the procedure and dropping the terms $O(\varepsilon^{r+1})$, we obtain the truncated normalized Hamiltonian $K^{(r)}$ to order r

$$K^{(r)} = K_0+\sum_{i=1}^{r}\varepsilon^i\overline{K}_i, \quad \{K_0,\overline{K}_i\}=0\ \forall i. \tag{12.2.2}$$

The normalized Hamiltonian $K^{(r)}$ is invariant under the dynamical flow generated by K_0, hence must be a function only of the first integrals of K_0, i.e., of $K_0, \vec{G}, \vec{A}$. Among these seven quantities the two algebraic relations (12.1.12) hold. Equivalently, the two vectors

$$\vec{\gamma} = \frac{1}{2}(\vec{G}+\vec{A}), \quad \vec{\alpha} = \frac{1}{2}(\vec{G}-\vec{A}), \tag{12.2.3}$$

with Poisson brackets

$$\{\gamma_i,\gamma_j\} = -\epsilon_{ijk}\gamma_k, \quad \{\alpha_i,\alpha_j\} = -\epsilon_{ijk}\alpha_k, \quad \{\gamma_i,\alpha_j\}=0, \tag{12.2.4}$$

satisfy the algebraic relations

$$\gamma^2 = \alpha^2 = \frac{1}{4} K_0^2 .$$

The phase space, $\overline{M}$ say, of the normalized truncated Hamiltonian $K^{(r)}$ is thus a symplectic leaf of the 6-dimensional Poisson manifold $\mathbb{R}^3 \times \mathbb{R}^3$ of rank 4, with global coordinates $\vec{\gamma}$ and $\vec{\alpha}$. $\overline{M}$ is clearly symplectomorphic to $S^2 \times S^2$.

REMARK 12.3 This last is a particular case of a more general statement, valid for the n-dimensional Kepler Problem: $\overline{M}$, which is the space of the orbits for a fixed energy, is homeomorphic (indeed, symplectomorphic) to $\frac{\mathrm{SO}(n+1)}{\mathrm{SO}(2)\times \mathrm{SO}(n-1)}$. In fact $\mathrm{SO}(n+1)$ acts transitively (besides on S^n) on $T_1^* S^n$ (the unit cotangent bundle), hence, being the isometry group of S^n, acts transitively also on the set of oriented orbits of its geodesic flow. But a homogeneous manifold may be viewed as the quotient of a group, acting transitively on it, by the isotropy subgroup of any of its points. The isotropy subgroup leaving an orbit (i.e., a big circle of S^n) invariant, is $\mathrm{SO}(2) \times \mathrm{SO}(n-1)$: from this the statement follows. For $n = 3$ the manifold $\overline{M}$ of the orbits is symplectomorphic to $\frac{\mathrm{SO}(4)}{\mathrm{SO}(2)\times\mathrm{SO}(2)} = S^2 \times S^2$ (since $\mathrm{SO}(4) = \mathrm{SO}(3) \times \mathrm{SO}(3)$) and for $n = 2$ to $\frac{\mathrm{SO}(3)}{\mathrm{SO}(2)} = S^2$.

REMARK 12.4 The Poisson brackets (12.2.4) give to the two sphere the usual symplectic structure of Example A.9 on page 351 and A.11 on page 353.

The normalization method does not lead, in general, to an integrable truncated Hamiltonian, since the Kepler Problem is completely degenerate, but allows lowering from 6 to 4 the dimensions of the phase space. For the complete integrability one needs another first integral: this case, frequently met in practice, will be investigated in the following chapter. But also in the generic case, from the existence of critical points[(4)] of the averaged Hamiltonian, some interesting conclusions can be drawn about the existence of periodic orbits.

Let us in fact consider the averaged truncated Hamiltonian $K^{(r)}$ which induces a dynamical flow on the space of orbits $S^2 \times S^2$. Every point of the space of orbits is, by definition, an orbit, i.e., an ellipse, in the original physical space, while a trajectory in the orbit space describes a deformation of this ellipse. It is obvious that to a critical point of $K^{(r)}$ corresponds a fixed ellipse, thus a periodic orbit. Although all the orbits of the unperturbed system are periodic, this need not be the case

[(4)] We remember that a critical point of a vector field f is a point for which $f = 0$. Analogously, a critical point for a Hamiltonian H is a point for which $dH = 0$, i.e., the Hamiltonian vector field vanishes.

for the perturbed truncated case and a periodic solution will branch off only from a (in general) finite number of points, namely from the critical points. In Moser (1970) it is proved that, if the critical points are nondegenerate[5] and if the neglected terms in the full perturbed Hamiltonian are small, this fact continues to be true for the full, i.e., not truncated, motion.

We now report the original proof of the Moser Theorem, which holds for generic vector fields, not only for the Hamiltonian ones; but first we need to state some facts and definitions.

Consider a flow ϕ_t (satisfying $\phi_t \circ \phi_T = \phi_{t+T}$) on a manifold M, such that all orbits are periodic with fixed period, say 2π. We assume that no point is stationary: if $f_0 = \left.\frac{d\phi_t}{dt}\right|_{t=0}$ is the vector generating ϕ_t, we have $f_0 \neq 0$. Our aim is to study the perturbed vector fields $f = f_0 + \varepsilon g + O(\varepsilon^2)$, in particular we are interested in finding periodic orbits for f whose periods are close to 2π. For any continuous vector field v on M, we define the averaged (along ϕ_t) vector field $\overline{v}$ by

$$\overline{v} \stackrel{\text{def}}{=} \frac{1}{2\pi} \int_0^{2\pi} \phi_t^* v \, dt.$$

Since obviously $\phi_t^* \overline{v} = \overline{v}$, any averaged vector field is determined along the orbits of f_0 if it is known at one point p of such an orbit. Indeed, we have (notation of the Appendix A)

$$\overline{v}(\phi_t(p)) = \mathbf{T}_{\phi_t} \overline{v}(p).$$

Therefore, we can consider any such averaged vector field $\overline{v}$ as defined on a manifold $\overline{M}$ obtained by identifying the points of M that lie on the same oriented orbit: in symbols, $\overline{M} = \frac{M}{S^1}$. Since $f \neq 0$, for the Proposition B.6 this is again a manifold, whose dimension is one less than that of M. We denote by Π the projection mapping an orbit $\phi_t(p) \in M$ into a point of $\overline{M}$.

THEOREM 12.1 (MOSER) *Let $f = f_0 + \varepsilon g + O(\varepsilon^2)$ with f_0 generating a flow such that all unperturbed orbits have period 2π. If the averaged vector field $\overline{g}$ admits a nondegenerate critical point at $p^* \in \overline{M}$, then the perturbed vector field f has a periodic orbit of period close to 2π in a pre-assigned neighbourhood of $\Pi^{-1}(p^*)$, provided ε is small enough.*

Proof. Pick a point p_0 on the orbit $\Pi^{-1}(p^*)$ and describe a neighbourhood of p_0 by the coordinates $x = (x_1, \dots, x_n)$. We choose the

[5] A critical point x^* of a vector field with components $f_k(x_1, \dots, x_n)$, $k = 1, \dots, n$ is said to be nondegenerate if $\det \left(\frac{\partial f_k}{\partial x_h} \right)_{x=x^*} \neq 0$.

coordinates so that f_0 has components $(0,\dots,0,1)$ in this neighbourhood of p_0, and such that p_0 corresponds to $x = 0$. Near $x = 0$, the surface $\sigma_0 : x_n = 0$ is intersected transversely by the orbits of f_0, which satisfy

$$\begin{cases} \dot{x}_\nu = 0, & \nu = 1,2,\dots,n-1, \\ \dot{x}_n = 1. \end{cases}$$

Now we can extend the validity of such coordinates to a neighbourhood of the whole orbit $\Pi^{-1}(p^*)$: for any point p in such a neighbourhood there exists a unique t with $0 \le t < 2\pi$ such that $\phi_{-t}(p) \in \sigma_0$. Setting $t = x_n$ and describing $\phi_{-t}(p) \in \sigma_0$ by the coordinates $x_1,\dots,x_{n-1},0$ gives rise to such an extension. In this way a neighbourhood of $\Pi^{-1}(p^*)$ can be mapped differentially into a tubular neighbourhood of the x_n-axis of Euclidean space, where points whose n-th coordinates differ by 2π are identified.

In these coordinates the vector field $f = f_0 + \varepsilon g + O(\varepsilon^2)$ takes the form

$$\begin{cases} \dot{x}_\nu = \phantom{1 +{}} \varepsilon g_\nu + O(\varepsilon^2), \\ \dot{x}_n = 1 + \varepsilon g_n + O(\varepsilon^2), \end{cases} \tag{12.2.5}$$

where the right-hand side has period 2π in x_n, and the averaged vector field $\overline{g}$ is given by

$$\overline{g}_\nu = \frac{1}{2\pi}\int_0^{2\pi} g_\nu(x_1,\dots,x_n)\,dx_n \quad \text{for } \nu = 1,2,\dots,n.$$

To establish the existence of a periodic orbit of f, we determine the mapping of σ_0 into itself generated by the flow of f by following the orbits from one intersection with σ_0 to the next one. We are seeking a fixed point for this mapping, which will give us a periodic orbit for f. To determine this mapping, we write the solution curves of f with initial values $x_\nu(0) = a_\nu$ in the form

$$x(t) = a + e_n t + \varepsilon \int_0^t g(a + e_n\tau)\,d\tau + O(\varepsilon^2), \tag{12.2.6}$$

where $e_n = (0,\dots,0,1)$. We restrict ourselves to $a \in \sigma_0$, i.e., $a_n = 0$, and determine the time $T = T(a,\varepsilon)$, elapsed between consecutive intersections with σ_0, by integrating the second of (12.2.5)

$$2\pi = T + \varepsilon \int_0^T g_n(a + e_n t)\,dt + O(\varepsilon^2),$$

which has a unique solution of the form

$$T = 2\pi + \varepsilon T_1(a) + O(\varepsilon^2).$$

Substituting this into (12.2.6) we obtain for the desired mapping of σ_0 into itself

$$x_\nu(T) = a_\nu + 2\pi\,\varepsilon\overline{g}_\nu(a) + O(\varepsilon^2) \quad \text{for } \nu = 1, 2, \ldots, n-1.$$

To find a fixed point of this mapping we have to look for a solution of

$$\varepsilon\overline{g}_\nu(a) + O(\varepsilon^2) = 0, \quad \nu = 1, 2, \ldots, n-1.$$

By assumption p^* is a critical point of the averaged vector field; because of our choice of coordinates, this means that

$$\overline{g}_\nu(0) = 0 \quad \text{for } \nu = 1, 2, \ldots, n-1,$$

while nondegeneracy means that

$$\det\left(\frac{\partial g_\nu}{\partial x_\mu}\right)_{x=0} \neq 0 \quad \text{for } \mu, \nu = 1, 2, \ldots, n-1,$$

i.e., the Jacobian of this mapping at $a = 0$ does not vanish: therefore, the equation has a unique solution with $a = O(\varepsilon)$. QED

COROLLARY 12.2 *Consider the perturbed Hamiltonian $H = H_0 + \varepsilon G + O(\varepsilon^2)$. Let M be the hypersurface H_0 = constant and ϕ_t the flow on it generated by the vector field $\Omega^\#(dH_0)$, where $\Omega^\#$ is the inverse matrix of the symplectic form. Suppose that the orbits of ϕ_t have all period 2π and let $\overline{M}$ be the quotient manifold with respect to the flow. Then to every nondegenerate critical point $p^* \in \overline{M}$ of the averaged Hamiltonian*

$$\overline{G} = \frac{1}{2\pi}\int_0^{2\pi} \phi_t^* G\, dt$$

corresponds a periodic solution of the vector field $\Omega^\#(dH)$, that branches off from the orbit representing p^* and has period close to 2π.

Proof. Note simply that, if $g = \Omega^\#(dG)$, then its average along ϕ_t

$$\overline{g} = \frac{1}{2\pi}\int_0^{2\pi} \phi_t^*\left(\Omega^\#(dG)\right)\, dt = \frac{1}{2\pi}\Omega^\#\left(d\int_0^{2\pi} \phi_t^* G\, dt\right) = \Omega^\#(d\overline{G})$$

is the vector field generated by the averaged Hamiltonian $\overline{G}$. In the above expression we have taken advantage of the property that ϕ_t leaves invariant the symplectic form, and that the pull-back and the exterior derivative commute. Since $\{H_0, \overline{G}\} = 0$, the vector field $\overline{g}$ is tangent to M and the statement of the corollary is reduced to that of the previous theorem. QED

The corollary can be applied to the normalized nontruncated Hamiltonian

$$K = K_0 + \sum_{i=1}^{r} \varepsilon^i \overline{K}_i + O(\varepsilon^{r+1})$$

noticing that $K_0 = \frac{1}{2}x(y^2+1)$ is the square root of the Hamiltonian of the geodesic motion on the sphere S^3 : it is clear that all the orbits have period 2π, with the "time" s denoting the angle of the position of the point on the great circle. For more details, see Section 5.1 and Chapter 6.

We are led hence to finding the critical points of the Hamiltonian $\sum \varepsilon^i \overline{K}_i$, defined on $\overline{M} = S^2 \times S^2$ (or $\overline{M} = S^2$ in the plane case). This gives a lower bound for the number of periodic orbits of the perturbed Kepler Problem. Indeed, the sum of the indices of the nondegenerate critical points of a vector field on a manifold $\overline{M}$ is equal to the Euler-Poincaré characteristic[(6)] of $\overline{M}$: see for example Milnor (1965) or Perko (1991). If the vector field is Hamiltonian, the nondegenerate critical points can be only elliptic (stable) with index $+1$ or hyperbolic (unstable) with index -1 : see Figure 12.1. The Euler-Poincaré characteristic of $\overline{M} = S^2 \times S^2$ being four, on $\overline{M}$ there exists at least four elliptic points (or six elliptic and two hyperbolic, and so forth). Therefore, the above topological argument ensures that, for a generic perturbation of the Kepler Problem, at least four periodic orbits of the averaged truncated Hamiltonian exist, while the Moser Theorem states that these orbits continue to persist for the full Hamiltonian.

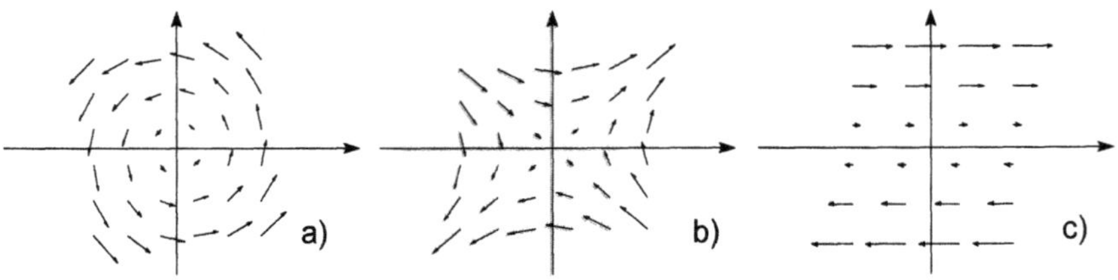

Figure 12.1: a) Elliptic point: following a closed curve which surrounds the critical point, the vector of the field rotates by $+2\pi$; hence the index is $+1$. b) Hyperbolic point: the vector of the field rotates by 2π in the opposite sense, hence the index is -1. c) The horizontal axis is made up of degenerate critical points.

We will now investigate, by means of Corollary 11.8 of Kolmogorov's

[(6)]The Euler-Poincaré characteristic of a two-dimensional compact manifold is given by $F - L + V$, where, for a given triangulation, F is the number of the triangles, L of the edges and V of the vertices. It can be shown that it is a topological invariant. It is immediate to check that for the sphere it is equal to 2.

Theorem, if the 3-dimensional orbits, corresponding to those orbits on $S^2 \times S^2$ surrounding one of the elliptic critical points, can survive for the full Hamiltonian also.

To this end, let us consider one of these elliptic critical points, which we may assume to be at the two North Poles of $S^2 \times S^2$; in fact, we can always rotate it to this location, and a 4-dimensional rotation is a symplectomorphism that leaves K_0 invariant. The Hamiltonian $\overline{K}_1(\vec{y}, \vec{\alpha})$ has thus an elliptic critical point for $\vec{y} = \vec{\alpha} = \left(0, 0, \frac{1}{2}K_0\right)$. We can choose local canonical coordinates[(7)] $\pi_1, \pi_2, \chi_1, \chi_2$ centered on this minimum of $\overline{K}_1$, by putting

$$y_1 = \frac{\chi_1}{2}\sqrt{2K_0 - (\chi_1^2 + \pi_1^2)}, \quad \alpha_1 = \frac{\chi_2}{2}\sqrt{2K_0 - (\chi_2^2 + \pi_2^2)},$$
$$y_2 = \frac{\pi_1}{2}\sqrt{2K_0 - (\chi_1^2 + \pi_1^2)}, \quad \alpha_2 = \frac{\pi_2}{2}\sqrt{2K_0 - (\chi_2^2 + \pi_2^2)},$$
$$y_3 = \frac{1}{2}K_0 - \frac{1}{2}(\chi_1^2 + \pi_1^2), \quad \alpha_3 = \frac{1}{2}K_0 - \frac{1}{2}(\chi_2^2 + \pi_2^2),$$

with

$$\{\chi_i, \chi_j\} = \{\pi_i, \pi_j\} = 0, \quad \{\pi_i, \chi_j\} = \delta_{ij}.$$

Substitute into $\overline{K}_1(\vec{y}, \vec{\alpha})$; this Hamiltonian has a minimum, say $a(K_0)$, in $\chi_i = \pi_j = 0$, and can be expanded into a Taylor series around this point. Consider the constant and the quadratic terms only; since the equilibrium point is elliptic, the quadratic form can be diagonalized by a linear symplectic transformation. Define the action-angle variables

$$I_k = \frac{1}{2}(\chi_k^2 + \pi_k^2), \quad \varphi_h = \arctan\frac{\chi_h}{\pi_h}, \quad h, k = 1, 2$$

and put $K_0 = I_0$. Thus

$$\overline{K}_1 = a(I_0) + \omega_1(I_0)I_1 + \omega_2(I_0)I_2 + \cdots,$$

where the dots stand for a polynomial in χ_h and π_k, whose first terms are cubic. The Birkhoff Theorem (Moser 1968), whose proof runs basically along the same lines of the general perturbation theory of the previous chapter, guarantees the existence of a formal canonical transformation that takes $\overline{K}_1$ into a power series in the variables I_1, I_2, with coefficients depending from I_0, provided ω_1 and ω_2 are rationally independent[(8)]. We are left finally with a total Hamiltonian of the type

$$K = I_0 + \varepsilon\left[a(I_0) + \omega_h(I_0)I_h + \frac{1}{2}\omega_{hk}(I_0)I_hI_k + \cdots\right] + O(\varepsilon^2),$$

(7) They have nothing to do with the coordinates $\vec{\chi}, \vec{\pi}$ of Chapter 10.

(8) More exactly, in order to obtain a power series in I_1, I_2 up to order r, ω_1 and ω_2 must not admit any resonamce $k \cdot \omega = 0$ with $k \in \mathbb{Z}^2 - \{0\}$, $|k| \le 2r$, where $|k| = \sum_i |k_i|$.

which is the sum of an integrable Hamiltonian

$$K_0' = I_0 + \varepsilon\left[a(I_0) + \omega_h(I_0)I_h + \frac{1}{2}\omega_{hk}(I_0)I_hI_k\right]$$

with perturbative terms. These last are of the two types: the terms $O(\varepsilon^2)$ and those, linear in ε, which come from $\overline{K}_1$, for which the role of perturbative parameter is played by the size of the neighbourhood of the origin. We are forced to consider this more complicated unperturbed Hamiltonian: indeed, $K_0 = I_0$ is clearly degenerate, since $\det\left(\dfrac{\partial^2 K_0}{\partial I_\mu \partial I_\nu}\right) = 0,\ \mu, \nu = 0, 1, 2.$

We can therefore conclude that the tori labeled by frequencies $\lambda_\mu = \dfrac{\partial K_0'}{\partial I_\mu}$, satisfying the Diophantine condition, survive to the full perturbation, provided the two other conditions of the Kolmogorov Theorem are satisfied, i.e., perturbations sufficiently small and nondegeneracy of K_0', which is clearly a generic condition.

12.3 Numerical Integration

Equations (12.1.20) are eight first order equations in normal form and one could pass directly to their numerical integration, but with its inevitable errors. To limit them we can do two things.

(i) The right-hand side of (12.1.20) is the sum of a "large" vector field, whose integral flow is exactly known, and a "small" one. We are in the situation described in the first appendix to this chapter, i.e., the method of variation of the constants can be applied. We are thus led to integrate the eight equations

$$\frac{dU_\alpha}{d\mathfrak{s}} = \varepsilon\{K_p(u(\mathfrak{s}), v(\mathfrak{s})), U_\alpha\}, \quad \frac{dV_\alpha}{d\mathfrak{s}} = \varepsilon\{K_p(u(\mathfrak{s}), v(\mathfrak{s})), V_\alpha\}, \tag{12.3.1}$$

where U, V satisfy the same Poisson brackets as u, v and $u(\mathfrak{s}), v(\mathfrak{s})$ is the general unperturbed solution (12.1.14). This is in the spirit of the action-angle variables (since we avoid integrating numerically the known and predominant part), but without its shortcomings: in fact, the dynamical variables U, V are *global* coordinates and simple *algebraic* functions of position and velocity.

(ii) U and V must satisfy the same constraints as u and v. We are thus in the situation described in the second appendix to this chapter, with

$$z = \begin{pmatrix} \frac{1}{2}(U^2 - V^2) \\ U \cdot V \end{pmatrix}, \quad Z = \begin{pmatrix} U^T & -V^T \\ V^T & U^T \end{pmatrix}.$$

The corrective term $W = -\gamma Z^T(ZZ^T)^{-1}z$ results in

$$W = -\gamma \frac{1}{U^2 + V^2} \begin{pmatrix} \frac{1}{2}(U^2 - V^2)U + (U \cdot V)V \\ -\frac{1}{2}(U^2 - V^2)V + (U \cdot V)U \end{pmatrix} \qquad (12.3.2)$$

and this is the term that must be added to the right-hand side of (12.3.1) to preserve the constraints.

Let us summarize *la marche à suivre* for the numerical integration.

- The starting point is the Hamiltonian (12.1.1) and the initial conditions. Compute its constant value h and, with the constants m and k, compute the scale factors (12.1.2).
- Rescale the initial conditions as in (12.1.3).
- Define two 4-vectors u, v according to (12.1.7). They satisfy the constraints (12.1.8) and have Poisson brackets (12.1.10). Compute their initial values using the rescaled conditions of the previous item.
- Integrate the Equations (12.3.1) with the initial values computed in the previous item. To preserve the constraints (this would be automatic for exact integration) add to the right-hand side of (12.3.1) the corrective term (12.3.2).
- Insert the computed values of $U(\mathfrak{s})$ and $V(\mathfrak{s})$ as varying initial values into the unperturbed equations of motion (12.1.14).
- Compute with (12.1.15) and (12.1.18) the physical rescaled coordinates as functions of $\mathfrak{s}$; integrate (12.1.6) to obtain the relation between $\mathfrak{s}$ and t.
- Finally return to the physical, nonrescaled coordinates.

Apart from the various rescaling processes, more conveniently performed "by hand", the above list has been implemented in the program KEPLER. Two stages are involved, namely the symbolic manipulation, which produces the eight equations to integrate starting from the perturbing Hamiltonian, and the numerical integration, which, however, must be preceded by the "compilation" of the equations.

12.3.1 Symbolic Manipulation

The perturbing Hamiltonian, introduced by the user, is manipulated using a special software module, which was developed to obtain the eight equations to be integrated. This manipulation is performed in a symbolic way and no numerical calculation is performed until the integration of the eight equations. The software is written in C language. This symbolic manipulation can be divided into five steps.

1. The perturbing Hamiltonian is multiplied by x.

2. The substitution of $\vec{x}, \vec{y}$ with corresponding values (in terms of $\vec{U}, \vec{V}, \vec{A}, K_0$ and of the independent variable s) is performed into the output of the previous item. Instead of U_α, V_β, we have chosen the variables $\vec{U}, \vec{V}, \vec{A}, K_0$ to simplify the output of the differentiations in the following item; consider indeed the expressions (12.1.15) and (12.1.18).

3. The resulting expression of the previous item is differentiated with respect to the $\vec{U}, \vec{V}, \vec{A}, K_0$ variables, obtaining 10 expressions in terms of the same variables. These expressions should be considered as a column vector.

4. The column vector, obtained in the previous item, is multiplied by an 8×10 matrix (whose entries are the mutual Poisson brackets among the variables) in order to obtain the eight right-hand sides of the equations in terms of $\vec{U}, \vec{V}, \vec{A}, K_0$.

5. Finally, the symbols $\vec{A}, K_0$ are substituted, inside the previous eight equations, by their values in terms of U_α, V_β. After this substitution, we get the eight equations in the variables U_α, V_β ready to be numerically integrated.

Substitutions are easy to implement in a computer program: just find the characters representing the name of the variable to be substituted and replace it with the characters representing its value.

The only problem to keep in mind is that substituting a variable with an expression can change the operation order: to avoid this problem, all substitutions are put inside round brackets. For example, multiplication by x at point 1 is solved just creating a new string by concatenating the string "$x*$" with the user-inserted Hamiltonian inside round brackets. The same approach is used for matrix multiplication: all factors are put inside round brackets and concatenated together with the addition symbol as separator.

The problem described at point 3 is more complicated. It can be summarized in this way: implement a computer algorithm that takes two strings as input (representing a formula and a variable name) and returns an output string, representing a formula that is the derivative of the input formula with respect to the given variable name.

To solve this problem, we divided it in two parts: the first analyses the input formula, checks the input syntax, parses it and produces a data structure that is examined by the second part. The second part takes this data structure and produces the required derivative function. We called the first part PARSER, and the second one DERIVE.

In fact, the PARSER software is a particular implementation of a parser, which checks the syntax and performs a lexical analysis. Instead of evaluating the input function, this parser produces a data structure representing the formula: a binary tree. In this tree, each node contains a token (i.e., a "formula unit" such as variable name, function name, operator or number) and an enumerative field containing its classification made by a lexical analyzer. The structure of this tree depends on the input formula.

This tree represents the formula in an easy-to-manipulate way; in fact, no brackets are needed, and operator priority is determined by the position of the node in the tree. Of course, only variables and numbers should be leaves; operators and functions must be nonterminal nodes.

We have implemented a syntax-driven parser because it is very easy to realize, and it maps the problem to solve in an excellent way. The standard parser implementation evaluates the input expression; we have modified it, in order to produce the binary tree.

The DERIVE software module uses this tree to produce a new binary tree representing the derivative expression. Then, converting the binary tree into the output string is an easy task: just visit the output tree in symmetrical order, copying the visited nodes into the string (into round brackets to avoid operator priority problems).

The next problem to solve is how to produce the derivative tree starting from the input tree, but this is not complicated because the recursive structure of the binary tree maps the recursive structure of the derivative rules. So, analyzing the root of an input subtree, it is possible to build the corresponding output subtree. This rule applies to subtrees of this one, and so on until the empty tree is reached. To implement this rule, we use, of course, recursion: this programming style is a very powerful tool and it is the best choice for this kind of problems.

The last consideration regards formula simplification since, after the derivative is performed, some expressions could be simplified. We gave a partial solution to this problem, simplifying only when this operation can be made in an easy way, so that the output expression may be not the best simplified solution. In fact, trying to simplify in all cases produces a complicated algorithm, very computer-time expensive, losing any advantage to get a smaller formula. Moreover, an operative definition of the terms "simplification" or "simplify" does not exist.

12.3.2 Compiling Equations

The numerical integration is performed with a standard method, for example a fourth order Runge-Kutta. During integration, the eight right-hand sides of the equations generated by symbolic manipulation are evaluated thousands of times, depending on the number of points required by the user. We need a parser to evaluate an expression, but a standard implementation of a parser has to check the syntax, evaluate variable values and solve priorities every time it is called. Since the equations are the same for all integration points, we can "compile" them in a simple binary language. During this phase, each expression is split into elementary operations, and a list of these operations is produced. Further, variable names are substituted with variable addresses to avoid finding their value at evaluation time. At evaluation time, the only operation required is to scan the operation list, performing the operations with the current variable values and storing results in a suitable temporary array. The address of the temporary array is contained in the operation list. In this way, the evaluation cycle is very short and very optimized.

As an example, consider the formula:

$$(U_1 + 3) * U_2 + U_3 * \sin(U_4).$$

This formula will be reduced to the following operation list:

$$\begin{aligned}
\mathsf{TMP}_1 &\leftarrow U_1 + 3 \\
\mathsf{TMP}_2 &\leftarrow \mathsf{TMP}_1 * U_2 \\
\mathsf{TMP}_3 &\leftarrow \sin(U_4) \\
\mathsf{TMP}_4 &\leftarrow U_3 * \mathsf{TMP}_3 \\
\mathsf{TMP}_5 &\leftarrow \mathsf{TMP}_2 + \mathsf{TMP}_4.
\end{aligned}$$

These operations are very easy to execute, because only "small" operations are involved and no priority problems arise. This evaluation will be very fast with respect to evaluating the starting expression. Further, no syntax checking is performed: this task is performed only once, when the equation is compiled. The problem is how to obtain the operation list starting from the string containing the formula. To do this, we use again the software already developed, dividing the problem into two parts, called PARSER and REDUCE.

PARSER is the same software described in the previous subsection, which takes an input string and produces a binary tree representing the formula itself.

The REDUCE module works on this tree, reducing it in this way: when a "small subtree" is found, it is detached from the tree and substituted with a temporary variable. For "small subtree" we mean a subtree representing a simple operation; it can be structured as a single binary or unary operation with the corresponding parameters. Once detached, the operation is copied into the operation list and the tree is simplified. This procedure will be repeated until the tree becomes empty. Temporary variables are chosen sequentially, so it is impossible to choose an already used variable. The number of these variables can be calculated by counting the number of operations in the tree. Before introducing new operations in the list, the operation list is scanned to verify if it is already in the list itself and, if so, the same temporary variable is used to avoid to perform the same operation twice. Finally, the last used temporary variable will contain the result.

Consider, for example, the previous formula:

$$(U_1 + 3) * U_2 + U_3 * \sin(U_4).$$

The corresponding tree is shown in the following picture.

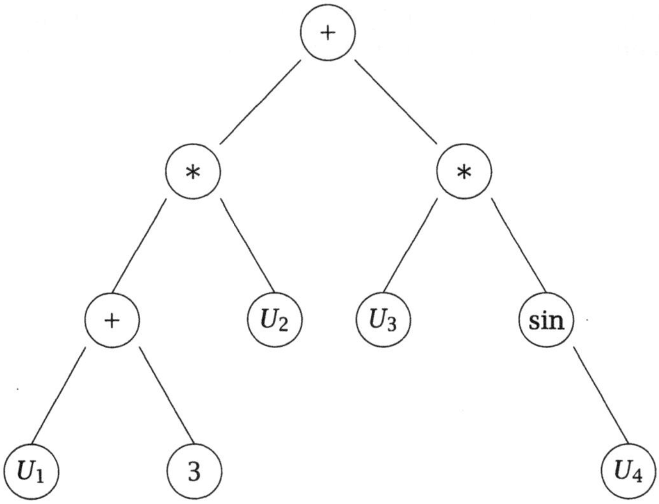

The first elementary operation found is $U_1 + 3$, so REDUCE will write the first operation in the operation list:

$$\mathsf{TMP}_1 \leftarrow U_1 + 3$$

and the tree will become the following.

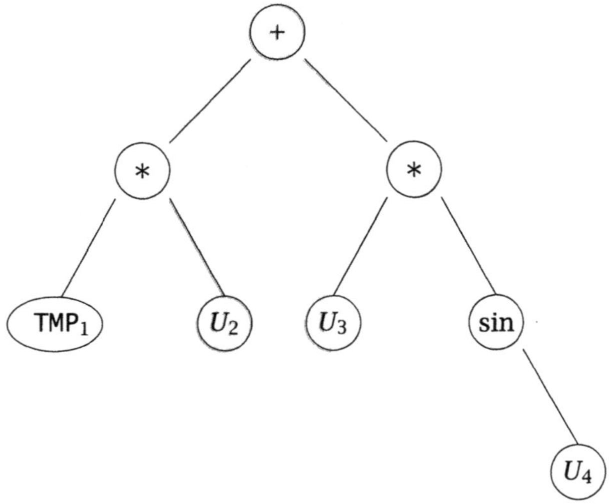

The next operation will be $\mathsf{TMP}_2 \leftarrow \mathsf{TMP}_1 * U_2$ and so on, until the tree will be reduced to a single temporary variable representing the entire tree.

APPENDICES

12.A Variation of the Constants

Let us consider the dynamical system

$$\frac{dx}{dt} = f_0(x) + \varepsilon f_p(x), \quad x \in \mathbb{R}^N, \quad f_0, f_p : \mathbb{R}^N \to \mathbb{R}^N,$$

and assume we know the general solution of the unperturbed part

$$x = x(t, X), \tag{12.A.1}$$

where $X \in \mathbb{R}^N$ is the vector of the integration constants. Suppose now that these integration constants are functions of the time in such a way that (12.A.1) becomes a solution of the perturbed equation. Thus, if the Jacobian matrix of the transformation (12.A.1) can be inverted,

$$\frac{dX}{dt} = \varepsilon \frac{\partial X}{\partial x} f_p(x(t, X)).$$

For the numerical integration this means that the errors do not affect the integration of the known (and predominant) part.

In particular, if the dynamical system is a generalized Hamiltonian one as

$$\frac{dx^\mu}{dt} = \{H_0(x) + \varepsilon H_p(x), x^\mu\},$$

we likewise obtain

$$\frac{dX^\mu}{dt} = \varepsilon \{H_p(x(t, X)), X^\mu\}.$$

We have exploited the fact that the Poisson structure is invariant under the transformation $x \mapsto X$, and this is assured because X is the initial value of x.

12.B The Stabilization Method

Let us consider the dynamical system

$$\frac{dx}{dt} = f(x), \quad x \in \mathbb{R}^N$$

and suppose that if the initial conditions satisfy

$$z(x) = 0, \quad z : \mathbb{R}^N \to \mathbb{R}^m, \; m < N, \tag{12.B.1}$$

this relation is preserved by the exact integration. The inevitable errors of the numerical integration will give rise to a progressive violation of this constraint and the solution will depart from the $(N - m)$-dimensional manifold (12.B.1). The stabilization method adds a term to the dynamical system, which has no effect if the constraint is respected but pulls the point back to the manifold (12.B.1) if it begins to depart from it. Define the $(m \times N)$ matrix

$$Z = \frac{\partial z}{\partial x}$$

and add to the dynamical system the term: $-\gamma Z^T (ZZ^T)^{-1} z$, where γ is a numerical constant whose value is chosen to avoid any instability in numerical integration. The dynamical evolution of the variable z *due only to this term* is the solution to the equation

$$\frac{dz}{dt} = -\gamma Z Z^T (Z Z^T)^{-1} z = -\gamma z,$$

i.e., the constraint function z tends exponentially to zero.

For the choice of the constant γ, we remark that, while it seems difficult to find its optimal value in a theoretical way, in practice, in our case, there are no problems at all: a choice of γ very greatly under the critical value (i.e., under that giving rise to possible instabilities) is sufficient to perform a total respect of the 6-dimensional constraint.

CHAPTER 13

Perturbations with Axial Symmetry

In this chapter we study the particular case of perturbations with axial symmetry. Important problems, like that of the Moon motion, that of a hydrogen atom in a constant electric plus crossed magnetic field or that of a satellite around an oblate primary are of (or can be reduced to) this type.

The main ideas of this chapter are essentially due (modulo some changes in the formalism) to Cushman (1991), Deprit (1981) and Kummer (1983), which rests ultimately on Conley (1963).

13.1 Reduction of Orbit Manifold

Unlike the general case of the previous chapter, the axisymmetric perturbations lead, after normalization, to a Liouville integrable problem; beside K and K_0, the projection of the angular momentum on the symmetry axis is a first integral, clearly in involution with the other two. This allows us to study the perturbed motion in a graphical and global way. The basic idea is to reduce the manifold of the orbits, which has $S^2 \times S^2$ topology, with the rotational symmetry, obtaining a symplectic manifold that turns out to be homeomorphic (but in general not diffeomorphic) to S^2. Then, the level surfaces of the twice reduced Hamiltonian (thus with one degree of freedom), intersecting this symplectic manifold, give rise to a family of curves describing the perturbed motion.

Our first task is therefore to obtain the symplectic reduction, $\overline{\overline{M}}$ say, of the manifold of the orbits $\overline{M} \simeq S^2 \times S^2$ with respect to the action of the axial symmetry. We may always assume the symmetry to be around the third axis, thus generated by the action of G_3. Following the Marsden-Weinstein Theorem, we should consider the moment map $\Gamma : \overline{M} \to so^*(2)$ given by G_3, choose some value[(1)], with $-K_0 \leq G_3 \leq K_0$, and, on the 3-dimensional manifold $\Gamma^{-1}(G_3)$, divide out the flow generated by the action of G_3. Unfortunately, for $G_3 = 0$ this flow has fixed points, as is clear from Figure 13.1, so that, in this case, the reduced space will be no longer a *differentiable* manifold.

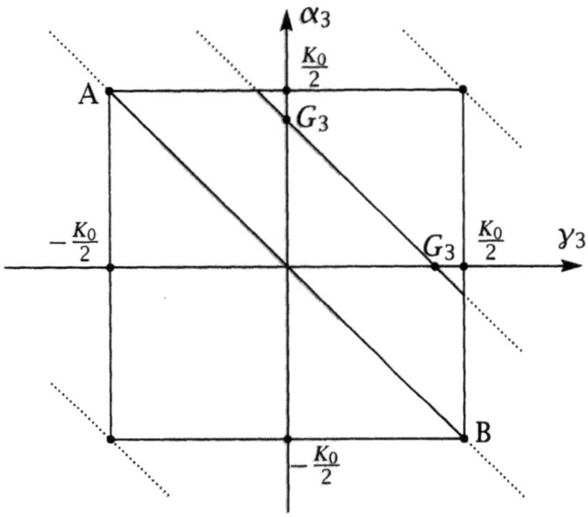

Figure 13.1: Once the value of G_3 is chosen, the admissible values of γ_3, α_3 (defined in (12.2.3)) are those lying on the straight line $\gamma_3 + \alpha_3 = G_3$ and falling on and internally to the square. The manifold $\Gamma^{-1}(G_3)$ is empty for $|G_3| > K_0$ and trivial (one point) for $|G_3| = K_0$. For $-K_0 < G_3 < K_0$, $G_3 \neq 0$ the rotational action on it has no fixed points, while for $G_3 = 0$ the two points $A \equiv \left(0, 0, -\frac{K_0}{2}, 0, 0, \frac{K_0}{2}\right)$ and $B \equiv \left(0, 0, \frac{K_0}{2}, 0, 0, -\frac{K_0}{2}\right)$ are fixed points.

This fact has been pointed out by Cushman (1991), who uses the general technique of "singular reduction" to construct $\overline{\overline{M}}$ for every value of G_3. To achieve the same goal, we prefer an *ad hoc* but very simple (and basically equivalent) procedure. We do not, indeed, reduce directly the symplectic manifold $S^2 \times S^2$, but, rather, the 6-dimensional Poisson manifold $so^*(4)$, of which $S^2 \times S^2$ is a symplectic leaf. In this way,

---

[(1)] To lighten the notation and at the cost of a slight abuse of language, we denote with G_3 (as for K_0) both the function and its value, the context making clear the sense.

the $G_3 = 0$ value is no longer a "bad" one. The Poisson reduction (see Appendix D) yields a 4-dimensional Poisson manifold of rank 2. The two Casimir invariants descend to the reduced manifold, and the two-fold reduced Kepler manifold $\overline{\overline{M}}$ appears as a symplectic 2-dimensional leaf.

Let us view the details. Under the S^1-action on so*(4) of G_3, this last and A_3 stay unchanged, the action being effective only on G_1, G_2, A_1, A_2 not all null, thus on $\mathbb{R}^4 - \{0\}$. To reduce the S^1-action, one uses the Hopf fibration:

$$\frac{\mathbb{R}^4 - \{0\}}{S^1} \simeq \frac{\mathbb{R}^+ \times S^3}{S^1} \simeq \mathbb{R}^+ \times S^2 \simeq \mathbb{R}^3 - \{0\}.$$

Explicitly, define the spinor, i.e., an element of $\mathbb{C}^2 - \{0\}$,

$$\psi = \begin{pmatrix} \psi_1 \\ \psi_2 \end{pmatrix} = \begin{pmatrix} G_1 + iG_2 \\ A_1 + iA_2 \end{pmatrix}, \quad \psi \neq 0,$$

on which the S^1-action of G_3 is given by $\psi \mapsto \psi e^{i\varphi}$. To reduce this action we must firstly find the 3-dimensional manifold that parametrizes the orbits of the action.

PROPOSITION 13.1 *The Hermitian matrices with null trace*

$$\psi\psi^\dagger - \frac{1}{2}(\psi^\dagger\psi)1_2 \overset{def}{=} \frac{1}{2}\begin{pmatrix} \xi_3 & \xi_4 + i\xi_2 \\ \xi_4 - i\xi_2 & -\xi_3 \end{pmatrix},$$
$$(\xi_2, \xi_3, \xi_4) \in \mathbb{R}^3 - \{0\},$$

are in 1-1 correspondence with the orbits of the S^1-action on $\mathbb{C}^2 - \{0\}$.

Proof. Put $\psi = \begin{pmatrix} \sqrt{\varrho_1}e^{i\theta_1} \\ \sqrt{\varrho_2}e^{i\theta_2} \end{pmatrix}$. The S^1-action lies invariant ϱ_1, ϱ_2 and $\theta_1 - \theta_2$, hence this triplet identifies univocally the orbit. Given this triplet, i.e., the orbit, the matrix is fixed unambiguously. Vice versa, given the matrix, i.e., ξ_2, ξ_3, ξ_4, by definition of ψ we write

$$\xi_3 = \varrho_1 - \varrho_2,$$
$$\xi_4 + i\xi_2 = \sqrt{\varrho_1\varrho_2}e^{i(\theta_1 - \theta_2)}.$$

From the latter equation we get $\theta_1 - \theta_2$, then we know the difference ξ_3 and the product $\xi_4^2 + \xi_2^2$ of ϱ_1 and ϱ_2, which are thus fixed unambiguously. QED

Putting $G_3 =$ constant, we obtain the Poisson reduction with respect to the S^1-action

$$\text{so}^*(4) \simeq \mathbb{R}^2 \times (\mathbb{C}^2 - \{0\}) \to \mathbb{R} \times (\mathbb{R}^3 - \{0\}) : (\vec{G}, \vec{A}) \mapsto (\xi_1, \xi_2, \xi_3, \xi_4),$$

where

$$\xi_1 = A_3, \qquad \xi_2 = 2(G_2A_1 - G_1A_2),$$
$$\xi_3 = G_1^2 + G_2^2 - A_1^2 - A_2^2, \quad \xi_4 = 2(G_1A_1 + G_2A_2). \tag{13.1.1}$$

The two Casimir invariants of so*(4) are

$$G^2 + A^2 = G_3^2 + \xi_1^2 + \sqrt{\xi_2^2 + \xi_3^2 + \xi_4^2},$$
$$\vec{G} \cdot \vec{A} = G_3\xi_1 + \frac{1}{2}\xi_4,$$

while the manifold $\overline{M} \simeq S^2 \times S^2$ is characterized by

$$G^2 + A^2 = K_0^2, \quad \vec{G} \cdot \vec{A} = 0.$$

The second relation allows us to eliminate $\xi_4 = -2G_3\xi_1$ in the first relation, so that the two-fold reduced Kepler manifold turns out to be a 2-dimensional algebraic variety imbedded in $\mathbb{R}^3$:

$$\left(K_0^2 + G_3^2 - \xi_1^2\right)^2 - \xi_2^2 - \xi_3^2 = 4K_0^2G_3^2. \tag{13.1.2}$$

This is thus the equation of $\overline{\overline{M}}$, that we call of the "A-type", since it is symmetric around the $\xi_1 = A_3$ axis. See Figure 13.2.

From the Definitions (13.1.1), and bearing in mind the Poisson brackets (12.1.11), one calculates

$$\{\xi_1, \xi_2\} = -2\xi_3, \quad \{\xi_1, \xi_3\} = 2\xi_2, \quad \{\xi_2, \xi_3\} = 4\xi_1(\xi_1^2 - K_0^2 - G_3^2). \tag{13.1.3}$$

Putting

$$C(\xi) = \left(K_0^2 + G_3^2 - \xi_1^2\right)^2 - \xi_2^2 - \xi_3^2, \tag{13.1.4}$$

we find that the Poisson brackets (13.1.3) can be written in the compact form

$$\{\xi_i, \xi_j\} = \epsilon_{ijh}\frac{\partial C(\xi)}{\partial \xi_h}.$$

Plainly, $C(\xi)$ is a Casimir function:

$$\{\xi_i, C(\xi)\} = \{\xi_i, \xi_h\}\frac{\partial C(\xi)}{\partial \xi_h} = \epsilon_{ijh}\frac{\partial C(\xi)}{\partial \xi_h}\frac{\partial C(\xi)}{\partial \xi_j} = 0.$$

Summing up, we have proved the following

PROPOSITION 13.2 *The symplectic reduction of the manifold $S^2 \times S^2$ with respect to the G_3-action is the 2-dimensional algebraic variety (13.1.2), i.e., a symplectic leaf of $\mathbb{R}^3$ equipped with the Poisson structure (13.1.3). Varying G_3, these symplectic leaves are diffeomorphic (for $G_3 \neq 0$) but only homeomorphic (for $G_3 = 0$) to the two-dimensional sphere S^2.*

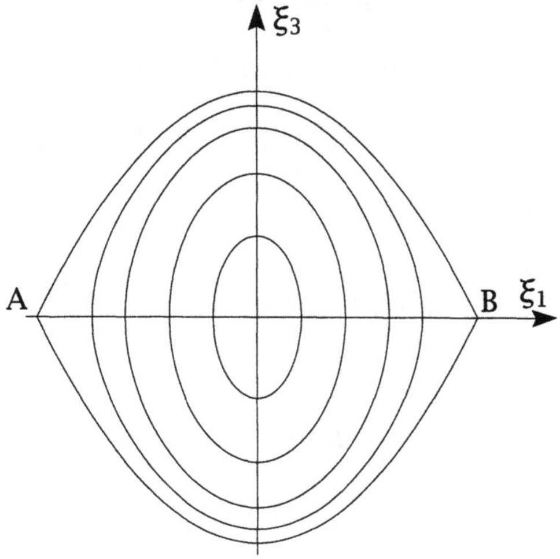

Figure 13.2: Section on the plane $\xi_1\xi_3$ of the algebraic variety $\overline{\overline{M}}$ of the A-type, for $K_0 = 1$ and various values of $G_3 = 0.8, 0.6, 0.4, 0.25, 0$. The variety has rotational symmetry around the axis ξ_1. Notice, for $G_3 = 0$, the two singular points A and B on axis ξ_1, which make the variety homeomorphic, but nondiffeomorphic, to the sphere.

We introduce now another parametrization, called of G-type, of $\overline{\overline{M}}$, characterized by the fact that it is symmetric about the third axis. Put

$$\eta_1 = \xi_1\sqrt{\frac{1}{2}(\xi_3 - \xi_1^2 + K_0^2 + G_3^2)},$$
$$\eta_2 = \frac{1}{2}\xi_2,$$
$$\eta_3 = \sqrt{\frac{1}{2}(\xi_3 - \xi_1^2 + K_0^2 + G_3^2)},$$

whose (formal) inverse relations are

$$\xi_1 = \frac{\eta_1}{\eta_3}, \quad \xi_2 = 2\eta_2, \quad \xi_3 = 2\eta_3^2 + \frac{\eta_1^2}{\eta_3^2} - (K_0^2 + G_3^2).$$

Clearly, this change of coordinates is a diffeomorphism for $\eta_3 \neq 0$. Recalling the Definitions (13.1.1), one easily verifies that $\eta_3 = G$, i.e., the norm of the total angular momentum: this explains the choice of the name "G-type". The condition $\eta_3 \neq 0$ holds if $G_3 \neq 0$, and in this case manifolds diffeomorphic to S^2 are be transformed into manifolds again diffeomorphic to S^2.

Let us examine how the algebraic variety and the Poisson structure change when passing from A-type to G-type. Substituting the above transformation into the Definition (13.1.4) we find

$$C(\eta) = -4\left[\eta_1^2 + \eta_2^2 + \eta_3^4 - (K_0^2 + G_3^2)\eta_3^2\right],$$

and the equation of $\overline{\overline{M}}$ is written as

$$\eta_1^2 + \eta_2^2 + \left[\eta_3^2 - \frac{1}{2}(K_0^2 + G_3^2)\right]^2 = \frac{1}{4}(K_0^2 - G_3^2)^2. \qquad (13.1.5)$$

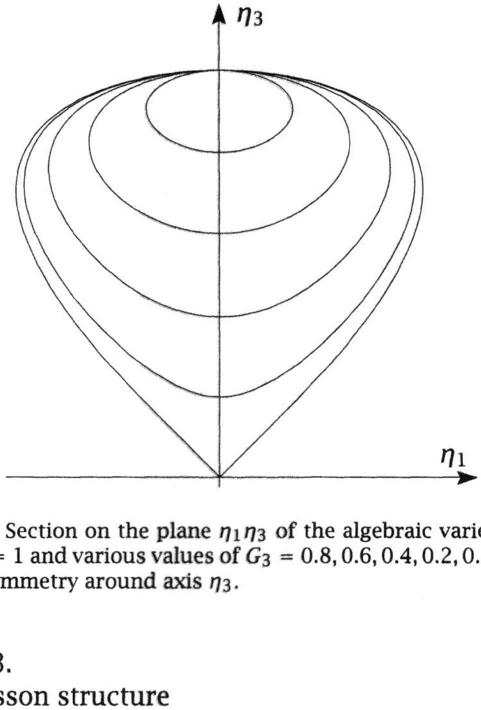

Figure 13.3: Section on the plane $\eta_1\eta_3$ of the algebraic variety $\overline{\overline{M}}$ of the G-type for $K_0 = 1$ and various values of $G_3 = 0.8, 0.6, 0.4, 0.2, 0$. The variety has rotational symmetry around axis η_3.

See Figure 13.3.

For the Poisson structure

$$\{\eta_r, \eta_s\} = \frac{\partial \eta_r}{\partial \xi_i}\frac{\partial \eta_s}{\partial \xi_h}\epsilon_{ihk}\frac{\partial C(\eta)}{\partial \eta_t}\frac{\partial \eta_t}{\xi_k} = \det\left(\frac{\partial \eta}{\partial \xi}\right)\epsilon_{rst}\frac{\partial C(\eta)}{\partial \eta_t},$$

and since

$$\det\left(\frac{\partial \xi}{\partial \eta}\right) = \det\begin{pmatrix} \dfrac{1}{\eta_3} & 0 & -\dfrac{\eta_1}{\eta_3^2} \\ 0 & 2 & 0 \\ 2\dfrac{\eta_1}{\eta_3^2} & 0 & 4\eta_3 - 2\dfrac{\eta_1^2}{\eta_3^3} \end{pmatrix} = 8$$

we find

$$\{\eta_r, \eta_s\} = \frac{1}{8}\epsilon_{rst}\frac{\partial C(\eta)}{\partial \eta_t}.$$

Making explicit these last relations, we write the transformed Poisson structure as

$$\begin{aligned}
\{\eta_1, \eta_2\} &= -2\eta_3\left[\eta_3^2 - \frac{1}{2}(K_0^2 + G_3^2)\right], \\
\{\eta_3, \eta_2\} &= \eta_1, \qquad\qquad (13.1.6)\\
\{\eta_3, \eta_1\} &= -\eta_2.
\end{aligned}$$

Owing to the rotational symmetry about the ξ_1 and η_3 axis respectively, we can immediately find action-angle coordinates in the two cases. In fact, from (13.1.3) and (13.1.6) we deduce

$$\left\{\xi_1, -\frac{1}{2}\arctan\frac{\xi_2}{\xi_3}\right\} = 1,$$

$$\left\{\eta_3, -\arctan\frac{\eta_1}{\eta_2}\right\} = 1.$$

In a slightly different form this result was already known. With regard to the first bracket, recalling the second of (3.2.7) on page 59 with $\vec{\gamma}, \vec{\alpha}$ replacing $\vec{M}, \vec{N}$, we write

$$\begin{aligned}
\arctan\frac{\gamma_2}{\gamma_1} - \arctan\frac{\alpha_2}{\alpha_1} &= \arctan\frac{\gamma_2\alpha_1 - \gamma_1\alpha_2}{\gamma_1\alpha_1 + \gamma_2\alpha_2} \\
&= \arctan\frac{2(G_1A_2 - G_2A_1)}{G_1^2 + G_2^2 - A_1^2 - A_2^2} = -\arctan\frac{\xi_2}{\xi_3},
\end{aligned}$$

while we recall that $\xi_1 = A_3$. With regard to the second bracket, bearing in mind the expression (2.4.1) on page 32 of the pericentre argument ω, with $\vec{A}$ replacing $\vec{E}$, we find

$$\tan\omega = \frac{A_3G}{-G_2A_1 + G_1A_2} = -\frac{\eta_1}{\eta_2},$$

while we recall that $\eta_3 = G$. We have thus found again the third pair of the action-angle variables for the Pauli and Delaunay parameters respectively.

Let us consider the G-type parametrization (but what we will say can be, with minor changes, repeated for the A-type). If the full Hamiltonian is axisymmetric, the perturbing term, once averaged with respect to the unperturbed evolution, turns out to be defined on the two-fold reduced Kepler manifold $\overline{\overline{M}}$. Therefore, the Hamiltonian can be written as

$$K = K_0 + \sum_{i=1}^{r} \varepsilon^i \overline{K}_i(K_0, G_3, \eta_1, \eta_2, \eta_3) + O(\varepsilon^{r+1}) = K^{(r)} + O(\varepsilon^{r+1})$$

with $\{\overline{K}_i, K_0\} = 0 = \{\overline{K}_i, G_3\}$. Disregarding the term $O(\varepsilon^{r+1})$, we are left with the truncated Hamiltonian $K^{(r)}$, which, with K_0 and G_3, forms a system of three first integrals in involution: the system is Liouville-integrable.

Considering K_0 and G_3 as parameters, the equation $K^{(r)}$ = constant describes, varying the constant, a family of two-dimensional surfaces in $\mathbb{R}^3$ that intersect the reduced Kepler variety $\overline{\overline{M}}$ in a family of curves. A point on one of these intersection curves represents an elliptic orbit rotating around the x_3-axis with frequency $\frac{\partial K^{(r)}}{\partial G_3}$, while the motion on the intersection curve describes variations of the eccentricity-inclination and rotations on its plane of the elliptic orbit. The interesting fact is that this description is *global*, and may reveal some features hidden in a *local* description, as is usual in the standard perturbation theory.

Anyway, the local description can be easily recovered; to this end, remembering (13.1.6) we put, for the G-type case,

$$\begin{aligned}
\eta_1 &= -\sin\omega\sqrt{\frac{1}{4}(K_0^2 - G_3^2)^2 - \left[G^2 - \frac{1}{2}(K_0^2 + G_3^2)\right]^2},\\
\eta_2 &= \cos\omega\sqrt{\frac{1}{4}(K_0^2 - G_3^2)^2 - \left[G^2 - \frac{1}{2}(K_0^2 + G_3^2)\right]^2},\\
\eta_3 &= G,
\end{aligned}$$

and substitute in $K^{(r)}$. We recall that $\{G, \omega\} = 1$.

Analogously, for the A-type case, put

$$\begin{aligned}
\xi_1 &= A_3,\\
\xi_2 &= -\sin 2\varphi\sqrt{(K_0^2 + G_3^2 - A_3^2)^2 - 4K_0^2G_3^2},\\
\xi_3 &= \cos 2\varphi\sqrt{(K_0^2 + G_3^2 - A_3^2)^2 - 4K_0^2G_3^2},
\end{aligned}$$

with $\{A_3, \varphi\} = 1$.

Before applying the general theory to the study of some interesting concrete case, let us show that the Kepler Problem with axially symmetric perturbation may have nontrivial monodromy, as pointed out by Cushman & Sadovskií (1999) or, equivalently, that A and B may be focus-focus singularities: see Subsection D.4.3 for the definitions.

The symplectic manifold $\overline{M} = S^2 \times S^2$, with an averaged and truncated Hamiltonian $K^{(r)}$, defines a Hamiltonian system with two degrees of freedom. Since $K^{(r)}$ has axial symmetry, the dynamics descends to the reduced space $\overline{\overline{M}}$. We are interested in what happens in a small neighbourhood of a singular point of the moment map $S^2 \times S^2 \to \mathbb{R}^2$ given by the two first integrals $K^{(r)}$ and G_3. We may therefore limit ourselves to Hamiltonians whose level surfaces, once projected on the reduced space

$\overline{\overline{M}}$, are planes. Moreover, because of the rotational symmetry around the ξ_1-axis of $\overline{\overline{M}}$, we may simply take

$$K^{(r)} = a\xi_3 + b\xi_1 + c, \quad \text{with } \xi_3 = 4(\alpha_1\gamma_1 + \alpha_2\gamma_2), \quad \xi_1 = \gamma_3 - \alpha_3,$$

which generates the equations of motion

$$\begin{aligned} \dot\alpha_1 &= 4a\alpha_3\gamma_2 + b\alpha_2, & \dot\gamma_1 &= 4a\alpha_2\gamma_3 - b\gamma_2, \\ \dot\alpha_2 &= -4a\alpha_3\gamma_1 - b\alpha_1, & \dot\gamma_2 &= -4a\alpha_1\gamma_3 + b\gamma_1, \\ \dot\alpha_3 &= 4a(\alpha_2\gamma_1 - \alpha_1\gamma_2), & \dot\gamma_3 &= 4a(\alpha_1\gamma_2 - \alpha_2\gamma_1). \end{aligned}$$

Clearly, the only equilibrium points candidate to support monodromy are $\mathrm{A} \equiv \left(0,0,-\frac{K_0}{2},0,0,\frac{K_0}{2}\right)$ and $\mathrm{B} \equiv \left(0,0,\frac{K_0}{2},0,0,-\frac{K_0}{2}\right)$ of Figure 13.1 and 13.2, all the other being "trivial" elliptic points. Let us linearize the above equations of motion around A (the arguments for B are similar), assuming $\alpha_1, \alpha_2, \gamma_1, \gamma_2$ as local coordinates. Thus $\dot\alpha_3 = \dot\gamma_3 = 0$ and

$$\begin{pmatrix} \dot\alpha_1 \\ \dot\alpha_2 \\ \dot\gamma_1 \\ \dot\gamma_2 \end{pmatrix} = \begin{pmatrix} 0 & b & 0 & 2aK_0 \\ -b & 0 & -2aK_0 & 0 \\ 0 & -2aK_0 & 0 & -b \\ 2aK_0 & 0 & b & 0 \end{pmatrix} \begin{pmatrix} \alpha_1 \\ \alpha_2 \\ \gamma_1 \\ \gamma_2 \end{pmatrix}.$$

The characteristic polynomial of the matrix is $\left[\lambda^2 - (4K_0^2a^2 - b^2)\right]^2$, which implies two double, real or imaginary, eigenvalues of opposite sign. Since the eigenvalues relative to the other first integral $G_3 = \gamma_3 + \alpha_3$ are $\pm i$ with double multiplicity, the final outcome is that there exists monodromy if $4K_0^2a^2 - b^2 > 0$, or, equivalently, if the plane $K^{(r)} =$ constant, and passing through A, intersects other pieces of $\overline{\overline{M}}$ also.

Another way to show that we are just in the situation of the focus-focus singularity is to verify that, for a suitable $K^{(r)}$ and when $G_3 = 0$, the Hamiltonian vector field around the singularity on the reduced space $\overline{\overline{M}}$ is topologically like that in Figure D.5. Assume, for example, $K^{(r)} = \xi_3$, then, in a small neighbourhood of the singularity, we have

$$\dot\xi_2 \approxeq 4\sqrt{\xi_2^2 + \xi_3^2}, \quad \dot\xi_3 = 0.$$

The relative vector field is shown in Figure 13.4. The angle conjugate to the action variable $\xi_1 = A_3$ is $-\frac{1}{2}\arctan\frac{\xi_2}{\xi_3}$, and the presence of the $\frac{1}{2}$ factor explains the topological equivalence between this field and that of the harmonic repulsor of Figure D.5: indeed, doubling the angle in Figure D.5, transforms it in Figure 13.4.

In general, taking Hamiltonians $K^{(r)}(\xi_1, \xi_3)$, whose level surfaces are as in left Figure 13.5, gives rise to monodromy with $k = 1$ (notation of Subsection D.4.3), while monodromy is trivial with Hamiltonians as in right Figure 13.5.

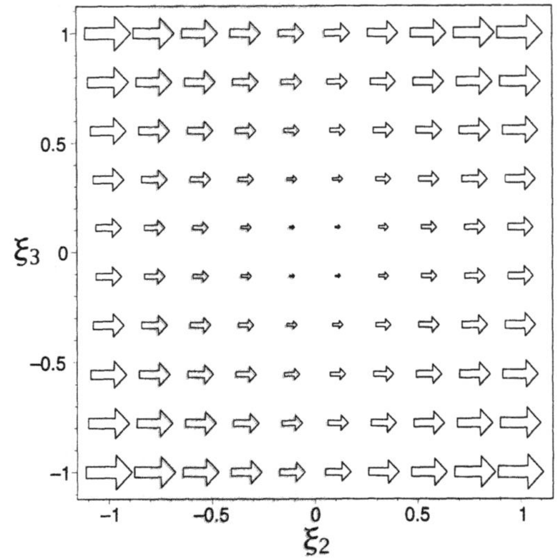

Figure 13.4: Vector field around the singularity in the axially reduced space.

13.2 Lunar Problem

As in Kummer (1983), let us consider the Sun (mass μ) and the Earth (mass ν, with $\mu + \nu = 1$) moving on a circle about their common center of mass with angular speed 1. For simplicity assume that the gravitational constant is equal to 1. We introduce rotating coordinates with the origin in the common center of mass and the x_1-axis passing through Sun and Earth. We want to study the motion of the Moon, having negligeable mass, in the hypothesis that the distance Moon-Earth is very small. In other words, we are modelling the situation with the help of the restricted (nonplanar) three body problem, but in distinction with the usual assumption $\mu << 1$ we keep μ arbitrary between 0 and 1 and instead make the above assumption on the distance Moon-Earth.

Elementary considerations show that the Hamiltonian of the Moon in this rotating system is

$$H = \frac{1}{2}y^2 + (x_1 y_2 - x_2 y_1) - \frac{\nu}{\|\vec{x} - \mu\vec{e}_1\|} - \frac{\mu}{\|\vec{x} + \nu\vec{e}_1\|},$$

where $\vec{e}_1$ is the unit vector along the x_1-axis. The term $x_1 y_2 - x_2 y_1$ encompasses centrifugal and Coriolis forces. Shift the Earth into the origin with the substitution $\vec{x} - \mu\vec{e}_1 \mapsto \vec{x}$. We obtain

$$H = \frac{1}{2}y^2 + (x_1 y_2 - x_2 y_1) - \mu y_2 - \frac{\nu}{x} - \frac{\mu}{\|\vec{x} + \vec{e}_1\|}.$$

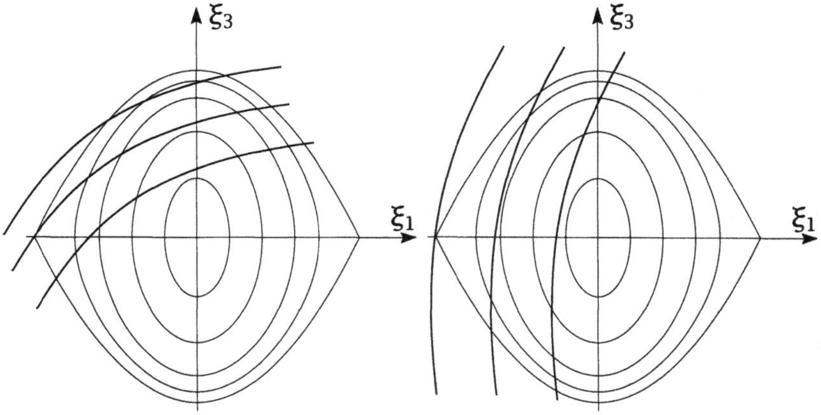

Figure 13.5: Nontrivial (left) and trivial (right) monodromy.

This last and the Hamiltonian

$$H = \frac{1}{2}y^2 + (x_1 y_2 - x_2 y_1) + \mu(1 - x_1) - \frac{\nu}{x} - \frac{\mu}{\left\|\vec{x} + \vec{e}_1\right\|}$$

induce flows in the phase space $(\mathbb{R}^3 - \{0\}) \times \mathbb{R}^3$ whose projections onto the $\vec{x}$ space agree. This is verified by eliminating $\vec{y}$ from the corresponding Hamiltonian Equations. Remembering that x stays small, we may expand the term $\frac{1}{\|\vec{x}+\vec{e}_1\|}$ with respect to $\vec{x}$

$$\frac{1}{\left\|\vec{x} + \vec{e}_1\right\|} = 1 - x_1 + \frac{1}{2}(3x_1^2 - x^2) + O(x^3).$$

Substituting into the Hamiltonian yields

$$H = \frac{1}{2}y^2 - \frac{\nu}{x} + (x_1 y_2 - x_2 y_1) - \frac{\mu}{2}(3x_1^2 - x^2) + O(x^3).$$

Our goal is to study the flow of this Hamiltonian on the so-called *Jacobi surface*

$$H = -\frac{1}{2\lambda^2}, \quad \lambda << 1.$$

We stretch the variables according to the recipe already discussed in the previous chapter

$$\vec{x} \mapsto \frac{\vec{x}}{\nu\lambda^2}, \quad \vec{y} \mapsto \lambda\vec{y}, \quad H \mapsto \lambda^2 H, \quad t \mapsto \frac{t}{\nu\lambda^3},$$

so that, putting $\varepsilon = \nu\lambda^3$, we obtain

$$H = \frac{1}{2}y^2 - \frac{1}{x} + \varepsilon(x_1y_2 - x_2y_1) + \varepsilon^2\frac{\mu}{2}(x^2 - 3x_1^2) + O(\lambda^8\nu^3).$$

We have succeeded in bringing the Hamiltonian into the form (12.1.4), to which the theory previously explained is applicable. Switch to the regularized truncated form (12.1.5)

$$K = K_0 + \varepsilon K_1 + \varepsilon^2\frac{\mu}{2}K_2$$

where

$$K_1 = x(x_1y_2 - x_2y_1), \quad K_2 = x(x^2 - 3x_1^2),$$

which we now normalize up to third order terms.

To this end, we express K_1 and K_2 as functions of the dynamical variables $u_\alpha, v_\beta, K_0, \vec{G}, \vec{A}$, so that the knowledge of their Poisson brackets can be exploited to perform the normalization process:

$$K_1 = (K_0 + v_4)G_3, \quad K_2 = (K_0 + v_4)\left[(K_0 + v_4)^2 - 3(u_1 - A_1)^2\right].$$

Start the normalization process, multiplying K by the operator

$$\exp(\varepsilon\mathcal{L}_{\chi_1}) = 1 + \varepsilon\mathcal{L}_{\chi_1} + \frac{1}{2}\varepsilon^2\mathcal{L}_{\chi_1}^2 + O(\varepsilon^3),$$

then rearrange the terms

$$\begin{aligned}\exp(\varepsilon\mathcal{L}_{\chi_1})K &= K_0 + \varepsilon\left[K_1 + \{\chi_1, K_0\}\right] \\ &+ \varepsilon^2\left[\frac{\mu}{2}K_2 + \{\chi_1, K_1\} + \frac{1}{2}\{\chi_1, \{\chi_1, K_0\}\}\right] + O(\varepsilon^3).\end{aligned}$$

K_0 and G_3 are first integrals of the unperturbed Hamiltonian K_0, while it is clear from (12.1.14) that the mean value of v_4 is zero, then $\overline{K}_1 = K_0G_3$. We must find χ_1 such that

$$\{K_0, \chi_1\} = K_1 - \overline{K}_1 = G_3v_4,$$

which is easily solved by $\chi_1 = G_3u_4$. The second order term becomes

$$\begin{aligned}&\frac{\mu}{2}K_2 + \{\chi_1, K_1\} + \frac{1}{2}\{\chi_1, \{\chi_1, K_0\}\} \\ &= \frac{\mu}{2}\left[\begin{array}{c} K_0^3 + 3K_0v_4^2 + 3K_0^2v_4 + v_4^3 - 3K_0u_1^2 - 3K_0A_1^2 \\ +6K_0A_1u_1 - 3u_1^2v_4 - 3A_1^2v_4 + 6A_1u_1v_4 \end{array}\right] \\ &-\frac{1}{2}G_3^2K_0 - G_3^2v_4.\end{aligned}$$

We must find the mean value of this term that, recalling (12.2.2), we call $\overline{K}_2$ (which thus does *not* denote the mean value of K_2). The evolution of u_α and v_α is given by (12.1.14), so that the odd monomials have vanishing mean value, while the terms containing only K_0, A_k, G_h are constant. Let us consider the quadratic monomials:

$$u_\alpha u_\beta = U_\alpha U_\beta \cos^2 s + V_\alpha V_\beta \sin^2 s + (U_\alpha V_\beta + U_\beta V_\alpha) \sin s \cos s,$$

whose mean value is clearly

$$\overline{u_\alpha u_\beta} = \frac{1}{2}(U_\alpha U_\beta + V_\alpha V_\beta);$$

analogously

$$\overline{v_\alpha v_\beta} = \frac{1}{2}(U_\alpha U_\beta + V_\alpha V_\beta),$$

so that, from the expression of the first integrals $G_{\alpha\beta} = \frac{1}{K_0}(v_\alpha u_\beta - v_\beta u_\alpha)$, it results that

$$\overline{u_\alpha u_\beta} = \overline{v_\alpha v_\beta} = \frac{1}{2}\sum_\gamma G_{\alpha\gamma} G_{\beta\gamma}.$$

A similar procedure yields

$$\overline{u_\alpha v_\beta} = -\frac{1}{2} K_0 G_{\alpha\beta}.$$

The final expression of the second order term in the normalized Hamiltonian is

$$\overline{K}_2 = \frac{\mu}{2} K_0 \left[K_0^2 + \frac{3}{2}(A_2^2 + A_3^2) - \frac{3}{2}(G_2^2 + G_3^2) - 6A_1^2 \right] - \frac{1}{2} K_0 G_3^2.$$

We have obtained the normalized truncated Hamiltonian

$$K^{(2)} = K_0 + \varepsilon \overline{K}_1 + \varepsilon^2 \overline{K}_2,$$

which is not, however, Liouville integrable: only K_0 and $\overline{K}_1 + \varepsilon \overline{K}_2$ are first integrals in involution. To obtain a third integral, Cushman (1991) has suggested to proceed as follows. Consider the system, with two degree of freedom, having phase space $S^2 \times S^2$ and Hamiltonian $\overline{K}_1 + \varepsilon \overline{K}_2$; the term $\overline{K}_1 = K_0 G_3$ generates a rotation of the two spheres around the third axis, so that, regarding $\overline{K}_1$ as unperturbed Hamiltonian, the normalization process can be applied again, calculating so the average $\overline{\overline{K}}_2$ over the orbits of $\overline{K}_1$. Arguing as in the previous case and recalling the Definitions (13.1.1), finally one finds

$$\overline{\overline{K}}_2 = \frac{3}{8} \mu K_0 (\xi_3 + 4\xi_1^2) - \frac{K_0}{4}(\mu K_0^2 + 2G_3^2).$$

We have thus arrived at the final normalized Moon Hamiltonian

$$\overline{K}^{(2)} = K_0 + \varepsilon\overline{K}_1 + \varepsilon^2\overline{\overline{K}}_2,$$

(the bar over $\overline{K}^{(2)}$ to remember the second normalization) that is Liouville integrable, since $K_0, \overline{K}_1, \overline{\overline{K}}_2$ are in involution.

Let us study its motion. The orbit rotates around the third axis with angular velocity

$$\dot{\Omega} = \frac{\partial \overline{K}^{(2)}}{\partial G_3} = \varepsilon K_0 - \varepsilon^2 K_0 G_3,$$

where Ω is the longitude of the ascending node. The second order term $\overline{\overline{K}}_2$ generates the motion of the pericentre and the variations of eccentricity and inclination of the orbit. In fact, $\overline{\overline{K}}_2$ is a first integral, so that the motion must stay on the intersection between $\overline{\overline{M}}$ and the level surfaces $\overline{\overline{K}}_2 = \text{constant}$.

In Figure 13.6, we have plotted the graphs of $\overline{\overline{M}}$, of G-type, for $G_3 = 0.9$ (here and in the subsequent figures K_0 is normalized to 1) and for the level surfaces $\overline{\overline{K}}_2 = \text{constant}$. The projection of an intersection point onto the horizontal plane gives the argument of the pericentre $\omega = -\arctan\frac{\eta_1}{\eta_2}$, while the projection on the vertical axis gives the $\eta_3 = G$ value, from which eccentricity $E = \sqrt{1 - \frac{G^2}{K_0^2}}$ and inclination $i = \arccos\frac{G_3}{G}$ are determined. Points in the neighbourhood of the north pole correspond to orbits with small eccentricity (null at north pole) and high inclination, while those in the neighbourhood of south pole correspond to orbits with high eccentricity and small inclination (null at south pole).

In Figure 13.7 the case $G_3 = 0.2$ is considered. As is clear, the motion is qualitatively somewhat different: while for high values of G_3 the pericentre always circulates, for low values its motion may be a circulation or a libration with a separatrix, as in a pendulum. The discriminant value G_3^* separating the two cases is calculated imposing that the two sections of $\overline{\overline{M}}$ and $\overline{\overline{K}}_2 = \text{constant}$ with the plane $\eta_1\eta_3$ have the same second derivative at their common maximum. One finds $G_3^* = \sqrt{\frac{3}{5}}K_0$.

For low values of G_3 the action-angle variables G, ω are not well suited, and it is preferable to pass to the A_3, φ variables, i.e., to the $\overline{\overline{M}}$ of A-type. In Figure 13.8 the case $G_3 = 0.9$ is considered: now the value of the angle φ is obtained by projecting onto the plane $\xi_2\xi_3$, while that of the action A_3 by projecting on the axis $\xi_1 = A_3$. In Figure 13.9 the case $G_3 = 0.55$ is depicted.

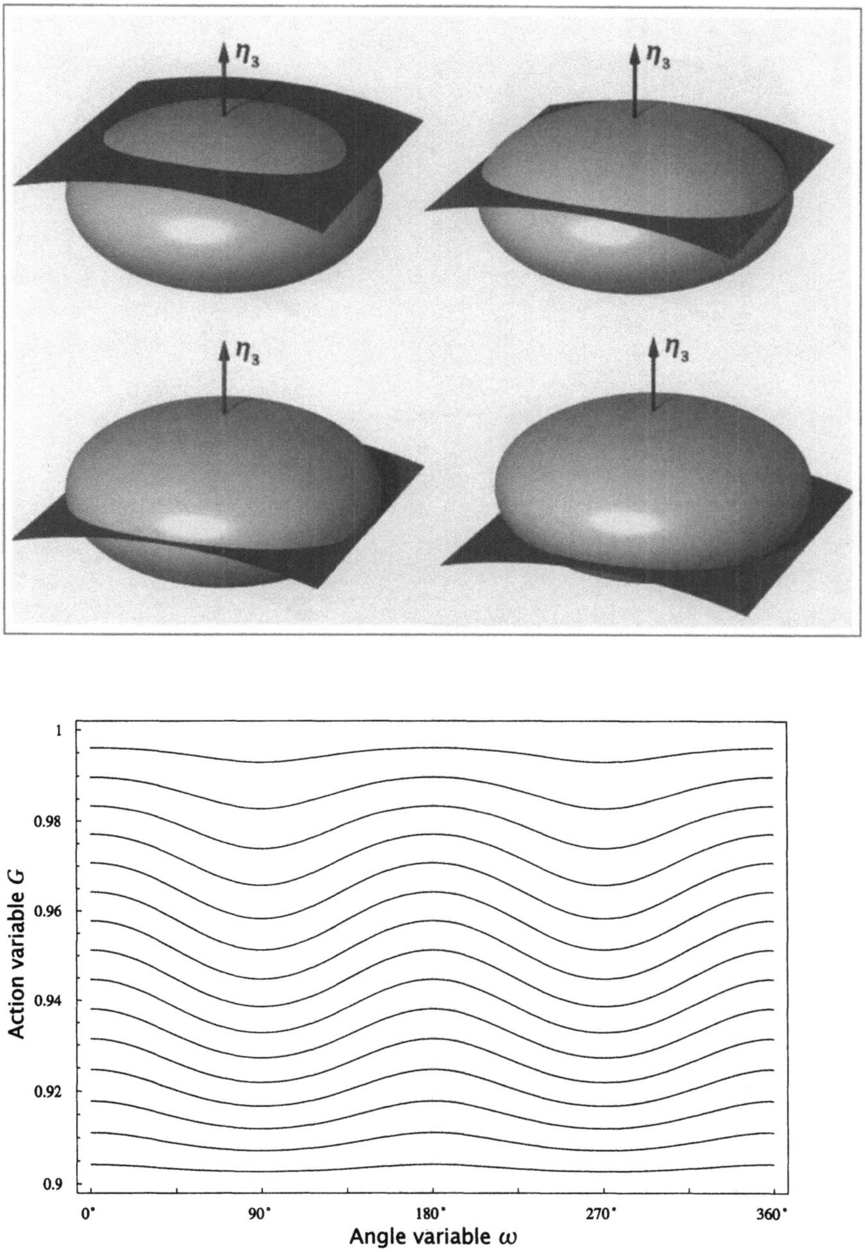

Figure 13.6: The level surfaces of the Moon Hamiltonian $\overline{\overline{K}}_2$ and the twofold reduced Kepler variety $\overline{\overline{M}}$ of G-type for $G_3 = 0.9$. Below, the action-angle variables.

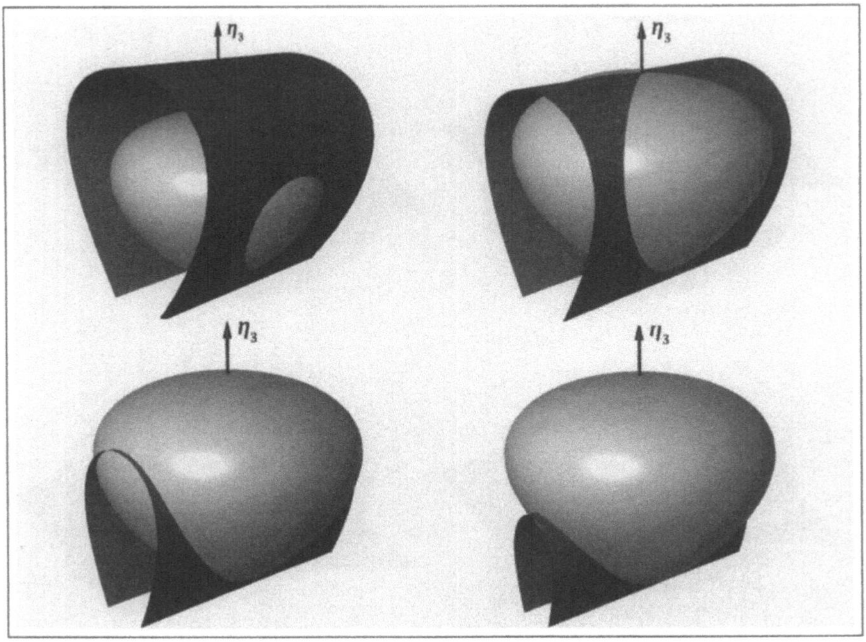

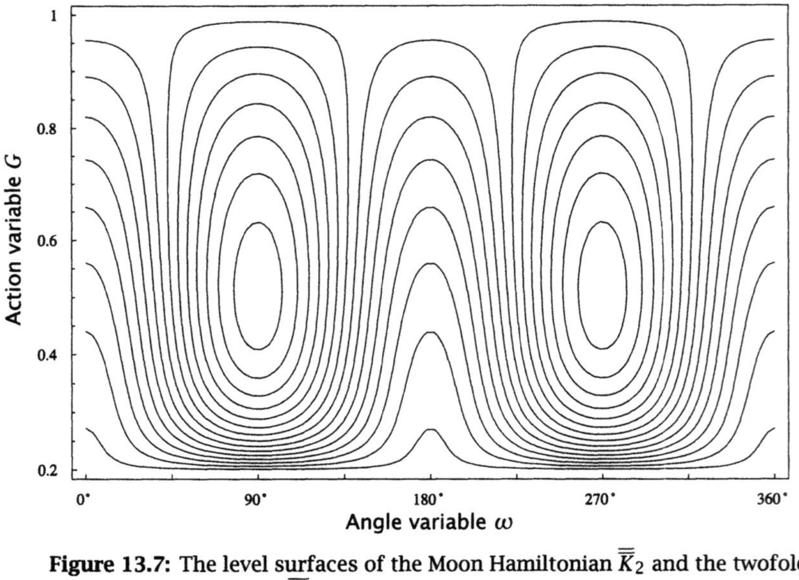

Figure 13.7: The level surfaces of the Moon Hamiltonian $\overline{\overline{K}}_2$ and the twofold reduced Kepler variety $\overline{\overline{M}}$ of G-type for $G_3 = 0.2$. Below, the action-angle variables.

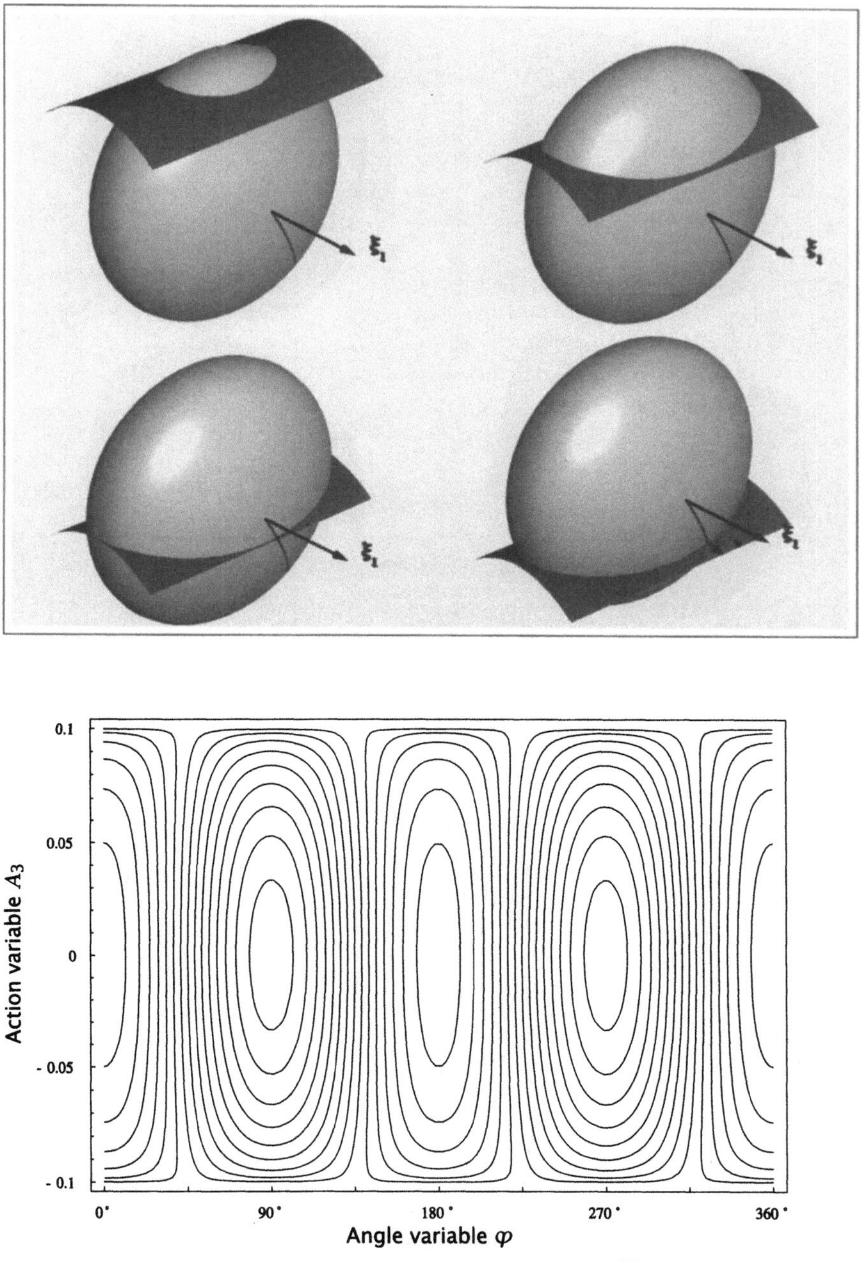

Figure 13.8: The level surfaces of the Moon Hamiltonian $\overline{\overline{K}}_2$ and the twofold reduced Kepler variety $\overline{\overline{M}}$ of A-type for $G_3 = 0.9$. Below, the action-angle variables.

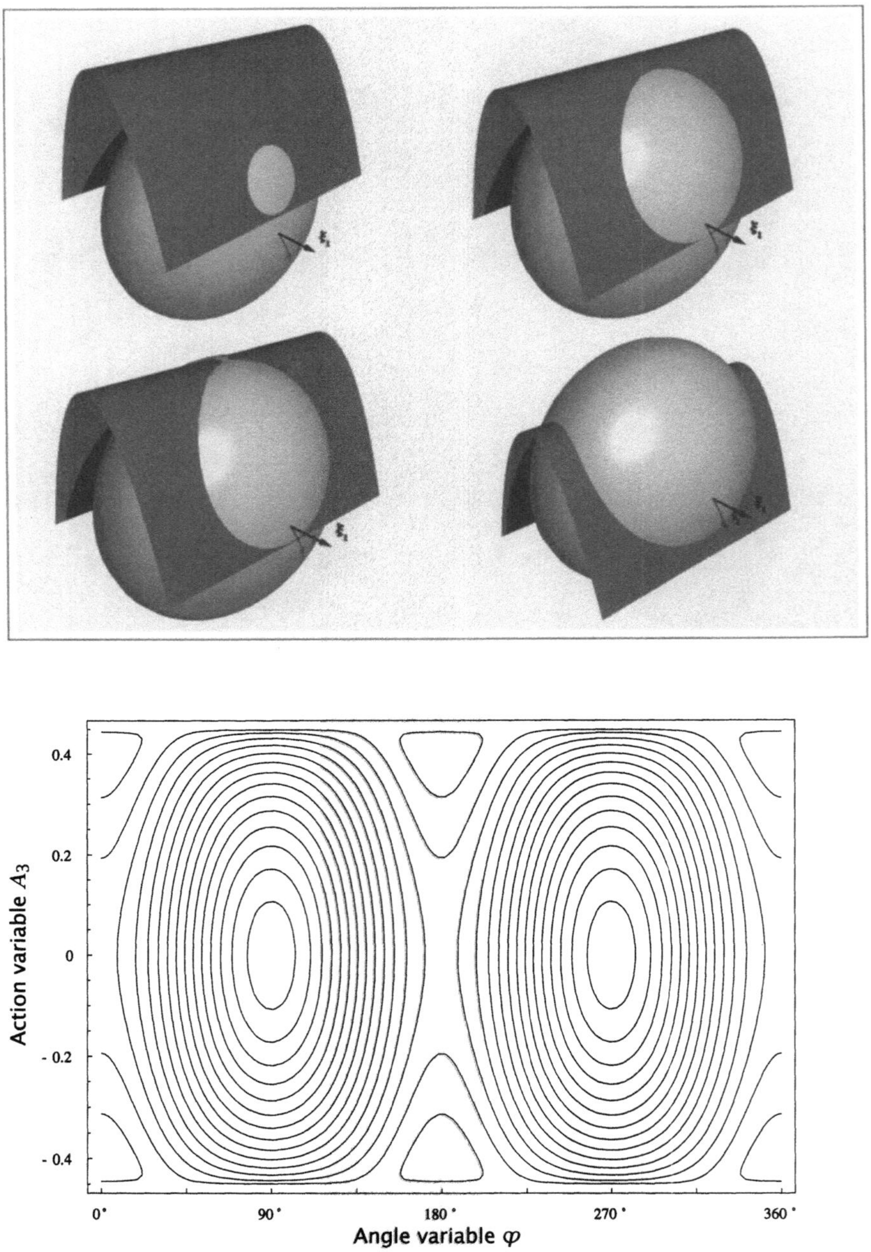

Figure 13.9: The level surfaces of the Moon Hamiltonian $\overline{\overline{K}}_2$ and the twofold reduced Kepler variety $\overline{\overline{M}}$ of A-type for $G_3 = 0.55$. Below, the action-angle variables.

13.3 Stark and Quadratic Zeeman Effect

Another interesting case is that of the hydrogen atom subject to the combined Stark (constant electric field) and quadratic Zeeman (constant magnetic field) effect. We suppose that the two fields are orthogonal one to the other. The corresponding Hamiltonian is

$$\begin{aligned} H &= \frac{1}{2}\left[(y_1 - \mathcal{B}x_2)^2 + (y_2 + \mathcal{B}x_1)^2 + y_3^2\right] - \frac{1}{x} - \mathcal{E}x_1 \\ &= \frac{1}{2}y^2 - \frac{1}{x} + \mathcal{B}(x_1y_2 - x_2y_1) + \frac{\mathcal{B}^2}{2}(x_1^2 + x_2^2) - \mathcal{E}x_1, \end{aligned}$$

where $2\mathcal{B}$ is the value of the magnetic field, directed along the x_3-axis, and $\mathcal{E}$ that of the electric field, directed along the x_1-axis. To treat the problem as a perturbative one, we suppose that $\mathcal{B}$ and $\mathcal{E}$ are very small, but not so small that $\mathcal{B}^2$ is negligible; the adjective "quadratic" refers to this fact.

The following reparametrization will prove to be convenient. Put

$$\mathcal{E} = \frac{2}{3}\varepsilon\alpha, \quad \mathcal{B} = \varepsilon\beta,$$

with $\alpha^2 + \beta^2 = 1$ and ε small, from which, switching to the regularized form of the Hamiltonian, we get

$$K = K_0 + \varepsilon K_1 + \varepsilon^2 K_2,$$

where

$$\begin{aligned} K_1 &= -\frac{2}{3}\alpha x x_1 + \beta x(x_1y_2 - x_2y_1), \\ K_2 &= \frac{\beta^2}{2}x(x_1^2 + x_2^2). \end{aligned}$$

Recalling (12.1.16) and (12.1.17), one finds the mean value

$$\overline{x(s)x_1(s)} = -\frac{3}{2}K_0A_1,$$

so that

$$\overline{K}_1 = K_0(\alpha A_1 + \beta G_3).$$

Since $u_4v_1 - u_1v_4 = K_0A_1$, we find

$$K_1 - \overline{K}_1 = -\frac{2}{3}\alpha K_0 u_1 + \left(\frac{2}{3}\alpha A_1 + \beta G_3\right)v_4 - \frac{1}{3}\alpha(u_1v_4 + u_4v_1).$$

Solving the equation $\{K_0, \chi_1\} = K_1 - \overline{K}_1$ yields

$$\chi_1 = \frac{2}{3}\alpha K_0 v_1 + \left(\frac{2}{3}\alpha A_1 + \beta G_3\right)u_4 - \frac{1}{3}\alpha u_1 u_4.$$

At this point, one can argue as in the previous case of the Lunar problem: the second order term is a sum of cubic monomials in the generators $K_0, G_{\alpha\beta}, u_\alpha, v_\beta$ so that, averaging, one obtains $\overline{K}_2$ as a function of $K_0, G_{\alpha\beta}$ only, i.e., of the generators of the invariance subgroup $SO(2) \times SO(4)$.

The Hamiltonian $K = K_0 + \varepsilon \overline{K}_1 + \varepsilon^2 \overline{K}_2$ is *not* Liouville integrable. However, the following "rotation" (thus a symplectomorphism) on the two-plane 2–4 of the Lie algebra $so^*(2,4)$

$$\begin{pmatrix} G_3' \\ A_1' \end{pmatrix} = \begin{pmatrix} \beta & \alpha \\ -\alpha & \beta \end{pmatrix} \begin{pmatrix} G_3 \\ A_1 \end{pmatrix}$$

gives $\overline{K}_1 = K_0 G_3'$, and we can perform, as for the Lunar problem, a second normalization: $\overline{K}_2 \mapsto \overline{\overline{K}}_2$. The calculations are somewhat lengthy but straightforward. Finally, an expression of the type

$$\overline{\overline{K}}_2 = a\xi_3 + b\xi_1^2$$

results, very similar to that of the Lunar problem. Analogous considerations follow.

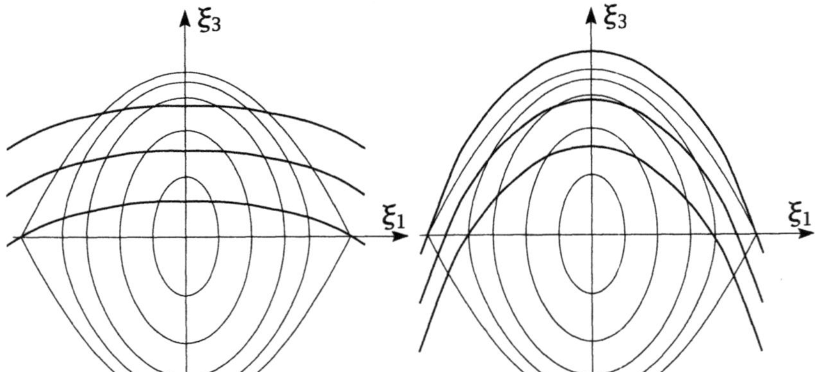

Figure 13.10: Nontrivial monodromy with double pinched torus (left) and trivial monodromy (right).

Notice that, varying the parameters a and b, i.e., $\mathcal{B}$ and $\mathcal{E}$, we can have basically the two configurations of Figure 13.10. In the left case, we have monodromy with matrix $\mathbf{M} = \begin{pmatrix} 1 & 2 \\ 0 & 1 \end{pmatrix}$, to which corresponds a double pinched torus, while in the right case the monodromy is trivial.

13.4 Satellite around Oblate Primary

For a satellite close to its primary, the principal perturbation arises from the nonsphericity of the planet. We want to study the so-called *main problem*, i.e., we will take into account only the second zonal harmonic; in particular the problem will be axially symmetric around the polar axis.

First, we rescale to the new variables $\vec{q}, \vec{p}$ by putting

$$\lambda_m = m, \quad \lambda_Q = k, \quad \lambda_T = \sqrt{mk^2},$$

so that

$$H = H_0 + \varepsilon H_p, \quad \text{with } H_0 = \frac{1}{2}p^2 - \frac{1}{q}.$$

The perturbation for the main problem is

$$H_p = -\frac{\rho_{\text{eq}}^2}{q^3} P_2(\cos\vartheta),$$

where $P_2(x) = \frac{3}{2}x^2 - \frac{1}{2}$ is the second Legendre polynomial and ρ_{eq} the equatorial radius of the primary. We use the previously defined coordinates of Figure 2.4 on page 30 and Figure 3.1 on page 46.

We do not switch to the regularized form K_0, since applying the averaging method, along the lines of the previous sections, leads to insurmountable difficulties in the integrations necessary to find the generators χ_k. Rather, it will prove convenient to use the Whittaker coordinates, defined in Subsection 3.2.1:

$$H_0 = \frac{1}{2}\left(p_r^2 + \frac{G^2}{r^2}\right) - \frac{1}{r},$$

$$H_p = -\frac{\rho_{\text{eq}}^2}{r^3} P_2(s\sin\theta) = \frac{\rho_{\text{eq}}^2}{r^3}\left(\frac{1}{2} - \frac{3}{4}s^2 + \frac{3}{4}s^2\cos 2\theta\right),$$

where $s = \sin i$. At this point one should start with the usual normalization procedure, averaging H_p with respect to the action of $H_0 = \dfrac{1}{-2\mathcal{L}^2}$, i.e., over the mean anomaly ℓ, and so on. Unfortunately, this direct method leads again to intricate calculations. Fortunately Deprit (1981), with the method of the *elimination of the parallax*, has radically unravelled the problem. His idea is to factorize the wanted canonical transformation into the product of two more simple: the first eliminates the angle θ, the second completes the procedure averaging over the mean anomaly ℓ. It turns out that to find the two transformations is strikingly simpler than to find directly the whole one. Moreover, the method can be implemented in a computer program and the average of the perturbative terms can be calculated to higher orders (we will treat here however only the first two terms); see Deprit (1981) and Coffey, Deprit & Deprit (1994).

The Deprit method rests on the simple observation that functions of the type $\frac{G}{r^2}f(\theta)$ stay on the image of the operator $\mathcal{L}_{H_0}$, i.e., for a generic function F,

$$\{H_0, F(\theta)\} = \frac{G}{r^2}F'(\theta),$$

and can, therefore, be completely eliminated. To this end, write the trajectory equation as

$$\frac{G^2}{r} = 1 + E\cos f = 1 + E\cos\theta\cos\omega + E\sin\theta\sin\omega,$$

and substitute into the perturbation Hamiltonian

$$H_p = \frac{\rho_{\rm eq}^2}{G^2}\frac{1}{r^2}\left(\frac{1}{2} - \frac{3}{4}s^2\right) + \frac{\rho_{\rm eq}^2}{G^3}\frac{G}{r^2}f(\theta),$$

where

$$f(\theta) = \frac{3}{4}s^2\cos 2\theta + E\cos\omega\left[\left(\frac{1}{2} - \frac{3}{8}s^2\right)\cos\theta + \frac{3}{8}s^2\cos 3\theta\right]$$
$$+ E\sin\omega\left[\left(\frac{1}{2} - \frac{9}{8}s^2\right)\sin\theta + \frac{3}{8}s^2\sin 3\theta\right].$$

The presence in $f(\theta)$ of the terms E, ω and s does not disturb anyway, since these dynamical variables are in involution with H_0. Multiply H from the left with the operator

$$\exp(\varepsilon\mathcal{L}_{w_1}) = 1 + \varepsilon\mathcal{L}_{w_1} + \frac{1}{2}\varepsilon^2\mathcal{L}_{w_1}^2 + O(\varepsilon^3)$$

obtaining

$$H = H_0 + \varepsilon\left[H_p - \{H_0, w_1\}\right]$$
$$+ \varepsilon^2\left[\{w_1, H_p\} + \frac{1}{2}\{w_1, \{w_1, H_0\}\}\right] + O(\varepsilon^3).$$

Choose $w_1 = \frac{\rho_{\rm eq}^2}{G^3}F(\theta)$ with $F'(\theta) = f(\theta)$. The first and second order terms become

$$H_p - \{H_0, w_1\} = \frac{\rho_{\rm eq}^2}{G^2}\frac{1}{r^2}\left(\frac{1}{2} - \frac{3}{4}s^2\right),$$

$$\{w_1, H_p\} + \frac{1}{2}\{w_1, \{w_1, H_0\}\}$$
$$= \frac{\rho_{\rm eq}^2}{G^3}F(\theta), \frac{\rho_{\rm eq}^2}{G^2}\frac{1}{r^2}\left(\frac{1}{2} - \frac{3}{4}s^2\right) + \frac{1}{2}\frac{\rho_{\rm eq}^2}{G^3}\frac{G}{r^2}f(\theta).$$

After development of the Poisson bracket and averaging in θ, the latter term becomes

$$\frac{\rho_{\text{eq}}^4}{G^6}\frac{1}{r^2}\left[-\frac{13}{16}+\frac{69}{32}s^2-\frac{159}{128}s^4 + \left(\frac{3}{16}-\frac{27}{32}s^2+\frac{105}{128}s^4\right)\frac{G^2}{\mathcal{L}^2}-\left(\frac{21}{16}-\frac{45}{32}s^2\right)s^2E^2\sin^2\omega\right].$$

Once θ is eliminated, we must average $\frac{1}{r^2}$ with respect to the action of $H_0 = \frac{1}{-2\mathcal{L}^2}$ in order to obtain

$$H = H_0 + \varepsilon\overline{H}_1 + \varepsilon^2\overline{H}_2 + O(\varepsilon^3).$$

Since, by (2.2.2) on page 23 and (2.3.2) on page 27,

$$\frac{d\ell}{r^2} = \frac{df}{\mathcal{L}^3 G}$$

holds, the averaged first order term is

$$\overline{H}_1 = \frac{\rho_{\text{eq}}^2}{\mathcal{L}^3 G^3}\left(\frac{1}{2}-\frac{3}{4}s^2\right),$$

with the corresponding generator

$$\chi_1 = \frac{\rho_{\text{eq}}^2}{G^3}\left(\frac{1}{2}-\frac{3}{4}s^2\right)(f-\ell).$$

Finally, after some, at this point obvious, calculations that can be found in Deprit (1981), the averaged second order term results

$$\overline{H}_2 = \frac{\rho_{\text{eq}}^4}{\mathcal{L}^3 G^7}\frac{1}{8}\left[-\frac{15}{2}+\frac{81}{4}s^2-\frac{195}{16}s^4+\left(-3+9s^2-\frac{27}{4}s^4\right)\frac{G}{\mathcal{L}} + \left(\frac{3}{2}-\frac{27}{4}s^2+\frac{75}{16}s^4\right)\frac{G^2}{\mathcal{L}^2}-\left(\frac{21}{2}-\frac{45}{4}s^2\right)s^2E^2\sin^2\omega\right].$$

Notice that $sE\sin\omega = E_3$.

We have reached the goal to write $\overline{H}_1$ and $\overline{H}_2$ as some functions on the two-fold reduced Kepler variety $\overline{\overline{M}}$. We may in fact repeat the reduction process $T^+S^3 \to \overline{\overline{M}}$, taking into account the simple substitutions $\vec{A} \mapsto \vec{R}$ and $K_0 \mapsto \mathcal{L}$, then the statement follows from the relations

$$E_3 = \frac{\xi_1}{\mathcal{L}} = \frac{1}{\mathcal{L}}\frac{\eta_1}{\eta_3}, \quad s^2 = 1-\frac{G_3^2}{G^2}, \quad G = \eta_3,$$

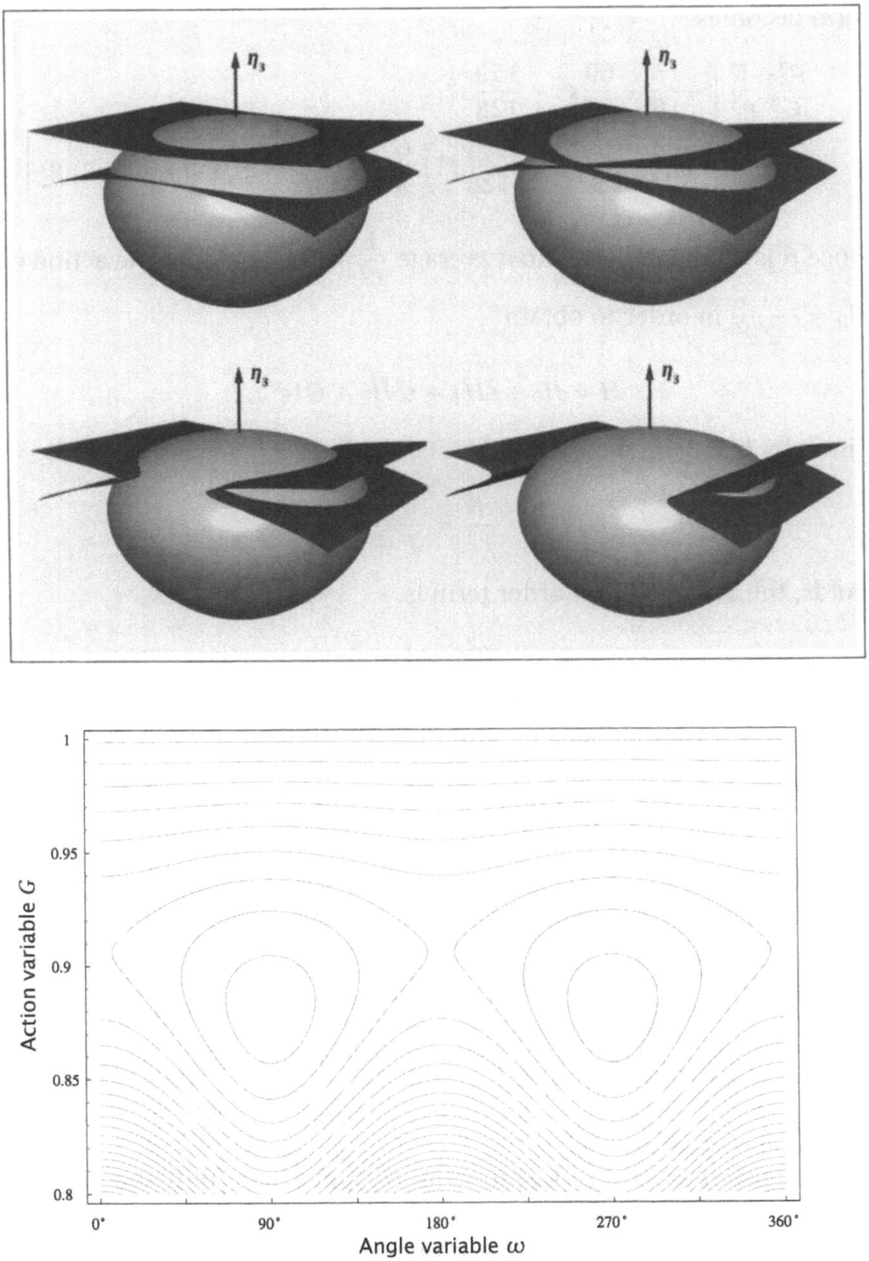

Figure 13.11: The level surfaces of the Satellite Hamiltonian $\overline{H}_1 + \varepsilon\overline{H}_2$ and the twofold reduced Kepler variety $\overline{\overline{M}}$ of G-type for $G_3 = 0.4$. Below, the action-angle variables.

so that $\overline{H}_1$ and $\overline{H}_2$ are expressed as some functions of η_1 and η_3. The level surfaces of $\overline{H}_1(\eta_1,\eta_3)+\varepsilon\overline{H}_2(\eta_1,\eta_3)$ are shown in Figure 13.11 for $G_3 = 0.4$ (K_0 normalized to 1).

If we limit ourselves to consider only the first order term of the averaged perturbation

$$\overline{H}_1 = \frac{\rho_{\mathrm{eq}}^2}{\mathcal{L}^3 G^3}\left(\frac{3}{4}\frac{G_3^2}{G^2} - \frac{1}{4}\right)$$

we find

$$\dot{\omega} = \varepsilon\frac{\partial \overline{H}_1}{\partial G} = \varepsilon\frac{\rho_{\mathrm{eq}}^2}{\mathcal{L}^3}\frac{3}{4G^4}\left(1 - 5\frac{G_3^2}{G^2}\right).$$

All the points belonging to the intersection between $\overline{\overline{M}}$ and the plane $\eta_3 = \sqrt{5}G_3$ are degenerate critical points, as in Figure 12.1c. Since the north pole of $\overline{\overline{M}}$ of the G-type has coordinate $\eta_3 = K_0$ for every value of G_3, the intersection is nonempty for $G_3 \leq \frac{K_0}{\sqrt{5}}$. The inclination $i_{\mathrm{crit}} = \arccos\frac{1}{\sqrt{5}}$ is called *critical*: the pericentre of the orbits with $i < i_{\mathrm{crit}}$ circulates in one sense, those with $i > i_{\mathrm{crit}}$ in the other. As G_3 increases toward $\frac{K_0}{\sqrt{5}}$, the critical circle collapses to the north pole of $\overline{\overline{M}}$. This very crude representation is broken up taking into account the second order term $\overline{H}_2$ also; the continuous critical circle now disappears, leaving only two hyperbolic and two elliptic points, provided that it is not too close to the north pole, i.e. $G_3 << \frac{K_0}{\sqrt{5}}$. Obviously, to these four critical points one must add the always present critical points at the poles. For $G_3 >> \frac{K_0}{\sqrt{5}}$ only these last two critical points are present.

To see what happens when G_3 is close to $\frac{K_0}{\sqrt{5}}$, one should proceed as follows. Solve the defining Equation (13.1.5) for η_3, then substitute into $\overline{H}_1 + \varepsilon\overline{H}_2$, writing the perturbation Hamiltonian in the form

$$Q(\eta_1,\eta_2) \stackrel{\mathrm{def}}{=} \overline{H}_1(\eta_1,\eta_3(\eta_1,\eta_2)) + \varepsilon\overline{H}_2(\eta_1,\eta_3(\eta_1,\eta_2));$$

then, $\eta_1 = \eta_2 = 0$ being a stationary point for $Q(\eta_1,\eta_2)$ (which is in fact an even function of both arguments) study the Hessian at north pole

$$(\mathbf{Q})_{hk} = \left.\frac{\partial^2 Q}{\partial\eta_h\partial\eta_k}\right|_{\eta_1=\eta_2=0}.$$

Obviously, $\mathbf{Q}$ will be diagonal. See Cushman (1983) and Coffey, Deprit & Miller (1986).

However, without performing these very lengthy calculations, one can reach a good qualitative understanding by plotting the graph of the function $Q(\eta_1,\eta_2)$, for $\eta_1 \geq 0$ and $\eta_2 \geq 0$ as in Figure 13.12 (left column) where the left-bottom edge of the boxes is the origin $\eta_1 = \eta_2 = 0$.

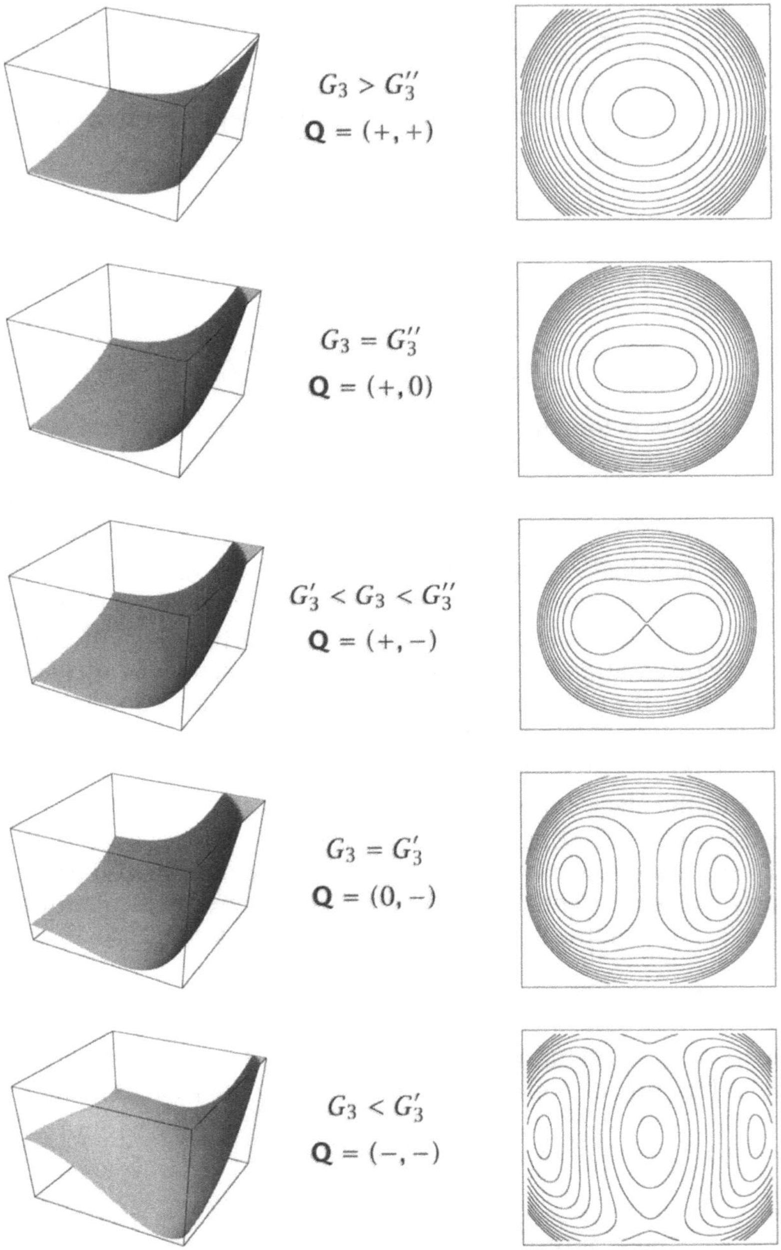

Figure 13.12: The graph and the level lines of $Q(\eta_1, \eta_2)$. The sign of the eigenvalues of the Hessian $\mathbf{Q}$ at the origin is shown.

It is clear that there exist two values of G_3, say $G_3' < \frac{K_0}{\sqrt{5}} < G_3''$, for which the first or the latter eigenvalue of $\mathbf{Q}$ is zero. In the right column of Figure 13.12 are plotted the corresponding level lines of $Q(\eta_1, \eta_2)$, clearly showing how, with growth of G_3, the external four critical points coalesce into the central one. The firstly vanishing are the hyperbolic ones, as it is also suggested by the fact that, in the action-angle diagram of Figure 13.11, they occur at a value of G higher than that of the elliptic ones, and therefore reach first the $G = K_0 (= 1)$ value.

The diagrams of the right column, moreover, describe the motion of the pericentre and the variations of the eccentricity-inclination of the orbit. In fact, taking into account that $\eta_3 = G$, we rewrite (13.1.5) as

$$\eta_1^2 + \eta_2^2 = (K_0^2 - G_3^2)(K_0^2 - G^2) - (K_0^2 - G^2)^2,$$

from which one sees that points very close to the origin represent orbits with $K_0^2 - G^2$ very small, i.e., with low eccentricity (null in the origin) and high inclination; going slightly away from the origin the eccentricity increases and the inclination decreases.

Part IV

Appendices

APPENDIX A

Differential Geometry

Differential geometry supplies the natural framework to treat analytical mechanics. In this chapter we shall review, in a very brief and schematic way, some basic notions, i.e., those of differentiable manifolds, tensors and forms, Riemann and symplectic structures, fibre bundles, etcetera. For a more serious study, many books have been written, for example Kobayashi & Nomizu (1963–1968), Auslander & MacKenzie (1977), Helgason (1978), Choquet-Bruhat (1968), Von Westenholz (1978), Sternberg (1983), Dubrovin, Novikov & Fomenko (1982–1987) and the very readable Crampin & Pirani (1986) or Nash & Sen (1983). For the topology, see Croom (1989).

A.1 Rudiments of Topology

The basic structure within which we work is called a *topological space*.

DEFINITION A.1 *A topological space T is a set of elements, called sometimes* points, *on which the notion of* open sets *has been defined. The open sets are subsets of T satisfying the following properties:*

(i) any union (finite or infinite) of open sets is an open set;

(ii) any finite intersection of open sets is an open set;

(iii) the total space T and the empty set are open sets.

EXAMPLE A.1 The prototype of topological space is $\mathbb{R}$, where the open sets are the unions of open intervals. This example makes evident the necessity of the term "finite intersection" in the item (ii) of the definition; in fact the intersection of the open intervals $(x - \frac{1}{n}, x + \frac{1}{n})$, $n = 1, 2, \ldots$ for $n \to \infty$ shrinks to the point x, which is not the union of open intervals. This topology is called the *usual topology* of $\mathbb{R}$. Similarly, the usual topology on $\mathbb{R}^n$ is given by the product topology, that is the open sets are the unions of $U_1 \times U_2 \times \cdots \times U_n$, where $U_i = (a_i, b_i)$, $1 \leq i \leq n$, are open intervals in $\mathbb{R}$.

The same set can be endowed with different topological structures, the two extreme being the *discrete topology*, where all the subsets are open sets, and the *trivial topology*, where only the whole and the empty set are open sets. Specifying "infinite intersection" in Definition A.1 is equivalent to giving to $\mathbb{R}$ the discrete topology.

A neighbourhood of a point $x \in T$ is a subset (not necessarily open) of T containing some open set to which x belongs. A topological space is *separable* (or Hausdorff) if any two distinct points have at least a neighbourhood with empty intersection.

Given a family of sets $\{F_\alpha\} = F$ say, then F is said to be a *cover* of another set V if $\bigcup F_\alpha$ contains V. If all the F_α are open sets, then the cover is called an *open cover.*

A set V is *compact* if, for every open covering $\{F_\alpha\}$ with $\bigcup F_\alpha \supset V$, there always exists a *finite* subcovering $\{F_1, \ldots, F_n\}$ of V with $F_1 \cup F_2 \ldots \cup F_n \supset V$. In a topological space the compact sets are the bounded and closed subsets, where *closed* is defined in a moment.

A subset of T is said to be *closed* if its complement in T is an open set. A dual definition of topological space can be given by substituting in Definition A.1 "open" with "closed" and exchanging "union" and "intersection".

A topological space T is said to be *connected* if it cannot be written as: $T = T_1 \cup T_2$, where T_1 and T_2 are both open and $T_1 \cap T_2 =$ empty set. Notice that there exist other connectivity definitions, for example topological spaces can be path or arcwise connected.

Let T' be a subset of a topological space T. We can endow T' with the so-called *induced topology* in the following way: a subset U' of T' is an open set for the induced topology if and only if there exists an open subset U of T such that $U' = U \cap T'$.

Let us define the *quotient topology*. If an equivalence relation $\sim$ is defined in T, then the points of T are divided into disjoint equivalence classes:

$$[[x]] = \{x' \in T : x' \sim x\}$$

is the equivalence class containing $x \in T$, and $F = T/\sim = \{[[x]] : x \in T\}$ denotes the set of these equivalence classes. The assignment

$x \mapsto [[x]]$ defines a map

$$\pi : T \to F = T/\sim$$

of T onto F called the *canonical* or *natural projection.* To the set F we can give a topology as follows: a set V in F is called open if and only if the inverse image $\pi^{-1}(V)$ is an open set in T.

A map from a topological space T to a topological space T' is said to be *continuous* if the inverse image of any open set in T' is an open set in T. The same map is called a *homeomorphism* if it is continuous and has an inverse that is also continuous. The notion of homeomorphism is central in topology, and two homeomorphic spaces are considered to be the "same" space from the topological point of view, since they can be continuously mapped one into the other. The homeomorphism is denoted by $T' \simeq T$. Notice that specifying "infinite intersection" in the Definition A.1, thus choosing the discrete topology for $\mathbb{R}$, would make any function in $\mathbb{R}$ a homeomorphism, even those clearly discontinuous for the common intuition.

A.2 Differentiable Manifolds

A.2.1 Definition

The concept of differentiable manifold generalizes the intuitive idea of surface, in the sense that we want to get rid of the constraints of the two dimensions and of the immersion in some $\mathbb{R}^n$. The central point is that every surface can be constructed assembling together some patches that are homeomorphic to open connected sets of $\mathbb{R}^2$.

DEFINITION A.2 *A topological space M is a differentiable manifold of dimension n if*

(i) M is provided with a family of pairs $\{(U_\alpha, \phi_\alpha)\}$, where the U_α are a family of open connected sets that cover M, while the ϕ_α are homeomorphisms from U_α to an open set of $\mathbb{R}^n$;

(ii) given U_α and U_β such that $U_\alpha \cap U_\beta \neq$ empty set, the map $\phi_\beta \circ \phi_\alpha^{-1}$ from the subset $\phi_\alpha(U_\alpha \cap U_\beta) \subset \mathbb{R}^n$ to the subset $\phi_\beta(U_\alpha \cap U_\beta) \subset \mathbb{R}^n$ is C^∞, that is infinitely differentiable (or smooth).

Item (i) asserts that M is a space that is locally like $\mathbb{R}^n$, that is M can be covered with patches U_α, which assign coordinates in $\mathbb{R}^n$ by the ϕ_α. Within one of these patches M looks like a subset of $\mathbb{R}^n$, but of course we do not expect, in general, M to be like $\mathbb{R}^n$ globally; this depends on how the patches fit together to form the whole M. Item (ii) asserts that, if two patches overlap, then in the overlap region $U_\alpha \cap U_\beta$ we have two sets of coordinates in $\mathbb{R}^n$, given by ϕ_α and ϕ_β, and that if we decide to change from one set of coordinates to the other, i.e., to use the function $\phi_\beta \circ \phi_\alpha^{-1}$, this can be done in a smooth manner.

EXAMPLE A.2 Let us consider the sphere $S^n = \{X \in \mathbb{R}^{n+1} : X_1^2 + X_2^2 + \cdots + X_{n+1}^2 = 1\}$, which is in a natural way a space whose topology is induced from the usual topology of $\mathbb{R}^{n+1}$. The sphere can be covered with the two charts obtained by stereographic projection, one from the North pole $\mathrm{N} \equiv (0,0\ldots,1)$ and the other from the South pole $\mathrm{S} \equiv (0,0\ldots,-1)$; see Figure A.1. The same point of the sphere is labelled by the two different coordinate sets $\{x_k^{\mathrm{N}}\}$ and $\{x_k^{\mathrm{S}}\}$, $k = 1,\ldots,n$, that is

$$\phi^{\mathrm{N}} : U_{\mathrm{N}} = S^n - \{\mathrm{N}\} \to \mathbb{R}^n, \quad \text{in coordinates } x_k^{\mathrm{N}} = \frac{X_k}{1 - X_{n+1}},$$
$$\phi^{\mathrm{S}} : U_{\mathrm{S}} = S^n - \{\mathrm{S}\} \to \mathbb{R}^n, \quad \text{in coordinates } x_k^{\mathrm{S}} = \frac{X_k}{1 + X_{n+1}}.$$

In the overlap region $U_{\mathrm{N}} \cap U_{\mathrm{S}} = S^n - \{\mathrm{N}\} - \{\mathrm{S}\}$ the transition functions

$$\phi^{\mathrm{S}} \circ (\phi^{\mathrm{N}})^{-1}, \quad \text{in coordinates } x_k^{\mathrm{S}} = \frac{x_k^{\mathrm{N}}}{(x_1^{\mathrm{N}})^2 + \ldots + (x_n^{\mathrm{N}})^2},$$
$$\phi^{\mathrm{N}} \circ (\phi^{\mathrm{S}})^{-1}, \quad \text{in coordinates } x_k^{\mathrm{N}} = \frac{x_k^{\mathrm{S}}}{(x_1^{\mathrm{S}})^2 + \ldots + (x_n^{\mathrm{S}})^2},$$

are clearly C^∞.

EXAMPLE A.3 Let us consider the 2-dimensional(1) real projective space

$$P_{\mathbb{R}}^2 = (\mathbb{R}^3 - \{0\})/\sim, \quad \vec{X} \sim \vec{X}' = \lambda\vec{X},\ \vec{X} \in \mathbb{R}^3 - \{0\},\ \lambda \in \mathbb{R} - \{0\}.$$

It is thus a space whose "points" are the straight lines through the origin of $\mathbb{R}^3$. A model for it is the sphere S^2 but with the diametrically opposed points identified; it is obvious that it cannot be immersed homeomorphically into $\mathbb{R}^3$. We make $P_{\mathbb{R}}^2$ a topological space with the quotient topology. $P_{\mathbb{R}}^2$ may be covered with three charts:

$$\phi_x : U_x = P_{\mathbb{R}}^2 - \{X_1 = 0\} \to \mathbb{R}^2, \quad \text{in coordinates } x_1 = \frac{X_2}{X_1},\ x_2 = \frac{X_3}{X_1},$$
$$\phi_y : U_y = P_{\mathbb{R}}^2 - \{X_2 = 0\} \to \mathbb{R}^2, \quad \text{in coordinates } y_1 = \frac{X_1}{X_2},\ y_2 = \frac{X_3}{X_2},$$
$$\phi_z : U_z = P_{\mathbb{R}}^2 - \{X_3 = 0\} \to \mathbb{R}^2, \quad \text{in coordinates } z_1 = \frac{X_1}{X_3},\ z_2 = \frac{X_2}{X_3},$$

(1)We limit ourselves to two dimensions for simplicity's sake, but all that is said can be straightforwardly extended to any dimensions.

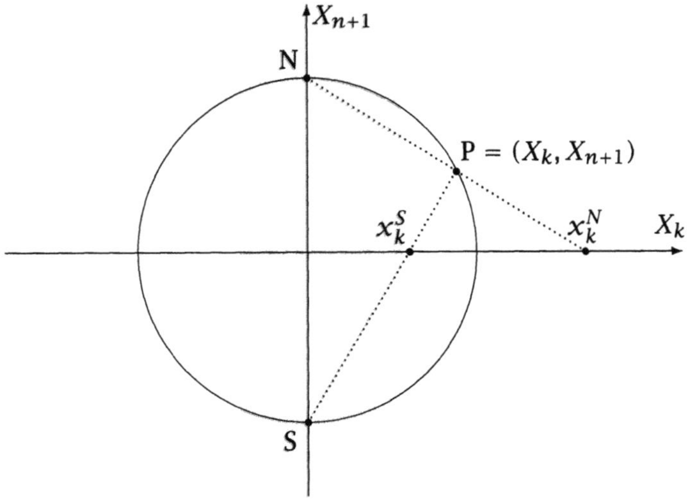

Figure A.1: Stereographic projection.

where X_1, X_2, X_3 are Cartesian coordinates on $\mathbb{R}^3$. Consider for example the overlap region $U_x \cap U_y$; here the transition functions

$$\begin{aligned} \phi_x \circ \phi_y^{-1}, &\quad \text{in coordinates } x_1 = \frac{1}{y_1},\ x_2 = \frac{y_2}{y_1}, \\ \phi_y \circ \phi_x^{-1}, &\quad \text{in coordinates } y_1 = \frac{1}{x_1},\ y_2 = \frac{x_2}{x_1}, \end{aligned}$$

are C^∞. Analogously for the other two overlap regions $U_x \cap U_z$ and $U_y \cap U_z$.

The pair (U_α, ϕ_α) for a fixed α is a *chart*. Two charts (U_α, ϕ_α) and (U_β, ϕ_β) are *compatible* if $\phi_\alpha \circ \phi_\beta^{-1} : \mathbb{R}^n \to \mathbb{R}^n$ and its inverse are smooth. A family of compatible charts $\{(U_\alpha, \phi_\alpha)\}$ such that $\bigcup_\alpha U_\alpha \supseteq M$ is an *atlas* for the differentiable manifold M.

Consider two differentiable manifolds M and N, respectively m and n-dimensional and let $f : M \to N$ be a map. Choosing local charts, ϕ for M and ψ for N say, the map f induces the map $\psi \circ f \circ \phi^{-1} : \mathbb{R}^m \to \mathbb{R}^n$, called *coordinate presentation* of f. A coordinate presentation of a map f is therefore a map of open subsets of real number spaces, given explicitly by n functions of m variables. The map $f : M \to N$ is called *smooth* if its coordinate presentation is given by smooth functions for all charts of an atlas of both its domain and codomain. For the differentiability of M and N, this smoothness definition does not clearly depend on the choice of the particular local chart.

An important special case of a map of manifolds is a smooth bijective map with a smooth inverse. Such a map is called a *diffeomorphism*, and two manifolds connected by a diffeomorphism are said to be *diffeomorphic*. From the point of view of the differential geometry, two diffeomorphic manifolds are considered to be the "same" manifold.

Consider the $n \times m$ Jacobian matrix of the coordinate presentation of the smooth map f. Its rank r at any point does not change under an invertible change of coordinates and is therefore a property of the map f itself: we call it the *rank* of f at the point. In particular, we are interested in those maps whose rank does not vary from point to point. Two extreme cases are relevant. First, when $r = n \leq m$, i.e., the rank is everywhere equal to the dimension of the codomain of f, we call the map a *submersion*. At the other extreme, we call a map an *immersion* when $r = m \leq n$, i.e., the rank is everywhere equal to the dimension of the domain of f. Notice that while an immersion is locally injective, this is not necessarily true globally; we thus define a smooth map to be an *embedding* if it is an (injective) immersion and is a homeomorphism onto its image.

A.2.2 Tangent and Cotangent Spaces

A particular but very important case of a map between manifolds is that defining a curve. A *curve* in an n-dimensional manifold is a map $\sigma : \mathbb{R} \to M$, that is in local coordinates

$$\phi \circ \sigma : x^k = x^k(t), \quad k = 1.2. \ldots, n \quad a < t < b, \; a, b \in \mathbb{R}.$$

Given a smooth function $f : M \to \mathbb{R}$, we define the derivative vf of the function along the curve σ

$$vf = \frac{d}{dt}(f \circ \sigma) = \partial_k f \frac{dx^k}{dt}, \quad \partial_k = \frac{\partial}{\partial x^k}.$$

We can view v as an operator that maps functions into functions; it has the two properties (g is another function and $\lambda, \mu \in \mathbb{R}$)

(i) $v(\lambda f + \nu g) = \lambda v f + \mu v g$,

(ii) $v(fg) = (vf)g + f(vg)$,

showing that v is a linear operator satisfying the Leibniz rule of the derivative of a product.

We define a *tangent vector* at a point p of a smooth manifold M to be an operator on smooth functions that satisfies properties (i) and (ii).

Denote the set of all tangent vectors to M at p by T_pM. We make this set a vector space defining the linear combination

$$(\lambda v + \mu w)f = \lambda v f + \mu w f, \quad v, w \in T_pM, \quad \lambda, \mu \in \mathbb{R}.$$

The linear space T_pM is called the *tangent space* to M at p. The definition of tangent space generalizes the intuitive idea of a plane tangent to a surface, but without to resort to the immersion of M into an ambient space.

A tangent vector v is written in coordinates as $v^k \partial_k$ and $\{\partial_k\}$ is a basis (more exactly, the *natural basis*) for the vector space T_pM. v^k are said to be the *contravariant components*. This shows that, as is intuitively clear, the dimension of the tangent space is equal to that of M. The adjective "contravariant" reminds that if T_pM is submitted to a linear change of basis and if we pretend (as is sensible) that the tangent vector, which is intrinsically defined, does not change, the components v^k must change with the inverse transformation of the basis(2).

As known from linear algebra, the set of the linear maps from a vector space V to $\mathbb{R}$ is again a vector space of the same dimension, which we denote V^* and call *dual space*. In other words, if $v \in V$ and $\omega \in V^*$ then $\omega(v) \in \mathbb{R}$. Often, one writes this *pairing* as $\langle \omega, v \rangle$ instead of $\omega(v)$ to emphasize the linearity. If $\{e_k\}$ is a basis of V, the dual basis $\{e^h\}$ of V^* is by definition that satisfying $\langle e^h, e_k \rangle = \delta^h_k$. Moreover, as it is easy to show, the dual of a dual space V^{**} is canonically, i.e., independently of the basis, isomorphic to V. The dual space T^*_pM of the tangent space T_pM is called *cotangent space* to M at p.

For fixed f and p, let us define a map $df : T_pM \to \mathbb{R}$ such that $v \mapsto (df)(v) = vf$. We call df the *differential* of f, where it is understood that every quantity is calculated at point p. It is immediate to verify that the linearity of the operator v entails that of df :

$$df(\lambda v + \mu w) = (\lambda v + \mu w)f = \lambda(df)(v) + \mu(df)(w).$$

The linearity ensures that $df \in T^*_pM$.

Take $f = x^k$, $k = 1, 2, \ldots, n$, where, as usual, $\{x^k\}$ are local coordinates on M, and let us examine the linear action of dx^k on the basis vectors ∂_h

$$dx^k : \partial_h \mapsto (dx^k)(\partial_h) = \frac{\partial x^k}{\partial x^h} = \delta^k_h,$$

which may be written as

$$\langle dx^k, \partial_h \rangle = \delta^k_h.$$

(2) As says the chief character of the Italian novel *Il Gattopardo* of Tomasi di Lampedusa: "All must change, in order that nothing changes".

Thus, $\{dx^k\}$ is the dual basis of $\{\partial_h\}$.

An element $\omega \in T_p^*M$ is called a *covector* and is written in coordinates $\omega = \omega_k\, dx^k$. The ω_k are called *covariant* components, since they change as the natural basis.

A.2.3 Push-forward and Pull-back

Given two vector spaces V and W, not necessarily of the same dimension, and a linear map $\mathbf{T} : V \to W$, one defines the dual map $\mathbf{T}^* : W^* \to V^*$ as that satisfying

$$\langle \mathbf{T}^*\omega, v\rangle = \langle \omega, \mathbf{T}v\rangle , \quad \forall v \in V \text{ and } \forall \omega \in W^*;$$

notice that the two pairings refer to two different couples of vector spaces, i.e., to V^*, V and W^*, W, respectively.

A map $\phi : M \to N$, where M and N are two differentiable manifolds, induces in a natural way a linear map $\mathbf{T}_\phi : T_pM \to T_{\phi(p)}N$, which is defined as that satisfying the (intrinsic) relation

$$(\mathbf{T}_\phi v)f = v(f \circ \phi), \quad \forall v \in T_pM \text{ and } \forall f : N \to \mathbb{R}.$$

Let us find the expression in local coordinates of the linear map $\mathbf{T}_\phi$. Let $x = (x^1, \dots, x^m)$ be the coordinates on M and $y = (y^1, \dots, y^n)$ those on N; the map ϕ has coordinate presentation $y = \phi(x)$. From elementary calculus

$$v(f \circ \phi) = v^k \frac{\partial}{\partial x^k} f(\phi(x)) = v^k \frac{\partial}{\partial y^h} f(y) \frac{\partial \phi^h}{\partial x^k} = (\mathbf{T}_\phi v)f, \tag{A.2.1}$$

from which

$$(\mathbf{T}_\phi)^h_k = \frac{\partial \phi^h}{\partial x^k},$$

thus the matrix representing $\mathbf{T}_\phi$ with respect to the standard basis is the Jacobian matrix of the map ϕ. We call $\mathbf{T}_\phi$ the *derivative map* of ϕ.

We can now define the image $\phi_* v$, or *push-forward*, of a vector v on M under a diffeomorphism ϕ as the vector on N that makes commutative the relative diagram in Figure A.2; then

$$\phi_* v = \mathbf{T}_\phi v \circ \phi^{-1},$$

which requires the inversion of the map ϕ. This explains our request that ϕ be a diffeomorphism.

Analogously, we define the inverse image, or *pull-back*, $\phi^*\omega$ of a covector on N. We start with a covector on N, instead of on M, since $\mathbf{T}^*$ reverses the direction of the map arrow with respect to $\mathbf{T}$. We have

$$\phi^*\omega = \mathbf{T}_\phi^*\omega \circ \phi.$$

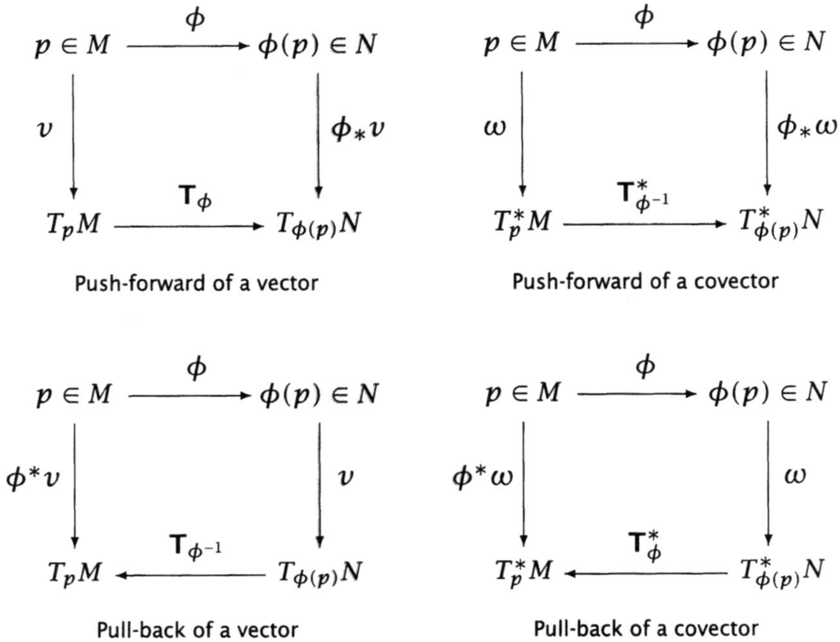

Figure A.2: Push-forward and pull-back of vectors and covectors.

Notice that the pull-back of a covector can be defined also if ϕ is not invertible. For example, if ω is a covector defined on N, it can be safely restricted to a covector on a submanifold $M \subset N$.

The definitions of the push-forward $\phi_* \omega$ of a covector and of the pull-back $\phi^* v$ of a vector is obvious[3]; see Figure A.2, where $\mathbf{T}_{\phi^{-1}} = \mathbf{T}_{\phi}^{-1}$. Also in these two other cases it is required that ϕ is invertible. In case ϕ is not invertible, we may however say that two vectors v and w are *ϕ-related* if

$$w \circ \phi = \mathbf{T}_{\phi} v, \tag{A.2.2}$$

which, in some sense, replaces the definition of pull-back and push-forward of a vector. The relation is natural with respect to Lie bracket, defined in (11.1.4), as elucidated in the following

PROPOSITION A.3 *If v_1, v_2 are ϕ-related to w_1, w_2 respectively, the Lie bracket $[v_1, v_2]$ is ϕ-related to $[w_1, w_2]$.*

[3] Admittedly, the usage of the asterisk in the various definitions is at this point somewhat inflated, but unfortunately this is the custom.

Proof. Taking into account (A.2.1), the definition (A.2.2) may be rewritten as

$$v(f \circ \phi) = (wf) \circ \phi, \quad \forall f : N \to \mathbb{R},$$

from which immediately follows

$$[v_1, v_2](f \circ \phi) = ([w_1, w_2]f) \circ \phi,$$

which is the claim. QED

A *vector field* on M is defined as the assignment to every point $p \in M$ of an element belonging to T_pM. Analogously for a *covector field*, or *1-form*. The definition of pull-back and push-forward is straightforwardly extended to fields.

A.3 Tensors and Forms

A.3.1 Tensors

An element of a dual space is a linear map from a vector space to the real numbers: the tensor definition is a natural generalization of this definition. First of all notice that, the dual of the dual of a vector space V being canonically isomorphic to V itself, we can define an element of V as a linear map from V^* to the real numbers. There is therefore a perfect symmetry, despite the notation, between V and V^*.

DEFINITION A.4 *We call* tensor of (r, s)*-type, or* r*-fold contravariant and* s*-fold covariant, a multilinear (i.e., linear in every argument) map* T *of the type*

$$T : \overbrace{V^* \times \cdots \times V^*}^{r \text{ times}} \times \overbrace{V \times \cdots \times V}^{s \text{ times}} \to \mathbb{R}.$$

The set of all these maps is, in a natural way, a vector space and is called $\mathcal{T}^r_s$. In particular $\mathcal{T}^1_0 = V$ and $\mathcal{T}^0_1 = V^*$.

DEFINITION A.5 *We define the* tensor product *of two tensors* $T_1 \in \mathcal{T}^r_s$ *and* $T_2 \in \mathcal{T}^p_q$ *the tensor* $T_1 \otimes T_2 \in \mathcal{T}^{r+p}_{s+q}$ *such that*

$$\begin{aligned} T_1 \otimes T_2(\omega_1, \dots, \omega_{r+p}, v_1, \dots, v_{s+q}) \\ = T_1(\omega_1, \dots, \omega_r, v_1, \dots, v_s) T_2(\omega_{r+1}, \dots, \omega_{r+p}, v_{s+1}, \dots, v_{s+q}), \\ \forall \omega_k \in V^* \text{ and } \forall v_h \in V. \end{aligned}$$

Tensor multiplication is associative and distributive with respect to addition, but not commutative.

It is easy to prove that, if $\{e_h\}$ is a basis of V and $\{e^k\}$ the dual basis of V^*, then

$$\left\{e_{h_1} \otimes \cdots \otimes e_{h_r} \otimes e^{k_1} \otimes \cdots \otimes e^{k_s}\right\}$$

is a basis of $\mathcal{T}^r_s$, which has thus dimension n^{r+s}, where $\dim V = n$. For the linearity, we can write for the generic tensor $T \in \mathcal{T}^r_s$

$$T = T^{h_1 \dots h_r}_{k_1 \dots k_s} e_{h_1} \otimes \cdots \otimes e_{h_r} \otimes e^{k_1} \otimes \cdots \otimes e^{k_s}$$

where

$$T^{h_1 \dots h_r}_{k_1 \dots k_s} = T(e^{h_1}, \dots, e^{h_r}, e_{k_1}, \dots, e_{k_s})$$

are the components of T in the chosen basis.

Let us consider a linear and invertible change of basis in V, that is $e_h \mapsto e'_k = A^h_k e_h$. It induces, if we demand that the transformed basis of V^* be still the dual basis, the transformation $e^h \mapsto e'^k = (A^{-1})^k_h e^h$. Therefore the, at first sight terrifying, expression

$$\begin{aligned} &e'_{i_1} \otimes \cdots \otimes e'_{i_r} \otimes e'^{j_1} \otimes \cdots \otimes e'^{j_s} \\ &=A^{h_1}_{i_1} \dots A^{h_r}_{i_r} (A^{-1})^{j_1}_{k_1} \dots (A^{-1})^{j_s}_{k_s} e_{h_1} \otimes \cdots \otimes e_{h_r} \otimes e^{k_1} \otimes \cdots \otimes e^{k_s} \end{aligned}$$

holds for the change of basis of $\mathcal{T}^r_s$, and

$$T'^{i_1 \dots i_r}_{j_1 \dots j_s} = (A^{-1})^{i_1}_{h_1} \dots (A^{-1})^{i_r}_{h_r} A^{k_1}_{j_1} \dots A^{k_s}_{j_s} T^{h_1 \dots h_r}_{k_1 \dots k_s}$$

for the corresponding change of the components. The basic point to keep in mind is that contravariant components (indices above) and covariant components (indices below) change with an inverse law, so that *contracting* (or *saturating*, i.e., summing over) a pair of such indices leads to a quantity invariant under change of basis.

Sometimes, with a slight abuse of language, one calls for brevity $T^{i_1 \dots i_r}_{j_1 \dots j_s}$ "a tensor", instead of "the components of a tensor", which would be precise.

A *tensor field* is the assignment to every point $p \in M$ of a tensor whose underlying vector space is $T_p M$, thus $e_h = \partial_h$ and $e^k = dx^k$. The definition of push-forward and pull-back is naturally extended to the tensor fields, with $A^h_k = (\mathbf{T}_\phi)^h_k$.

A.3.2 Forms and Exterior Derivatives

Let us define the symmetry properties of a tensor. We say that a tensor is *symmetric* or *skew symmetric* in a pair of indices (of the same type) if

$$T_{\dots i_r \dots i_s \dots} = T_{\dots i_s \dots i_r \dots} \quad \text{symmetric,}$$
$$T_{\dots i_r \dots i_s \dots} = -T_{\dots i_s \dots i_r \dots} \quad \text{skew symmetric,}$$

and *completely* symmetric or skew symmetric if the property holds for every pair of indices. Analogously for contravariant indices.

Given a generic tensor, we can extract from it the symmetric or skew symmetric part. Let P_r be the group of permutations of the r integers $1, \dots, r$ and let $\pi \in P_r$. Define

$$(\pi T)_{i_1 \dots i_r} = T_{\pi(i_1) \dots \pi(i_r)}.$$

A symmetrization operator Sym and a skew symmetrization operator Sk can be defined as

$$\mathrm{Sym}(T) = \frac{1}{r!} \sum_{P_r} \pi T, \quad \mathrm{Sk}(T) = \frac{1}{r!} \sum_{P_r} \mathrm{sign}_\pi \pi T,$$

where $\mathrm{sign}_\pi = \pm 1$ according to whether the permutation π is even or odd. $\mathrm{Sym}(T)$ is a completely symmetric and $\mathrm{Sk}(T)$ a completely skew symmetric tensor. Analogously for contravariant indices.

DEFINITION A.6 *The covariant completely skew symmetric tensors of order r are called r-*covectors. *The set of all r-covectors in an n-dimensional vector space V_n forms a vector space called $\bigwedge^r V_n$, $r \leq n$.*

DEFINITION A.7 *The* exterior product $\wedge$ *(or* wedge product, *or* Grassman product*) of an r-covector with an s-covector is a map*

$$\wedge : \left(\bigwedge^r V_n, \bigwedge^s V_n \right) \to \bigwedge^{r+s} V_n$$

defined by

$$\alpha \wedge \beta = \frac{(r+s)!}{r!s!} \mathrm{Sk}(\alpha \otimes \beta), \quad \alpha \in \bigwedge^r V_n, \ \beta \in \bigwedge^s V_n.$$

The exterior product is associative and distributive with respect to the sum, but in general not commutative

$$\alpha \wedge \beta = (-1)^{rs} \beta \wedge \alpha.$$

As one easily checks, a basis for $\bigwedge^r V_n$ is given by the set

$$\{ e^{i_1} \wedge \cdots \wedge e^{i_r}, \ i_1 < i_2 < \cdots < i_r \},$$

and $\dim \bigwedge^r V_n = \binom{n}{r}$.

DEFINITION A.8 *The contravariant completely skew symmetric tensors of order r are called r-vectors; they form a vector space denoted $\bigwedge_r V_n$ and the exterior product on it is defined in an analogous way.*

The definition of *field* is naturally extended to r-vectors and r-co-vectors. In the second case, we speak of *r-forms* or simply of *forms.*

DEFINITION A.9 *The* exterior derivative *is a map from the space of the r-forms to the space of the $(r+1)$-forms, given in local coordinates by*

$$d\omega = \frac{\partial \omega_{i_1 \dots i_r}}{\partial x^j} dx^j \wedge dx^{i_1} \wedge \cdots \wedge dx^{i_r}, \quad \forall j \text{ and } i_1 < i_2 < \cdots < i_r. \tag{A.3.1}$$

where

$$\omega = \omega_{i_1 \dots i_r} dx^{i_1} \wedge \cdots \wedge dx^{i_r}, \quad \text{with } i_1 < i_2 < \cdots < i_r,$$

is the expression of an r-form.

One checks that this definition is independent of the choice of coordinates. This is basically due to the fact that the skew symmetrization among j and the other indices kills the symmetric terms containing the second partial derivatives.

The exterior derivative satisfies the following properties:

(i) if f is a 0-form, the 1-form df is the differential already defined;

(ii) if $\omega = \omega_1 + \omega_2$, then $d\omega = d\omega_1 + d\omega_2$;

(iii) $dd\omega = 0, \ \forall \omega$;

(iv) $d(\omega_1 \wedge \omega_2) = d\omega_1 \wedge \omega_2 + (-1)^r \omega_1 \wedge d\omega_2$, ω_1 being an r-form and ω_2 any form.

The check of (i) and (ii) is immediate, that of (iii) is due to the symmetry of the second derivatives $\frac{\partial^2}{\partial x^i \partial x^k} = \frac{\partial^2}{\partial x^k \partial x^i}$, that of (iv) to the skew symmetry of the exterior product of two 1-forms.

Vice versa, one can take the properties (i) – (iv) as defining the exterior derivative, then show that the coordinate expression (A.3.1) follows uniquely.

Another useful property is given by

PROPOSITION A.10 *Pull-back and exterior derivative commute*

$$\phi^*(d\omega) = d(\phi^* \omega).$$

Proof. An arbitrary form is a finite sum of exterior products of functions and differentials of functions, hence we have to prove the theorem only when ω is a 0-form or the differential of a function.

(i) If ω is a 0-form g, then

$$\phi^* dg = d(g \circ \phi) = d(\phi^* g).$$

(ii) If $\omega = dg$, then for property (iii) above

$$\phi^*(ddg) = 0,$$

but, on the other hand, for item (i)

$$d(\phi^* dg) = dd(\phi^* g) = 0,$$

from which the proposition follows. QED

A.3.3 Lie Derivative

In Subsection 11.1.1 we have already defined the flow Φ_τ on a manifold M and the Lie derivative $\mathcal{L}_v f = vf$ of $f : M \to \mathbb{R}$ with respect to the vector v (previously X), the generator of the flow. It is clear that

$$\mathcal{L}_v f = \lim_{\tau \to 0} \frac{1}{\tau}(\Phi_\tau^* f - f),$$

where $\Phi_\tau^* f = f \circ \Phi_\tau$. In other words, we compare the value taken by the function f at the point $\Phi_\tau(p)$ (which is the image of the point p under the diffeomorphism Φ_τ) with that taken at the point p itself (i.e., at the position corresponding to $\tau = 0$).

We define the Lie derivative of a tensor field T basically in the same way

$$\mathcal{L}_v T = \lim_{\tau \to 0} \frac{1}{\tau}(\Phi_\tau^* T - T).$$

The only difference, intimated in the definition of Φ_τ^*, is that now we can no longer compare directly the values of T at p and at $\Phi_\tau(p)$, since they belong to different (co)tangent spaces, but must before "pull-back" the tensor.

From the definition it is immediate to find the relation

$$\begin{aligned} \frac{d}{d\tau}(\Phi_\tau^* T) &= \lim_{\triangle\tau \to 0} \frac{1}{\triangle\tau}(\Phi_{\tau+\triangle\tau}^* T - \Phi_\tau^* T) \\ &= \lim_{\triangle\tau \to 0} \frac{1}{\triangle\tau}[\Phi_\tau^*(\Phi_{\triangle\tau}^* T - T)] = \Phi_\tau^*(\mathcal{L}_v T), \end{aligned} \tag{A.3.2}$$

which is sometimes taken as a definition of Lie derivative.

To express the Lie derivative of a tensor field in local coordinates is technically slightly more complicated than for a scalar field. We first prove the following two propositions.

PROPOSITION A.11 *The Lie derivative is a derivation on the algebra of the differentiable tensor fields:*

$$\mathcal{L}_v(T_1 + T_2) = \mathcal{L}_v T_1 + \mathcal{L}_v T_2, \qquad \textit{(linearity)},$$
$$\mathcal{L}_v(T_1 \otimes T_2) = \mathcal{L}_v T_1 \otimes T_2 + T_1 \otimes \mathcal{L}_v T_2, \qquad \textit{(Leibniz rule)}.$$

Proof. The additive property can be verified readily. The Leibniz rule rests on the obvious fact that the transform of a tensor product is the tensor product of the transforms of its factors. Therefore

$$\begin{aligned}\mathcal{L}_v(T_1 \otimes T_2) &= \lim_{\tau\to 0}\frac{1}{\tau}(\Phi_\tau^* T_1 \otimes \Phi_\tau^* T_2 - T_1 \otimes T_2 + \Phi_\tau^* T_1 \otimes T_2 - \Phi_\tau^* T_1 \otimes T_2) \\ &= \lim_{\tau\to 0}\frac{1}{\tau}\left[\Phi_\tau^* T_1 \otimes (\Phi_\tau^* T_2 - T_2) + (\Phi_\tau^* T_1 - T_1) \otimes T_2\right],\end{aligned}$$

and the thesis follows. QED

PROPOSITION A.12 *Lie and exterior derivative commute:*

$$\mathcal{L}_v(d\omega) = d(\mathcal{L}_v\omega).$$

Proof. This is an immediate consequence of Proposition A.10 and of the definition of $\mathcal{L}_v$. QED

An arbitrary tensor is a finite sum of tensor products of functions and of elements of the basis and the dual basis, hence we have to find the Lie derivative of dx^h and $\frac{\partial}{\partial x^k}$. Firstly

$$\mathcal{L}_v(dx^h) = d(\mathcal{L}_v x^h) = d\left(v^k\frac{\partial x^h}{\partial x^k}\right) = \frac{\partial v^h}{\partial x^i}dx^i. \tag{A.3.3}$$

Secondly, to find $\mathcal{L}_v\frac{\partial}{\partial x^k}$, we demand that the transforms of $\frac{\partial}{\partial x^k}$ and dx^h be still elements of the dual bases:

$$\left\langle dx^h + \tau\mathcal{L}_v dx^h + O(\tau^2), \frac{\partial}{\partial x^k} + \tau\mathcal{L}_v\frac{\partial}{\partial x^k} + O(\tau^2)\right\rangle = \delta_k^h + O(\tau^2),$$

which entails

$$\left\langle \mathcal{L}_v dx^h, \frac{\partial}{\partial x^k}\right\rangle + \left\langle dx^h, \mathcal{L}_v\frac{\partial}{\partial x^k}\right\rangle = \mathcal{L}_v\left\langle dx^h, \frac{\partial}{\partial x^k}\right\rangle = 0.$$

Substituting the calculated expression of $\mathcal{L}_v dx^h$ and writing $\mathcal{L}_v \frac{\partial}{\partial x^k}$ as a linear combination $\lambda^i_k \frac{\partial}{\partial x^i}$ of the basis vectors, we find

$$\mathcal{L}_v \frac{\partial}{\partial x^k} = -\frac{\partial v^h}{\partial x^k} \frac{\partial}{\partial x^h}.$$

At this point we can calculate the Lie derivative of a generic tensor.

EXAMPLE A.4 Let us calculate the Lie derivative of a vector u :

$$\mathcal{L}_v \left(u^h \frac{\partial}{\partial x^h} \right) = \left(v^k \frac{\partial}{\partial x^k} u^h - u^k \frac{\partial}{\partial x^k} v^h \right) \frac{\partial}{\partial x^h} = [v, u],$$

where by definition $[v, u] = vu - uv$ is the commutator, or *Lie bracket*, of the two vectors.

Finally, let us view a formula connecting Lie and exterior derivative. Let us denote with the symbol $i_v \omega$ the contraction of a vector v with the first index of an r-form ω, that yields an $(r-1)$-form. Then

$$\mathcal{L}_v \omega = i_v \, d\omega + d(i_v \omega) \tag{A.3.4}$$

holds. The proof consists in a direct but somewhat lengthy check.

A.3.4 Integration of Differential Forms

Two coordinate systems $\{x^i\}$ and $\{y^j\}$ on an open set of $\mathbb{R}^r$ are said to define the same *orientation* if the determinant of the Jacobian $\det \left(\frac{\partial x^i}{\partial y^j} \right)$ is positive at all points of the set. A chart (U, ϕ) on a manifold defines an orientation of U by means of the orientation provided by the ordering of the coordinates on $\phi(U) \subset \mathbb{R}^r$. Indeed, exchanging the order of two consecutive coordinates changes the orientation. A differentiable manifold is said to be *orientable* if there exists an atlas such that, on the overlap of any two charts, the determinant of the Jacobian is positive. This is like saying that a basis of the tangent space can be propagated coherently.

To define a chain c and its frontier ∂c (in the sense of the algebraic topology) called its boundary, we start by considering some elementary sets of $\mathbb{R}^r$, the generalized rectangles.

An r-dimensional *chain* element of an n-dimensional manifold M is defined by giving an oriented generalized rectangle R, $a^i \le x^i \le b^i$ $(i = 1, \dots, r)$ and a differentiable map $\phi : U \to M$, where $U \subset \mathbb{R}^r$ is an open neighbourhood of R. The integral of a differential r-form ω of M over the chain element c is by definition

$$\int_c \omega \stackrel{\text{def}}{=} \int_R \phi^* \omega.$$

Changing over to the opposite orientation of c, i.e., $-c$, the integral changes sign

$$\int_c \omega = -\int_{-c} \omega.$$

The linear formal combination of elements of chain $c = \sum_i k_i c_i$, $k_i \in \mathbb{R}$ (but one may consider more general cases with k_i elements of a group) is called an r-*chain*, or simply a *chain* when there is no risk of confusion. The integral of ω over c is by definition

$$\int_c \omega = \sum k_i \int_{c_i} \omega.$$

The $2r$ generalized rectangles of $\mathbb{R}^{r-1} \subset \mathbb{R}^r$ that are the faces $x^i = a^i$ or $x^i = b^i$ are called the *boundary* ∂R of the generalized rectangle $R \subset \mathbb{R}^r$. A point of a face is defined by the $r-1$ coordinates

$$(x^1, \ldots, \hat{x}^i, \ldots, x^r), \ a^j \le x^j \le b^j, \ j = 1, \ldots, \hat{\imath}, \ldots, r.$$

The symbol ˆ denotes that the corresponding element is missing.

The boundary ∂R is said to be *coherently oriented* with R when the following orientations are given to the faces:

$$(x^1, \ldots, \hat{x}^i, \ldots, x^r) \quad \text{for} \quad \begin{cases} x^i = a^i & \text{and } i \text{ even} \\ x^i = b^i & \text{and } i \text{ odd,} \end{cases}$$

$$\text{the opposite orientations} \quad \text{for} \quad \begin{cases} x^i = a^i & \text{and } i \text{ odd} \\ x^i = b^i & \text{and } i \text{ even.} \end{cases}$$

If c is the image in M of R, then the image of the boundary ∂R is an $(r-1)$-chain ∂c, called still the boundary of c. It is defined by summing over the images of the $2r$ faces, with coefficients $+1$ or -1 to take into account the orientations. A very important property of the boundary operator is that it is nilpotent:

$$\partial \partial c = 0 \quad \forall c,$$

as one may easily convince oneself by noticing, in low dimensions, that the boundary of a chain has no boundary.

THEOREM A.13 (GENERALIZED STOKES) *For every* $(r-1)$*-form* ω *and* r*-chain* c *the beautiful formula holds,*

$$\int_c d\omega = \int_{\partial c} \omega.$$

Proof. We limit ourselves to proving the theorem for a 1-form and a rectangle; the case of generalized rectangles involves only the labelling of an arbitrary number of coordinates, and the extension to a chain is straightforward. Let

$$\omega = \omega_1(x^1, x^2)\, dx^1 + \omega_2(x^1, x^2)\, dx^2,$$

then

$$d\omega = \left(\frac{\partial \omega_2}{\partial x^1} - \frac{\partial \omega_1}{\partial x^2}\right) dx^1 \wedge dx^2,$$

and

$$\begin{aligned}\int\int_R d\omega &= \int\int_R \frac{\partial \omega_2}{\partial x^1}\, dx^1\, dx^2 - \int\int_R \frac{\partial \omega_1}{\partial x^2}\, dx^1\, dx^2 \\ &= \int_{a^1}^{b^1} \omega_1(x^1, a^2)\, dx^1 + \int_{a^2}^{b^2} \omega_2(b^1, x^2)\, dx^2 \\ &+ \int_{b^1}^{a^1} \omega_1(x^1, b^2)\, dx^1 + \int_{b^2}^{a^2} \omega_2(a^1, x^2)\, dx^2 = \int_{\partial R} \omega,\end{aligned}$$

which is the claim. QED

The sets of all the r-forms and of all the r-chains on a manifold are two (infinite-dimensional) vector spaces that can be put in duality. Indeed, if Ω is an r-form and C an r-chain, put

$$\langle C, \Omega \rangle \stackrel{\text{def}}{=} \int_C \Omega.$$

The statement of the Generalized Stokes Theorem may be rewritten as

$$\langle c, d\omega \rangle = \langle \partial c, \omega \rangle\,.$$

The boundary operator ∂ appears to be the dual of the exterior derivative, which is thus called *coboundary operator* also, to emphasize the duality. This pairing is however not satisfying, since it is degenerate. Indeed, if $C = \partial B \neq 0$ is any boundary and $\Omega = d\Theta \neq 0$ is the derivative of any form, then

$$\langle \partial B, d\Theta \rangle = \langle B, dd\Theta \rangle = 0.$$

This suggests that a nondegenerate pairing may be obtained by passing judiciously to some quotient, coming in this way to the deRahm Theorem. We state in advance some definitions.

(i) A form ω such that $d\omega = 0$ is called a *cocycle* or *closed form.*

(ii) A finite chain C such that $\partial C = 0$ is called a *cycle.*

(iii) A form ω such that $\omega = d\theta$ is called a *coboundary* or *exact form.*

(iv) A finite chain C such that $C = \partial B$ is called a *boundary*.

Notice that

$$\omega = d\theta \Rightarrow d\omega = 0 \quad \text{but usually} \quad d\omega = 0 \not\Rightarrow \omega = d\theta,$$
$$C = \partial B \Rightarrow \partial C = 0 \quad \text{but usually} \quad \partial C = 0 \not\Rightarrow C = \partial B.$$

Let Z^r or Z_r denote the vector space of cocycles or cycles, respectively of degree or dimension r. Let B^r or B_r denote the vector space of coboundaries or boundaries, respectively of degree or dimension r. Since $dd = 0$, $B^r \subset Z^r$. Since $\partial\partial = 0$, $B_r \subset Z_r$.

The space $H^r(M) = Z^r/B^r$ is the *r-cohomology class,* whose elements are equivalence sets of cocycles. Two cocycles ω_1 and ω_2 belong to the same equivalence set $[[\omega_1]] = [[\omega_2]]$, or are *homologous* ($\omega_1 \sim \omega_2$), if and only if they differ by a coboundary: $\omega_1 - \omega_2 = d\theta$.

Similarly $H_r(M) = Z_r/B_r$ is the *r-homology class,* whose elements are equivalence sets of cycles. Two cycles c_1 and c_2 belong to the same equivalence set $[[c_1]] = [[c_2]]$, or are *homologous* ($c_1 \sim c_2$), if and only if they differ by a boundary: $c_1 - c_2 = \partial b$.

$H^r(M)$ and $H_r(M)$ are vector spaces. If in general the coefficients of the elementary chains are elements of a fixed group, then $H^r(M)$ and $H_r(M)$ are groups, not necessarily vector spaces. For this one generally speaks of (co)homology groups.

THEOREM A.14 (DERAHM) *The map*

$$\mathcal{H} : H_r \times H^r \to \mathbb{R}, \quad ([[c]],[[\omega]]) \mapsto \langle c, \omega \rangle \quad \textit{for} \quad c \in Z_r,\ \omega \in Z^r$$

is a bilinear nondegenerate map.

Sketch of the Proof. It easy to prove that $\mathcal{H}$ is a bilinear map $H_r \times H^r \to \mathbb{R}$, i.e., the integral of a closed form over a cycle depends only on the homology class of the cycle and the cohomology class of the form. Indeed, applying the Stokes Theorem,

$$c \in B_r \Rightarrow \langle c, \omega \rangle = 0 \quad \text{because} \quad c = \partial b,\ d\omega = 0,$$
$$\omega \in B^r \Rightarrow \langle c, \omega \rangle = 0 \quad \text{because} \quad \omega = d\theta,\ \partial c = 0.$$

To prove that $\mathcal{H}$ is non degenerate is to prove the converse of the above two statements:

$$\langle c, \omega \rangle = 0 \quad \forall \omega \in Z^r \Rightarrow c \sim 0,$$
$$\langle c, \omega \rangle = 0 \quad \forall c \in Z_r \Rightarrow \omega \sim 0,$$

but this goes beyond the aim of this appendix; see the original proof in DeRahm (1960). QED

A.4 Distributions and Frobenius Theorem

Let M be an n-dimensional differentiable manifold.

DEFINITION A.15 *An r-dimensional (with $1 \le r \le n$)* distribution Δ *on M is a map $\Delta : p \mapsto \Delta_p$, where $\Delta_p \subset T_pM$ is an r-dimensional subspace of T_pM, $\forall p \in M$.*

We will consider only distributions whose dimension is constant for any $p \in M$. A set $\{X_1, \dots, X_r\}$ of r vector fields[(4)] on $U \subset M$ is called a local basis for the distribution Δ in a neighbourhood of $p \in M$ if the r differentiable vector fields $X_1, \dots, X_r$ on U defines a basis of Δ for every point of U.

If, for any $p \in M$, Δ has a local basis of smooth vector fields, then Δ is called a smooth distribution. In the sequel, "distribution" stands for "smooth distribution". A vector field X is said to belong to Δ ($X \in \Delta$) if, for any $p \in M$, $X|_p \in \Delta_p$.

DEFINITION A.16 *An r-dimensional submanifold N of M is an* integral manifold *of an r-dimensional distribution $p \mapsto \Delta_p$ if $T_pN = \Delta_p$ $\forall p \in N$.*

Not all the distributions admit an integral manifold; if this is the case, the distribution is said to be *integrable*. An integrable distribution admits, for every point $p \in M$, a neighbourhood U such that Δ is defined on U by the equations

$$dx^{r+1} = 0,\ dx^{r+2} = 0,\ \dots\ ,\ dx^n = 0, \tag{A.4.1}$$

where $x^{r+1}, x^{r+2}, \dots, x^n$ are $n-r$ independent functions defined on U. The functions x^i are called *first integrals* of the distribution.

Let us add to these first integrals other r independent functions $x^1, \dots, x^r$, so that $x^1, \dots, x^n$ make up a local coordinate system on U, called *adapted system* to the distribution Δ. In an adapted coordinate system the integral manifolds are the r-dimensional planes spanned by the vectors $X_1 = \frac{\partial}{\partial x^1}, \dots, X_r = \frac{\partial}{\partial x^r}$.

Now, a natural and fundamental question arises: when is a given distribution integrable? The answer is given by the Frobenius Theorem, but we need first a definition.

DEFINITION A.17 *A distribution Δ is said to be* involutive *if*

$$X, Y \in \Delta \Rightarrow [X, Y] \in \Delta,$$

or, in other words, if $\{X_1, \dots, X_r\}$ is a basis of Δ on U, then on U are defined the functions c^i_{hk}, $i, h, k = 1, \dots r$, such that $[X_h, X_k] = c^i_{hk} X_i$.

---

(4) In the sequel, X_i^h denotes the h-th component of the i-th vector, and ω_i^h the i-th component of the h-th 1-form.

THEOREM A.18 (FROBENIUS) *A distribution Δ is integrable on a neighbourhood U if and only if it is involutive on U.*

Proof. The condition is necessary. If the distribution is integrable, an adapted coordinate system $\{x^i\}$, $i = 1,\dots,n$ exists on U, such that Δ is defined by the r independent vector fields $\frac{\partial}{\partial x^1},\dots,\frac{\partial}{\partial x^r}$. If X and Y are two generic vector fields belonging to the distribution, then they and their Lie bracket $[X,Y]$ are some linear combination of the vectors of this basis with coefficients depending on $x^1,\dots,x^r$.

The condition is sufficient. If the distribution is involutive, we want to prove that, for every point $p \in U$, an adapted coordinate system $\{x^i\}$, $i = 1,\dots,n$ exists such that the manifold x^h = constant, $h = r+1,\dots,n$ is an integral manifold and the set of the vectors $\frac{\partial}{\partial x^1},\dots,\frac{\partial}{\partial x^r}$ is a local basis of Δ in $U(p)$.

The proof proceeds by induction on the dimension of Δ. For $r = 1$ the theorem reduces to the classical existence theorem of ordinary first order differential equations. Let us now suppose that the theorem is true for dimension $r-1$ and prove it for dimension r. To this end, take a generic basis $X_1,\dots,X_r$ of the distribution and construct another basis $Y_1,\dots Y_r$, say, relative to a coordinate system $y^1,\dots,y^n$

$$Y_1 = X_1 = \frac{\partial}{\partial y^1},$$

$$Y_a = X_a - (X_a y^1)X_1, \quad a = 2,\dots,r.$$

Also the independent vector fields $Y_1,\dots,Y_r$ span Δ. Since $X_1 y^1 = 1$, then $Y_a y^1 = 0$, hence

$$[Y_i, Y_k]y^1 = Y_i Y_k y^1 - Y_k Y_i y^1 = 0.$$

In other words, the vector fields $[Y_i, Y_k]$ have null component with respect to $\frac{\partial}{\partial y^1}$. We can write

$$[Y_i, Y_k] = c^a_{ik} Y_a, \quad a = 2,\dots,r.$$

If (i,N), $i : N \to M$, is the submanifold of M defined by $y^1 = 0$, then $Y_a|_p \in T_pN$, thus there exist $(r-1)$ vector fields Z_a on N satisfying $i_* Z_a = Y_a$, with $[Z_b, Z_c] = c^a_{bc}\big|_N Z_a$. Thus $Z_2,\dots,Z_r$ span an $(r-1)$-dimensional involutive distribution on N. Since by the inductive hypothesis the assertion is true for $r-1$, there exists, in a neighbourhood of p in N, a set of local coordinates $\{z^2,\dots,z^n\}$, with $z^i = z^i(y^2,\dots,y^n)$ such that the vector fields $\left\{\frac{\partial}{\partial z^a}\right\}$, $2 \le a \le r$, belong to the basis of Δ. Now, we can define local coordinates in some neighbourhood $U(p)$ in M by $x^1 = y^1$, $x^i = z^i(y^2,\dots,y^n)$, $2 \le i \le n$, and by construction

$$Y_1 = \frac{\partial}{\partial y^1} = \frac{\partial}{\partial x^1}, \quad Y_1 x^i = 0 \text{ for } i \ge 2.$$

Clearly, $\{x^k\}$, $k = 1\ldots,n$ is an adapted coordinate system for the distribution, which is therefore integrable. QED

The Frobenius Theorem gives a necessary and sufficient condition for the integrability of an r-dimensional distribution Δ in terms of the r vector fields of the basis. As is obvious, for example by (A.4.1), the same distribution can be described by $(n-r)$ 1-forms $\omega^{r+1},\ldots,\omega^n$, which are defined as those annihilating in every point $p \in U$ all the vectors X belonging to the distribution:

$$\langle \omega^\alpha, X_\mu \rangle = 0, \quad \alpha = r+1,\ldots,n, \quad \mu = 1,\ldots,r.$$

The condition on the integrability of the distribution can be given in terms of these 1-forms. This is the content of the

THEOREM A.19 (FROBENIUS DUAL VERSION) *A distribution Δ is integrable if and only if in an open $U \subset M$ there exist some functions $c^\alpha_{\beta\gamma}$ with $\alpha,\beta,\gamma = r+1,\ldots,n$ such that*

$$d\omega^\alpha = c^\alpha_{\beta\gamma}\omega^\beta \wedge \omega^\gamma.$$

Proof. For every vector field X_i and 1-form ω^h with $h,i,j = 1,\ldots,n$ the formula

$$\left\langle d\omega^h, X_i \wedge X_j \right\rangle = X_i \left\langle \omega^h, X_j \right\rangle - X_j \left\langle \omega^h, X_i \right\rangle - \left\langle \omega^h, [X_i, X_j] \right\rangle$$

holds, as one can check with a direct computation. Take the $n-r$ 1-forms defining the distribution Δ and add besides r 1-forms so that the set $\{\omega^h\}$ forms a basis of T^*_pM, $\forall p \in U$. Let X_i be the vectors of the dual basis: $\left\langle \omega^h, X_i \right\rangle = \delta^h_i$, thus the set $\{X_\mu\}$, $\mu = 1,\ldots,r$, is a basis of the distribution Δ. By the above formula

$$\langle d\omega^\alpha, X_\mu \wedge X_\nu \rangle = -\langle \omega^\alpha, [X_\mu, X_\nu] \rangle,$$
$$\mu,\nu = 1,\ldots,r, \quad \alpha = r+1,\ldots,n.$$

Since the right-hand side is zero if and only if Δ is involutive, the same thing holds for the left-hand side, which is equivalent to the thesis.

QED

The following corollary gives an equivalent useful formulation.

COROLLARY A.20 *The integrability condition of the Dual Frobenius Theorem can be expressed also as*

$$d\omega^\alpha \wedge \omega^{r+1} \wedge \cdots \wedge \omega^n = 0, \quad \forall \alpha = r+1,\ldots,n.$$

Proof. It is sufficient to take into account that $\omega^h \wedge \omega^h = 0$. QED

EXAMPLE A.5 As an application, let us verify that a disk, of radius R, rolling on a plane is a system with nonholonomic constraints. Let x, y be the Cartesian coordinates of the contact point, θ the angle between the axle of the disk and the axis x, and φ the proper rotation angle of the disk. The velocity of the center of the disk is equal to $R\dot{\varphi}$, and its direction is perpendicular to the axle. Projecting on the axes

$$\dot{x} = R\dot{\varphi}\sin\theta, \quad \dot{y} = -R\dot{\varphi}\cos\theta.$$

The constraints are given by the two 1-forms in four dimensions

$$\omega^1 \equiv dx - R\sin\theta\, d\varphi = 0,$$
$$\omega^2 \equiv dy + R\cos\theta\, d\varphi = 0.$$

They are not integrable because the two expressions

$$d\omega^1 \wedge \omega^1 \wedge \omega^2 = -R\cos\theta\, d\theta \wedge d\varphi \wedge dx \wedge dy,$$
$$d\omega^2 \wedge \omega^1 \wedge \omega^2 = -R\sin\theta\, d\theta \wedge d\varphi \wedge dx \wedge dy$$

are not identically zero. In the 4-dimensional configuration manifold there are no integral manifolds of dimension $4 - 2 = 2$. The meaning of this result should be clear: starting from any configuration, the disk can reach whatever other, but only with suitable manoeuvering (this is the famous "second parking problem").

A.5 Riemannian, Symplectic and Poisson Manifolds

Up to now, both tangent and cotangent spaces have looked like separate worlds: a vector cannot be transformed into a covector, and vice versa. If a manifold is endowed with an additional, Riemannian or symplectic, structure, this becomes possible. These two structures differ for their symmetry properties. The Riemannian manifolds are typically, as far as we are concerned, the configuration spaces of mechanical systems, while the symplectic manifolds, with their Poisson generalization, are the phase spaces.

A.5.1 Riemannian Manifolds

A *Riemannian manifold* is a smooth manifold M together with a covariant tensor field $g = g_{hk}dx^h \otimes dx^k$ of order two, called a *metric tensor*, such that

(i) g is symmetric,

(ii) for every point $p \in M$, the bilinear form $g|_p$ is nondegenerate, that is $g|_p(u,v) = 0$ for all $v \in T_pM$ if and only if $u = 0$.

A Riemannian manifold is called *proper* if

$$g|_p(u,u) > 0, \quad \forall u \in T_pM, \quad u \neq 0.$$

Otherwise the manifold is called *pseudo-Riemannian* or is said to have an *indefinite metric.* The tensor g endows every vector space T_pM with an inner, or scalar product

$$u \cdot v \stackrel{\text{def}}{=} g|_p(u,v) \quad \forall u, v \in T_pM.$$

For the linearity of the definition of the tensor map it follows that in components

$$u \cdot v = g_{hk}u^h v^k, \quad g_{hk} \stackrel{\text{def}}{=} g(e_h, e_k).$$

If we take a point $p' \in M$ "infinitesimally close" to p, the vector joining the two points belongs to T_pM, and the scalar product allows one to compute the distance between p and p'. One usually writes

$$ds^2 = g_{hk}dx^h dx^k,$$

without $\otimes$, for the symmetry of the tensor.

Given a curve

$$\gamma : \mathbb{R} \to M \quad \text{by } \tau \mapsto x^k = \gamma^k(\tau), \quad a \le \tau \le b,$$

its length is given by

$$\int_\gamma ds \stackrel{\text{def}}{=} \int_a^b \sqrt{\gamma^*(ds^2)}.$$

The curve, joining two fixed points of M, of minimal length is called a *geodesic.*

An inner product on any vector space defines a canonical isomorphism between the space and its dual. For a fixed u the mapping

$$g|_p(u,\cdot) : T_pM \to \mathbb{R} \quad \text{by } v \mapsto u \cdot v$$

is by definition an element of T_p^*M. The canonical isomorphism is the mapping

$$T_pM \to T_p^*M \quad \text{by } u \mapsto g|_p(u,\cdot).$$

The same vector can be written as $u = u^h e_h$ or $u = u_k e^k$, the relation between contravariant u^h and covariant u_k components being given by

$$u_k = g_{kh} u^h, \quad u^h = g^{hk} u_k, \quad \text{where } g^{hk} g_{ki} = \delta_i^h.$$

The g^{ij} are the contravariant components of the metric tensor. One says that the indices of a generic tensor are raised or lowered by means of the metric tensor.

EXAMPLE A.6 The most obvious example of Riemannian structure is the ordinary Euclidean space. Taking Cartesian coordinates, the metric tensor is represented by the unit matrix

$$(g_{hk}) = \begin{pmatrix} 1 & & \\ & \ddots & \\ & & 1 \end{pmatrix}.$$

EXAMPLE A.7 A less obvious example is that of the induced metric. Let $f : M \to N$, $\dim M = m < \dim N = n$, be an immersion of rank m. When N (the ambient space) is endowed with a Riemannian metric g, the immersed manifold M acquires an induced metric $f^* g$, the pull-back of the ambient metric. If, for example, $N = \mathbb{R}^3$ with Euclidean metric and M is the sphere S^2 of radius R, then the induced metric is

$$(g_{hk}) = \begin{pmatrix} R^2 & 0 \\ 0 & R^2 \sin^2 \vartheta \end{pmatrix},$$

where the local coordinates $q^1 = \vartheta$, $q^2 = \varphi$ on the sphere are respectively colatitude and longitude.

As one verifies, the partial derivatives $\partial_h v^k$ of (the components of) a vector do not transform like (the components of) a tensor $T \in \mathcal{T}_1^1$. We are thus led to the following

DEFINITION A.21 *A* connection *on a manifold is a mapping* $v \mapsto \nabla v$ *from the set of vectors into the set of tensors of type* $\mathcal{T}_1^1$ *such that*

$$\nabla(v + w) = \nabla v + \nabla w,$$
$$\nabla(f v) = df \otimes v + f \nabla v$$

for any function f.

The tensor ∇v is called the *covariant derivative* of v.

To assign a connection, one must define the connection coefficients Γ^j_{hk}, with which one is able to differentiate the vectors of the natural basis, that is

$$\nabla \partial_k = \Gamma^j_{hk} \partial_j \otimes dx^h.$$

The general expression

$$\nabla v = (\partial_k v^i + \Gamma^i_{kj} v^j) \partial_i \otimes dx^k \stackrel{\text{def}}{=} (\nabla_k v^i) \partial_i \otimes dx^k,$$

immediately follows. With a slight abuse of language one also refers to $\nabla_k v^i$ the "covariant derivative". Since we demand that $\nabla_k v^i$ be a tensor, the connection coefficients can not be a tensor, rather their "lack of tensoriality" must make up for that of $\partial_k v^i$.

For the covariant derivative of a covariant vector, one requires, in addition to the properties of the above definition, the extension of the Leibniz rule to the product of tensors. Since

$$\nabla_h (v^k v_k) = \partial_h (v^k v_k),$$

it follows that

$$\nabla_h v_k = \partial_h v_k - \Gamma^i_{hk} v_i.$$

The extension to the covariant derivative of a generic tensor is straightforward.

With the coefficients of a connection one can construct the two tensors

$$\begin{aligned} T^i_{hk} &= \Gamma^i_{hk} - \Gamma^i_{kh}, \\ R^j_{ikh} &= \partial_k \Gamma^j_{hi} - \partial_h \Gamma^j_{ki} + \Gamma^j_{km} \Gamma^m_{hi} - \Gamma^j_{hm} \Gamma^m_{ki}, \end{aligned}$$

the first being the *torsion tensor*, the latter the *curvature* (or *Riemann*) *tensor*. They appear naturally when calculating the effect, on functions and vectors, respectively, of the commutator of two covariant derivatives along two basis vectors:

$$\begin{aligned} [\nabla_k, \nabla_h] f &= -T^i_{kh} \nabla_i f, \\ [\nabla_k, \nabla_h] v^j &= R^j_{ikh} v^i - T^i_{kh} \nabla_i v^j. \end{aligned}$$

In principle, the connection on a manifold can be given independently of its Riemannian structure. Usually, however, connection and metric tensor are linked together, by demanding that this last be covariantly constant, that is

$$\nabla_j g_{hk} = 0 = \nabla_j g^{hk},$$

which implies that

$$\Gamma^m_{ik} = \frac{1}{2} g^{jm} (\partial_k g_{ij} + \partial_i g_{kj} - \partial_j g_{ik}) \quad \text{and } \Gamma^m_{ik} = \Gamma^m_{ki}. \tag{A.5.1}$$

A connection of this type is called a *metric connection.*

Let M and M' be two smooth manifolds with Riemannian structures g and g', respectively. The mapping $f : M \to M'$ is called an *isometry* if f is a diffeomorphism and $f^* g' = g$. Two manifolds are said to be isometric if there exists an isometry of one onto the other. A manifold is said to be *flat* or *(pseudo)-Euclidean* if it is locally isometric to a manifold with a metric of the type $(g_{hk}) = \mathrm{diag}(-1, \ldots, -1, 1, \ldots, 1)$. We enunciate without proof the following theorem, fundamental in Riemannian geometry and general relativity.

THEOREM A.22 *A manifold with symmetric (i.e., $\Gamma^m_{ik} = \Gamma^m_{ki}$) connection is flat if and only if its curvature tensor is identically zero.*

A vector field v on M generates a one-parameter group of local isometries if and only if $\mathcal{L}_v g = 0$. For example, if M is the usual 3-dimensional Euclidean space, a family of continuous isometries is a composition of translations and rotations. If, on the contrary, the symmetric tensor $\mathcal{L}_v g$ of order two is different from zero, it is a measure of the deformation of the continuum, and is called, in the theory of elasticity, a *strain tensor.*

A.5.2 Symplectic Manifolds

A *symplectic manifold* is a $2n$-dimensional smooth manifold P endowed with a 2-form $\Omega = \Omega_{\mu\nu} dx^\mu \wedge dx^\nu$, called *symplectic form,* such that

(i) Ω is closed: $d\Omega = 0$,

(ii) for every point $x \in P$ the symplectic form is nondegenerate, i.e., $\underbrace{\Omega \wedge \ldots \wedge \Omega}_{n \text{ times}} \neq 0$.

Notice differences and similarities with the Riemannian case: skew symmetry of the tensor $\Omega_{\mu\nu}$ versus symmetry of the tensor g_{hk}, but nondegeneracy in both cases. This last property ensures that a bivector $\Omega^{\alpha\beta}$, called a *Poisson bivector*, exists such that $\Omega_{\mu\nu}\Omega^{\nu\beta} = \delta^\beta_\mu$. As in the Riemannian case, the bilinear nondegenerate form Ω defines an inner product in $T_x P$, and thus a canonical isomorphism between $T_x P$ and its dual T_x^*P. This isomorphism is sometimes denoted with the symbols[(5)] $\sharp$ and $\flat$

$$\sharp : T_x^*P \to T_xP \quad \text{by } \omega \mapsto \omega^\sharp = v \quad \text{or } v^\mu = \Omega^{\mu\nu}\omega_\nu,$$
$$\flat : T_xP \to T_x^*P \quad \text{by } v \mapsto v^\flat = \omega \quad \text{or } \omega_\mu = \Omega_{\mu\nu}v^\nu.$$

To the isometry of the Riemannian case corresponds now the concept of *symplectomorphism.* Let P and P' be two $2n$-dimensional manifolds with symplectic forms Ω and Ω' respectively. A map $f : P \to P'$ is called a symplectomorphism if f is a diffeomorphism and $f^*\Omega' = \Omega$. Two manifolds are said to be symplectomorphic if there exists a symplectomorphism of one onto the other.

Let v be a vector field on M. The vector field v generates a one-parameter group of local symplectomorphisms if and only if $\mathcal{L}_v\Omega = 0$.

The closure property corresponds to the flatness of the Riemannian case, as will be clear in a moment by the Darboux Theorem. In coordinates, the closure property reads as

$$\begin{aligned} \partial_\lambda\Omega_{\mu\nu} + \partial_\mu\Omega_{\nu\lambda} + \partial_\nu\Omega_{\lambda\mu} &= 0, \\ \Omega^{\lambda\alpha}\partial_\alpha\Omega^{\mu\nu} + \Omega^{\mu\alpha}\partial_\alpha\Omega^{\nu\lambda} + \Omega^{\nu\alpha}\partial_\alpha\Omega^{\lambda\mu} &= 0. \end{aligned} \tag{A.5.2}$$

THEOREM A.23 (DARBOUX) *For any $p \in U \subset P$ there is a chart such that, if $(x^1, \dots, x^{2n})$ are local coordinates (called* canonical coordinates*) in U with $p = (0, \dots, 0)$, the symplectic form acquires the canonical form*

$$\Omega|_U = \sum_{i=1}^{n} dx^{n+i} \wedge dx^i. \tag{A.5.3}$$

Proof. The following proof, suggested by Moser (1965), was exposed in detail by Weinstein (1971). We prove in the order that (i) there is a local chart in which Ω is constant, and (ii) every $2n \times 2n$ invertible skew symmetric constant matrix can be put in the canonical form

$$(\Omega_{\mu\nu}) = \begin{pmatrix} 0_n & -\mathbf{1}_n \\ \mathbf{1}_n & 0_n \end{pmatrix}$$

with a linear transformation. Clearly, to prove (i) and (ii) is equivalent to proving the theorem.

Let us prove (i). Let Ω^1 be the constant form which equals $\Omega|_p$. We construct a coordinate change transforming Ω to the constant form Ω^1. Define $\tilde{\Omega} = \Omega^1 - \Omega$ and $\Omega^\tau = \Omega + \tau\tilde{\Omega}$, $0 \le \tau \le 1$. For every τ,

[(5)]The usage of these symbols, exhibiting the great learning of the mathematicians, is inherited from music notation, where $\sharp$ (diesis or sharp) raises while $\flat$ (bemolle or flat) lowers a note by one-half tone.

$\Omega^\tau|_p = \Omega|_p$ is nondegenerate, hence there is a neighbourhood of p on which Ω^τ is nondegenerate for all $0 \le \tau \le 1$. We can assume that this neighbourhood is connected and simply connected, thus $d\tilde{\Omega} = 0 \Rightarrow \tilde{\Omega} = d\alpha$, for a 1-form α. We can suppose $\alpha|_p = 0$. Define a smooth vector field X_τ by $i_{X_\tau}\Omega^\tau = -\alpha$, which is possible since Ω^τ is nondegenerate. Since $X_\tau|_p = 0$, by the local existence theory there is a ball about zero on which the flow of the vector field X_τ is defined for a τ at least one. Call this flow F_τ, with initial condition F_0 = identity. Then, by (A.3.2) and bearing in mind that Ω^τ is τ-dependent,

$$\begin{aligned}\frac{d}{d\tau}(F_\tau^*\Omega^\tau) &= F_\tau^*(\mathcal{L}_{X_\tau}\Omega^\tau) + F_\tau^*\frac{d}{d\tau}\Omega^\tau \\ &\overset{(A.3.4)}{=} F_\tau^* d(i_{X_\tau}\Omega^\tau) + F_\tau^*\tilde{\Omega} = F_\tau^*(-d\alpha + \tilde{\Omega}) = 0.\end{aligned}$$

Therefore, $F_\tau^*\Omega^\tau = F_0^*\Omega^0 = \Omega$, so F_τ provides the coordinate change transforming Ω to the constant form Ω^1.

Let us prove (ii). Let e_1 and e_{n+1} be two basis elements of $E = \mathbb{R}^{2n}$ such that $\Omega(e_1, e_{n+1}) \ne 0$. Dividing by a constant we can assume $\Omega(e_1, e_{n+1}) = -1$. Since $\Omega(e_1, e_1) = \Omega(e_{n+1}, e_{n+1}) = 0$ and Ω is skew, the matrix of Ω in the plane Π_1, spanned by e_1, e_{n+1}, is $\begin{pmatrix} 0 & -1 \\ 1 & 0 \end{pmatrix}$. Let E_2 be the Ω-orthogonal complement of Π_1, that is

$$E_2 = \{z,\ \Omega(z, z_1) = 0\ \forall z_1 \in \Pi_1\}.$$

Clearly $E_2 \cap \Pi_1 = \{0\}$; also, $\forall z \in E$,

$$z + \Omega(z, e_{n+1})e_1 - \Omega(z, e_1)e_{n+1} \in E_2,$$

as one verifies with a direct computation. Thus $E = \Pi_1 \oplus E_2$. Then, we continue inductively. QED

The Darboux Theorem is a basic result, showing that all the symplectic manifolds are locally the "same" manifold, and look like $\mathbb{R}^{2n}$ endowed with the canonical form. They thus differ only for global, topological properties. Moreover, the canonical form (A.5.3) is the natural one for the phase spaces of mechanical systems, as the following example shows.

EXAMPLE A.8 Let Q be an n-dimensional manifold, which in the applications will be the configuration space of a mechanical system. Define the *tangent* and *cotangent bundle* respectively as

$$TQ = \bigcup_{q\in Q} T_qQ, \quad T^*Q = \bigcup_{q\in Q} T_q^*Q.$$

They are $2n$-dimensional manifolds with local coordinates[(6)] $(q,v) = (q^1,\ldots,q^n,v^1,\ldots,v^n)$ and $(q,p) = (q^1,\ldots,q^n,p_1,\ldots,p_n)$, respectively; the first n are local coordinates on Q, the others n are the components, with respect to the natural bases $\left\{\frac{\partial}{\partial q^k}\right\}$ and $\{dq^h\}$, of a tangent and a cotangent vector, respectively.

Any cotangent bundle carries a canonical 1-form Θ defined as follows. Let $\pi : T^*Q \to Q$, $x = (q,p) \mapsto q$ be the canonical projection and let $w \in T_x(T^*Q)$, $\Theta|_x \in T^*_x(T^*Q)$; then, taking advantage of the fact that a point $x \in T^*Q$ identifies a cotangent vector $p = p_k dq^k$ to Q, we define

$$\langle \Theta, w\rangle|_x = \langle p, \pi_* w\rangle\big|_q, \quad \forall w \in T_x(T^*Q).$$

Notice that the pairing in the left-hand side is between $T^*_x(T^*Q)$ and $T_x(T^*Q)$, while that in the right-hand side is between $T^*_q Q$ and $T_q Q$. In terms of local coordinates

$$\left\langle \Theta, \frac{\partial}{\partial q^k}\right\rangle = p_k, \quad \left\langle \Theta, \frac{\partial}{\partial p_k}\right\rangle = 0,$$

and therefore $\Theta = p_k dq^k$, from which $\Omega = d\Theta = dp_k \wedge dq^k$. We have reached the fundamental result that the cotangent bundle of a configuration space is endowed in a natural way with the canonical symplectic structure.

Obviously, not all the symplectic manifolds are cotangent bundles, as the following example shows.

EXAMPLE A.9 Let us consider the sphere $S^2 = \{\vec{x} \in \mathbb{R}^3 : \|\vec{x}\| = R\}$. We can make the sphere a symplectic manifold defining on it the area 2-form

$$\Omega|_{\vec{x}}(\vec{u},\vec{v}) = \vec{x}\cdot\vec{u}\times\vec{v}, \quad \vec{u},\vec{v} \in T_{\vec{x}}S^2.$$

This form is closed, since an $(n+1)$-form on an n-dimensional manifold vanishes identically. This definition is clearly a global one, and no point on the sphere plays a privileged role. Now pick a polar axis and choose local coordinates ϑ and φ as in Example A.7. Thus $\Omega = R^2 \sin\vartheta d\varphi \wedge d\vartheta$, and the symplectic form is undefined at the poles. Put

$$p = R\cos\vartheta, \; q = R\varphi \;\Rightarrow\; \Omega = dp \wedge dq,$$

i.e., q, p are local canonical coordinates.

(6) Hereafter p will denote a covector of T^*Q, no longer a generic point on a manifold.

A.5.3 Poisson Manifolds

The Poisson manifolds are a generalization of the symplectic manifolds in the following sense. On a symplectic manifold P let us define the *Poisson bracket* $\{f, g\}$ between two functions $f, g : P \to \mathbb{R}$ by

$$\{f, g\} = \left\langle dg, df^{\sharp} \right\rangle = \frac{\partial g}{\partial x^{\mu}} \Omega^{\mu\nu} \frac{\partial f}{\partial x^{\nu}}. \tag{A.5.4}$$

In canonical coordinates,

$$\{f, g\} = \frac{\partial f}{\partial p_h} \frac{\partial g}{\partial q^h} - \frac{\partial g}{\partial p_h} \frac{\partial f}{\partial q^h}.$$

The closure property (A.5.2) of Ω entails (h is another function)

$$\{\{f, g\}, h\} + \{\{h, f\}, g\} + \{\{g, h\}, f\} = 0,$$

which is known as the *Jacobi identity*.

From this basic identity a useful property follows, relating Lie and Poisson bracket. Let us indicate the vector $(df)^{\sharp}$ with X_f, then, recalling that a vector is a derivation operator, we can write

$$\{f, g\} = X_f g = -X_g f.$$

Taking a third generic function h, one obtains

$$\begin{aligned} [X_f, X_g]h = X_f X_g h - X_g X_f h &= \{f, \{g, h\}\} - \{g, \{f, h\}\} \\ \text{(for Jacobi identity)} &= \{\{f, g\}, h\} = X_{\{f,g\}} h, \end{aligned}$$

therefore

$$[X_f, X_g] = X_{\{f,g\}}. \tag{A.5.5}$$

A *Poisson manifold* is a manifold on which is defined a Poisson bracket, or, equivalently, on which is defined a bivector $\Omega^{\sharp} = \Omega^{\mu\nu} \partial_{\mu} \partial_{\nu}$ satisfying the closure property. It is obvious that any symplectic manifold is also a Poisson manifold. If however the matrix $(\Omega^{\mu\nu})$ is not invertible, we have at hand only a Poisson structure, without the corresponding symplectic structure. Notice that a Poisson manifold can be of odd dimension. Hereafter, "Poisson manifold" will denote in general a "true" one, that is not derived from a symplectic one.

At any point $x \in P$ only the map $\sharp$ still exists, and it is no longer surjective. The dimension r of its image, which we suppose constant when varying x, is called the *rank*. The image of $\sharp$ is a subspace of $T_x P$ of constant dimension r, thus defines a distribution Δ.

THEOREM A.24 *The distribution Δ is integrable.*

Proof. Exploiting the closure property (A.5.2) one can check that, for any two 1-forms α and β, the identity

$$[\alpha^\sharp, \beta^\sharp] = \left(\mathcal{L}_{\alpha^\sharp}\beta - \mathcal{L}_{\beta^\sharp}\alpha - d\left\langle \alpha, \beta^\sharp \right\rangle\right)^\sharp, \quad \forall \alpha, \beta \in T^*P,$$

holds, asserting that the Lie bracket of any two vectors belonging to the image of $\sharp$ is a vector that belongs itself to the image of $\sharp$. The distribution Δ is therefore involutive, and the claim follows from the Frobenius Theorem. QED

COROLLARY A.25 *Any integral manifold M of Δ is a symplectic manifold.*

Proof. We would like to restrict the bivector $\Omega^\sharp$ from the ambient space P to the submanifold M, but in general the pull-back of a k-vector is not defined. In this case, however, we exploit the fact that the tangent space to M coincides with the image of $\Omega^\sharp$, and so define the 2-form

$$\Omega^\flat(X_f, X_g) \overset{\text{def}}{=} \{f, g\}, \quad \forall f, g : P \to \mathbb{R}.$$

This 2-form is well defined on M (the vectors X_f and X_g are tangent to M) where it is obviously nondegenerate, moreover it is closed for the Jacobi identity: therefore it endows M with a symplectic structure. QED

REMARK A.10 We have supposed constant the rank of the Poisson structure. The general case is more complicated but, as proved in Marsden & Ratiu (1994, page 302), the same conclusion can be reached.

The integral manifolds of the distribution Δ are called *symplectic leaves* and are described by $(n - r)$ equations $C_a(x)$ = constant (n is the dimension of P). The functions $C_a(x)$ are called *Casimir functions* and have zero Poisson bracket with any functions, since

$$\{C_a(x), x^h\} = 0, \quad a = 1, \ldots, n - r, \quad h = 1, \ldots, n,$$

as is clear bearing in mind that dC_a belongs, by definition, to the kernel of $\sharp$.

EXAMPLE A.11 Take $P = \mathbb{R}^3 - \{0\}$ with the Poisson structure

$$\{x^1, x^2\} = x^3, \quad \{x^2, x^3\} = x^1, \quad \{x^3, x^1\} = x^2,$$

or equivalently

$$(\Omega^{\mu\nu}) = \begin{pmatrix} 0 & -x^3 & x^2 \\ x^3 & 0 & -x^1 \\ -x^2 & x^1 & 0 \end{pmatrix}.$$

The kernel of $\sharp$ is $\begin{pmatrix} x^1 & x^2 & x^3 \end{pmatrix}$, the rank is 1, the Casimir function is $C(x) = (x^1)^2 + (x^2)^2 + (x^3)^2$ and the symplectic leaves are the concentric spheres. Put $p = x^3$, $q = \arctan \frac{x^1}{x^2}$, from which $\{p, q\} = 1$, so that q, p are local canonical coordinates on a sphere. This symplectic structure on the sphere is clearly equal to that of Example A.9.

A.6 Fibre Bundles

A fibre bundle is, in intuitive terms, a topological space that is locally, but not necessarily globally, a product of two spaces. For example, take a rectangular strip and join the ends to form a cylinder, which is thus the product of a circle and a line segment. Now, instead, take the same rectangular strip but twist one end through 180° relative to the other end before gluing; you so obtain a Moebius strip, which is locally like the cylinder, i.e., the product of two line segments, but not globally. The idea in a fibre bundle is to study spaces that are locally products but not, in general, globally. If the space is also globally a product, the fibre bundle is said to be *trivial*. Following Nash & Sen (1983) , let us see the exact definition.

A.6.1 Definition

DEFINITION A.26 *A fibre bundle* (E, π, B, F, G), *or* $\pi : E \xrightarrow{F} B$, *is defined by the following requirements.*

(i) A topological space E called the total space.

(ii) A topological space B called the base space *and a* projection $\pi : E \to B$.

(iii) A topological space F called the fibre.

(iv) A group G of homeomorphisms of the fibre F, called the structural group.

(v) A set of coordinate neighbourhoods U_α covering B. With every U_α there is given a homeomorphism

$$\phi_\alpha : \pi^{-1}(U_\alpha) \to U_\alpha \times F$$

where ϕ_α^{-1} satisfies

$$\pi\, \phi_\alpha^{-1}(x, f) = x, \quad x \in U_\alpha,\ f \in F.$$

For two open U_α and U_β such that $U_\alpha \cap U_\beta \neq 0$, then $\phi_\alpha \circ \phi_\beta^{-1}$ is a continuous invertible map of the form

$$\phi_\alpha \circ \phi_\beta^{-1} : (U_\alpha \cap U_\beta) \times F \to (U_\alpha \cap U_\beta) \times F.$$

For a fixed $x \in B$ the map $\phi_\alpha \circ \phi_\beta^{-1} \stackrel{def}{=} g_{\alpha\beta}(x)$ is a homeomorphism $F \to F$. The set of all these homeomorphisms for all choices of U_α, ϕ_α form just the group G, while $g_{\alpha\beta}(x)$ are called the transition functions.

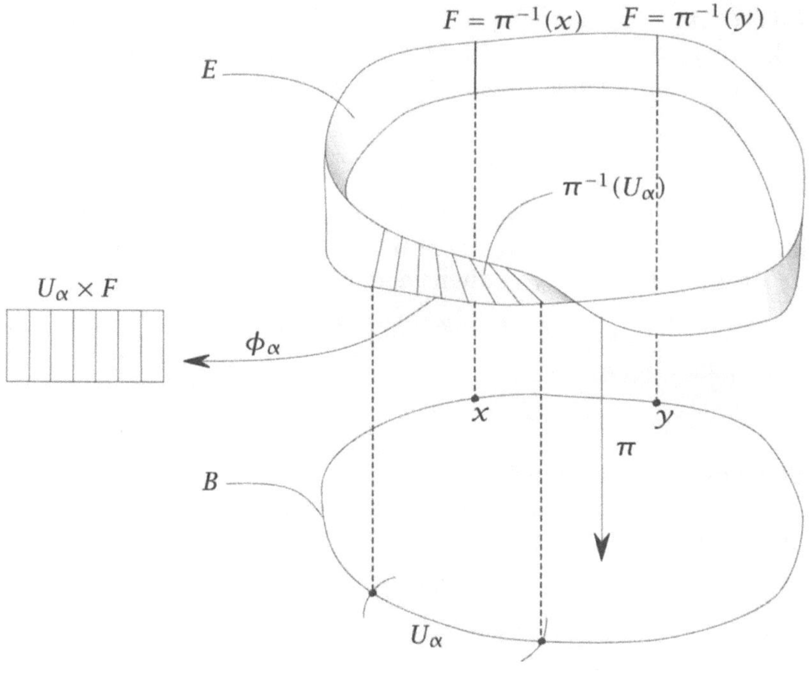

Figure A.3: The Moebius ring.

Let us comment on this definition. We may say that a fibre bundle consists of a collection of "trivial" pieces $U_\alpha \times F$ (α is a classification index varying in some set) that are glued together above the intersections $U_\alpha \cap U_\beta$, but with the possible addition of a warp or a twist to the F-factors in the gluing process. The transition functions codify this twisting, by saying how to proceed before gluing.

EXAMPLE A.12 (MOEBIUS RING) See Figure A.3. We take for B the circle S^1, for F the line segment $[-a, a]$ and for G the two-element group

$\{-1, 1\}$, acting on F by multiplication. We cover the base circle with two open subsets U_1, U_2, each homeomorphic to an open interval, so that $U_1 \cap U_2$ is the union of two disjoint open intervals, say I and II. We give instructions how to assemble the Moebius ring by setting the transition functions to

$$g_{12}(x) = 1 \quad \text{if } x \in \text{I},$$
$$g_{12}(x) = -1 \quad \text{if } x \in \text{II}.$$

Given B, the transition functions $g_{\alpha\beta}(x)$, the fibre F and the group G, then one can reconstruct the bundle E to which these objects belong. Let us see how this happens. Clearly, we need to find ϕ_α, π and E. First of all, E will be constructed by considering the "trivializing set" $\widetilde{E} = \bigcup_\alpha U_\alpha \times F$ and applying to it an equivalence relation: let $(x, f) \in U_\alpha \times F$ and $(x', f') \in U_\beta \times F$, then

$$(x, f) \sim (x', f') \quad \text{if } x = x',\ g_{\alpha\beta}(x) f = f';$$

E is therefore the set of all equivalence classes under this equivalence relation: $E = \widetilde{E}/\sim$. If we denote the equivalence class containing $(x, f) \in U_\alpha \times F$ by $[[x, f]]$, then we define the projection π by

$$\pi : E \to B, \quad [[x, f]] \mapsto x.$$

Lastly, define the function ϕ_α by

$$\phi_\alpha^{-1} : U_\alpha \times F \to \pi^{-1}(U_\alpha), \quad (x, f) \mapsto [[x, f]],$$

which automatically satisfies $\pi\phi_\alpha^{-1}(x, f) = x$, as it should.

It is worth noticing that, for a given fibre bundle, the transition functions satisfy the properties

$$g_{\alpha\alpha}(x) = \text{identity} \quad x \in U_\alpha,$$
$$g_{\alpha\beta}(x) = g_{\beta\alpha}^{-1}(x) \quad x \in U_\alpha \cap U_\beta,$$
$$g_{\alpha\beta}(x) g_{\beta\gamma}(x) = g_{\alpha\gamma}(x) \quad x \in U_\alpha \cap U_\beta \cap U_\gamma,$$

as one easily verifies. They are called *compatibility relations.*

The transition functions are not unambiguously defined. To see why, let us take two fibre bundles E and E' with the same base space, fibre and group. Let their coordinates and covering be given by (ϕ_α, U_α) and (ψ_α, U_α) respectively. Further, require that the map $\phi_\alpha \circ \psi_\alpha^{-1} : U_\alpha \times F \to U_\alpha \times F$ gives rise, for every x, to a homeomorphism $\lambda_\alpha(x) \in G$. Since, by definition,

$$\lambda_\alpha(x) = \phi_\alpha \circ \psi_\alpha^{-1}(x),$$
$$g_{\alpha\beta}(x) = \phi_\alpha \circ \phi_\beta^{-1}(x),$$
$$g'_{\alpha\beta}(x) = \psi_\alpha \circ \psi_\beta^{-1}(x),$$

it is immediate to find that the relation

$$g'_{\alpha\beta}(x) = \lambda_\alpha^{-1}(x) \circ g_{\alpha\beta}(x) \circ \lambda_\beta(x)$$

holds. It is thus natural to identify the two fibre bundles E and E', which differ only in their assignments ϕ_α and ψ_α of coordinates, but are topologically the same.

An immediate consequence is that the condition $g_{\alpha\beta}$ = identity is only sufficient for the triviality of a fibre bundle, while the general, necessary and sufficient, condition is

$$g_{\alpha\beta}(x) = \lambda_\alpha^{-1}(x) \circ \lambda_\beta(x), \quad x \in U_\alpha \cap U_\beta.$$

A.6.2 Principal and Associated Fibre Bundles

An important class of fibre bundles is that of the principal bundles.

DEFINITION A.27 *A* principal bundle *P is a fibre bundle such that $F = G$, with G acting on F by group multiplication.*

EXAMPLE A.13 A very interesting and important example is that of the Hopf fibration $\pi : S^3 \xrightarrow{S^1} S^2$, which we now examine. Let

$$\psi = \begin{pmatrix} \psi_1 \\ \psi_2 \end{pmatrix} \in \mathbb{C}^2, \quad \psi^\dagger \psi = \overline{\psi}_1 \psi_1 + \overline{\psi}_2 \psi_2 = 1$$

be, by definition, a *spinor*; it defines a unit sphere S^3 imbedded in $\mathbb{C}^2 = \mathbb{R}^4$. $\psi\psi^\dagger$ is a 2×2 Hermitian matrix with $\operatorname{Tr} \psi\psi^\dagger = \psi^\dagger\psi = 1$ and $\det \psi\psi^\dagger = 0$, hence we may write

$$\psi\psi^\dagger = \frac{1}{2} \begin{pmatrix} 1 - X_3 & X_1 + iX_2 \\ X_1 - iX_2 & 1 + X_3 \end{pmatrix}, \quad \det \psi\psi^\dagger = 0 \Rightarrow X_1^2 + X_2^2 + X_3^2 = 1,$$

that is $(X_1, X_2, X_3) \in S^2$. The Hopf fibration is $\pi : \psi \mapsto (X_1, X_2, X_3)$. Since ψ and $\psi e^{i\chi}$, $0 \le \chi < 2\pi$ have the same projection onto S^2, the fibre is $\mathrm{U}(1) = S^1$.

The fibre bundle structure is given as follows. Let us consider the stereographic projection (Example A.2) from the North and South poles of S^2 onto $\mathbb{C} = \mathbb{R}^2 : (X_1, X_2, X_3) \mapsto (x_1, x_2)$

$$x_1 - ix_2 = \frac{X_1 - iX_2}{1 - X_3} = \frac{\psi_2 \overline{\psi}_1}{\psi_1 \overline{\psi}_1} = \frac{\psi_2}{\psi_1} \quad \text{in } U_{\mathrm{N}} = S^2 - \{\mathrm{N}\},$$

$$x_1 + ix_2 = \frac{X_1 + iX_2}{1 + X_3} = \frac{\psi_1 \overline{\psi}_2}{\psi_2 \overline{\psi}_2} = \frac{\psi_1}{\psi_2} \quad \text{in } U_{\mathrm{S}} = S^2 - \{\mathrm{S}\}.$$

Bearing in mind item (v) of Definition A.26, we choose the local trivialisations

$$\phi_{\mathrm{N}} : \pi^{-1}(U_{\mathrm{N}}) \to U_{\mathrm{N}} \times \mathrm{U}(1) \quad \text{by} \quad \begin{pmatrix} \psi_1 \\ \psi_2 \end{pmatrix} \mapsto \left(\frac{\psi_2}{\psi_1}, \frac{\psi_1}{|\psi_1|} \right),$$

$$\phi_{\mathrm{S}} : \pi^{-1}(U_{\mathrm{S}}) \to U_{\mathrm{S}} \times \mathrm{U}(1) \quad \text{by} \quad \begin{pmatrix} \psi_1 \\ \psi_2 \end{pmatrix} \mapsto \left(\frac{\psi_1}{\psi_2}, \frac{\psi_2}{|\psi_2|} \right).$$

Once one has fixed the complex value $\frac{\psi_2}{\psi_1}$ (respectively $\frac{\psi_1}{\psi_2}$) and the unitary complex value $\frac{\psi_1}{|\psi_1|}$ (respectively $\frac{\psi_2}{|\psi_2|}$), the spinor ψ is determined up to a real positive factor, hence a point on S^3 is fixed unambiguously. Notice that these local trivialisations are well defined in the respective two charts. On the equator $X_3 = 0$, we have $\overline{\psi}_1\psi_1 = \overline{\psi}_2\psi_2 = \frac{1}{2}$. Accordingly, the local trivialisations *on the equator* are

$$\phi_{\mathrm{N}} : \begin{pmatrix} \psi_1 \\ \psi_2 \end{pmatrix} \mapsto \left(\frac{\psi_2}{\psi_1}, \sqrt{2}\psi_1 \right),$$

$$\phi_{\mathrm{S}} : \begin{pmatrix} \psi_1 \\ \psi_2 \end{pmatrix} \mapsto \left(\frac{\psi_1}{\psi_2}, \sqrt{2}\psi_2 \right).$$

The transition function *on the equator* is

$$g_{\mathrm{NS}}(X) = \frac{\sqrt{2}\psi_1}{\sqrt{2}\psi_2} = X_1 + iX_2 \in \mathrm{U}(1).$$

If we circumnavigate the equator, $g_{\mathrm{NS}}(X)$ traverses the unit circle in the complex plane once, and the transition function is not reducible with a change of coordinates to the identity. So, the Hopf fibration is not trivial.

EXAMPLE A.14 The real projective space $P^n_{\mathbb{R}}$ is the set of lines through the origin in $\mathbb{R}^{n+1}$. The multiplicative group of nonzero real numbers $\mathbb{R}^+$ acts on $\mathbb{R}^{n+1} - \{0\}$. Thus $\mathbb{R}^{n+1} - \{0\}$ is the bundle space of a principal bundle over $P^n_{\mathbb{R}}$ with structural group $\mathbb{R}^+$.

EXAMPLE A.15 The n-dimensional torus $\mathbb{T}^n$ is the orbit space of the action of $\mathbb{Z}^n = \underbrace{\mathbb{Z} \oplus \cdots \oplus \mathbb{Z}}_{n \text{ times}}$ on $\mathbb{R}^n$ by $\{x^1, x^2, \ldots, x^n\} \mapsto \{x^1 + m_1, x^2 + m_2, \ldots, x^n + m_n\}$. Thus $\mathbb{R}^n$ is the bundle space of a principal bundle over $\mathbb{T}^n$ with group $\mathbb{Z}^n$.

Given any bundle E with fibre F, structural group G and base space B, together with its transition functions $g_{\alpha\beta}(x)$, one can always construct from E a principal bundle $P(E)$: simply replace the fibre F with G but keep the same transition functions, then apply the previous procedure of construction. For example, the principal bundle associated to the Moebius ring is a double (since G is a two-element group) covering of S^1; see Figure 7.1 on page 144, but reading "principal bundle" instead of "SU(2)" and "S^1" instead of "SO(3)".

The procedure can be reversed: given a principal bundle $\pi : P \xrightarrow{G} B$ and a manifold F on which the structural group G acts, one can construct an *associated bundle* $\pi : E \xrightarrow{F} B$ by substituting, in some sense, the fibre G with the fibre F. The mechanism is the following. On the direct product $P \times F$ the structural group G acts in a natural way $G : P \times F \to P' \times F'$. Suppose that, for some reason, $P \times F$ and $P' \times F'$ can be considered equivalent, then one can assume the

DEFINITION A.28 *An* associated bundle *is the set of all the equivalence classes* $E = (P \times F)/G$.

An example will clarify things.

EXAMPLE A.16 (THE FRAME BUNDLE) A *linear frame* F_xM at a point x of an n-dimensional manifold M is a basis $\{e_1, \ldots, e_n\}$ of T_xM. The *frame bundle* FM is defined by

$$FM = \bigcup_{x \in M} F_xM,$$

that is, as the collection of all linear frames at all points of M. FM is a fibre bundle whose base space is M and whose projection is the map which sends every frame F_xM to the point x.

If $\{e_k\}$ is a local field of linear frames on a neighbourhood U and $\{v_k\}$ is a frame at $x \in U$, then $v_k = \lambda_k^h e_h \big|_x$, where the numbers λ_k^h are the entries in a nonsingular $n \times n$ matrix $\boldsymbol{\lambda}$, with h labelling the rows and k the columns. Thus, relative to a local field, each linear frame determines an element of the general linear group GL(n), and each element of GL(n) determines a linear frame. We therefore take GL(n) as the fibre for the frame bundle. Put for brevity $\underline{e} = (e_1, \ldots, e_n)$; two subsequent changes of basis, first $\boldsymbol{\lambda}_1$, then $\boldsymbol{\lambda}_2$, transform $\underline{e} \mapsto \underline{e}' = \underline{e}\boldsymbol{\lambda}_1\boldsymbol{\lambda}_2$. One sees that the structural group is GL(n), acting on the fibre, i.e., on itself, to the right. Therefore, the frame bundle FM is a principal bundle with base space M and fibre-structural group GL(n).

Now we show that the tangent bundle TM, already defined as the collection of all the tangent spaces T_xM, is an associated bundle to FM. A basis $\underline{e}$ and an element $\underline{\xi} = \{\xi_k\} \in \mathbb{R}^n$, considered as a column vector, determine a vector $\underline{e}\underline{\xi}$ (matrix multiplication of a 1-row matrix with vector entries by a 1-column matrix) of T_xM. But the same vector is determined by any other pair $(\underline{e}', \underline{\xi}')$ such that $\underline{e}' = \underline{e}\boldsymbol{\lambda}$ and $\underline{\xi}' = \boldsymbol{\lambda}^{-1}\underline{\xi}$, $\boldsymbol{\lambda} \in \mathrm{GL}(n)$, hence by the equivalence class $[[\underline{e}, \underline{\xi}]]$, where the equivalence relation is $(\underline{e}, \underline{\xi}) \sim (\underline{e}', \underline{\xi}')$. Therefore

$$TM = (FM \times \mathbb{R}^n)/\mathrm{GL}(n).$$

With a slight adjustment on the action of $\boldsymbol{\lambda}$, the same reasoning leads to

$$T^*M = (FM \times \mathbb{R}^n)/\mathrm{GL}(n).$$

Take indeed the dual basis $\underline{e}^*$ (a 1-column matrix) and the components $\underline{\omega}$ of a covector (a 1-row matrix), then the covector is $\underline{\omega}\,\underline{e}^*$, and $\underline{e}'^* = \boldsymbol{\lambda}^{-1}\underline{e}^*$, $\underline{\omega}' = \underline{\omega}\boldsymbol{\lambda}$. The generalization to a tensor bundle is straightforward.

The principal bundle of the frames appears therefore as the prolific parent of several other fibre bundles, to which it transmits the genetic content of its transition functions and base space.

Given a fibre bundle, to detect if it is a trivial one is not in general an easy task. The subsequent theorem is sometimes useful.

DEFINITION A.29 *A* global section *or simply a* section *of a bundle E is a continuous map $s : B \to E$ satisfying $\pi s(x) = x$, $\forall x \in B$.*

In other words a section is a homeomorphic "lifting" of the whole base into the fibre space.

THEOREM A.30 *A bundle E and the associated principal bundle $P(E)$ are trivial if and only if $P(E)$ has a section*

Proof. E having the same transition functions of P (which ensures that E is trivial if and only if P also is trivial), we can limit ourselves to proving the theorem for the principal bundle.

Let us suppose that a section does exist and prove that $P(E)$ is trivial by constructing a homeomorphism ϕ between $P(E)$ and $B \times G$. Because $P(E)$ is a principal bundle, $s(x) \in G$, and for $g \in G$, then $gs(x)$ belongs to the fibre over the point $x \in B$. Further, all elements of the fibre over x are of the form $gs(x)$, for some $g \in G$. If we let x vary, since a bundle is the union of all its fibres, then we see that all the elements of $P(E)$ can be written as $gs(x)$ for some $g \in G$ and some $x \in B$. Thus the desired homeomorphism is given by

$$\phi : P(E) \to B \times G \quad \text{by } gs(x) \mapsto (x, g).$$

Let us suppose that $P(E)$ is trivial. It is obvious that $B \times G$ has a section, namely

$$B \to B \times G \quad \text{by } x \mapsto (x, s(x)),$$

where $s : B \to G$ is some continuous map. QED

REMARK A.17 A nonprincipal bundle *can* have a section *without* being trivial. Let us consider, for example, the tangent bundle TM to an n-dimensional manifold. It admits always the zero section, that is to every $x \in M$ we may attach the zero vector, but this does not imply that TM is always trivial. Rather TM is trivial if there exists a section of its associated bundle FM, that is if $\forall x \in M$ we can find a continuous basis for T_xM. In other words, TM is trivial if one can find n continuous vector fields, independent at every point of M, thus in particular never vanishing. But this is not always possible; on the sphere S^2, for example, it is intuitive (but not easy to prove) that a smooth and nowhere vanishing vector field does not exist. Surely, we can say that if M can be covered with only one chart, then TM is trivial. Another example is the Moebius ring, which obviously admits a section, while the associated principal bundle does not.

A typical example of a trivial bundle is, as one easily verifies, a bundle whose base is continuously contractible to a point. Another example is the (co)tangent bundle to a Lie group manifold.

APPENDIX B

Lie Groups and Lie Algebras

A Lie group is a group whose elements are labelled by one or more continuously variable parameters; taking these parameters as coordinates, a Lie group can thus be seen also as a manifold, while the tangent space at the identity, once equipped with a natural composition law, becomes the corresponding Lie algebra. Lie groups are the fundamental tool when investigating the symmetries of mechanical systems.

For a deeper study the classical reference text is Chevalley (1946), while for a high readability, with applications to physics, see for example Cornwell (1989) or Gilmore (1974).

B.1 Definition and Properties

DEFINITION B.1 *A Lie group $\mathfrak{G}$ is a manifold on which an analytic composition law $\mathfrak{G} \times \mathfrak{G} \to \mathfrak{G}$ is defined, satisfying the group multiplication law.*

This means that there exist in a neighbourhood of the identity the so-called *composition functions*

$$F : \mathfrak{G} \times \mathfrak{G} \to \mathfrak{G}, \quad (x, y) \mapsto F(x, y) = xy,$$

where xy is the group product, which are analytic and satisfy

(i) $F(x, 1) = F(1, x) = x, \quad \text{where } 1x = x;$

(ii) $F(x, x^{-1}) = F(x^{-1}, x) = 1, \qquad$ where $x^{-1}x = 1$;

(iii) $F(x, F(y, z)) = F(F(x, y), z), \quad$ i.e., the associative law.

These properties, in particular the third, endow the group manifold with a rich structure, that we want to study. We stress however that, once given the composition functions in a neighbourhood of the identity, the group is determined only locally. Indeed the group manifold has in general a nontrivial topology, and in this case can not be covered with a single chart.

Let $x^h, y^k, z^i,\ h, k, i = 1, \dots, N$ be local coordinates of the points $x, y, z \in \mathfrak{G}$, and the group multiplication have coordinate presentation $z^i = F^i(x, y)$. The most important actors on the scene are the $N + N$ vector fields defined by

$$R_a^k = \left.\frac{\partial F^k(x,y)}{\partial x^a}\right|_{x=1}, \quad L_a^k = \left.\frac{\partial F^k(y,x)}{\partial x^a}\right|_{x=1}. \tag{B.1.1}$$

As it is clear from the definition, the vectors $R_a,\ a = 1, \dots, N$ are the generators of the N flows of the *left translations*:

$$\lambda_x : \mathfrak{G} \to \mathfrak{G} \quad \text{by } \lambda_x y = xy,$$

while the vectors L_a are the generators of the N flows of the *right translations*:

$$\rho_x : \mathfrak{G} \to \mathfrak{G} \quad \text{by } \rho_x y = yx.$$

The maps λ_x and ρ_x, which are by assumption analytic, satisfy

$$\lambda_x \lambda_y = \lambda_{xy}, \quad \rho_x \rho_y = \rho_{yx}$$

and these relations may be taken as a definition of *left* and *right action*, respectively.

PROPOSITION B.2 *The vectors R_a are invariant with respect to the right translations (or simply* right invariant*) while the vectors L_a are invariant with respect to the left translations (or simply* left invariant*).*

Proof. Differentiating the associative law with respect to x and z we obtain respectively

$$\frac{\partial F^h(x,yz)}{\partial x^k} = \frac{\partial F^h(xy,z)}{\partial (xy)^i}\frac{\partial F^i(x,y)}{\partial x^k},$$
$$\frac{\partial F^h(x,yz)}{\partial (yz)^i}\frac{\partial F^i(y,z)}{\partial z^k} = \frac{\partial F^h(xy,z)}{\partial z^k}.$$

Putting $x = 1$ in the first equation and $z = 1$ in the latter, we obtain

$$R_a^h(yz) = \frac{\partial F^h(y,z)}{\partial y^i} R_a^i(y), \qquad \frac{\partial F^h(x,y)}{\partial y^i} L_a^i(y) = L_a^h(xy), \tag{B.1.2}$$

which may be rewritten as

$$R_a \circ \rho_z = \mathbf{T}_{\rho_z} R_a, \quad \text{that is } \rho_{z*} R_a = R_a,$$
$$\mathbf{T}_{\lambda_x} L_a = L_a \circ \lambda_x, \quad \text{that is } \lambda_{x*} L_a = L_a.$$

This is just the definition of invariance: the vector fields R_a and L_a are transformed into themselves by any right or left translation. QED

The set of left (right) invariant vector fields constitutes a vector space, because of the linearity of $\lambda_{x*}(\rho_{x*})$. Moreover, a left (right) invariant vector field can be constructed taking some vector at the point 1 and left (right) translating it at the generic point of the group manifold. The vector space of left (right) invariant vector fields may thus be identified with $T_1\mathfrak{G}$, the tangent space to the group manifold at the identity, and denoted by $\mathfrak{g}$.

PROPOSITION B.3 *The Lie bracket* $[\cdot,\cdot]$ *of two left (right) invariant vector fields is still a left (right) invariant vector field.*

Proof. The claim is a straightforward consequence of the following general property of the push-forward of vectors:

$$\phi_*[v,w] = [\phi_* v, \phi_* w]$$

(an analogous property holds for the pull-back). In turn, this simply derives from the naturalness of the ϕ-relation with respect to the Lie bracket: see Proposition A.3. QED

As a basis for the vector space of the left (right) invariant vector fields we may take the N vectors $\{L_a\}(\{R_a\})$ previously defined. For the last proposition, the Lie bracket is still a left (right) invariant vector field and can thus be decomposed on the respective bases:

$$[L_a, L_b] = c_{ab}^c L_c, \quad [R_a, R_b] = d_{ab}^c R_c, \quad a,b,c = 1,\ldots,N, \tag{B.1.3}$$

where c_{ab}^c and d_{ab}^c are some constants. To find their value, substitute the Definition (B.1.1) into the Lie brackets (B.1.3) and put $x = 1$. Taking into account that, for item (i) of the properties of the composition functions, the relations $R^k_a(1) = L^k_a(1) = \delta^k_a$ hold, we obtain

$$\begin{aligned} c^c_{ab} &= \left(\frac{\partial^2 F^c}{\partial x^a \partial x^b} - \frac{\partial^2 F^c}{\partial x^b \partial x^a} \right)_{x=y=1}, \\ d^c_{ab} &= -c^c_{ab}. \end{aligned} \tag{B.1.4}$$

Therefore, the two structures essentially coincide.

To complete the list of the Lie brackets among left and right invariant vector fields, we notice that left and right translations commute, so that

$$[L_a, R_b] = 0, \quad \forall a, b.$$

Therefore, to any local Lie group are associated some constants c^c_{ab}, $a, b, c = 1, \dots, N$, called *structure constants*. They can not be completely arbitrary, but must satisfy

(i) skew symmetry: $[L_a, L_b] + [L_b, L_a] = 0 \Rightarrow c^c_{ab} + c^c_{ba} = 0.$

(ii) Jacobi identity:

$$\begin{aligned} &[[L_a, L_b], L_c] + [[L_b, L_c], L_a] + [[L_c, L_a], L_b] = 0 \\ &\qquad \Rightarrow c^e_{cd} c^d_{ab} + c^e_{ad} c^d_{bc} + c^e_{bd} c^d_{ca} = 0. \end{aligned}$$

We have reached the fundamental result that the N-dimensional vector space of the left or right invariant vector fields on the group manifold $\mathfrak{G}$ is naturally endowed with the respective composition law (B.1.3). As seen, this vector space is identified with $\mathfrak{g} = T_1\mathfrak{G}$, which thus is equipped with the same composition law.

DEFINITION B.4 *A vector space* $\mathfrak{g}$ *with a bilinear composition law*

$$\mathfrak{g} \times \mathfrak{g} \to \mathfrak{g} : (u, v) \mapsto [u, v]$$

satisfying the two properties

(i) $[u, v] = -[v, u]$,

(ii) $[[u.v], w] + [[v, w], u] + [[w, u], v] = 0$

is said to be a Lie algebra.

We can rephrase the content of this section in terms of left (right) invariant dual forms. Consider the vector space $\mathfrak{g}^*$, the dual space of $\mathfrak{g}$, and let $\{\omega^a\}$ be the dual basis of $\{L_b\}$, that is $\langle \omega^a, L_b \rangle = \delta^a_b$, then (as in the proof of the dual version of the Frobenius Theorem)

$$\langle d\omega^a, L_b \wedge L_c \rangle = - \langle \omega^a, [L_b, L_c] \rangle = -c^a_{bc} \tag{B.1.5}$$

which can be rewritten as

$$d\omega^a + c^a_{bc}\omega^b\omega^c = 0, \quad a,b,c = 1,\dots,N.$$

These equations are called the *Maurer-Cartan Equations* for $\mathfrak{G}$. The ω^a are left invariant forms, as one easily deduces by applying the general relation $\langle \alpha, \phi_* v\rangle = \langle \phi^*\alpha, v\rangle$:

$$\langle \omega^a, L_b\rangle = \langle \omega^a, \lambda_{x*}L_b\rangle = \langle \lambda_x^*\omega^a, L_b\rangle \Rightarrow \lambda_x^*\omega^a = \omega^a.$$

Analogously, one could introduce the right invariant forms π^a, the dual forms of R_b, which satisfy

$$d\pi^a - c^a_{bc}\pi^b\pi^c = 0, \quad a,b,c = 1,\dots,N.$$

B.2 Adjoint and Coadjoint Representation

Besides left and right translations, we may define a third action of a Lie group on itself, i.e., the *conjugation* or *inner automorphism*

$$\gamma_x : \mathfrak{G} \to \mathfrak{G} \quad \text{by } y \mapsto \gamma_x y = xyx^{-1}.$$

Since $\gamma_z\gamma_x = \gamma_{zx}$, the conjugation is a left action. The derivative map $\mathbf{T}_{\gamma_x}$ at the identity is a linear transformation of $T_1\mathfrak{G} = \mathfrak{g}$ into itself, denoted $\mathbf{Ad}_x$, which satisfies $\mathbf{Ad}_z\mathbf{Ad}_x = \mathbf{Ad}_{zx}$. It is a linear representation of the group, called *adjoint representation.* Let us find its explicit expression in coordinates. By definition

$$(\mathbf{Ad}_x)^a_b = \left.\frac{\partial F^a(xy, x^{-1})}{\partial y^b}\right|_{y=1} = \left.\frac{\partial F^a(xy, x^{-1})}{\partial (xy)^i}\frac{\partial F^i(x,y)}{\partial y^b}\right|_{y=1}.$$

Putting $z = y^{-1}$ in the first of (B.1.2), we find $\frac{\partial F^a(y,y^{-1})}{\partial y^i} = \pi^a_i(y)$, the right invariant forms. Hence

$$(\mathbf{Ad}_x)^a_b = \pi^a_i(x)L^i_b(x), \quad \text{and } (\mathbf{Ad}_x^{-1})^a_b = \omega^a_i(x)R^i_b(x).$$

Besides the adjoint representation $\mathbf{Ad}_x : \mathfrak{g} \to \mathfrak{g}$ we can define by duality the *coadjoint representation* $\mathbf{Ad}^*_x : \mathfrak{g}^* \to \mathfrak{g}^*$ by

$$\langle \mathbf{Ad}^*_x\alpha, v\rangle = \left\langle \alpha, \mathbf{Ad}_x^{-1}v\right\rangle, \quad \forall \alpha \in \mathfrak{g}^*, \ \forall v \in \mathfrak{g}, \tag{B.2.1}$$

which satisfies $\mathbf{Ad}^*_x\mathbf{Ad}^*_y = \mathbf{Ad}^*_{xy}$, as required.

REMARK B.1 When we do not need to stress the particular group element x, we write $\mathbf{Ad}_{\mathfrak{G}}$ and $\mathbf{Ad}^*_{\mathfrak{G}}$, or even $\mathbf{Ad}$ and $\mathbf{Ad}^*$, if there is no risk of confusion.

The derivative map **ad** of the adjoint representation at the identity is strictly related to the structure constants. To show it, differentiate at the identity the relation $\pi^a_k(x)R^k_b(x) = \delta^a_b$, finding $\frac{\partial \pi^a_b}{\partial x^c}\Big|_{x=1} + \frac{\partial R^a_b}{\partial x^c}\Big|_{x=1} = 0$. Then, differentiate $\mathbf{Ad}_x$ at the identity and substitute. Remembering (B.1.4), one finds

$$\frac{\partial}{\partial x^c}(\mathbf{Ad}_x)^a_b\Big|_{x=1} = c^a_{cb}. \tag{B.2.2}$$

The derivative map of $\mathbf{Ad}_x$ at the identity is therefore

$$(\mathbf{ad}_{L_c})^a_b = c^a_{cb},$$

which may be written intrinsically as

$$\mathbf{ad}_u v = [u, v], \quad \forall u, v \in \mathfrak{g}.$$

$\mathbf{ad}_u$ is a linear operator $\mathfrak{g} \to \mathfrak{g}$, $\forall u \in \mathfrak{g}$, called the *adjoint map*.

In the linear space $\mathfrak{g}^*$ let us consider the submanifolds generated by the *coadjoint action* of $\mathfrak{G}$ over an element $\mu \in \mathfrak{g}^*$. They are called *coadjoint orbits*

$$\mathcal{O}_\mu \stackrel{\text{def}}{=} \{\mu' \in \mathfrak{g}^* : \mu' = \mathbf{Ad}^*_{\mathfrak{G}}\mu\},$$

and are, in a natural way, symplectic manifolds. The symplectic structure descends by the restriction to the symplectic leaves of a Poisson structure on $\mathfrak{g}^*$. Let us view it in detail.

We endow $\mathfrak{g}^*$ with a Poisson structure assigning a Poisson bivector, that is, for every $\mu \in \mathfrak{g}^*$, a linear map $\Omega^\sharp_{(\mu)} : T^*_\mu\mathfrak{g}^* \times T^*_\mu\mathfrak{g}^* \to \mathbb{R}$ satisfying the properties of skew symmetry and closure. For the linearity of the manifold $\mathfrak{g}^*$, we can identify $T^*_\mu\mathfrak{g}^*$ with $(\mathfrak{g}^*)^* = \mathfrak{g}$. The Equation (B.1.5) suggests defining the Poisson structure

$$\Omega^\sharp_{(\mu)}(u, v) \stackrel{\text{def}}{=} -\langle \mu, [u, v]\rangle, \quad u, v \in \mathfrak{g},$$

(the minus sign is not essential and is inserted only for convenience). Its skew symmetry is obvious, bearing in mind the skew symmetry of the Lie bracket. The closure property is easily checked: writing explicitly the Poisson bivector in components

$$(\Omega^\sharp_{(\mu)})^{ab} = -\mu_c c^c_{ab},$$

the second of (A.5.2) follows from the Jacobi identity on the structure constants.

The rank of this Poisson structure is not constant on $\mathfrak{g}^*$, but only on each coadjoint orbit, as we now see. We show indeed that the image of $\Omega^\sharp_{(\mu)}$, considered for each fixed μ as a linear operator $\mathfrak{g} \to \mathfrak{g}^*$, coincides with $T_\mu\mathcal{O}_\mu$, the tangent space in μ to the coadjoint orbit through μ. To this end, let us consider, in the definition of coadjoint orbit, elements of $\mathfrak{G}$ very close to the identity, then, remembering that the derivative map at the identity of **Ad** is **ad**, we may write

$$\mathbf{Ad}^*_{\mathfrak{G}}\mu = \mu - \langle \mu, \mathbf{ad}_{\mathfrak{g}} \cdot \rangle + \cdots$$

while, by definition of Poisson bivector,

$$\Omega^{\sharp}_{(\mu)}(u, \cdot) = -\langle \mu, \mathbf{ad}_u \cdot \rangle, \quad u \in \mathfrak{g}.$$

This proves the above statement. We know that the image of a Poisson bivector is an integrable distribution, therefore we can conclude that the symplectic integral manifold through μ coincides with $\mathcal{O}_\mu$. The coadjoint orbits are therefore symplectic manifolds: in some sense, the symplectic form is the inverse of the restriction, to the orbit, of the Poisson bivector (see Corollary A.25) and is called the *Kirillov form*, or the *Kirillov-Kostant-Souriau form*,

Let $\mathfrak{G}_\mu$ be the isotropy subgroup of μ. Since

$$\mathbf{Ad}^*_{\mathfrak{G}_\mu}\mu = \mu = \mu - \left\langle \mu, \mathbf{ad}_{\mathfrak{g}_\mu} \cdot \right\rangle + \cdots,$$

the Poisson bivector is null when evaluated on the elements of $\mathfrak{g}_\mu$

$$\Omega^{\sharp}_{(\mu)}(u, v) = 0 \quad \text{if } u \text{ or } v \in \mathfrak{g}_\mu.$$

In other words, $\mathfrak{g}_\mu$ is the kernel of the linear operator $\Omega^{\sharp}_{(\mu)} : \mathfrak{g} \to \mathfrak{g}^*$.

Assigning the Poisson structure $\Omega^{\sharp}_{(\mu)}$ makes $\mathfrak{g}^*$ a Poisson manifold, endowed with the Poisson bracket

$$\{\nu_a, \nu_b\} = c^c_{ab}\nu_c, \quad \nu \in \mathfrak{g}^*, \ \nu_a = \langle \nu, L_a \rangle.$$

The functions $F : \mathfrak{g}^* \to \mathbb{R}$ satisfying the relation

$$\{F(\nu), \nu_b\}|_\mu = 0 \quad \forall \mu \text{ such that } F(\mu) = 0, \tag{B.2.3}$$

have an interesting property. This relation implies that $dF|_\mu$ belongs to the kernel of $\Omega^{\sharp}_{(\mu)}$. The pairing between the kernel of the dual of a linear operator and the image of the same operator is zero; moreover, $\Omega^{\sharp}_{(\mu)}$ is skew symmetric, thus coincides with its dual up to the sign; therefore, we have

$$\left\langle \nu, dF|_\mu \right\rangle = 0, \quad \forall \nu \in \text{image of } \Omega^{\sharp}_{(\mu)}.$$

This means that the image of $\Omega^{\sharp}_{(\mu)}$ is tangent to the manifold $F(\nu) = 0$. But the image of $\Omega^{\sharp}_{(\mu)}$ is the tangent space $T_\mu\mathcal{O}_\mu$, and we reach the conclusion that the manifold $F(\nu) = 0$ contains the coadjoint orbit $\mathcal{O}_\mu$. A coadjoint orbit can therefore be described by the zero's of the independent functions $F_k : \mathfrak{g}^* \to \mathbb{R}$, $k = 1, \ldots, d$ satisfying (B.2.3), where d is the codimension of the orbit.

EXAMPLE B.2 In so*$(2, n+1)$ take

$$\{v_a\} = \{G_{AB}\} = \{K, U_\alpha, V_\beta, G_{\alpha\beta}\},$$
$$A, B = -1, 0, \ldots, n+1, \quad \alpha, \beta = 1, \ldots, n+1.$$

The Kepler manifold is the coadjoint orbit, symplectomorphic to T^+S^n, described by the zero's of the functions

$$K - \|U\|, \quad K - \|V\|, \quad U \cdot V, \quad V_\alpha U_\beta - V_\beta U_\alpha - K G_{\alpha\beta}.$$

B.3 Action of a Lie Group on a Manifold

For what concerns us, Lie groups arise as transformation groups of manifolds and describe the symmetries of mechanical systems, whose configuration or phase space is just the manifold on which the group acts. On this manifold some vector fields are naturally defined, which are, in some sense, the velocity fields of the incipient motion caused by the group action. Not surprisingly, the Lie algebra of the group turns out to be homomorphic to the Lie algebra of these fields.

Indeed, let us consider an n-dimensional manifold M with local coordinates $q = \{q^\mu\}$ and let

$$\Phi_x : M \to M \quad \text{by } q \mapsto \Phi_x q, \quad \text{in coordinates } q^\mu \mapsto \Phi^\mu(x, q),$$

be a left action of the N-dimensional Lie group $\mathfrak{G}$:

(i) $\Phi_1 q = q,$ i.e., $\Phi^\mu(1, q) = q^\mu,$

(ii) $\Phi_x \Phi_y q = \Phi_{xy} q,$ i.e., $\Phi^\mu(x, \Phi(y, q)) = \Phi^\mu(F(x, y), q).$

We define N vector fields on M by $X^\mu_a(q) = \left.\dfrac{\partial \Phi^\mu}{\partial x^a}\right|_{x=1}$. They are called *infinitesimal generators* of the action.

PROPOSITION B.5 *The map $L_a \mapsto -X_a$ is a Lie algebra homomorphism.*

Proof. Differentiate (ii) above with respect to x and put $x = 1$, then with respect to y and put $y = 1$. Antisymmetrizing in b and c, and making use of the Maurer–Cartan Equations for the right invariant forms we find

$$[X_b, X_c] = -c^a_{bc} X_a.$$

If none of the vectors X_a is identically zero (thus if the action of $\mathfrak{G}$ is effective: see below) the map is an isomorphism. QED

Starting instead with a right action: $\Phi_x\Phi_y q = \Phi_{yx}q$, one finds that the homomorphism is given by $L_a \mapsto X_a$.

Let us view some general features of a group action. The set of the elements of $\mathfrak{G}$, leaving fixed a chosen point $q \in M$, is a subgroup of $\mathfrak{G}$, called *isotropy (sub)group* of q and is denoted by $\mathfrak{G}_q$. Thus $\mathfrak{G}_q = \{x \in \mathfrak{G} : \Phi_x q = q\}$. If the isotropy group of every point of M is the identity, then $\mathfrak{G}$ is said to act *freely* on M; in this case no element except the identity leaves *any* point fixed, i.e., *all* points are moved by the group action. Less strongly, $\mathfrak{G}$ is said to act *effectively* on M if no element, except the identity, leaves *every* point fixed, i.e., not all but surely *some* points are moved by the group action. The orbit $\mathcal{O}_q$ is defined as the set of points of M that can be reached, starting from q, by the action of the whole group $\mathfrak{G}$; thus $\mathcal{O}_q = \{\Phi_x q,\ x \in \mathfrak{G}\}$. If $\mathcal{O}_q = M$, the group $\mathfrak{G}$ is said to act *transitively*, and M is called a *homogeneous manifold* of $\mathfrak{G}$.

As an example, let us consider the action of the rotation and translation group on the 3-dimensional Euclidean space. The group of translations acts freely and transitively. The group of rotations, around a given point O, acts effectively but neither freely nor transitively, the orbit for a point P, different from O, being the sphere with center O and radius $\overline{\mathrm{OP}}$; the restriction of the action of the rotation group to this sphere is a homogeneous nonfree action, the isotropy subgroup for a point P being the subgroup of the rotations around the axis OP.

An action Φ of $\mathfrak{G}$ on a manifold M defines an equivalence relation on M by the relation of belonging to the same orbit. Explicitly, for $q, q' \in M$, we write $q \sim q'$ if there exists an $x \in \mathfrak{G}$ such that $\Phi_x q = q'$, that is if $q' \in \mathcal{O}_q$. We let $M/\mathfrak{G}$ be the set of these equivalence classes, that is the set of orbits, sometimes called the *orbit space*. We give to $M/\mathfrak{G}$ the quotient topology. To guarantee however that the orbit space has a smooth manifold structure also, further conditions on the action are required. An action Φ is called *proper* if the mapping $\widetilde{\Phi} : \mathfrak{G} \times M \to M \times M$ defined by $\widetilde{\Phi}(x, q) = (q, \Phi_x q)$ is proper, that is, for $K \subset M \times M$ compact, then $\widetilde{\Phi}^{-1}(K)$ also is compact. We notice that if $\mathfrak{G}$ is compact, the property is automatically satisfied. The next proposition gives a useful sufficient condition for $M/\mathfrak{G}$ to be a smooth manifold.

PROPOSITION B.6 *If* $\Phi : \mathfrak{G} \times M \to M$ *is a free and proper action, then* $M/\mathfrak{G}$ *is a smooth manifold and* $\pi : M \to M/\mathfrak{G}$ *a smooth submersion.*

For the proof see Abraham & Marsden (1978, Proposition 4.1.23).

EXAMPLE B.3 Take M to be the punctured plane and $\mathfrak{G}$ the rotation group around the puncture point. The action of $\mathfrak{G}$ is proper (the rotation group is compact) and free (the point fixed under rotations has been removed).

The orbits are concentric circles, and the value of the radius $R > 0$ parametrizes the set of the orbits. Then $M/\mathfrak{G} = \mathbb{R}^+$.

An important particular case is the following. Consider the action on $\mathfrak{G}$ of a closed Lie subgroup $\mathfrak{H}$ of $\mathfrak{G}$ itself, given by restriction of the left action. The orbits of $\mathfrak{G}$ under this action are called *right cosets* of $\mathfrak{H}$ in $\mathfrak{G}$; they are the sets $\mathfrak{H}x = \{hx,\, h \in \mathfrak{H},\, x \in \mathfrak{G}\}$ (unfortunately some authors call them left cosets). One verifies that this action is proper and free, see Abraham & Marsden (1978, Proposition 4.1.23). Therefore $\mathfrak{G}/\mathfrak{H}$ is a smooth manifold, M say.

One may reverse the argument. Let us consider a transitive nonfree action of a Lie group $\mathfrak{G}$ on a manifold M, and let $\mathfrak{G}_q$ be the isotropy subgroup relative to any point $q \in M$ (notice that all $\mathfrak{G}_q$, $\forall q \in M$, are isomorphic for the homogeneity of M): thus $M = \mathfrak{G}/\mathfrak{G}_q$. The closure of any isotropy subgroup is always assured, since it is the preimage of a point q (which is a closed set) of the smooth map $\Phi_q : \mathfrak{G} \to M$, given by $\Phi_q(x) = \Phi(x, q)$.

EXAMPLE B.4 The 3-dimensional rotation group SO(3) acts transitively on the sphere S^2. Choose a point of S^2 : its isotropy subgroup is the group of the rotations around the axis through this point. Therefore $S^2 = \mathrm{SO}(3)/\mathrm{SO}(2)$.

The above construction gives to $\mathfrak{G}$ the structure of a principal fibre bundle over M with structural group $\mathfrak{G}_q$.

As seen, to any local Lie group are associated some structure constants satisfying the two properties of skew symmetry and Jacobi identity. Vice versa, it is possible to prove that, given some constants satisfying the two properties, or, in other words, given an abstract Lie algebra $\mathfrak{g}$, there is a local Lie group $\mathfrak{G}$ whose Lie algebra is isomorphic to $\mathfrak{g}$. The group is determined by $\mathfrak{g}$ to within a local isomorphism. This is the content of the converse of the third Lie Theorem.

In general there are homomorphic Lie groups with isomorphic Lie algebras. We remember that a subgroup $\mathfrak{D}$ of $\mathfrak{G}$ is *discrete* if every $d \in \mathfrak{D}$ has a neighbourhood in $\mathfrak{G}$ that contains no element of $\mathfrak{D}$, apart from d. $\mathfrak{D}$ is *normal* if $xdx^{-1} \in \mathfrak{D}$, $\forall d \in \mathfrak{D}$, $\forall x \in \mathfrak{G}$. It is thus possible to prove that the general situation is the following: if $\mathfrak{G}$ is a simply connected Lie group (that is every closed curve on the group manifold can be smoothly shrunk to a point), then any other Lie group $\mathfrak{G}'$, such that $\mathfrak{g}' = \mathfrak{g}$, is covered by $\mathfrak{G}$, that is $\mathfrak{G}' = \mathfrak{G}/\mathfrak{D}$. A typical example (see the following section for the definitions) is the simply connected Lie group SU(2), which is the twofold covering of SO(3), thus $\mathfrak{D} = \mathbb{Z}_2$.

B.4 Classification of Lie Groups and Lie Algebras

We recall some definitions.

Let $\mathfrak{G}_1$ and $\mathfrak{G}_2$ be any two groups. Consider the set of pairs (x_1, x_2), where $x_1 \in \mathfrak{G}_1$ and $x_2 \in \mathfrak{G}_2$, and define the product of two such pairs (x_1, x_2) and (y_1, y_2) by $(x_1, x_2)(y_1, y_2) = (x_1 y_1, x_2 y_2)$. Then the set of pairs (x_1, x_2) forms a group, as one easily checks, called the *direct product* and denoted by $\mathfrak{G}_1 \times \mathfrak{G}_2$. The corresponding Lie algebras satisfy

$$[\mathfrak{g}_1, \mathfrak{g}_1] \subseteq \mathfrak{g}_1, \quad [\mathfrak{g}_2, \mathfrak{g}_2] \subseteq \mathfrak{g}_2, \quad [\mathfrak{g}_1, \mathfrak{g}_2] = 0.$$

A group $\mathfrak{G}$ is said to be a *semidirect product* group if it possesses two subgroups $\mathfrak{G}_1$ and $\mathfrak{G}_2$ such that

(i) $\mathfrak{G}_1$ is an invariant subgroup of $\mathfrak{G}$, i.e., $g^{-1} g_1 g \subseteq \mathfrak{G}_1$;

(ii) $\mathfrak{G}_1$ and $\mathfrak{G}_2$ have only the identity element in common;

(iii) any element of $\mathfrak{G}$ can be written as the product of an element of $\mathfrak{G}_1$ by an element of $\mathfrak{G}_2$.

In this case, we will write $\mathfrak{G} = \mathfrak{G}_2 \times_S \mathfrak{G}_1$. The corresponding Lie algebras satisfy

$$[\mathfrak{g}_1, \mathfrak{g}_1] \subseteq \mathfrak{g}_1, \quad [\mathfrak{g}_2, \mathfrak{g}_2] \subseteq \mathfrak{g}_2, \quad [\mathfrak{g}_1, \mathfrak{g}_2] \subseteq \mathfrak{g}_1.$$

One says that $\mathfrak{g}_1$ is an *invariant subalgebra* of $\mathfrak{g} = \mathfrak{g}_1 \oplus \mathfrak{g}_2$, or that it is an *ideal.*

EXAMPLE B.5 The Euclidean group of the continuous isometries of the 3-dimensional Euclidean space is the semidirect product of $\mathfrak{G}_1$ = translations and $\mathfrak{G}_2$ = rotations.

A nonabelian (that is a noncommutative) Lie algebra is said to be *simple* if it does not posses an invariant subalgebra. Hence $[\mathfrak{g}, \mathfrak{g}] = \mathfrak{g}$, that is the Lie brackets among all the elements generate the whole algebra. A Lie group is said to be *simple* if its Lie algebra is simple. A Lie group is said to be *semisimple* if it is the direct product of simple groups.

A Lie algebra is said to be *solvable* if the k-derived algebra $\mathfrak{g}^{(k)}$ is zero for some $k \geq 0$, where the *k-derived algebra* is defined recursively as

$$\mathfrak{g}^{(1)} = [\mathfrak{g}, \mathfrak{g}], \quad \mathfrak{g}^{(i+1)} = [\mathfrak{g}^{(i)}, \mathfrak{g}^{(i)}].$$

A Lie group is *solvable* if its algebra is solvable.

Simple and solvable Lie algebras are the two basic constituents of any Lie algebra. Indeed, any Lie algebra $\mathfrak{g}$ can be decomposed into the sum of a semisimple $\mathfrak{s}$ and a solvable $\mathfrak{h}$ Lie algebra: $\mathfrak{g} = \mathfrak{s} \oplus \mathfrak{h}$ (this is known as the *Levi-Malcev decomposition*), satisfying the relations

$$[\mathfrak{s}, \mathfrak{s}] = \mathfrak{s}, \quad [\mathfrak{s}, \mathfrak{h}] \subseteq \mathfrak{h}, \quad [\mathfrak{h}, \mathfrak{h}] = \mathfrak{h}^{(1)} \subset \mathfrak{h}.$$

Clearly, the classification of all Lie groups is a fundamental problem. Unfortunately, an exhaustive classification of the solvable groups does not exist, while, fortunately, that of the simple groups does: it has been performed by Killing and Cartan; see the books quoted at the beginning of this appendix. Without entering into the beautiful but lengthy analysis (which, however, is after all easily understandable, requiring only linear algebra), we jump to the conclusions.

The result of this classification is that a compact simple Lie group is necessarily one of the so-called "classical groups".

(i) The orthogonal or rotation groups $SO(n)$ with $n = 3, 5, 6, 7 \ldots$, which are the groups of the $n \times n$ matrices $\mathbf{R}$ such that

$$\mathbf{R}^t\mathbf{R} = \mathbf{1}_n, \quad \det \mathbf{R} = 1 \quad (\mathbf{R}^t = \text{transposed matrix}).$$

To find the corresponding Lie algebras $so(n)$, we write

$$\mathbf{R} = \mathbf{1}_n + \mathbf{r} + O(\mathbf{r}^2),$$

then

$$(\mathbf{1}_n + \mathbf{r}^t)(\mathbf{1}_n + \mathbf{r}) = \mathbf{1}_n + \mathbf{r} + \mathbf{r}^t + O(\mathbf{r}^2) = \mathbf{1}_n,$$

showing that $\mathbf{r} \in so(n) \Rightarrow \mathbf{r}$ is skew symmetric. The case $n = 2$ is excluded since $SO(2)$ is abelian, as well as the case $n = 4$ since $SO(4) = SO(3) \times SO(3)$.

(ii) The special unitary groups $SU(n)$ with $n = 2, 3, 4, 5 \ldots$, which are the groups of the $n \times n$ matrices $\mathbf{U}$ such that

$$\mathbf{U}^\dagger\mathbf{U} = \mathbf{1}_n, \quad \det \mathbf{U} = 1 \quad (\mathbf{U}^\dagger = \text{Hermitian conjugate}).$$

To find the corresponding Lie algebras $su(n)$, we write

$$\mathbf{U} = \mathbf{1}_n + \mathbf{u} + O(\mathbf{u}^2),$$

then $\det \mathbf{U} = 1 \Rightarrow \operatorname{Tr} \mathbf{u} = 0$, and

$$(\mathbf{1}_n + \mathbf{u}^\dagger)(\mathbf{1}_n + \mathbf{u}) = \mathbf{1}_n + \mathbf{u} + \mathbf{u}^\dagger + O(\mathbf{u}^2) = \mathbf{1}_n,$$

showing that $\mathbf{u} \in su(n) \Rightarrow \mathbf{u}$ is skew Hermitian.

(iii) The symplectic groups $\mathrm{Sp}(n)$, which are the groups of the $2n \times 2n$ matrices $\mathbf{S}$ such that

$$\mathbf{S}^t \mathbf{E}_n \mathbf{S} = \mathbf{E}_n, \quad \mathbf{E}_n = \begin{pmatrix} \mathbf{0}_n & \mathbf{1}_n \\ -\mathbf{1}_n & \mathbf{0}_n \end{pmatrix}.$$

A matrix $\mathbf{s}$ belongs to the Lie algebra $\mathrm{sp}(n)$ if $\mathbf{s}^t\mathbf{E} + \mathbf{E}\mathbf{s} = \mathbf{0}_{2n}$.

To the groups of these three series one must add the five "exceptional groups", which are isolated; see the quoted books.

Besides the above compact simple groups, there exist some other noncompact forms. They are obtained by complexifying the compact forms (that is, allowing to the group coordinates x^k to assume complex values), then finding all the possible real structures. One finds so, for example, the noncompact simple Lie groups $\mathrm{SO}(p,q)$, which are the isometry groups of pseudo-Euclidean metric, or the special (that is with $\det = 1$) linear groups $\mathrm{SL}(n)$: their corresponding compact forms are $\mathrm{SO}(p+q)$ and $\mathrm{SU}(n)$, respectively.

The simple groups obtained in this way are all distinct, but the same statement does not hold for the corresponding algebras. For example, we now show that $\mathrm{SO}(3) = \frac{\mathrm{SU}(2)}{\mathbb{Z}_2}$, $\mathbb{Z}_2 = \{1, -1\}$, from which $\mathrm{so}(3) = \mathrm{su}(2)$. Coming back to Example A.13, we see that if $\psi' = g\psi$, $g \in \mathrm{SU}(2)$, then $\psi'\psi'^{\dagger}$ is still a Hermitian matrix with trace 1 and null determinant, thus $\mathrm{SU}(2)$ induces an action on $\mathbb{R}^3$ which leaves invariant the Euclidean metric. A homomorphism between $\mathrm{SU}(2)$ and $\mathrm{SO}(3)$ is so established, whose kernel is discrete because the dimension of the two groups is equal. Let us investigate the topology of the two group manifolds. Any matrix $g \in \mathrm{SU}(2)$ can be written as

$$g = \begin{pmatrix} \alpha & \beta \\ -\overline{\beta} & \overline{\alpha} \end{pmatrix}, \quad \det g = \alpha\overline{\alpha} + \beta\overline{\beta} = 1, \quad \alpha, \beta \in \mathbb{C},$$

thus $\mathrm{SU}(2) \simeq S^3$, the 3-dimensional sphere. On the other hand, all the elements of $\mathrm{SO}(3)$ are rotations around the origin in $\mathbb{R}^3$. There is a correspondence between rotations and vectors $\vec{x} \in \mathbb{R}^3$, $\left\|\vec{x}\right\| \leq \pi$, which is 1-1 when $\left\|\vec{x}\right\| < \pi$, while one must make the identification $\vec{x} = -\vec{x}$ when $\left\|\vec{x}\right\| = \pi$. Therefore $\mathrm{SO}(3) \simeq P^3_{\mathbb{R}} \simeq \frac{S^3}{\mathbb{Z}_2}$.

A very important property of the semisimple Lie groups is that the corresponding group manifold is endowed with a natural (pseudo-)Riemannian metric, invariant under right and left translation. Indeed it is possible to prove that the symmetric matrix $F_{ab} = -c^c_{ad}c^d_{bc}$ is nonsingular if and only if the group is semisimple, and positive definite if it is compact. Then one defines

$$F_{hk}(x) = F_{ab}\omega^a_h\omega^b_k = F_{ab}\pi^a_h\pi^b_k, \tag{B.4.1}$$

bringing around on the group manifold the metric defined at the identity. That one obtains the same metric using in (B.4.1) left or right invariant forms indifferently, is due to the fact that the structure constants are invariant under the action of the adjoint representation, and that one passes from right to left invariant forms (and vice versa) just with the action of the adjoint representation. The metric defined in (B.4.1) is called the *Cartan-Killing metric.*

B.5 Connection on a Principal Bundle

In a fibre bundle there is no canonical way to "lift" a curve, lying on the base, up to the bundle manifold; a connection provides a convenient way.

DEFINITION B.7 *A* connection *on a principal bundle* $\pi : P \overset{\mathfrak{G}}{\to} M$*, where* $\mathfrak{G}$ *is the structural group, is defined by assigning, to each point* p *of the bundle space* P*, a subspace* H_p *of* T_pP *that*

(i) is complementary to the vertical subspace tangent to the fibre,

(ii) defines a smooth distribution on P*, and*

(iii) is invariant under the left action of $\mathfrak{G}$ *on* P*, in the sense that* $H_{xp} = \lambda_{x*}H_p$ *for all* $x \in \mathfrak{G}$*.*

The subspace H_p *is called* horizontal subspace *of the connection at* p.

Let a connection be given on $\pi : P \overset{\mathfrak{G}}{\to} M$. Then, to each tangent vector $v \in T_{\pi(p)}M$ there corresponds a unique horizontal vector at p that projects onto v, and is called the *horizontal lift* of v at p. Similarly, to each vector field V on M there corresponds a unique horizontal vector field on P that projects onto V, and is called the *horizontal lift* of V. Moreover, for each curve γ in M there is a unique horizontal curve (that is, whose tangent vector at a preassigned point p belongs to H_p) that projects onto γ; it is called the *horizontal lift* of γ through p. The horizontal lifts of curves define the so-called *parallel transport* as follows. Let γ be a curve joining points a and b in M. A map $\tau_\gamma : \pi^{-1}(a) \to \pi^{-1}(b)$ is defined: if $p \in \pi^{-1}(a)$, then $\tau_\gamma(p)$ is the point where the horizontal lift of γ through p meets $\pi^{-1}(b)$. The map τ_γ is called the *parallel transport* map from a to b along γ, determined by the connection.

Let us choose trivializing charts on the bundle, with q^μ coordinates on the base manifold M and x^k coordinates on the group manifold. The horizontal subspace H_p, where $p \equiv \{q^\mu, x^k\}$, is spanned by vectors of the type

$$\nabla_\mu \stackrel{\text{def}}{=} \frac{\partial}{\partial q^\mu} - A^a_\mu(q) L^k_a(x) \frac{\partial}{\partial x^k},$$

where $A^a_\mu(q)$ are arbitrary functions on the base manifold M. These vectors define univocally the connection and satisfy the requirements of Definition B.7. Sometimes one says that a connection on a principal bundle is defined by giving $A^a_\mu(q)\, dq^\mu$, that is a 1-form on the base manifold with values on the Lie algebra of the structural group.

The Frobenius Theorem suggests calculating the Lie bracket of the horizontal vectors. Since $[L_a, L_b] = c^c_{ab} L_c$ we find

$$[\nabla_\mu, \nabla_\nu] = F^c_{\mu\nu} L^k_c \frac{\partial}{\partial x^k}, \quad F^c_{\mu\nu} = \frac{\partial A^c_\mu}{\partial q^\nu} - \frac{\partial A^c_\nu}{\partial q^\mu} + c^c_{ab} A^a_\mu A^b_\nu,$$

which is a vertical vector. Therefore the distribution describing the connection is integrable if and only if $F^c_{\mu\nu}$, the so-called *curvature 2-form*, vanishes.

The horizontal subspace of a connection on a principal bundle may be defined (as for any distribution) in terms of 1-forms. One has merely to find dim $\mathfrak{G}$ linearly independent 1-forms, whose zero set is the horizontal subspace at each point. The 1-forms

$$\Omega^a \stackrel{\text{def}}{=} A^a_\mu\, dq^\mu + \omega^a_k\, dx^k$$

satisfy the requirement, since

$$\langle \Omega^a, \nabla_\mu \rangle = \left\langle A^a_\nu\, dq^\nu + \omega^a_k\, dx^k, \partial_\mu - A^b_\mu L^i_b \partial_i \right\rangle = A^a_\mu - A^a_\mu = 0,$$

and

$$\left\langle \Omega^b, L_a \right\rangle = \left\langle A^b_\mu\, dq^\mu + \omega^b_k\, dx^k, L^j_a \partial_j \right\rangle = \delta^b_a.$$

Taking into account the Maurer–Cartan Equations, we find

$$d\Omega^c + c^c_{ab} \Omega^a \Omega^b = F^c_{\mu\nu}\, dq^\mu \wedge dq^\nu.$$

Finally, we establish the transformation properties of the connection under the action of $\mathfrak{G}$, which leads to the important notion (for physical applications) of *gauge transformation*. We define the right action of $\mathfrak{G}$ on P as the one that: (i) preserves every fibre, and (ii) right translates along the fibre by an element $y \in \mathfrak{G}$, which smoothly depends on the point $q \in M$ onto which the fibre itself is projected. Hence

$$R_{y(q)} : q \mapsto \overline{q} = q, \quad x \mapsto \overline{x} = x y(q).$$

Given the function $y(q)$, and recalling that $\overline{x}^k = F^k(x, y(q))$, one finds

$$\begin{aligned} R_{y(q)*}\nabla_\mu &= \frac{\partial \overline{q}^\nu}{\partial q^\mu}\frac{\partial}{\partial \overline{q}^\nu} + \frac{\partial \overline{x}^h}{\partial q^\mu}\frac{\partial}{\partial \overline{x}^h} + A^a_\mu L^i_a \frac{\partial \overline{x}^h}{\partial x^i}\frac{\partial}{\partial \overline{x}^h} \\ &= \frac{\partial}{\partial \overline{q}^\mu} + \frac{\partial F^h(x, y(q))}{\partial y^i}\frac{\partial y^i}{\partial q^\mu}\frac{\partial}{\partial \overline{x}^h} + A^a_\mu L^i_a \frac{\partial F^h(x, y(q))}{\partial x^i}\frac{\partial}{\partial \overline{x}^h} \\ &= \frac{\partial}{\partial \overline{q}^\mu} + \left[E^a_\mu + \left(\mathbf{Ad}^{-1}_{y(q)}\right)^a_b A^b_\mu\right] L^h_a \frac{\partial}{\partial \overline{x}^h}, \end{aligned}$$

where we have defined

$$E^a_\mu = \omega^a_i(y(q))\frac{\partial y^i}{\partial q^\mu} \quad \Rightarrow \quad \frac{\partial E^a_\mu}{\partial q^\nu} - \frac{\partial E^a_\nu}{\partial q^\mu} + c^a_{bc}E^b_\mu E^c_\nu = 0.$$

We can thus infer that

$$\overline{A} = \mathbf{Ad}^{-1}_{y(q)}A + E, \quad \overline{F} = \mathbf{Ad}^{-1}_{y(q)}F,$$

which is by definition a gauge transformation. In this formula, A and E are 1-forms and F a 2-form living on the base space M and with values in $\mathfrak{g}$.

APPENDIX C

Lagrangian Dynamics

By Lagrange's Equations we mean the equations of dynamics derived from a function L, the Lagrangian, which is a function of generalized coordinates q^μ and generalized velocities $\dot{q}^\mu$ (and in general explicitly of time). We shall interpret q^μ as local coordinates on a manifold Q, the so-called configuration space of the mechanical system, and $\dot{q}^\mu$ as the corresponding fibre coordinates on TQ. A Lagrangian is then a function on TQ, the tangent bundle of the configuration space.

Many books have been written on analytical mechanics. We quote Whittaker (1917), Goldstein (1950), Sommerfeld (1964), and Landau & Lifchitz (1960) or, for those who like the conceptual aspects, the beautiful Lanczos (1970), all written in traditional language. But in the last decades the matter has changed its language, employing more and more the tools of differential geometry; see Arnold (1989), Abraham & Marsden (1978), Souriau (1997), Libermann & Marle (1987), Cushman & Bates (1997) and Marsden & Ratiu (1994).

C.1 Lagrange Equations

The Lagrange Equations are a powerful tool that allows one to write, in an automatic way, as many pure (i.e., without forces of reaction) equations of the second order as degrees of freedom.

Let $\vec{r}_1, \ldots, \vec{r}_N$ be the position vectors of the N particles of our system[(1)] of mass $m_1, \dots, m_N$. Let $\vec{v}_k = \frac{d\vec{r}_k}{dt}$ and $\vec{a}_k = \frac{d^2\vec{r}_k}{dt^2}$ be velocity and acceleration vectors of the generic k-th particle. The total force acting on this particle is divided into the sum of two forces: $\vec{F}_k$, the active force, and $\vec{\Phi}_k$, the reaction force due to the constraints,

$$\vec{F}_k + \vec{\Phi}_k = m_k \vec{a}_k.$$

We state the following basic postulate concerning $\vec{\Phi}_k$: the virtual work of the forces of reaction is always zero for any virtual displacement that is in harmony with the given kinematic constraints. We recall that a virtual displacement is infinitesimal and that we suppose the constraints frozen at the time when the virtual displacement is imposed. Plainly, the postulate generalizes the idea that the forces of reaction are orthogonal to the constraints. Let $\delta\vec{r}_k$ be the virtual displacement of the generic point; for the above postulate we can write

$$\sum_{k=1}^{N} \vec{F}_k \cdot \delta\vec{r}_k = \sum_{k=1}^{N} m_k \vec{a}_k \cdot \delta\vec{r}_k,$$

that is an equation from which the forces of reaction have disappeared. If the constraints are holonomic, we can obtain as many equations as degrees of freedom by projecting the above equation along all the independent virtual displacements.

Let $q^1, \dots, q^n$ be local coordinates on the configuration manifold Q of the mechanical system, hence the position vector of any point can be expressed as a function $\vec{r}_k(q,t)$. If the time t does not appear explicitly, the constraints are said to be *fixed*. Any virtual displacement impresses to the generic point the displacement

$$\delta\vec{r}_k = \frac{\partial\vec{r}_k}{\partial q^\mu}\delta q^\mu, \quad \mu, \nu \ldots = 1, \dots, n.$$

Notice that instead for a *real* displacement: $d\vec{r}_k = \delta\vec{r}_k + \frac{\partial\vec{r}_k}{\partial t}dt$. While the vectors $\delta\vec{r}_k$ are not in general independent (they take into account the constraints of the system), the δq^μ are independent, if the constraints are holonomic as we suppose. We can thus write the n independent equations

$$\sum_{k=1}^{N} \vec{F}_k \cdot \frac{\partial\vec{r}_k}{\partial q^\mu} = \sum_{k=1}^{N} m_k \vec{a}_k \cdot \frac{\partial\vec{r}_k}{\partial q^\mu}.$$

[(1)] We suppose the material system discrete. Otherwise, it is understood that the subsequent sums over the particles are substituted by integrals.

The left-hand side is the sum of the projections of all the active forces onto the vectors of the natural basis and is denoted by Q_μ, or (covariant components of the) *generalized force.* The right-hand side is rearranged as follows. Write

$$m_k \vec{a}_k \cdot \frac{\partial \vec{r}_k}{\partial q^\mu} = \frac{d}{dt}\left(m_k \vec{v}_k \cdot \frac{\partial \vec{r}_k}{\partial q^\mu}\right) - m_k \vec{v}_k \cdot \frac{d}{dt}\frac{\partial \vec{r}_k}{\partial q^\mu}.$$

Then, from the definition of real velocity

$$\vec{v}_k = \frac{\partial \vec{r}_k}{\partial q^\mu}\dot{q}^\mu + \frac{\partial \vec{r}_k}{\partial t},$$

we derive the two identities

$$\frac{\partial \vec{r}_k}{\partial q^\mu} = \frac{\partial \vec{v}_k}{\partial \dot{q}^\mu} \quad \text{and} \quad \frac{d}{dt}\frac{\partial \vec{r}_k}{\partial q^\mu} = \frac{\partial^2 \vec{r}_k}{\partial q^\nu \partial q^\mu}\dot{q}^\nu + \frac{\partial^2 \vec{r}_k}{\partial t \partial q^\mu} = \frac{\partial \vec{v}_k}{\partial q^\mu}.$$

Defining the total kinetic energy $T = \sum_k \frac{1}{2} m_k \vec{v}_k \cdot \vec{v}_k$, we can write

$$\sum_{k=1}^{N} m_k \vec{a}_k \cdot \frac{\partial \vec{r}_k}{\partial q^\mu} = \frac{d}{dt}\frac{\partial T}{\partial \dot{q}^\mu} - \frac{\partial T}{\partial q^\mu},$$

from which the Lagrange Equations follow

$$\frac{d}{dt}\frac{\partial T}{\partial \dot{q}^\mu} - \frac{\partial T}{\partial q^\mu} = Q_\mu.$$

If the active forces admit a potential energy V, that is if $Q_\mu = -\frac{\partial V}{\partial q^\mu}$, then the Lagrange Equations may be written as

$$\frac{d}{dt}\frac{\partial L}{\partial \dot{q}^\mu} - \frac{\partial L}{\partial q^\mu} = 0,$$

where $L = T - V$ is the *Lagrangian* of the mechanical system.

The general expression of the kinetic energy is

$$T = \frac{1}{2} g_{\mu\nu}\dot{q}^\mu \dot{q}^\nu + A_\mu \dot{q}^\mu + V,$$

$$g_{\mu\nu}(q,t) = \sum_{k=1}^{N} m_k \frac{\partial \vec{r}_k}{\partial q^\mu} \cdot \frac{\partial \vec{r}_k}{\partial q^\nu},$$

$$A_\mu(q,t) = \sum_{k=1}^{N} m_k \frac{\partial \vec{r}_k}{\partial q^\mu} \cdot \frac{\partial \vec{r}_k}{\partial t},$$

$$V(q,t) = \frac{1}{2}\sum_{k=1}^{N} m_k \frac{\partial \vec{r}_k}{\partial t} \cdot \frac{\partial \vec{r}_k}{\partial t}.$$

If the constraints are fixed, the kinetic energy is a quadratic homogeneous function of the velocities

$$T = \frac{1}{2} g_{\mu\nu}(q) \dot{q}^\mu \dot{q}^\nu,$$

and endows the configuration manifold Q with a Riemannian metric $g_{\mu\nu}$.

Exploiting the Riemannian structure, to Lagrange Equations can be given a more expressive form. For simplicity, let us limit ourselves to the case of fixed constraints. Let $v \in TM$ be the velocity vector, then $v = \dot{q}^\mu \partial_\mu$. It is natural to define the acceleration $a \in TM$ as

$$a = \frac{d}{dt}(\dot{q}^\mu \partial_\mu) = (\ddot{q}^\mu + \dot{q}^\nu \dot{q}^\varrho \Gamma^\mu_{\varrho\nu}) \partial_\mu \overset{\text{def}}{=} a^\mu \partial_\mu,$$

that can be written also as

$$a^\mu = \dot{q}^\nu (\partial_\nu \dot{q}^\mu + \Gamma^\mu_{\varrho\nu} \dot{q}^\varrho) = \dot{q}^\nu \nabla_\nu \dot{q}^\mu \overset{\text{def}}{=} \frac{D}{Dt} \dot{q}^\mu .$$

$\frac{D}{Dt}$ is called the *absolute derivative.* Recalling that the metric tensor is covariantly constant, we may write also

$$a_\varrho = g_{\varrho\mu} a^\mu = \dot{q}^\nu \nabla_\nu (g_{\varrho\mu} \dot{q}^\mu) = \frac{D}{Dt} p_\varrho,$$
$$\text{where } p_\varrho = g_{\varrho\mu} \dot{q}^\mu = \frac{\partial T}{\partial \dot{q}^\varrho}.$$

In other words, the acceleration is the (absolute) derivative of the (covariant expression of the) velocity.

Taking into account the expression (A.5.1) for the coefficients of a metric connection, the identity

$$g_{\varrho\nu} \Gamma^\nu_{\mu\sigma} \dot{q}^\varrho \dot{q}^\sigma = \frac{1}{2} \partial_\mu g_{\varrho\sigma} \dot{q}^\varrho \dot{q}^\sigma .$$

holds. Then

$$\begin{aligned} \frac{d}{dt} \frac{\partial T}{\partial \dot{q}^\mu} - \frac{\partial T}{\partial q^\mu} &= \frac{dp_\mu}{dt} - g_{\varrho\nu} \Gamma^\nu_{\mu\sigma} \dot{q}^\varrho \dot{q}^\sigma = \dot{q}^\sigma (\partial_\sigma p_\mu - \Gamma^\nu_{\mu\sigma} p_\nu) \\ &= \frac{Dp_\mu}{Dt} = Q_\mu . \end{aligned} \tag{C.1.1}$$

The Lagrange Equations can therefore be written in the form of the fundamental equation of dynamics (Newton's law).

C.2 Hamilton Principle

Before discussing the very important fact that the Lagrange Equations can be derived from a variational principle[(2)], we state a basic analytic identity.

Let us consider a generic smooth function $F : TQ \times \mathbb{R} \to \mathbb{R}$ by $(q, v, t) \mapsto F(q, v, t)$. Define the *action integral*

$$S = \int_{t_1}^{t_2} F(q(t), \dot{q}(t), t)\, dt,$$

which is a functional, i.e., in correspondence to a curve $q(t)$, $t_1 \leq t \leq t_2$ in the configuration space it yields a number. We want to calculate the "infinitesimal" variation of this action integral when passing from a curve γ_0 to other curves γ_α "infinitesimally close". These curves are labelled with a real parameter α :

$$\gamma_\alpha : t \mapsto q(t, \alpha).$$

Substituting into the action integral, this becomes a function of the parameter α. The usual methods of elementary calculus, in particular the derivation under the integral sign, can be applied, obtaining

$$\delta S[\alpha] \stackrel{\text{def}}{=} \frac{\partial S[\alpha]}{\partial \alpha} \delta\alpha = \int_{t_1}^{t_2} \left(\frac{\partial F}{\partial q^\mu} \frac{\partial q^\mu}{\partial \alpha} + \frac{\partial F}{\partial \dot{q}^\mu} \frac{\partial \dot{q}^\mu}{\partial \alpha} \right) \delta\alpha\, dt. \tag{C.2.1}$$

We can safely invert the order of the differentiations with respect to the two parameters t and α, then perform an integration by parts on the second term:

$$\int_{t_1}^{t_2} \frac{\partial F}{\partial \dot{q}^\mu} \frac{\partial \dot{q}^\mu}{\partial \alpha} \delta\alpha\, dt = \left. \frac{\partial F}{\partial \dot{q}^\mu} \frac{\partial q^\mu}{\partial \alpha} \delta\alpha \right|_{t_1}^{t_2} - \int_{t_1}^{t_2} \frac{d}{dt} \frac{\partial F}{\partial \dot{q}^\mu} \frac{\partial q^\mu}{\partial \alpha} \delta\alpha\, dt.$$

Now put $\alpha = 0$, so that δS measures the variation of the integral action when passing from γ_0 to γ_α, as required. Put, for brevity, $\delta q^\mu = \left. \frac{\partial q^\mu}{\partial \alpha} \delta\alpha \right|_{\alpha=0}$. The final formula we seek is

$$\delta S = \int_{t_1}^{t_2} \left(\frac{\partial F}{\partial q^\mu} - \frac{d}{dt} \frac{\partial F}{\partial \dot{q}^\mu} \right) \delta q^\mu\, dt + \left. \frac{\partial F}{\partial \dot{q}^\mu} \delta q^\mu \right|_{t_1}^{t_2}, \tag{C.2.2}$$

which, as we stress again, is a pure analytical identity, i.e., not involving any physical principle.

(2) In this context, they are sometimes called *Euler-Lagrange Equations.*

We can now state the Hamilton principle. Let us restrict ourselves to the class of curves γ_α that start from a point P_1 at time t_1 and reach another point P_2 at time t_2.Thus $\delta q^\mu = 0$ at time t_1 and t_2, and the last term in (C.2.2) vanishes. Therefore the fundamental identity (C.2.2) shows that

(i) if we choose γ_0 to be the real trajectory of the mechanical system, i.e., that satisfying the physical laws, then the Lagrange Equations hold, and we can infer that $\delta S = 0$: the natural motion of a mechanical system makes stationary the action integral;

(ii) if we assume as a primitive physical law that the action integral is stationary and suppose the constraints holonomic, then the Lagrange Equations follow.

Therefore, the Hamilton principle may be summarized as

Stationary action integral $\Leftrightarrow$ *Lagrange Equations.*

C.3 Noether Theorem

The several conservation laws of mechanics (for example of the linear or angular momentum and of the energy) are particular cases of the same general theorem, due to Noether: to each one-parameter group of diffeomorphisms of the configuration space, leaving invariant the Lagrangian, corresponds a first integral of the equations of motion. We recall that a first integral is a function $I : TQ \mapsto \mathbb{R}$, whose total time derivative vanishes when calculated along a solution of the Lagrange Equations. In this section we consider a time independent Lagrangian.

Before proving the theorem, we must define exactly what the invariance condition of the Lagrangian means. Let us consider a one-parameter group of diffeomorphisms acting on the configuration space: $\Phi_\alpha : q \mapsto \Phi_\alpha q$, and let $\Phi_{\alpha*} v$ be the induced push-forward of a tangent vector field to the manifold. We say that a Lagrangian $L(q, \dot{q})$ is invariant under the action of the group if

$$L(\Phi_\alpha q, \Phi_{\alpha*} v) = L(q, v), \quad \forall \alpha.$$

One may also say that the Lagrangian has a symmetry.

To prove the theorem, we start from the fundamental identity (C.2.2), where we put $F = L$. Let γ_0 be the real curve, i.e., that satisfying the Lagrange Equations. As γ_α take the transformed curve under the action of the symmetry group. Therefore $\delta q^\mu = \left. \dfrac{\partial \Phi^\mu_\alpha}{\partial \alpha} \delta\alpha \right|_{\alpha=0} = X^\mu(q)\, \delta\alpha$, where

X^μ is the vector field named previously as the infinitesimal generator of the action of the group. In (C.2.2), $\delta S = 0$ holds, for the assumption on the invariance of the Lagrangian, while the integral is zero, since only curves satisfying Lagrange Equations are considered. Therefore, also the difference $\left.\frac{\partial L}{\partial \dot{q}^\mu} X^\mu\right|_{t_2} - \left.\frac{\partial L}{\partial \dot{q}^\mu} X^\mu\right|_{t_1}$ vanishes. Since t_2 and t_1 are generic, we infer that

$$I(q.\dot{q}) \stackrel{\text{def}}{=} \frac{\partial L}{\partial \dot{q}^\mu} X^\mu$$

is a first integral.

EXAMPLE C.1 If the Lagrangian does not depend on some coordinate, say q^n, it is invariant under the group of the translations along this coordinate. Then $X = (0, \ldots, 0, 1)$ and $I = \frac{\partial L}{\partial \dot{q}^n}$.

EXAMPLE C.2 Let Q be the 2-dimensional Euclidean plane, with Cartesian coordinates x and y, and let the Lagrangian be invariant under rotations around the origin. Thus

$$\Phi_\alpha \begin{pmatrix} x \\ y \end{pmatrix} = \begin{pmatrix} \cos\alpha & \sin\alpha \\ -\sin\alpha & \cos\alpha \end{pmatrix} \begin{pmatrix} x \\ y \end{pmatrix} \Rightarrow X = \begin{pmatrix} y \\ -x \end{pmatrix}.$$

If $L = T - V$ with $T = \frac{1}{2}m(\dot{x}^2 + \dot{y}^2)$, then $I = m(\dot{x}y - \dot{y}x)$, that is the angular momentum with respect to the origin.

C.4 Reduced Lagrangian and Maupertuis Principle

As seen in the above example, if the Lagrangian is invariant under translations along a coordinate, say q^n, a first integral does exist. It is possible to eliminate completely from the Lagrangian the coordinate q^n, which is called *ignorable*, reducing the problem to one with $n - 1$ degrees of freedom. This is a first elementary example of *reduction by symmetry*. In the subsequent appendix, the idea of eliminating degrees of freedom, exploiting the symmetries of the system, will be amply generalized.

Let us indeed consider the differential (C.2.1) of the action integral and, *before* performing the integration by parts, take into account that now $\frac{\partial L}{\partial q^n} = 0$ and $\frac{\partial L}{\partial \dot{q}^n} = c_n$ hold. Obviously $c_n \delta \dot{q}^n = \delta(c_n \dot{q}^n)$, and we may write, after an integration by parts on the first $n - 1$ coordinates only,

$$\delta \int_{t_1}^{t_2} (L - c_n \dot{q}^n) dt = \int_{t_1}^{t_2} \left(\frac{\partial L}{\partial q^\mu} - \frac{d}{dt} \frac{\partial L}{\partial \dot{q}^\mu} \right) \delta q^\mu \, dt + \left. \frac{\partial L}{\partial \dot{q}^\mu} \delta q^\mu \right|_{t_1}^{t_2},$$

$$\mu = 1, \dots, n - 1.$$

We can now argue as in the previous section, and deduce that the equations of motion for the first $n - 1$ coordinates are derived from the reduced Lagrangian $L_R = L - c_n \dot{q}^n$.

We sum up the process that one must follow for the elimination of an ignorable variable q^n.

(i) Write the Lagrange Equation for the ignorable variable: $\frac{\partial L}{\partial \dot{q}^n} = c_n$.

(ii) Consider the reduced Lagrangian $L_R = L - c_n \dot{q}^n$.

(iii) Eliminate the ignorable velocity $\dot{q}^n$ by solving the equation in (i) for $\dot{q}^n$ and substituting in (ii). Then the new Lagrangian L_R does not depend on the ignorable variable.

EXAMPLE C.3 Let $L = \frac{1}{2} m (\dot{r}^2 + r^2 \dot{\theta}^2) - V(r)$ be the Lagrangian of a central motion on the plane with polar coordinates r, θ. The angle θ is ignorable, $\frac{\partial L}{\partial \dot{\theta}} = m r^2 \dot{\theta}$ is a constant c and $\dot{\theta} = \frac{c}{m r^2}$, then $L_R = \frac{1}{2} m \dot{r}^2 - V(r) - \frac{c^2}{2 m r^2}$: the motion is now 1-dimensional, but a new force with potential energy $\frac{c^2}{2 m r^2}$ appears.

The most interesting application of this machinery arises when the Lagrangian does not contain the time explicitly. Let us imagine that we do not use the time t as the independent variable, but that all the $n + 1$ variables $q^1, \dots, q^n$ and t are given as functions of some parameter τ. The system has now $n + 1$ degrees of freedom. Denoting derivatives with respect to τ by a prime, the action integral appears in the form

$$S = \int_{\tau_1}^{\tau_2} L \left(q, \frac{q'}{t'} \right) t' \, d\tau,$$

which defines a modified Lagrangian $\overline{L} = L \left(q, \frac{q'}{t'} \right) t'$. The time is now an ignorable coordinate, since only t' appears in $\overline{L}$, not t itself.

Owing to the fact that to an ignorable variable is associated a first integral, we deduce that

$$\frac{\partial \overline{L}}{\partial t'} = - \frac{\partial L}{\partial \dot{q}^\mu} \frac{q'^\mu}{t'^2} t' + L = L - \frac{\partial L}{\partial \dot{q}^\mu} \dot{q}^\mu$$

is a constant of the motion, which we call $-H(q, \dot{q})$. If, as we suppose, the system is *natural*, that is of the form

$$L = T - V = \frac{1}{2} g_{\mu\nu}(q) \dot{q}^\mu \dot{q}^\nu - V(q),$$

then $H(q, \dot{q}) = T + V$ is just the total energy. Thus the conservation of the total energy $H(q, \dot{q}) = E$ is due to the invariance of the Lagrangian with respect to the time translations.

But more than this can be said. We know that an ignorable variable can be eliminated, reducing the original problem by one degree of freedom. In the present case, t can be eliminated from the variational problem, obtaining a new variational problem that determines the motion in space but says nothing as to how the path is described with respect to the time t. This last information is obviously enclosed in the energy conservation law.

According to the general reduction process we first modify the Lagrangian. In the present case we put

$$\overline{L}_R = Lt' + Et' = (T - V + T + V)t' = 2Tt'$$

and the action integral becomes

$$S = \int_{\tau_1}^{\tau_2} 2Tt' \, d\tau = \int_{t_1}^{t_2} 2T \, dt.$$

But the last expression is unsatisfactory, because the time t cannot be used as independent variable in the variational problem. We want to completely eliminate the time in the action integral. Introducing the Riemannian square length $ds^2 = g_{\mu\nu} dq^\mu dq^\nu$, we can write, for the energy conservation,

$$\frac{1}{2}\left(\frac{ds}{dt}\right)^2 + V = E \quad \Rightarrow \quad dt = \frac{ds}{\sqrt{2(E - V)}}.$$

Therefore, by substituting in the action integral, we finally obtain

$$\int_{t_1}^{t_2} 2T \, dt = \int_{\mathrm{P}_1}^{\mathrm{P}_2} 2(E - V) \frac{ds}{\sqrt{2(E - V)}} = \int_{\mathrm{P}_1}^{\mathrm{P}_2} \sqrt{2(E - V)} \, ds.$$

The elimination process is completed, obtaining so the action integral of the reduced system. The time t does not appear in the action integral, which is moreover independent of the parameter τ.

We can now state the *least action* or *Maupertuis principle*: among all the trajectories from P_1 to P_2 of a natural system, covered with the same fixed energy E, the "true one" is the one making stationary the integral

$$\int_{\mathrm{P}_1}^{\mathrm{P}_2} \sqrt{2(E - V)} \, ds,$$

where ds is the Riemannian length element given by the quadratic form of the kinetic energy.

An immediate consequence is that if there are no active forces, the representative point follows a geodesic on the Riemannian configuration space.

APPENDIX D

Hamiltonian Dynamics

Paraphrasing the beginning of the previous appendix, we say that Hamilton's Equations are the equations of dynamics derived from a function H, the Hamiltonian, which is a function of generalized coordinates q^μ and generalized momenta p_μ (and in general explicitly of the time). We interpret q^μ as local coordinates on a manifold Q, the configuration space of the mechanical system, and p_μ as the corresponding fibre coordinates on T^*Q. A Hamiltonian is then a function on T^*Q, the cotangent bundle of the configuration space.

To pass from Lagrangian to Hamiltonian formulation is not simply a formal recasting of the matter, since new lands will appear to our eyes, indeed "now we are in the rarefied atmosphere of theories of excessive beauty and we are nearing a high plateau on which geometry, optics, mechanics, and wave mechanics meet on common ground.[(1)]"

D.1 From Lagrange to Hamilton

A Lagrangian[(2)] is a function $L : TQ \to \mathbb{R}$, and it is tempting to write the n Lagrange Equations of the second order (n is the dimension of the configuration space) as $2n$ equations of the first order on the tangent

[(1)] Lanczos (1970, page 229)

[(2)] Where not specified, in this appendix Lagrangian and Hamiltonian functions are supposed nonexplicitly depending on the time.

bundle TQ to the configuration space. The natural local coordinates on TQ are $\{q^\mu, v^\nu\}$, so that we may write

$$\frac{d}{dt}q^\mu = v^\mu, \quad \frac{d}{dt}\frac{\partial L(q,v)}{\partial v^\nu} = \frac{\partial L(q,v)}{\partial q^\nu}, \quad \mu,\nu = 1,\dots,n, \qquad \text{(D.1.1)}$$

as required. It is a very remarkable fact that TQ is naturally endowed with a symplectic structure and that to the system (D.1.1) can be given a Hamiltonian form. What this means, is immediately defined.

DEFINITION D.1 *A vector field on a symplectic manifold P is said to be* Hamiltonian *if it is the symplectic gradient of a Hamiltonian function $H : P \to \mathbb{R}$, that is has the form $\Omega^\sharp(dH)$, where Ω is the symplectic 2-form of P. A dynamical system is said to be Hamiltonian if the relative vector field is Hamiltonian.*

Let u^α, $\alpha,\beta,\gamma = 1,\dots,2n$ be local coordinates on P and

$$\Omega = \Omega_{\alpha\beta}\, du^\alpha \wedge du^\beta$$

the symplectic form; a Hamiltonian system can therefore be written as

$$\dot{u}^\alpha = \Omega^{\alpha\beta}\frac{\partial H}{\partial u^\beta}, \quad \Omega^{\alpha\beta}\Omega_{\beta\gamma} = \delta^\alpha_\gamma.$$

To show that (D.1.1) is a Hamiltonian system, firstly we endow TQ with the symplectic form $\Omega_L = d\Theta_L$, where $\Theta_L = \dfrac{\partial L}{\partial v^\mu}dq^\mu$. This ensures that $d\Omega_L = 0$. Then, we suppose that $\det\left(\dfrac{\partial^2 L}{\partial v^\mu \partial v^\nu}\right) \neq 0$, which ensures the invertibility of the matrix $(\Omega_L^{\alpha\beta})$: a Lagrangian satisfying this property is said to be *regular.* Explicitly

$$\Omega_L = \frac{\partial^2 L}{\partial q^\mu \partial v^\nu}dq^\mu \wedge dq^\nu + \frac{\partial^2 L}{\partial v^\mu \partial v^\nu}dv^\mu \wedge dq^\nu.$$

Finally, putting $(u^1,\dots,u^{2n}) = (q^1,\dots,q^n,v^1,\dots,v^n)$, define the Hamiltonian

$$H(u) \stackrel{\text{def}}{=} \frac{\partial L}{\partial v^\mu}v^\mu - L(q,v).$$

Then, it is a simple matter of calculation to check that the Equations (D.1.1) are equivalent to

$$\Omega_L(\dot{u}) = dH,$$

which, for the invertibility of the matrix of Ω_L, proves the statement.

From the Darboux Theorem we know that there exists a chart in which a symplectic structure is put in canonical form. In the Lagrangian case it is immediate to find it by simply inspecting Θ_L; it is indeed natural to define the map Leg : $TQ \to T^*Q$ by $(q, v) \mapsto (q, p)$ where $p_\mu = \frac{\partial L}{\partial v^\mu}$, so that T^*Q becomes equipped with the canonical 1-form $\Theta = p_\mu dq^\mu$. That $\{p_\mu\}$ can be assumed as coordinates on the fibres of T^*Q is obvious by thinking Θ as the pairing that puts in duality the tangent space, spanned by $\{dq^\mu\}$, with its dual space, spanned by $\{p_\mu\}$. In other words, $\frac{\partial L}{\partial v^\mu} v^\mu$ is a scalar invariant under coordinate transformations of the configuration space.

The Leg map, called traditionally *Legendre transformation*, is therefore a symplectomorphism between TQ, equipped with the symplectic structure Θ_L, and T^*Q, equipped with the canonical structure $\Omega = dp_\mu \wedge dq^\mu$:

$$\mathrm{Leg}^*\Omega = \Omega_L.$$

Let us sum up, in the traditional language, how to pass from Lagrangian to Hamiltonian formulation.

(i) Invert the definition of the momenta $p_\mu = \frac{\partial L}{\partial \dot{q}^\mu}$, finding the generalized velocities as functions of the momenta; this is possible if the Lagrangian is regular.

(ii) Define the Hamiltonian $H(q, p) = p_\mu \dot{q}^\mu - L(q, \dot{q})$; then the equations of motion are the Hamilton Equations in canonical form

$$\dot{q}^\mu = \frac{\partial H(q,p)}{\partial p_\mu}, \quad \dot{p}_\nu = -\frac{\partial H(q,p)}{\partial q^\nu}.$$

A particular but important case is that of the natural systems: the momenta $p_\mu = g_{\mu\nu}\dot{q}^\nu$, are the covariant components of the velocity, and the Hamiltonian $H = \frac{1}{2} g^{\mu\nu} p_\mu p_\nu + V$ is the total energy.

D.2 The Hamilton–Jacobi Integration Method

D.2.1 Canonical Transformations

A chart in T^*Q, with coordinates $u = (u^1, \ldots, u^{2n})$, is said to be *canonical* if the symplectic 2-form has the expression $\Omega = \sum_\mu du^{n+\mu} \wedge du^\mu$. A change of chart is said to be *canonical* if it preserves the canonical form. A canonical transformation is thus a (in general local) symplectomorphism of T^*Q, equipped with the canonical 2-form, into itself.

Making the identifications[3] $u^\mu = q^\mu,\ u^{n+\mu} = p_\mu$, the change of chart is given by assigning the $2n$ functions

$$Q^\mu = Q^\mu(q,p),\ P_\nu = P_\nu(p,q).$$

These functions are not completely arbitrary: taking into account the Definition (A.5.4) of Poisson bracket, then $\Omega^{\alpha\beta} = -\{u^\alpha, u^\beta\},\ \alpha,\beta = 1,\dots,2n$ holds, so that the change of chart defines a canonical transformation $\Gamma : (q,p) \mapsto (Q,P)$ if and only if

$$\{Q^\mu, Q^\nu\} = 0, \quad \{P_\mu, P_\nu\} = 0, \quad \{P_\mu, Q^\nu\} = \delta^\nu_\mu. \tag{D.2.1}$$

Given a Hamiltonian $H(q,p)$, put $K = \Gamma_* H$, then, for the canonicity of Γ, the Hamilton Equations in the new variables are

$$\dot{Q}^\mu = \frac{\partial K(Q,P)}{\partial P_\mu}, \quad \dot{P}_\nu = -\frac{\partial K(Q,P)}{\partial Q^\nu}.$$

Sometimes this property, i.e., to leave invariant in form *any* Hamiltonian system, is taken as the definition of canonical transformation.

The conditions (D.2.1) allow us to check if a given transformation is canonical, but do not tell us how to construct one. To this end, we notice that a transformation $(q,p) \mapsto (Q,P)$ such that

$$p_\mu dq^\mu - P_\mu dQ^\mu = dF(q,Q)$$

is surely canonical, since $ddF = 0$ implies that $dp_\mu \wedge dq^\mu = dP_\mu \wedge dQ^\mu$. In particular, F may be identically zero, then $p_\mu = \frac{\partial Q^\nu}{\partial q^\mu} P_\nu$, which is the transformation rule of a covector under a coordinate transformation $q \mapsto Q$. Such transformations are called *punctual*, and exhaust the class of transformations allowed in the Lagrangian formalism.

A function $F(q,Q)$, satisfying $\det\left(\frac{\partial^2 F}{\partial q^\mu \partial Q^\nu}\right) \neq 0$, defines, at least locally, an invertible canonical transformation, and is called a *generating function*. Indeed, expanding its differential, one finds

$$p_\mu = \frac{\partial F(q,Q)}{\partial q^\mu}, \quad P_\mu = -\frac{\partial F(q,Q)}{\partial Q^\mu}.$$

Owing to the condition on the Hessian these two relations can be inverted, expressing the new as functions of the old canonical coordinates and vice versa.

(3) Unfortunately, there is a certain problem with the position of the indices: we put them above, as for the u's coordinates, thinking of the symplectic structure, while we put them below, as for the p's, thinking of these as the components of the 1-form Θ *defined on the configuration space.*

The most simple canonical transformation of this type is generated by $F = \sum_\mu q^\mu Q^\mu$:

$$p_\mu = Q^\mu, \quad P_\mu = -q^\mu,$$

that is the exchange between coordinates and momenta. However, the identity transformation can not be generated by $F(q, Q)$.

We may consider other classes of generating functions; for example

$$p_\mu dq^\mu - P_\mu dQ^\mu = dF(q, P) - d(P_\mu Q^\mu)$$

gives surely a canonical transformation

$$p_\mu = \frac{\partial F(q,P)}{\partial q^\mu}, \quad Q^\mu = \frac{\partial F(q,P)}{\partial P_\mu}, \tag{D.2.2}$$

if the condition $\det\left(\frac{\partial^2 F}{\partial q^\mu \partial P_\nu}\right) \neq 0$ is satisfied. The identity is generated by $F = P_\mu q^\mu$. The punctual, or *extended point transformation*, can be generated by $F(q,P) = f^\mu(q)P_\mu$, with f^μ arbitrary functions. Indeed

$$Q^\mu = f^\mu(q), \quad p_\mu = \frac{\partial f^\nu}{\partial q^\mu} P_\nu.$$

One may consider also generating functions of the type $F(p.P)$ or $F(p, Q)$, although they are never used.

D.2.2 Hamilton-Jacobi Equation

The basic idea of the integration method of Hamilton-Jacobi is to find a canonical transformation Γ such that, in the new coordinates, the transformed Hamiltonian $K = \Gamma_* H$ is easily integrable.

As a typical example, let us seek a Γ such that $K(Q, P) \equiv P_1$, so that

$$\begin{aligned}
\dot{P}_\mu &= -\frac{\partial K}{\partial Q^\mu} = 0 && \Rightarrow P_\mu = \alpha_\mu, \quad \mu = 1, \dots, n, \\
\dot{Q}^1 &= \frac{\partial K}{\partial P_1} = 1 && \Rightarrow Q^1 = t + \beta^1, \\
\dot{Q}^\nu &= \frac{\partial K}{\partial P_\nu} = 0 && \Rightarrow Q^\nu = \beta^\nu, \quad \nu = 2, \dots, n,
\end{aligned}$$

where α_μ, β^μ are $2n$ integration constants, in particular $\alpha_1 = E$, the total energy. After inversion of Γ, the general integral can be written as

$$q^\mu = q^\mu(t, \alpha, \beta), \quad p_\mu = p_\mu(t, \alpha, \beta),$$

and the dynamical problem is solved.

Of course, there is nothing magic in all that, the question being now deferred to find such a canonical transformation. Let $W(q, P)$ (but one could take any of the other three types) be the unknown generating function of Γ and consider, taking into account the first of (D.2.2), the partial differential equation

$$H\left(q^1, \ldots, q^n, \frac{\partial W}{\partial q^1}, \ldots, \frac{\partial W}{\partial q^n}\right) = \alpha_1. \tag{D.2.3}$$

If one is able to find a so-called *complete integral* of this equation, that is a solution $W(q, \alpha)$ such that $\det\left(\frac{\partial^2 W}{\partial q^\mu \partial \alpha_\nu}\right) \neq 0$, $\mu, \nu = 1, \ldots, n$, the dynamical problem is solved. The partial differential equation of first order (D.2.3) is called a *(reduced) Hamilton–Jacobi Equation*, the full name (without "reduction") being reserved to time-dependent Hamiltonians.

The meaning of (D.2.3) should be clear: it imposes that the canonical transformation Γ we seek be such that the transformed Hamiltonian $K = \Gamma_* H$ coincides with the first of the new momenta.

D.2.3 Geometric Description

Let us expound the Hamilton–Jacobi theory from a geometric point of view.

DEFINITION D.2 *A* Lagrangian submanifold *of a $2n$-dimensional symplectic manifold, with Ω the symplectic 2-form, is an n-dimensional submanifold Λ such that $\Omega|_\Lambda = 0$.*

With $\Omega|_\Lambda$ we mean the evaluation of Ω on all the pairs of tangent vectors to Λ, that is the pull-back of Ω to Λ. Notice that n is the maximal dimension of a submanifold for which the defining property holds.

Consider the cotangent bundle T^*Q (the generalization to a generic symplectic manifold is immediate owing to the Darboux Theorem) and let $\pi : T^*Q \to Q$, $u \mapsto q$ be the canonical projection. Let Λ be a section of T^*Q, that is an n-dimensional submanifold of T^*Q with a one-to-one projection on the base manifold Q. But, by definition, a section of T^*Q is also a 1-form (i.e., a covector field) λ on Q, so that Λ and λ are two different symbols for the same object, according to whether we consider it as a submanifold of T^*Q or as a 1-form on Q. Moreover, λ may be regarded also as a map $Q \to \Lambda$.

We need the following basic property of the canonical 1-form Θ of T^*Q:

$$\lambda^*\Theta = \lambda, \quad \forall \lambda.$$

In the left-hand side, $\lambda : Q \to \Lambda \subset T^*Q$ is viewed as a map, while in the right-hand side λ is viewed as a 1-form on Q. In canonical coordinates, if $\lambda = \lambda_\mu(q)dq^\mu$ and $\Theta = p_\mu dq^\mu$, then $\lambda^*\Theta$ is obtained by simply replacing $p_\mu = \lambda_\mu(q)$ into Θ, which gives back λ, so $\lambda^*\Theta = \lambda$. To prove the property intrinsically, we recall (Example A.8) that the canonical 1-form Θ is defined by

$$\langle\Theta, w\rangle|_u = \langle\lambda, \pi_* w\rangle|_q, \quad \Theta|_u \in T_u^*(T^*Q), \quad \forall w \in T_u(T^*Q),$$

therefore, $\forall v \in T_qQ$ and taking into account that $\pi \circ \lambda$ = identity in Q,

$$\begin{aligned}\langle\lambda^*\Theta, v\rangle|_q &= \langle\Theta, \lambda_* v\rangle|_u = \langle\lambda, \pi_*\lambda_* v\rangle|_q \\ &= \langle\lambda, (\pi \circ \lambda)_* v\rangle|_q = \langle\lambda, v\rangle|_q,\end{aligned}$$

from which $\lambda^*\Theta = \lambda$.

We can now prove the

THEOREM D.3 *A section Λ of T^*Q is Lagrangian if and only if the corresponding 1-form λ is closed.*

Proof. From the above property, and bearing in mind that pull-back and exterior derivative commute, one finds

$$d\lambda = d\lambda^*\Theta = \lambda^* d\Theta = \lambda^*\Omega = \Omega|_\Lambda,$$

from which the theorem follows. QED

An immediate consequence is that Λ is Lagrangian if and only if there exists a scalar function $W : Q \to \mathbb{R}$, called the *generating function* of the Lagrangian section Λ, such that, at least locally, $\lambda = dW$. But we may argue in another way: since $\Omega|_\Lambda = d\Theta|_\Lambda = 0$, thus, at least locally, $\Theta|_\Lambda = dW_\Lambda$, where $W_\Lambda : \Lambda \to \mathbb{R}$ is a scalar function defined on the Lagrangian submanifold Λ. The relation between the two functions is, up to an additive constant,

$$W = \lambda^* W_\Lambda = W_\Lambda \circ \lambda,$$

indeed

$$d(\lambda^* W_\Lambda) = \lambda^* dW_\Lambda = \lambda^*\Theta = \lambda = dW.$$

Therefore, W is the projection of W_Λ on the base Q : even if W_Λ is a well defined function on Λ, W fails to be so on the critical points of the projection.

The Hamilton–Jacobi Equation can be written as

$$H \circ \lambda = E,$$

which shows that *one* solution of the Hamilton–Jacobi Equation is a Lagrangian section of T^*Q, constrained to belong to the level hypersurface $H = E$. However, for the general solution of the dynamical problem described by the Hamiltonian H, we need a complete integral, that is a family of solutions continuously depending on n parameters.

To get a geometric understanding of the meaning of complete integral, let us first consider a *Lagrangian foliation* of a symplectic manifold[(4)], that is, by definition, a family λ_α of 1-forms continuously depending on n parameters $\alpha = \alpha_1, \ldots, \alpha_n$. (sometimes a Lagrangian foliation is called *polarization*). There is another way to describe a Lagrangian foliation. Consider indeed n functions $\Phi_k : T^*Q \to \mathbb{R}$, $k = 1, \ldots, n$; if their differentials $d\Phi_k$ are independent, i.e., if the matrix $\frac{\partial(\Phi_1,\ldots,\Phi_n)}{\partial(q^1,\ldots,p_n)}$ has rank n, than the n equations

$$\Phi_1(q,p) = \alpha_1, \ldots, \Phi_n(q,p) = \alpha_n \tag{D.2.4}$$

describe, when varying the parameters α, an n-dimensional foliation Λ_α of T^*Q.

THEOREM D.4 *The foliation Λ_α is Lagrangian if and only if the functions Φ_k are in* involution, *that is* $\{\Phi_h, \Phi_k\} = 0$, $h, k = 1, \ldots, n$.

Proof. From Definition (A.5.4) of Poisson bracket, one immediately deduces

$$\{f, g\} = \Omega(df^\sharp, dg^\sharp).$$

If the two functions are in involution, then the vector $df^\sharp$ is tangent to the manifold g = constant and $dg^\sharp$ to f = constant, since

$$\{f, g\} = \mathcal{L}_{df^\sharp} g = -\mathcal{L}_{dg^\sharp} f = 0.$$

Therefore, the theorem follows from

$$\{\Phi_h, \Phi_k\} = \Omega(d\Phi_h^\sharp, d\Phi_k^\sharp)$$

and from the independence of $d\Phi_1^\sharp, \ldots, d\Phi_n^\sharp$, which span the tangent space to the level manifold Λ_α of (D.2.4). QED

The relation between the functions Φ_k, the 1-forms λ_α and the complete integral $W(q,\alpha)$ should now be clear. If $\det\left(\frac{\partial \Phi_k}{\partial p_h}\right) \neq 0$, one can invert, at least locally, the n Equations (D.2.4) to find $p_k = \not{p}_k(q,\alpha)$, with which one constructs the 1-forms $\lambda_\alpha = \not{p}_k(q,\alpha)\, dq^k$. Varying the parameters α_k, these 1-forms describe a foliation Λ_α of sections of T^*Q that are Lagrangian, since by hypothesis the functions Φ_k are in involution.

[(4)] A *foliation* of a manifold M is a decompositiion of M into disjoint connected immersed submanifolds of constant dimension less than that of M.

Thus a generating function $W_\alpha \equiv W(q,\alpha)$ exists such that $\lambda_\alpha = dW_\alpha$. To find explicitly the complete integral W_α one must integrate

$$W_\alpha = \int_{q_0}^{q} \lambda_\alpha = \int_{q_0}^{q} \lambda_\alpha^* \Theta = \int_{q_0}^{q} \not{p}_k(q',\alpha)\, dq'^k \qquad \text{(D.2.5)}$$

along any path joining q_0 with q in the configuration manifold Q. That W_α is a complete integral, generating the canonical transformation

$$(p,q) \mapsto (\alpha,\beta):\ p_k = \frac{\partial W(q,\alpha)}{\partial q^k} = \not{p}_k(q,\alpha), \quad \beta^h = \frac{\partial W(q,\alpha)}{\partial \alpha_h},$$

is assured by the fact that the first relation is invertible with respect to the α's (indeed $\alpha_h = \Phi_h(q,p)$), hence $\det\left(\frac{\partial \not{p}_k}{\partial \alpha_h}\right) = \det\left(\frac{\partial^2 W(q,\alpha)}{\partial q^k \partial \alpha_h}\right) \neq 0$.

Since, in general, $\Phi_h(q,p) = \alpha_h$ can not be everywhere inverted with respect to the momenta, W_α is a local object. We can also consider the function $W_{\Lambda_\alpha} : \Lambda_\alpha \to \mathbb{R}$ given by

$$W_{\Lambda_\alpha} = \int_{\mathrm{P}_0}^{\mathrm{P}} \Theta,$$

where the integration path from P_0 to P lies completely on Λ_α. W_{Λ_α} is in general a multivalued function defined on Λ_α. The multivaluedness of W_{Λ_α} is due to the fact that Θ is closed on Λ_α but, if the first cohomology group of Λ_α is not trivial, fails to be exact. As seen

$$W_\alpha = W_{\Lambda_\alpha} \circ \lambda_\alpha,$$

that is W_α is the projection of W_{Λ_α} on the base manifold Q.

THEOREM D.5 (LIOUVILLE) *Given a Hamiltonian $H(q,p)$, let us suppose to know n first integrals $\Phi_k(q,p)$ in involution, with $d\Phi_k$ all independent and* $\det\left(\frac{\partial \Phi_k}{\partial p_h}\right) \neq 0$. *Then the general solution of the Hamilton Equations is found through quadratures, i.e., integrations and inversions.*

Proof. After the previous geometric analysis, the proof of the theorem is almost obvious. The first integrals in involution define a Lagrangian foliation, and (D.2.5) allows us to find the generating function of a canonical transformation for which the new momenta coincide with the first integrals. The Hamiltonian is in involution with the first integrals since

$$0 = \frac{d\Phi_k}{dt} = \frac{\partial \Phi_k}{\partial p_h}\dot{p}_h + \frac{\partial \Phi_k}{\partial q^h}\dot{q}^h = \{H, \Phi_k\},$$

which implies that the transformed Hamiltonian K does not depend on the new coordinates. The relative Hamilton Equations are

$$\frac{dP_k}{dt} = -\frac{\partial K(P)}{\partial Q^k} = 0, \quad \frac{dQ^h}{dt} = \frac{\partial K(P)}{\partial P_h},$$

which are trivially integrable. QED

D.2.4 The Time-dependent Case

In this subsection we consider Hamiltonians that are explicitly time-dependent.

We try to reduce this case to the time-independent one, by adding to the coordinates and momenta of the phase space also time t and conjugate momentum p_t. The enlarged phase space T^*M becomes $(2n+2)$-dimensional, but one must take into account that p_t can not be assumed as an independent coordinate: the previous work on Lagrangian dynamics shows that

$$p_t + H(q, p, t) = 0.$$

This relation appears as a constraint in the enlarged phase space, so that the dynamics is forced to take place on a $(2n+1)$-dimensional hypersurface, called the *evolution space.* Let i be the inclusion map. The pull-back of the canonical 1-form and 2-form of the enlarged phase space to this constrained hypersurface is

$$\begin{aligned} i^*\Theta &= p_\mu dq^\mu - H(q,p,t)\, dt, \\ i^*\Omega &= dp_\mu \wedge dq^\mu - dH(q,p,t) \wedge dt. \end{aligned}$$

$i^*\Theta$ is called the *Poincaré-Cartan 1-form.*

The matrix of the 2-form $i^*\Omega$ is

$$\begin{pmatrix} \mathbf{0}_n & -\mathbf{1}_n & -\partial_q H \\ \mathbf{1}_n & \mathbf{0}_n & -\partial_p H \\ \partial_q H & \partial_p H & 0 \end{pmatrix}$$

and has rank $2n$, thus has a 1-dimensional kernel. The lines, whose tangent vector belongs to this 1-dimensional subspace, are called the *characteristic lines* of the Poincaré-Cartan 1-form, and, as it is easy to check, satisfy the equations

$$\frac{dq^\mu}{dt} = \frac{\partial H}{\partial p_\mu}, \quad \frac{dp_\nu}{dt} = -\frac{\partial H}{\partial q^\nu},$$

that is the Hamiltonian Equations. The space of the characteristic lines is the *space of motions.*

A transformation $(q, p, H) \mapsto (Q, P, K)$ (notice that the time is not involved) is canonical if

$$p_\mu dq^\mu - H\, dt = P_\mu dQ^\mu - K\, dt + dF(q, Q, t).$$

Expanding the differential of the generating function we find

$$p_\mu = \frac{\partial F}{\partial q^\mu}, \quad P_\mu = -\frac{\partial F}{\partial Q^\mu}, \quad K = H + \frac{\partial F}{\partial t},$$

which define, if the usual condition on the Hessian is satisfied, a time-dependent invertible canonical transformation. Alternatively, one may consider generating functions of the type $F(q, P, t)$.

Let us consider the partial differential equation of the first order in the unknown function $S(q, P, t)$

$$H\left(q, \frac{\partial S}{\partial q}, t\right) + \frac{\partial S}{\partial t} = 0,$$

which is called the *Hamilton-Jacobi Equation.* If one is able to find a complete integral, thus a canonical transformation sending the Hamiltonian H into $K \equiv 0$, the dynamical problem is solved. Indeed, $P_\mu = \alpha_\mu$ and $Q^\nu = \beta^\nu$, where α_μ, β^ν are integration constants, integrates the trivial Hamilton Equations. Inverting the time-dependent canonical transformation, one finds

$$q^\mu = q^\mu(t, \alpha, \beta), \quad p_\mu = p_\mu(t, \alpha, \beta),$$

that is the general integral of the Hamilton Equations. The dynamical evolution is therefore viewed as the continuous unfolding of a family of canonical transformations, parametrized by the time.

The time-independent case is recovered simply by writing

$$S(q, \alpha, t) = W(q, \alpha) - \alpha_1 t, \quad \alpha_1 = E.$$

REMARK D.1 In point of fact, one can consider more general transformations than those viewed in this subsection and that are reported in any textbook. More exactly, one can define the *extended canonical transformations* as those leaving invariant the 2-form $dp_\mu \wedge dq^\mu + dp_t \wedge dt$, and enforce the constraint $p_t + H = 0$ only *after* the transformation: in this way, the time does not play the role of a simple parameter, but enters into the transformation like all the other coordinates. A transformation of this type has been used in Chapter 6 to show the equivalence between the Kepler Problem and the geodesic motion on the sphere, on the Euclidean space and on the hyperboloid for the various values of the energy, respectively.

D.3 Symmetries and Reduction

In this section we want to generalize the Noether Theorem, in a time independent Hamiltonian environment, to the case of an N-dimensional Lie group $\mathfrak{G}$ of symmetries.

D.3.1 The Moment Map

Let $\Phi : M \to M$ be a symplectomorphism of a $2n$-dimensional symplectic manifold (M, Ω). Obviously, Φ lies invariant the Poisson brackets:

$$\Phi^*\Omega = \Omega \Leftrightarrow \{\Phi^* f, \Phi^* g\} = \Phi^*\{f, g\}. \tag{D.3.1}$$

Let us now consider the left action $u \mapsto \Phi_x u$, $u \in M$, $x \in \mathfrak{G}$ of $\mathfrak{G}$ on M. Let

$$X_a = \left.\frac{\partial \Phi_x}{\partial x^a}\right|_{x=1}, \quad a = 1, \ldots, N$$

be a basis for the vector space of the infinitesimal generators of the action. We know from Proposition B.5 that they form a Lie algebra that is anti-homomorphic to the Lie algebra $\mathfrak{g}$. In other words, if $\xi = \xi^a L_a$, $\eta = \eta^a L_a$ are two elements of $\mathfrak{g}$, and $X_\xi = \xi^a X_a$, $X_\eta = \eta^a X_a$ the corresponding vector fields on M, we have

$$[X_\xi, X_\eta] = -X_{[\xi,\eta]}, \quad \xi, \eta \in \mathfrak{g}. \tag{D.3.2}$$

Now suppose that the $\mathfrak{G}$-action on M is a symplectomorphism for every $x \in \mathfrak{G}$. Recalling the relation (A.3.4), we can write

$$\mathcal{L}_{X_\xi}\Omega = i_{X_\xi} d\Omega + d(i_{X_\xi}\Omega).$$

But $\mathcal{L}_{X_\xi}\Omega = 0$, which is the "infinitesimal version" of (D.3.1), moreover $d\Omega = 0$, by definition of symplectic manifold. Therefore the 1-form $i_{X_\xi}\Omega$ is closed, hence locally exact, i.e., for a fixed ξ, there exists locally a function $J_\xi : M \to \mathbb{R}$ such that

$$i_{X_\xi}\Omega = -dJ_\xi, \quad \text{or } X_\xi = dJ_\xi^\sharp.$$

Thus, the infinitesimal generators of the action are local Hamiltonian vectors, with Hamiltonian J_ξ. If, as we suppose hereafter, the vector fields X_ξ are globally Hamiltonian, the action Φ is said to be *Hamiltonian*.

For a fixed point $u \in M$, the dependence of $J_\xi(u)$ from the element of the Lie algebra is linear. Indeed, since the basis vector fields X_a on M are Hamiltonian, $X_a = dJ_a^\sharp$ holds for some functions J_a, $a = 1, \ldots, N$. If $\xi = \xi^a L_a$, then X_ξ is Hamiltonian, with Hamiltonian $J_\xi = \xi^a J_a$. Since the dual of a vector space is the space of the linear functions, we can define a function $J : M \to \mathfrak{g}^*$ such that

$$J_\xi(u) = \langle J(u), \xi \rangle .$$

J is called a *moment map*.

Let us find the basic property of the moment map. Take, in expression (A.5.5), $f = J_\xi$ and $g = J_\eta$, then (since $X_\xi = X_{J_\xi}$ and $X_\eta = X_{J_\eta}$) we obtain, comparing with (D.3.2), that

$$X_{\{J_\xi, J_\eta\}} = -X_{J_{[\xi,\eta]}}.$$

The operator $\sharp$ is an isomorphism, and the above relation implies that the two functions $-J_{[\xi,\eta]}$ and $\{J_\xi, J_\eta\}$ have the same gradient, thus differ by a quantity constant on M

$$\{J_\xi, J_\eta\} = -J_{[\xi,\eta]} + C(\xi, \eta).$$

Projecting this relation on the basis $\{L_a\}$, we can write

$$\{J_a, J_b\} = -c^c_{ab} J_c + C_{ab}, \quad a, b, c = 1, \ldots, N.$$

The skew symmetric bilinear function $C(\xi_1, \xi_2)$ is called a *2-cocycle of the Lie algebra*, and must satisfy the relation

$$C([\xi_1, \xi_2], \xi_3) + C([\xi_2, \xi_3], \xi_1) + C([\xi_3, \xi_1], \xi_2) = 0,$$

equivalent to the Jacobi identity. When the cocycle is zero, there is an isomorphism between the Lie algebra of the infinitesimal generators X_a and the Lie algebra (with respect to the Poisson brackets) of the Hamiltonians J_a : the moment map is said to be *Ad*-equivariant*. This name is due to the fact that the equation $\{J_a, J_b\} = -c^c_{ab} J_c$ is the "infinitesimal" version of the statement that the following diagram is commutative

$$\begin{array}{ccc} M & \xrightarrow{\Phi_x} & M \\ J \big\downarrow & & \big\downarrow J \\ \mathfrak{g}^* & \xrightarrow{\mathbf{Ad}^*_x} & \mathfrak{g}^* \end{array} \qquad \text{i.e., } J(\Phi_x u) = \mathbf{Ad}^*_x J(u).$$

Indeed, projecting this commutativity relation on the basis $\{L_a\}$, we obtain, by Definition (B.2.1) of coadjoint representation,

$$J_a(\Phi_x u) = (\mathbf{Ad}_x^{-1})^c_a J_c(u). \tag{D.3.3}$$

Differentiating with respect to x^b and putting $x = 1$, then taking into account (B.2.2) and the skew symmetry of Poisson bracket and structure constants, we find

$$\{J_a, J_b\} = -c^c_{ab} J_c. \tag{D.3.4}$$

This shows that, if the diagram is commutative, (D.3.4) holds.

For the connected component of the identity of $\mathfrak{G}$, also the vice versa holds. Indeed, Equation (D.3.4) asserts that both members of (D.3.3) have the same derivative at the identity, moreover the two members of (D.3.3) agree at the identity: for the homogeneity of the group, the statement follows. Hereafter, we will consider only Ad*-equivariant maps.

EXAMPLE D.2 An important example is that arising when a group acts on the configuration space Q and the symplectomorphism Φ is an extended point transformation. The infinitesimal generator X_ξ is a vector field on Q and the relative Ad*-equivariant moment map is $J_\xi = \left\langle \Theta, X_\xi \right\rangle$, in coordinates $J_\xi(q,p) = p_\mu X_\xi^\mu(q)$. Indeed, a trivial calculation shows that

$$\{J_\xi, J_\eta\} = [X_\xi, X_\eta]^\mu p_\mu,$$

which ensures that (D.3.4), that is the infinitesimal Ad*-equivariance, holds.

If the Φ action on M is transitive, for the Ad*-equivariance also the coadjoint action of $\mathfrak{G}$ on $J(M)$ is so, and M can be identified with (a cover of) a coadjoint orbit of $\mathfrak{G}$ on $\mathfrak{g}^*$. This is the content of the Kirillov-Kostant-Souriau Theorem (Guillemin & Sternberg 1990).

Let us now suppose that on the symplectic manifold M a dynamics is given, which is invariant with respect to the group $\mathfrak{G}$, i.e., a Hamiltonian $H : M \to \mathbb{R}$ is assigned such that

$$H \circ \Phi_x = H, \quad \forall x \in \mathfrak{G}.$$

Differentiating with respect to the elements of $\mathfrak{G}$ at the identity, and noticing that the right-hand side does not depend on $\mathfrak{G}$, we find

$$\frac{\partial}{\partial x^a} H(\Phi_x u)\bigg|_{x=1} = \left\langle dH, dJ_a^\sharp \right\rangle = \{J_a, H\} = 0 \quad a = 1, \ldots, N.$$

The components of the moment map relative to the symmetry group $\mathfrak{G}$ are therefore first integrals. The expression

$$\{J_a, H\} = \mathcal{L}_{dJ_a^\sharp} H = -\mathcal{L}_{dH^\sharp} J_a = 0.$$

can be interpreted in two "complementary" ways. In the first, the Hamiltonian is invariant along the flows generated by the $\mathfrak{G}$-action: hence the Hamiltonian is invariant with respect to the group action. In the latter, the components of the moment map are invariant along the time flow generated by the Hamiltonian: hence the J_a's are first integrals. This relation between symmetries and conservation laws is the *Generalized Noether Theorem.*

But we can go even further. In the Lagrangian environment a one-parameter group of symmetry allows us to reduce by one the number of freedom degrees of a mechanical system. Also in the Hamiltonian environment an N-dimensional Lie group of symmetries allows us to proceed to a reduction. This is the content of the Marsden-Weinstein Theorem (Marsden & Weinstein 1974), concerning the reduction in the symplectic case, and of its generalization to the Poisson case (Marsden & Ratiu 1986).

D.3.2 Reduction of Symplectic Manifolds

Let us define

$$J^{-1}(\mu) = \{u \in M : J(u) = \mu,\ \mu \in \mathfrak{g}^*\}.$$

We say that μ is a *regular value* if $dJ|_u$ is surjective for every $u \in J^{-1}(\mu)$; in such a case $J^{-1}(\mu)$ is a differentiable submanifold of M. The action of $\mathfrak{G}$ on M in general moves this submanifold, while that of the isotropy subgroup

$$\mathfrak{G}_\mu \stackrel{\text{def}}{=} \{x \in \mathfrak{G} : \mathbf{Ad}_x^* \mu = \mu\},$$

leaves it invariant. Indeed, if $J(u) = \mu$ and $x \in \mathfrak{G}_\mu$, for the Ad*-equivariance we have $J(\Phi_x u) = \mathbf{Ad}_x^* J(u) = \mathbf{Ad}_x^* \mu = \mu$. If the action of $\mathfrak{G}_\mu$ on $J^{-1}(\mu)$ is free and proper, then the space of the orbits $M_\mu = J^{-1}(\mu)/\mathfrak{G}_\mu$ is a differentiable manifold. The restriction of the symplectic 2-form Ω to $J^{-1}(\mu)$ is still a closed 2-form, which is however in general degenerate, i.e., the induced linear map $\flat : TM \to T^*M$ has a nontrivial kernel. The content of the Marsden-Weinstein Theorem is essentially that this kernel is spanned by the tangent vectors to the orbit of the $\mathfrak{G}_\mu$-action on $J^{-1}(\mu)$, so that, passing to the quotient, we obtain that M_μ is symplectic. M_μ is said to be the *symplectic reduced manifold* of M. Moreover, if the Hamiltonian on M is invariant with respect to $\mathfrak{G}$-action, the dynamics projects down to M_μ.

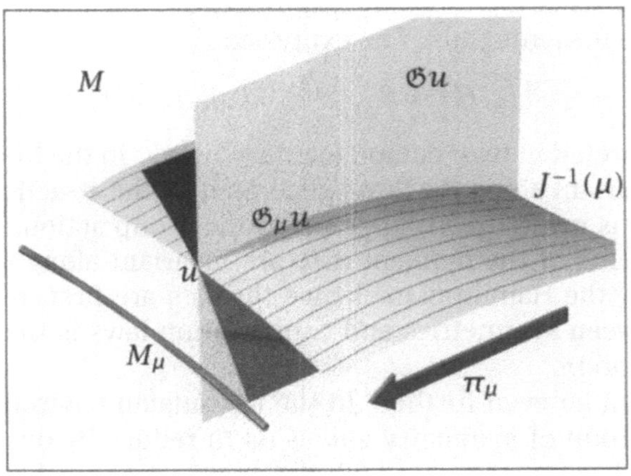

Figure D.1: Symplectic reduced manifold.

Before the theorem, we state and prove the following lemma; see Figure D.1.

LEMMA D.6 *Let $u \in M$ be such that $J(u) = \mu$, and $\mathfrak{G}u, \mathfrak{G}_\mu u$ be the orbits in M of the full group $\mathfrak{G}$ and of its isotropy subgroup $\mathfrak{G}_\mu$, respectively. Then*

(i) *The two linear subspaces $T_u(J^{-1}(\mu))$ and $T_u(\mathfrak{G}u)$ of T_uM are Ω-orthogonal complements of each other;*

(ii) *$T_u(\mathfrak{G}_\mu u) = T_u(\mathfrak{G}u) \cap T_u(J^{-1}(\mu))$.*

Proof. (i) In the relation

$$\Omega(v, X_\xi) = \mathcal{L}_v J_\xi, \quad \xi \in \mathfrak{g}, \; v \in T_uM$$

the right-hand side vanishes if and only if v has a direction along which the moment map is constant, that is if $v \in T_u(J^{-1}(\mu))$. Taking into account that in the left-hand side the vectors $\{X_\xi, \xi \in \mathfrak{g}\}$ span the tangent space $T_u(\mathfrak{G}u)$ to the orbit of the action of $\mathfrak{G}$ over M, we see that the Ω-orthogonal space to $T_u(\mathfrak{G}u)$ is $T_u(J^{-1}(\mu))$. Vice versa, in the relation

$$\Omega(w, X_F) = \mathcal{L}_w F, \quad w \in T_u(J^{-1}(\mu)), \; F : M \to \mathbb{R}$$

the right-hand side vanishes if and only if the function F is constant on the manifold $J^{-1}(\mu)$, that is if F is a function of the components J_a's of the moment map. In this case, when varying F, X_F spans $T_u(\mathfrak{G}u)$ and $T_u(\mathfrak{G}u)$ turns out to be Ω-orthogonal to $T_u(J^{-1}(\mu))$.

(ii) We must prove that $X_\xi \in T_u(J^{-1}(\mu))$ if and only if $\xi \in \mathfrak{g}_\mu$. This is an immediate consequence of the Ad*-equivariance of the moment map. Indeed X_ξ is tangent to $J^{-1}(\mu)$ if and only if it is generated by the action of those elements of $\mathfrak{G}$ that do not move the submanifold $J^{-1}(\mu)$, thus of those belonging to $\mathfrak{G}_\mu$. QED

THEOREM D.7 (MARSDEN-WEINSTEIN) *Let (M, Ω) be a symplectic manifold on which the Lie group $\mathfrak{G}$ acts symplectically and let $J : M \to \mathfrak{g}^*$ be an Ad*-equivariant moment map for this action. Assume $\mu \in \mathfrak{g}^*$ is a regular value of J and that the action of the isotropy subgroup $\mathfrak{G}_\mu$ on $J^{-1}(\mu)$ is free and proper. Then $M_\mu = J^{-1}(\mu)/\mathfrak{G}_\mu$ has a unique symplectic form Ω_μ with the property*

$$\pi_\mu^* \Omega_\mu = i_\mu^* \Omega, \tag{D.3.5}$$

where $\pi_\mu : J^{-1}(\mu) \to M_\mu$ is the canonical projection and $i_\mu : J^{-1}(\mu) \to M$ the inclusion.

The restriction of a $\mathfrak{G}$-invariant Hamiltonian H to $J^{-1}(\mu)$ induces a Hamiltonian H_μ on the reduced manifold M_μ, called the reduced Hamiltonian, *defined by*

$$\pi_\mu^* H_\mu = i_\mu^* H.$$

Then, the vector field X_H on $J^{-1}(\mu)$ is π_μ-related to the reduced vector field X_{H_μ} on M_μ, that is

$$\mathbf{T}\pi_\mu X_H = X_{H_\mu} \circ \pi_\mu .$$

Proof. A consequence of the previous lemma is that the kernel of the 2-form $\Omega|_{J^{-1}(\mu)} = i_\mu^* \Omega$ is $T_u(\mathfrak{G}_\mu u)$ for every point $u \in J^{-1}(\mu)$. Indeed, for (i) a vector of $T_u M$ is Ω-orthogonal to a vector of $T_u(J^{-1}(\mu))$ if and only if it belongs to $T_u(\mathfrak{G}u)$. For (ii), a vector of $T_u(\mathfrak{G}u)$ belongs to $T_u(J^{-1}(\mu))$ if and only if it belongs also to $T_u(\mathfrak{G}_\mu u)$. Therefore, $i_\mu^*\Omega(w,v) = i_\mu^*\Omega(w,v')$ if and only if $(v - v') \in T_u(\mathfrak{G}_\mu u)$, so that a well defined closed 2-form Ω_μ, satisfying (D.3.5), is defined on M_μ. The first assertion of the theorem is proved.

J is an integral for X_H. It follows that $J^{-1}(\mu)$ is invariant under the flow generated by H, so we can restrict ourselves to $J^{-1}(\mu)$. The relation between H and X_H is, by the mere definition of Hamiltonian vector field,

$$\Omega(X_H) = dH, \tag{D.3.6}$$

and analogously

$$\Omega_\mu(X_{H_\mu}) = dH_\mu . \tag{D.3.7}$$

It is immediate to check that the π_μ pull-back of (D.3.7) yields the restriction to $J^{-1}(\mu)$ of (D.3.6) if and only if the two vector fields X_H and X_{H_μ} are π_μ-related. The second assertion is so proved. QED

As for the dimension of the reduced manifold, $\dim M_\mu = \dim M - \dim \mathfrak{G} - \dim \mathfrak{G}_\mu$. The completely integrable systems may be viewed as a particular case, with $\mathfrak{G} = \mathfrak{G}_\mu = \mathbb{R}^n$, while the moment map is given by the first integrals in involution and $\dim M_\mu = 0$.

D.3.3 Reduction of Poisson Manifolds

In order to generalize the result of the previous subsection to the more general context of the Poisson manifolds, we now report almost verbatim the proof of Marsden & Ratiu (1986). The symplectic reduction turns out to be a particular, more intuitive case, which we have put before for didactic purpose.

First of all, let us recast the basic formula (D.3.5) of the symplectic reduction in the language of Poisson structure. Let f and g be real valued functions on the reduced manifold M_μ and let F and G be $\mathfrak{G}$-invariant extensions off $J^{-1}(\mu)$ of $f \circ \pi_\mu$ and $g \circ \pi_\mu$. Then (D.3.5) is equivalent to

$$\{F,G\}_M \circ i_\mu = \{f,g\}_{M_\mu} \circ \pi_\mu , \tag{D.3.8}$$

where obviously $\{.,.\}_M$ and $\{.,.\}_{M_\mu}$ are the Poisson brackets relative to the Poisson bivectors $\Omega^\#$ and $\Omega^\#_\mu$, respectively.

Let $(P, \{.,.\}_P)$ be a Poisson manifold, $N \subset P$ a submanifold and $i : N \to P$ the inclusion. Let $\mathfrak{B} \subset TP|_N$ be a distribution of the tangent bundle of P restricted to N.

We make the following three assumptions.

(A1) $\mathfrak{B} \cap TN$ is an integrable distribution of TN, so defines a foliation Ψ on N.

(A2) The foliation Ψ is regular, so the space of leaves N/Ψ is a manifold with projection $\pi : N \to N/\Psi$ a submersion.

(A3) The bundle $\mathfrak{B}$ leaves $\{.,.\}_P$ invariant in the sense that if K, L are smooth functions on P with differentials vanishing on $\mathfrak{B}$, then also $d\{K, L\}_P$ vanishes on $\mathfrak{B}$.

Then, we give the definition of Poisson reducibility.

Definition D.8 *Bearing in mind (D.3.8), we say that* $(P, N, \mathfrak{B})$ *is* Poisson reducible *if* N/Ψ *has a Poisson structure* $\{.,.\}_{N/\Psi}$ *such that, for any smooth functions* f, g *on* N/Ψ, *and smooth extensions* F, G *of* $f \circ \pi$, $g \circ \pi$ *with differentials vanishing on* $\mathfrak{B}$, *we have*

$$\{F, G\}_P \circ i = \{f, g\}_{N/\Psi} \circ \pi. \tag{D.3.9}$$

The following table shows the correspondence between the terms in the two cases of symplectic and Poisson reduction.

Symplectic reduction	Poisson reduction
M	P
$J^{-1}(\mu)$	N
$T(\mathfrak{G}u)$	$\mathfrak{B}$
Orbits of $\mathfrak{G}_\mu$ on $J^{-1}(\mu)$	Ψ
$M_\mu = J^{-1}(\mu)/\mathfrak{G}_\mu$	N/Ψ
F, G $\mathfrak{G}$-invariant extensions of $f \circ \pi_\mu$ and $g \circ \pi_\mu$	F, G extensions of $f \circ \pi, g \circ \pi$ whose differentials vanish on $\mathfrak{B}$
$\{F, G\}_M \circ i_\mu = \{f, g\}_{M_\mu} \circ \pi_\mu$	$\{F, G\}_P \circ i = \{f, g\}_{N/\Psi} \circ \pi$

Now we state and prove the central theorem of this subsection.

THEOREM D.9 (POISSON REDUCTION) *Let assumptions (A1) - (A3) hold and regard the Poisson structure on P as a map $\Omega^\# : T^*P \to TP$. The triplet $(P, N, \mathcal{S})$ is* Poisson reducible *if and only if*

$$\Omega^\# \left(\mathcal{S}^0\right) \subset TN + \mathcal{S}, \tag{D.3.10}$$

where $\mathcal{S}_u^0 = \{\alpha_u \in T_u^*P : \langle \alpha_u, \mathcal{S}_u \rangle = 0\}$ is the annihilator of $\mathcal{S}$ in $u \in P$.

Proof. First assume that $(P, N, \mathcal{S})$ is Poisson reducible. Let $\beta_u \in \mathcal{S}_u^0$, where $u \in N$ and let F be C^∞ in a neighbourhood of u such that dF vanishes on $\mathcal{S}$ and equals β_u at u. Let $\alpha_u \in \mathcal{S}_u^0 \cap (T_uN)^0 = (\mathcal{S}_u + T_uN)^0$ and choose an extension K of the zero function on N such that $dK|_u = \alpha_u$ and dK vanishes on $\mathcal{S}$. Thus, if f is the function on N/Ψ induced from $F|_N$,

$$\left\langle \alpha_u, \Omega^\#(\beta_u) \right\rangle = \{K, F\}_P|_u = \{0, f\}_{N/\Psi}|_{\pi u} = 0,$$

where the second equality holds since we have assumed (D.3.9) by hypothesis. Thus $\Omega^\#(\beta_u) \in (TN + \mathcal{S})|_u$.

Conversely, assume $\Omega^\# \left(\mathcal{S}^0\right) \subset TN + \mathcal{S}$. Let $f, g \in C^\infty(N/\Psi)$ and let F, G be extensions of $f \circ \pi, g \circ \pi$ whose differentials vanish on $\mathcal{S}$. By condition (A3), $\{F, G\}_P$ is constant on the leaves of Ψ and so induces a function on N/Ψ. We now show that this induced function is independent of the extension, thereby defining a function $\{f, g\}_{N/\Psi}$ satisfying (D.3.9). Let G' be another extension of $g \circ \pi$ whose differential vanishes on $\mathcal{S}$. Thus, $G - G'$ vanishes on N, therefore $\langle d(G - G'), TN \rangle = 0$ and $\langle d(G - G'), TN + \mathcal{S} \rangle = 0$. Since $dF \in \mathcal{S}^0$, by (D.3.10) we have $\Omega^\#(dF) \subset TN + \mathcal{S}$, from which

$$\left\langle d(G - G'), \Omega^\#(dF) \right\rangle \Big|_u = 0 \quad \Rightarrow \quad \{F, G\}_P|_u = \{F, G'\}_P|_u .$$

Thus. $\{F, G\}_P|_u$ is independent of how $g \circ \pi$ is extended, as long as the differential of the extension vanishes along $\mathcal{S}$. By skew symmetry of the bracket, it is also independent of the $f \circ \pi$ extension. Thus $\{f, g\}_{N/\Psi}$ is well defined and is uniquely determined by (D.3.9). It remains to show that N/Ψ is a Poisson manifold with this bracket, i.e., Jacobi identity holds. To this end, notice that if F, G are extensions of $f \circ \pi, g \circ \pi$ whose differentials vanish on $\mathcal{S}$, then so does $d\{F, G\}_P$ by (A3), so $\{F, G\}_P$ is an extension of $\{f, g\}_{N/\Psi} \circ \pi$ that we can use in (D.3.9) to give the identity

$$\{\{f, g\}_{N/\Psi}, k\}_{N/\Psi} \circ \pi = \{\{F, G\}_P, K\}_P \circ i$$

and the Jacobi identity on N/Ψ is inherited from that on P. QED

Moreover, we can define the reduction of a Poisson map between reducible Poisson manifolds and show that it is still a Poisson map.

Corollary D.10 *Let* $(P_i, N_i, \mathscr{G}_i)$, $i = 1,2$ *be Poisson reducible, and assume that* $\Phi : P_1 \to P_2$ *is a Poisson map such that* $\Phi(N_1) \subset N_2$, $\mathbf{T}_\Phi \mathscr{G}_1 \subset \mathscr{G}_2$, *and* Φ *maps the leaves of* Ψ_1 *into leaves of* Ψ_2. *Then* Φ *induces a unique Poisson map* $\phi : N_1/\Psi_1 \to N_2/\Psi_2$, *called* the reduction of Φ, *making the diagram*

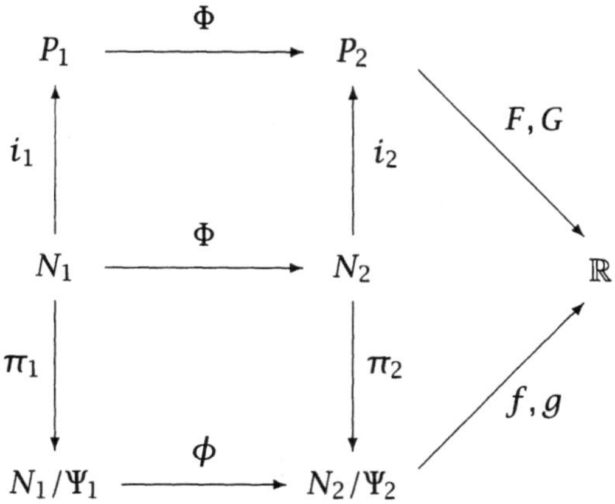

commutative, i.e., satisfying $\pi_2 \circ \Phi = \phi \circ \pi_1$.

Proof. Since Φ maps leaves into leaves, the map ϕ exists and is unique with the commutativity property. To show that it is Poisson, take $f, g : N_2/\Psi_2 \to \mathbb{R}$ and let $F, G : P_2 \to \mathbb{R}$ be extensions of $f \circ \pi_2$ and $g \circ \pi_2$ whose differentials vanish on $\mathscr{G}_2$. For any $v \in \mathscr{G}_1$ we have

$$\langle d(F \circ \Phi), v \rangle = \langle dF, \mathbf{T}_\Phi v \rangle = 0,$$

since $\mathbf{T}_\Phi \mathscr{G}_1 \subset \mathscr{G}_2$. Therefore, $F \circ \Phi$ and $G \circ \Phi$ are extensions of $f \circ \phi$ and $g \circ \phi$ whose differentials vanish on $\mathscr{G}_1$, and

$$\begin{aligned}\{f, g\}_{N_2/\Psi_2} \circ \phi \circ \pi_1 &= \{f, g\}_{N_2/\Psi_2} \circ \pi_2 \circ \Phi = \{F, G\}_{P_2} \circ i_2 \circ \Phi \\ &= \{F, G\}_{P_2} \circ \Phi \circ i_1 = \{F \circ \Phi, G \circ \Phi\}_{P_1} \circ i_1 \\ &= \{f \circ \phi, g \circ \phi\}_{N_1/\Psi_1} \circ \pi_1.\end{aligned}$$

The first three equalities holds for the commutativity of the diagram and since we have chosen F, G extensions of f, g with $(P_i, N_i, \mathscr{G}_i)$ Poisson reducible, thus (D.3.9) holds. The fourth equality holds since by assumption Φ is a Poisson map. The last equality is deduced from (D.3.9), bearing in mind that $F \circ \Phi$ and $G \circ \Phi$ are extensions of $f \circ \phi$ and $g \circ \phi$. Comparing the first and last term and taking into account that π_1 is surjective, shows that ϕ is a Poisson map. QED

Next we study the dynamical counterpart of the Poisson reduction Theorem. If $H : P \to \mathbb{R}$ is a Hamiltonian, we say that the submanifold $N \subset P$ is *conserved* for H if the relative Hamiltonian vector field is tangent to N in every point. For example, if $\Omega^\sharp\left(\mathcal{S}^0\right) \subset TN$, then N is conserved for all Hamiltonians whose differentials annihilate $\mathcal{S}$.

THEOREM D.11 (DYNAMIC POISSON REDUCTION) *Let $(P, N, \mathcal{S})$ be Poisson reducible and $H : P \to \mathbb{R}$ be a Hamiltonian for which N is conserved and that preserves the foliation Ψ. Then the flow Φ_t of X_H induces Poisson diffeomorphisms ϕ_t on N/Ψ. The vector field on N/Ψ whose flow is ϕ_t equals X_h, where $h : N/\Psi \to \mathbb{R}$ is uniquely determined by H via $h \circ \pi = H \circ i$, where $i : N \to P$ is the inclusion. In addition, the vector fields $X_H \circ i$ and X_h are π-related.*

Proof. Since N is conserved for H, the flow Φ_t of X_H leaves N invariant, moreover Ψ is preserved, so by the previous corollary ϕ_t is a flow of Poisson diffeomorphisms on N/Ψ.

Now let $f : N/\Psi \to \mathbb{R}$ be an arbitrary map and let $F : P \to \mathbb{R}$ be an extension of $f \circ \pi$ to P such that $dF \in \mathcal{S}^0$. Then for any $u \in N$ we have

$$\begin{aligned}\langle df, X_h\rangle \circ \pi &= \{h, f\}_{N/\Psi} \circ \pi = \{H, F\}_P \circ i\\ &= \langle dF, X_H\rangle \circ i = \langle df, \mathbf{T}_\pi X_H\rangle \circ i\end{aligned}$$

and so $X_h \circ \pi = \mathbf{T}_\pi X_H \circ i$, that is $X_H \circ i$ and X_h are π-related. QED

EXAMPLE D.3 Let $J : P \to \mathfrak{g}^*$ an Ad*-equivariant moment map for the action of a Lie group preserving the Poisson structure. As in the symplectic case, let μ be a regular value of J, $N = J^{-1}(\mu)$ and $\mathcal{S}$ the tangent space to the $\mathfrak{G}$-orbits. Then $(P, N, \mathcal{S})$ is Poisson reducible. To prove it, we observe that (A1) is satisfied since $\mathcal{S} \cap TN$ consists of the tangent spaces to the $\mathfrak{G}_\mu$-orbits by equivariance of J. (A2) holds if $\mathfrak{G}_\mu$ acts freely and properly, so $N/\Psi = J^{-1}(\mu)/\mathfrak{G}_\mu$ is a manifold. (A3) holds since the $\mathfrak{G}$-action preserves the Poisson structure and $\Omega^\sharp(\mathcal{S}^0) \subset T(J^{-1}(\mu))$ (so (D.3.10) holds) since J is conserved for $\mathfrak{G}$-invariant Hamiltonians. Therefore, $N/\Psi = J^{-1}(\mu)/\mathfrak{G}_\mu$ inherits a unique Poisson structure satisfying (D.3.9).

D.4 Action-Angle Variables

Liouville's Theorem D.5 states that if n independent first integrals Φ_k in involution of a Hamiltonian $H(p, q)$ are known, the problem (which is called *completely integrable*) is reduced to the quadratures. This is a local statement, as is obvious from the fact that the inversion of the relations $\Phi_k(p,q) = \alpha_k$ with respect to the momenta is required.

But more can be said. The n first integrals are the Hamiltonians of n commuting flows and this fact imposes a drastic condition on the topology of the surfaces of the foliation Φ_k = constant: these surfaces are the product of real lines and circles, hence, if they are compact and connected as we shall suppose, are n-dimensional tori $\mathbb{T}^n$. This is the content of the Arnold Theorem, which moreover states that "semi-globally" (the meaning of the term will be clear in the sequel) canonical coordinates can be found, called *action-angle*, such that the action coordinates parametrize the various tori, while the canonically conjugate angles locate a point on a torus.

Globally, the phase space (minus the subset where the differentials $d\Phi_k$ fail to be independent) has the structure of a fibre bundle, where the base, i.e., the space of the first integrals, is an open set in $\mathbb{R}^n$ and the fibre the torus $\mathbb{T}^n$. Thus the problem arises if this fibre bundle is trivial, or, in other words, if it has nontrivial *monodromy*. This is strictly related to the problem of univaluedness of the actions as functions of the first integrals.

D.4.1 Arnold Theorem

Before stating the Arnold Theorem, we prove a series of lemmas, following Arnold (1989) and Fasano & Marmi (1994). Let Λ_α be one of the n-dimensional differentiable compact manifolds of the foliation (D.2.4) of the $2n$-dimensional symplectic manifold M. On Λ_α act n commuting vector fields $X_{\Phi_k} = d\Phi_k^\sharp$, inducing n commuting flows $\phi_1^{t_1}, \dots, \phi_n^{t_n}$. The commutativity of the vector fields X_{Φ_k} is with respect to the Lie brackets and is inherited from the Poisson involutivity of the first integrals. We can therefore define an action on Λ_α of the n-dimensional abelian group $\mathbb{R}^n$ by

$$\mathbb{R}^n \times \Lambda_\alpha \to \Lambda_\alpha : (t_1, \dots, t_n, u) \mapsto \phi_1^{t_1} \circ \dots \circ \phi_n^{t_n} u,$$

which will be simply denoted by $u \mapsto \phi^t u,\ t \in \mathbb{R}^n$.

LEMMA D.12 *For any fixed $u_0 \in \Lambda_\alpha$ the map $\phi : \mathbb{R}^n \to \Lambda_\alpha$ given by $\phi^U u_0 = V$ is a local diffeomorphism; here U is a neighbourhood of* $0 \in \mathbb{R}^n$ *and V a neighbourhood of u_0.*

Proof. For the independence of the differentials $d\phi_k$ and the invertibility of $\sharp$, the n commuting vectors $X_{\Phi_k} = d\Phi_k^\sharp$ are independent. Integrating along their directions, it is possible to parametrize all points of V through an element of U (note that $\phi^0 u_0 = u_0$). Thus the transformation $t \mapsto u(t)$ is defined, and the rows of the Jacobian matrix of this transformation, evaluated at $t = 0$, are just the n vectors generating the flows. Therefore the Jacobian matrix is nonsingular, and the transformation is invertible for U and V sufficiently small. QED

REMARK D.4 The map $\phi : \mathbb{R}^n \to \Lambda_\alpha$ can not be bijective since Λ_α is compact, contrary to $\mathbb{R}^n$.

LEMMA D.13 *The action ϕ of $\mathbb{R}^n$ over Λ_α is transitive, that is*

$$\forall u_0, u \in \Lambda_\alpha,\ \exists t \in \mathbb{R}^n : \phi^t u_0 = u.$$

Proof. Connect u to u_0 with a curve on Λ_α. For the compactness of Λ_α the curve can be covered with a finite number of neighbourhoods of the previous lemma, then define t as the sum of the various translations corresponding to the neighbourhoods of the covering. QED

The above lemma says that Λ_α is a homogeneous manifold under the action of the Lie group $\mathbb{R}^n$. But, in general, if a Lie group $\mathfrak{G}$ acts transitively on a manifold M, the isotropy subgroups $\mathfrak{G}_u$, relative to any point $u \in M$, are all isomorphic (owing to the homogeneity of the $\mathfrak{G}$-action) and M can be identified with $\mathfrak{G}/\mathfrak{G}_u$. We are therefore led to study the isotropy subgroup, say Γ, of the $\mathbb{R}^n$-action on Λ_α, that is the set of the inverse images of $u \in \Lambda_\alpha$ under the map $\phi^t : \mathbb{R}^n \to \Lambda_\alpha$.

LEMMA D.14 *The isotropy subgroup Γ of $\mathbb{R}^n$ is a discrete group.*

Proof. Since ϕ^t is a local diffeomorphism, the origin $0 \in \mathbb{R}^n$ is an isolated point of Γ, so that there is a neighbourhood $U \subset \mathbb{R}^n$ of the origin such that $\Gamma \cap U = \{0\}$. This imposes that any other point of Γ is isolated. Indeed, let us suppose on the contrary that $t \neq 0$ is an accumulation point of Γ. Then

$$t + U = \{t + s : s \in U\}$$

is a neighbourhood of t, thus an element $s \in (t + U) \cap \Gamma$, $s \neq t$ should exist. But $s - t \neq 0$ and $s - t \in \Gamma \cap U$, which is a contradiction. QED

The following lemma gives a classification of all the discrete subgroups of $\mathbb{R}^n$.

LEMMA D.15 *Let Γ be a discrete subgroup of $\mathbb{R}^n$. Then there exist k linearly independent vectors $e_1, \dots, e_k \in \Gamma$, $0 \le k \le n$ such that Γ is generated by them, that is Γ is the set of all their linear combinations with integer coefficients:*

$$\Gamma : \{m_1 e_1 + \cdots + m_k e_k,\ (m_1, \dots, m_k) \in \mathbb{Z}^k\}.$$

The vectors $e_1, \ldots, e_k$ *form a* basis *of* Γ, *which is therefore contained in some subspace* $\mathbb{R}^k \subseteq \mathbb{R}^n$.

Proof. We first look for a characterization of a basis of Γ.

Take k independent vectors of Γ : the point is that in general they do not form a basis, since they do not generate the whole Γ but only a subgroup. To a set $\{e_1, \ldots, e_k\} \subset \Gamma$ of independent vectors we associate the k-dimensional parallelogram $D(e_1, \ldots, e_k)$ defined as

$$D(e_1, \ldots, e_k) = \{\mu_1 e_1 + \cdots + \mu_k e_k,\ 0 \le \mu_j < 1,\ j = 1, \ldots, k\}.$$

We claim that this set is a basis of Γ if and only if

$$D(e_1, \ldots, e_k) \cap \Gamma = \{0\}. \tag{D.4.1}$$

Indeed, if the condition is not satisfied, the vectors belonging to $D \cap \Gamma$ can not be generated (with integer coefficients) by the set, which therefore is not a basis of Γ. Vice versa, if the condition is satisfied all the elements of Γ are generated, since any element of Γ is external to D and by assumption the set $\{e_1, \ldots, e_k\}$ is a basis of the vector space $\mathbb{R}^k$ containing Γ.

Giving to $\mathbb{R}^k$ a Euclidean metric, we can speak of the area (or volume) of D. If the condition (D.4.1) is not satisfied, one can substitute an element of $\{e_1, \ldots, e_k\}$ with an element belonging to D, in such a way that the new set is still a basis of $\mathbb{R}^k$ and, moreover, the area of the new parallelogram is strictly smaller. With a finite number of steps we can construct a set satisfying the condition (D.4.1). QED

LEMMA D.16 *A basis* $\{e_1, \ldots, e_k\}$ *of* Γ *is not unique. Let* $\mathbf{M}$ *be a unimodular matrix, i.e., a square matrix with integer entries and* $\det \mathbf{M} = \pm 1$. *Then any other set* $\{e'_1, \ldots, e'_k\} \subset \Gamma$ *of k independent vectors is a basis of* Γ *if and only if there exists a* $k \times k$ *unimodular matrix* (M_{ij}) *such that*

$$e'_i = \sum_{j=1}^{k} M_{ij} e_j, \quad i = 1, \ldots k. \tag{D.4.2}$$

Proof. Let us suppose that the condition (D.4.2) is satisfied. Then the set $\{e'_i\}$ generates a subgroup $\Gamma' \subseteq \Gamma$, since its vectors are independent and generated with integer coefficients by a basis of Γ. Inverting (D.4.2) and noticing that the inverse of a unimodular matrix is still a unimodular matrix, we can also infer, with the same argument, that $\Gamma \subseteq \Gamma'$. Therefore $\Gamma' = \Gamma$.

Let us suppose that $\{e'_1, \ldots, e'_k\}$ is a basis of Γ. Then there exists an invertible matrix $\mathbf{M}$ transforming the basis $\{e_1, \ldots, e_k\}$ to the basis

$\{e'_1, \dots, e'_k\}$. Since $e'_j \in \Gamma$, the matrix $\mathbf{M}$ must have integer entries. On the other hand, since $\{e'_1, \dots, e'_k\}$ is a basis of Γ, also the inverse matrix must have integer entries. That implies that both $\mathbf{M}$ and $\mathbf{M}^{-1}$ have integer determinant, and by $\det \mathbf{M} \cdot \det \mathbf{M}^{-1} = 1$ we get $\det \mathbf{M} = \pm 1$. QED

REMARK D.5 If $\det \mathbf{M} = -1$, one can recover the case $\det \mathbf{M} = 1$ by simply exchanging two basis vectors. Hereafter, we will thus always take $\mathbf{M} \in SL(n, \mathbb{Z})$.

From the previous lemmas we deduce that $\Lambda_\alpha \simeq \mathbb{R}^n/\Gamma \simeq \mathbb{R}^{n-k} \times \mathbb{T}^k$. If Λ_α is compact, $k = n$ and we reach the basic conclusion that Λ_α is diffeomorphic to the n-dimensional torus $\mathbb{T}^n$. Moreover, we can now make a semiglobal statement about the topology of the symplectic manifold M.

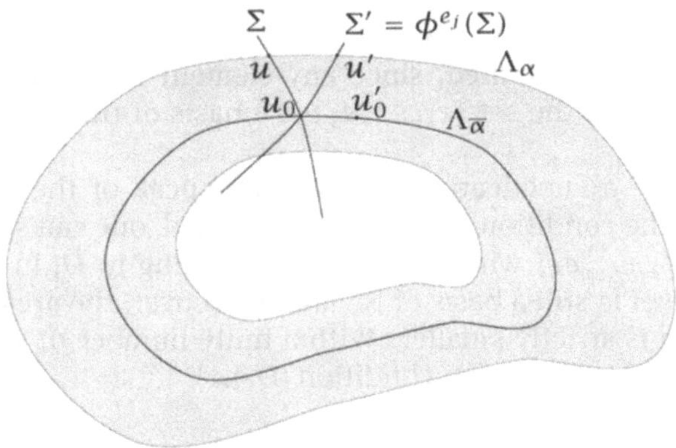

Figure D.2: Construction of the coordinates α, ψ. The shaded area is the neighbourhood $V \subset M$ of $\Lambda_{\overline{\alpha}}$.

LEMMA D.17 *There exists an open neighbourhood $V \subset M$ of any Λ_α that is diffeomorphic to the product of an open neighbourhood $U \subset \mathbb{R}^n$ and a torus, i.e., $V \simeq U \times \mathbb{T}^n$, and can be covered with coordinates $\alpha_1, \dots, \alpha_n$ on U and angular coordinates $\psi^1, \dots, \psi^n$ on $\mathbb{T}^n$.*

Proof. See Figure D.2. For a fixed $\overline{\alpha}$ take any point $u_0 \in \Lambda_{\overline{\alpha}}$ and construct a differentiable manifold Σ transversal to $\Lambda_{\overline{\alpha}}$ in u_0 : by definition this implies that the tangent space to Σ is complementary in $T_{u_0}M$ to that of $\Lambda_{\overline{\alpha}}$, thus $\dim \Sigma = n$. Consider the basis vectors e_j of the isotropy subgroup of u_0, i.e., $\phi^{e_j} u_0 = u_0$, $j = 1, \dots, n$. We can write, for any $u_0' \in \Lambda_{\overline{\alpha}}$, that $u_0' = \phi^t u_0$, $t \in \mathbb{R}^n$ and give to this point angular coordinates ψ, by putting $t = \sum_{i=1}^n \frac{\psi^i}{2\pi} e_i$. Let the surface Σ evolve under the action of ϕ^{e_j}. Another surface $\Sigma' = \phi^{e_j}(\Sigma)$ is obtained, different in general from Σ but containing surely u_0. If U is sufficiently small, Σ' is transversal to Λ_α, $\forall \alpha \subset U$, therefore for every $u' \in \Sigma'$, uniquely determined by a value α, there exists a differentiable function $\tau_j(\alpha)$ such that $\tau_j(\overline{\alpha}) = 0$ and, if $u \in \Sigma \cap \Lambda_\alpha$, such that $\phi^{\tau_j(\alpha)} u' = u$. Hence, $\phi^{e_j + \tau_j(\alpha)} u = u$. For every $\alpha = (\alpha_1, \ldots, \alpha_n)$ in U, the periods $e_j(\alpha) = e_j + \tau_j(\alpha)$ can thus be defined that determine on any Λ_α the isotropy subgroup of u, and allow us to give to any point of Λ_α (as above to u_0') angular coordinates ψ. QED

The previous lemma is global with regard to the parametrization of points on a torus, but only local with regard to the parametrization of the tori themselves. The lemma says that M is a fibre bundle whose base is covered by a family of open sets like U and the fibre is an n-dimensional torus. In this sense we say that it is a semiglobal statement, since nothing is said about the global topology of the bundle, which in general may be nontrivial.

The coordinates $\alpha_1, \ldots, \alpha_n, \psi^1, \ldots, \psi^n$ are *not* in general canonical coordinates. In other words, the vector X_{Φ_j} generates a flow that preserves Λ_α but neither is periodic in the angle ψ^j nor keeps constant the other angles ψ^k, $k \neq j$. It is possible however to construct n vectors with this property. Indeed, let us define the n vectors

$$Y_i = \sum_{j=1}^n e_i^j X_{\Phi_j}, \quad i = 1, \ldots n, \tag{D.4.3}$$

where e_i^j is the j-th component of the vector $e_i \in \Gamma$. On each $\Lambda_\alpha \simeq \mathbb{R}^n/\Gamma$ the vector X_{Φ_j} generates a linear flow ϕ_j^τ, $\tau \in \mathbb{R}$, thus the vector Y_i generates the flow $\phi_1^{\tau e_i^1} \circ \cdots \circ \phi_n^{\tau e_i^n} = \phi^{\tau e_i}$. By definition of isotropy subgroup, ϕ^{e_i} is the identity for every i and the vectors $Y_1, \ldots, Y_n$ generate n independent periodic flows with period 1. To prove that these n vectors are locally Hamiltonian, we construct explicitly their (locally defined) Hamiltonians, which are by definition the action variables $I_1, \ldots, I_n$ (up to the multiplicative factor 2π).

Let γ_i be the cycle on the torus generated by Y_i, and φ^i the corresponding angle. If we pretend that I_j, φ^k, $j, k = 1, \ldots, n$ are canonical coordinates, then the two 1-forms $p_k dq^k$ and $I_k d\varphi^k$ must differ for a closed 1-form whose integral along a cycle is zero, if the cycle is also a boundary. Thus

$$\oint_{\gamma_i} I_k d\varphi^k = \oint_{\gamma_i} p_k dq^k.$$

In the left-hand side $d\varphi^k = 0,\ \forall k \neq i$, and I_i is constant along the cycle since, by definition, it is just the Hamiltonian generating γ_i. Therefore

$$I_i = \frac{1}{2\pi} \oint_{\gamma_i} p_k dq^k. \tag{D.4.4}$$

The action variables are therefore functions of the first integrals Φ's only. As it is obvious from (D.4.3), on any fixed Λ_α these functions are linear invertible, so that the action variables are a system of n independent first integrals in involution.

The expressions (D.4.4) are usually taken, in text books, as definition of the action variables. Notice that they do not depend, to some extent, on the particular cycle γ. If we take, indeed, any other cycle $\gamma' = \gamma + \partial\sigma,\ \sigma \subset \Lambda_\alpha$, homologous to γ, the value of the integral does not change, owing to the Stokes Theorem and to the fact that $\Omega|_{\Lambda_\alpha} = d\Theta|_{\Lambda_\alpha} = 0$:

$$\oint_{\gamma'} \Theta = \oint_{\gamma} \Theta + \oint_{\partial\sigma} \Theta = \oint_{\gamma} \Theta + \oint_{\sigma} d\Theta = \oint_{\gamma} \Theta.$$

This definition, however, must be taken with some care when (in practice: always) the first integrals are generated by separation of variables, so that the cycle γ_k is assumed to be the closed curve, on the phase plane $q_k p_k$, described by $\Phi_k(q_k, p_k) =$ constant. Varying the value of the first integrals, also γ_k does vary: the subtle point, as explained below in the subsection devoted to the monodromy, is that, coming back to the starting value of the first integrals, the cycle γ_k does *not* in general agree with the initial one. Rather, if the torus bundle is not trivial,

$$\gamma_k \mapsto \gamma'_k = \sum_h M_{kh}\gamma_h, \quad (M_{kh}) \in \mathrm{SL}(n, \mathbb{Z}).$$

This implies that the Definition (D.4.4) is only local, i.e., it is meaningful as long as the value of the first integrals varies in a simply connected open set of the base space. Globally, the action variables are in general multivalued functions of the first integrals.

LEMMA D.18 *The canonical transformation $(p, q) \mapsto (I, \varphi)$ to action-angle variables is generated by*

$$W(I, q) = \int_{q_0}^{q} \mathfrak{p}_k(I, q')\, dq'^k,$$

where the functions $\mathfrak{p}(I, q)$ are obtained by inverting $I(p, q) =$ constant with respect to the momenta.

Proof. The formula descends from (D.2.5). We want to check that the new coordinates φ are really angles. The variation $\Delta_i W$, that affects the generating function W by integrating along a cycle (homologous to) γ_i, is $2\pi I_i$ for the Definition (D.4.4). Then the variation of φ^k on the same cycle is

$$\Delta_i \varphi^k = \Delta_i \frac{\partial W}{\partial I_k} = \frac{\partial}{\partial I_k} \Delta_i W = 2\pi \delta_i^k,$$

so that $\varphi^k \in \mathbb{T}^n$, $\forall k$. QED

We have therefore proved that the canonical transformation generated by $W(I, q)$ gives new canonical coordinates (I, φ), such that $Y_i = 2\pi \, dI_i^\sharp$. This canonical transformation transforms the given Hamiltonian $H(p, q)$ into a new Hamiltonian $K(I)$, which depends only on the actions. The Hamilton Equations are

$$\dot{I}_k = -\frac{\partial K(I)}{\partial \varphi^k} = 0, \quad \dot{\varphi}^k = \frac{\partial K(I)}{\partial I^k},$$

i.e., the actions are first integrals (as already known) and the angles evolve linearly in the time.

The chain of previous lemmas proves the

THEOREM D.19 (ARNOLD) *Given a completely integrable n-dimensional Hamiltonian system, the compact and connected components of the level surfaces of the first integrals are diffeomorphic to an n-dimensional torus. Moreover, there exist (locally) canonical coordinates called action-angle, such that the action variables parametrize the set of the tori and the angles the points on a torus. The Hamiltonian, expressed as a function of these coordinates, depends only on the actions, so that the dynamical evolution is a uniform rotation on an invariant torus.*

The action-angle variables are not defined in a unique way. New coordinates on the same torus bundle can be generated by the following canonical transformations.

(i) Translations of the action variables:

$$I'_k = I_k + c_k, \quad \varphi'^h = \varphi^h, \quad c_k \in \mathbb{R}^n;$$

the actions enter indeed in the formulas only through their gradients.

(ii) Translations of the origin of the angles by quantities depending on the torus:

$$I'_k = I_k, \quad \varphi'^h = \varphi^h + \frac{\partial f(I)}{\partial I_h},$$

where f is an arbitrary scalar function; this reflects the arbitrariness in the choice of the point u_0 and of the surface Σ.

(iii) A linear transformation by a unimodular matrix:

$$\varphi' = \mathbf{M}^{-1}\varphi, \quad I' = \mathbf{M}^t I,$$

which reflects the arbitrariness in the choice of the basis of the isotropy subgroup.

D.4.2 Degenerate Systems

Let us consider an n-dimensional integrable system, and suppose that the Hamiltonian is a function of $n-d$ action variables only. The dynamics takes place, therefore, on $(n-d)$-dimensional tori, and the system is said to be *d-fold degenerate*. If $d = n - 1$, then all the orbits are periodic, and the system is *completely degenerate*: it is the case of the Kepler Problem with negative energy.

The geometry of the degenerate systems is richer than that of the nondegenerate ones. As pointed out in Fassò (1996), the general situation is the following. The phase space of a d-fold degenerate Hamiltonian system (minus, in case, a subset of zero measure, comprising equilibrium points and separatrices) is fibered by $(n-d)$-dimensional invariant tori, filled by quasi-periodic motions, while the base space is an $(n+d)$-dimensional Poisson manifold of rank $2d$. The point is that the $2d$-dimensional symplectic leaves of this Poisson manifold are not in general homeomorphic to $\mathbb{R}^d \times \mathbb{T}^d$.

For example, in the 3-dimensional Kepler Problem the fibre is the circle S^1 and the base (the space of the orbits) the 5-dimensional Poisson manifold spanned by the two orthogonal vectors $\vec{G}$ and $\vec{R}$; the symplectic leaves (the space of the orbits for a fixed energy) have topology $S^2 \times S^2$.

It is clear that, while in the nondegenerate case the Lagrangian fibration in n-dimensional tori is intrinsic, in the degenerate case only the fibration in $(n-d)$-dimensional tori is so. To complete this last to a Lagrangian fibration, one selects a piece of the symplectic leaf, homeomorphic to a cylinder $\mathbb{R}^d \times \mathbb{T}^d$, and attaches the factor $\mathbb{T}^d$ to the $(n-d)$-dimensional invariant tori. This operation is in general, i.e., when the symplectic leaves are not homeomorphic to $\mathbb{R}^d \times \mathbb{T}^d$, neither global nor unique, and every such choice of action-angle variables develops, inevitably, singularities.

In the case of the Kepler Problem, one must add to $\mathfrak{L}$ and ℓ two other pairs of action-angle variables parametrizing (thus, only locally) the symplectic leaf $S^2 \times S^2$. Two of the many possible choices are those studied in Chapter 3 and called Delaunay's and Pauli's variables respectively.

There is a strict relation between degeneration and over-integrability. We say that a Hamiltonian is *over-integrable* when it admits several systems of n inequivalent first integrals in involution. The level surfaces of these inequivalent systems are, by definition, different families of n-dimensional tori, whose intersection is just the family of $(n - d)$-dimensional tori on which the dynamics takes place. Whatever is the n-ple of the first integrals in involution, they will, in general, lose their independence in a subset of the phase space of zero measure (think, typically, of the Delaunay parameters): this is obviously due to the fact that one is looking for an object (a Lagrangian fibration in tori) that does not exist globally.

D.4.3 Monodromy

The concept of monodromy was introduced by Duistermaat (1980) in the study of Hamiltonian integrable systems. Monodromy describes the twisting of the tori when one moves along a circle on the base of the Liouville-Arnold fibration. "Monodromy can *not* be described by examining a single trajectory since it is a property of the way in which trajectories fit together" (Cushman & Bates 1997). Monodromy has now been found in several integrable systems, see Cushman & Knörrer (1985), Bates (1991), Cushman & Sadovskií (2000).

Let us therefore consider an n-dimensional integrable system (M, H), and let

$$J = (\Phi_1, \dots, \Phi_n) : M \to \mathbb{R}^n$$

be a moment map, where Φ's are the n first integrals in involution. If we suppose that the level set $\Lambda_\alpha = J^{-1}(\alpha)$, with α a regular value, is compact and connected, then it is a torus $\mathbb{T}^n = \mathbb{R}^n/\Gamma_\alpha$, where Γ_α is the lattice of the periods $\{e_1(\alpha), \dots, e_n(\alpha)\}$ constructed in Lemma D.17. The union of all the isotropy subgroups is the *period bundle* over the base $B \subset \mathbb{R}^n$ of all the regular values α of J. Lemma D.17 asserts that this bundle is locally trivial. The point is that it can be globally non-trivial: if we let $\alpha = \{\alpha_1, \dots, \alpha_n\}$ evolve along a closed path γ in the base B, then when we return to the initial position, say $\overline{\alpha}$, we obtain a basis $\{e'_1(\overline{\alpha}), \dots, e'_n(\overline{\alpha})\}$ that may be different from the original one $\{e_1(\overline{\alpha}), \dots, e_n(\overline{\alpha})\}$. But the two bases generate the same lattice, so they must be connected by a unimodular matrix $\mathbf{M} \in \mathrm{SL}(n, \mathbb{Z})$. This matrix is called the *monodromy matrix*. Obviously, the bundle is trivial if and only if $\mathbf{M} = \mathbf{1}_n$. If the bundle is not trivial, the basis vectors $\{e_1(\alpha), \dots, e_n(\alpha)\}$ are not uniquely defined as functions of the values α of the first integrals. Then, Definition D.4.3 shows that the same holds for the vectors $Y_i = dI_i^\sharp$, and the actions $I_1, \dots, I_n$ also are multivalued functions of the first integrals.

An equivalent way of stating the thing is the following. Let us consider the fibre bundle, associated to the period bundle, obtained by substituting for the fibre $\{e_i(\alpha)\}$ the fibre $H_1(\Lambda_\alpha, \mathbb{Z})$ (i.e. the first homology group of $\mathbb{T}^n$, which is isomorphic to $\mathbb{Z}^n = \mathbb{Z} \oplus \cdots \oplus \mathbb{Z}$) but taking the same transition functions, with values into the group $\mathrm{SL}(n, \mathbb{Z})$. The advantage is that, the fibre being discrete, we have the notion of parallel transport above the closed path $\gamma \subset B$. Then the monodromy matrix $\mathbf{M}$ is defined as that transforming the fibre over a point of γ into the fibre over the same point after that γ has been covered once. The matrix $\mathbf{M}$ does not depend on the particular initial point, but only on the path. Moreover, two paths that can be continuously deformed one into the other, i.e., homotopic paths, give rise to the same monodromy matrix. Therefore, the monodromy of an integrable system appears as a map from the homotopy group $\pi_1(B)$ into $\mathrm{SL}(n, \mathbb{Z})$.

Let us study in more detail the 2-dimensional case: see Nguyên Tiên (1997). If $B \subset \mathbb{R}^2$ is simply connected, then the bundle is obviously trivial. The interesting case arises when the moment map has at least an isolated singularity, which we can think of as placed at the origin of B, thus $d\Phi_1|_0 = d\Phi_2|_0 = 0$. The origin is therefore an equilibrium point and, linearizing the Hamiltonian flows of Φ_1 and Φ_2 around the singularity, one can classify all the normal forms of the quadratic Hamiltonians. In a canonical chart, let $\underline{x}$ be the vector of components (q_1, q_2, p_1, p_2), then the quadratic part of each component of the moment map is written as

$$\Phi_i = \frac{1}{2} \langle \underline{x}, \mathbf{D}_i \underline{x} \rangle , \quad i = 1, 2,$$

where $\mathbf{D}_i$ is a symmetric 4×4 matrix. The Hamiltonian flow generated by Φ_i is $\dot{\underline{x}} = \mathbf{\Omega} \mathbf{D}_i \underline{x}$, where $\mathbf{\Omega}$ is the canonical symplectic matrix. The classification is pursued by considering the eigenvalues of the two matrices $\mathbf{\Omega D}_i$, finding so 15 inequivalent cases: see Williamson (1936), Lerman & Umanskii (1991) or also Passoni (2000). If however one considers only the *nondegenerate* (or *stable*) singular points (i.e., those for which the eigenvalues are generic), only four types survive. Up to a linear symplectic transformation they are

(i) $\Phi_1 = q_1^2 + p_1^2$, $\Phi_2 = q_2^2 + p_2^2$,

(elliptic case with eigenvalues $\pm i$ and $\pm i$)

(ii) $\Phi_1 = q_1^2 + p_1^2$, $\Phi_2 = q_2 p_2$,

(elliptic–hyperbolic case with eigenvalues $\pm i$ and ± 1)

(iii) $\Phi_1 = q_1 p_1$, $\Phi_2 = q_2 p_2$,

(hyperbolic case with eigenvalues ± 1 and ± 1)

(iv) $\Phi_1 = q_1p_1 + q_2p_2$, $\Phi_2 = q_1p_2 - q_2p_1$,
(focus-focus case with eigenvalues $\pm 1, \pm 1$ and $\pm i, \pm i$).

It easy to verify that only the focus-focus type implies an isolated singularity, so that it is the unique candidate for nontrivial monodromy. Physically it corresponds to a linear isotropic repulsor, as is clear after a $\frac{\pi}{4}$ rotation in the two planes q_1p_1 and q_2p_2 (which is a canonical transformation). A typical example is the spherical pendulum, where the focus-focus singularity is at the unstable equilibrium point; indeed, developing in series the potential of the gravitational force, one finds just the potential of a linear isotropic repulsor.

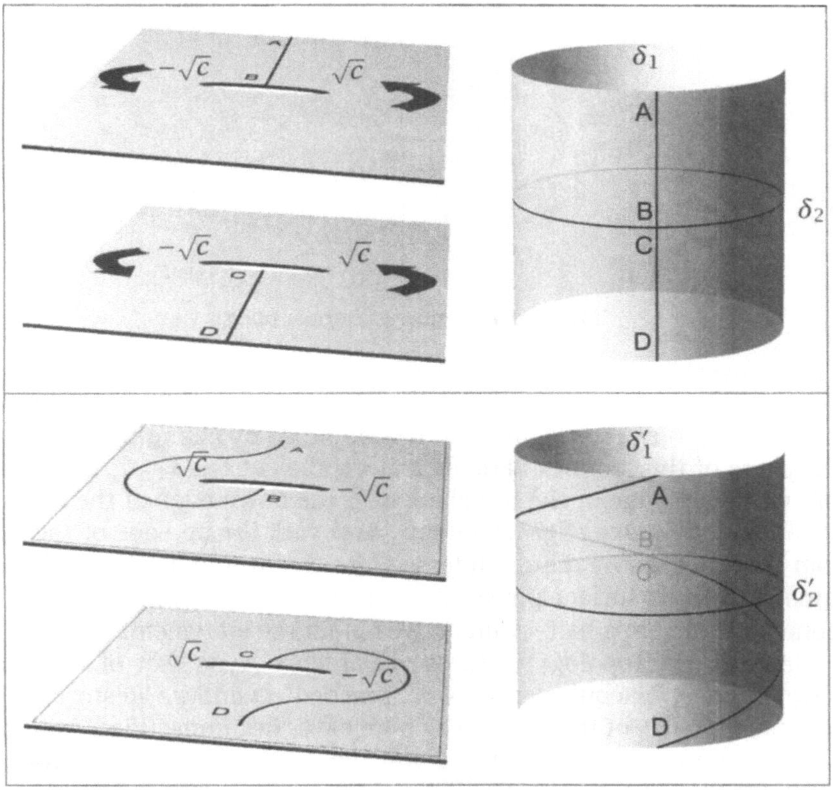

Figure D.3: Monodromy of the focus-focus singularity.

To study the topology of the focus-focus singularity, it is useful to pass to the complex coordinates

$$z = \frac{1}{2}(p_1 - ip_2 + q_1 + iq_2), \quad w = \frac{i}{2}(p_1 - ip_2 - q_1 - iq_2),$$

so that the moment map J is expressed as

$$J : \mathbb{C}^2 \to \mathbb{C}, \quad (z, w) \mapsto z^2 + w^2 = \Phi_1 + i\Phi_2.$$

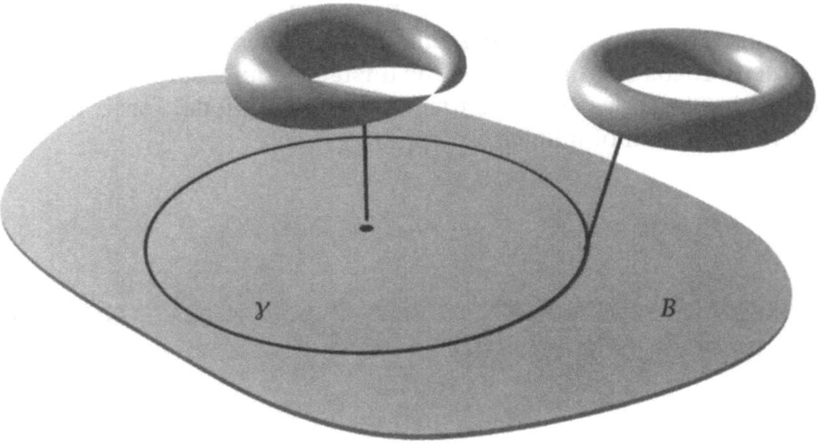

Figure D.4: Singular vs. regular fibers

The topology of the generic fibre $\Lambda_c = J^{-1}(c)$, $0 \neq c \in \mathbb{C}$ is that of a cylinder $S^1 \times \mathbb{R}$. Indeed, let us consider the Riemann surface of the function $w = \sqrt{c - z^2}$: see Figure D.3. It is obtained by gluing two copies of the plane of the complex variable z along the cut line $(-\sqrt{c}, \sqrt{c})$, identifying the up edge of the top sheet with the down edge of the bottom sheet, and the down edge of the top sheet with the up edge of the bottom sheet (we ignore the self intersections that would be obtained if we constructed our surface physically). Since each copy of the cut plane is homeomorphic to a half-cylinder, we obtain the statement.

Instead, the topology of the singular fibre Λ_0 is that of a double sheeted cone, or equivalently of a "pinched" cylinder. Restoring the original topology of the fibres (which we have "decompactified" with the linearization) we obtain a torus $\mathbb{T}^2$ for the regular fibres and a pinched torus for the singular one. See Figure D.4.

The monodromy matrix $\mathbf{M} \in \mathrm{SL}(2, \mathbb{Z})$ must have one of the two rows equal to that of the unit matrix $\mathbf{1}_2$, as is clear from the fact that Φ_2, which generates a periodic flow, is already an action. Therefore it must be that

$$\mathbf{M} = \begin{pmatrix} 1 & k \\ 0 & 1 \end{pmatrix}, \quad k \in \mathbb{Z}. \tag{D.4.5}$$

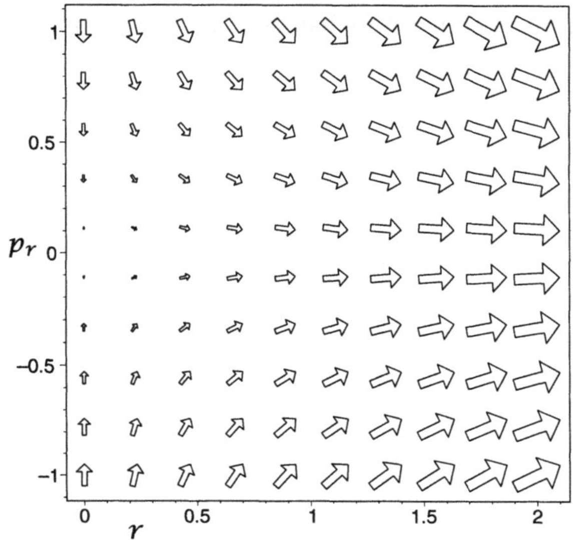

Figure D.5: Hamiltonian vector field of a harmonic repulsor in the reduced space.

Let us show that, for the focus-focus singularity, $k = 1$ holds: see Figure D.3. Consider on the regular fibre over $c \neq 0$ two generators δ_1 and δ_2 of the first cohomology group, then move along a circle $\gamma \subset B$, parametrized by $c(\tau) = ce^{i\tau}$, $0 \leq \tau \leq 2\pi$. When the circle is covered once, the cut on the two sheets is rotated by π, so that $\delta_1 \mapsto \delta_1' = \delta_1 + \delta_2$ while $\delta_2' = \delta_2$, as is clear from Figure D.3. This proves that the monodromy matrix has the form D.4.5 with $k = 1$.

Another useful way to characterize a focus-focus singularity is to analyze the Hamiltonian vector field, generated by Φ_1, on the 2-dimensional phase space obtained reducing out the S^1 symmetry generated by Φ_2. Passing to polar coordinates r, θ we get

$$\Phi_1 = r p_r, \quad \Phi_2 = p_\theta.$$

The reduced space is $\mathbb{R}_{\geq 0} \times \mathbb{R}$ and the Φ_1 flow: $\dot{r} = r$, $\dot{p}_r = -p_r$ is depicted in Figure D.5

A more general situation is that of k focus-focus singularities. A path $\gamma \subset B$ enclosing all these singularities can then be viewed as the product of k paths, all beginning and ending at the same point and each enclosing one singularity only. The monodromy matrix of γ is the product of k matrices

$$\mathbf{M} = \begin{pmatrix} 1 & 1 \\ 0 & 1 \end{pmatrix} \cdots \begin{pmatrix} 1 & 1 \\ 0 & 1 \end{pmatrix} = \begin{pmatrix} 1 & k \\ 0 & 1 \end{pmatrix},$$

i.e., the general matrix (D.4.5). Letting all these singularities collapse into a point, the singular fibre has in this case the topology of a k-times pinched torus.

Bibliography

ABRAHAM, R. & MARSDEN, J. E. *Foundations of Mechanics.* Benjamin Cummings, Reading, Mass., (1978).

AKYILDIZ, Y. On the dynamical symmetries of the Kepler problem. *J. Math. Phys.*, **21**, pp. 665–670, (1980).

APOSTOL, T. M. *Calculus.* John Wiley & Sons, New York, (1969).

ARNOLD, V. I. Small denominators and problems of stability of motion in classical and celestial mechanics. *Russ. Math. Surv.*, **18**, pp. 85–191, (1963).

ARNOLD, V. I. *Mathematical Methods of Classical Mechanics.* Springer-Verlag, Berlin, 2nd edn., (1989).

ASCHER, U. M. Stabilization of invariants of discretized differential systems. *Numer. Algorithms*, **14**, pp. 1–24, (1997).

ATIYAH, M. & HITCHIN, N. *The Geometry and Dynamics of Magnetic Monopoles.* Princeton University Press, Princeton, New Jersey, (1988).

AUSLANDER, L. & MACKENZIE, R. E. *Introduction to Differentiable Manifolds.* Dover Publications, New York, (1977).

BACRY, H. The de Sitter group $L_{4,1}$ and the bound states of hydrogen atom. *Nuovo Cimento*, **41A**, pp. 222–234, (1966).

BANDER, M. & ITZYKSON, C. Group theory and the Hydrogen Atom. *Rev. Mod. Phys.*, **38**, pp. 330–345, (1966).

BARGMANN, V. Irreducible unitary representations of the Lorentz group. *Ann. Math.*, **48**, pp. 568–640, (1946).

BARUT, A. O. & BORNZIN, G. L. SO(4,2)-formulation of the symmetry breaking in relativistic Kepler problems with or without magnetic charges. *J. Math. Phys.*, **12**, pp. 841–846, (1971).

BARUT, A. O. & RACZKA, R. *Theory of Group Representations and Applications.* PWN - Polish Scientific Publishers, Warszawa, (1980).

BATES, L. Symmetry preserving deformations of the Kepler problem. *Rep. Math Phys.*, **26**, pp. 413–428, (1988).

BATES, L. Monodromy in the champagne bottle. *J. Appl. Math. and Phys. (ZAMP)*, **42**, pp. 837–847, (1991).

BELAVIN, A. A., POLYAKOV, A. M., SCHWARZ, A. S. & TYUPKIN, Y. S. Pseudoparticle solutions of the Yang-Mills equations. *Phys. Lett.*, **59B**, pp. 85–87, (1975).

BENENTI, S. Separation of variables in the geodesic Hamilton-Jacobi equation. In *Symplectic geometry and mathematical physics (Aix-en-Provence, 1990)*, pp. 1–36. Birkhäuser, Boston, MA, (1991).

BENENTI, S. Orthogonal separable dynamical systems. In Kowalski, O. & Krupka, D., eds., *Differential geometry and its applications (Opava, 1992)*, pp. 163–184. Silesian Univ. Opava., Opava, (1993). Electronic edition ELibEMS, http://www.emis.de/proceedings/.

BENENTI, S. Intrinsic characterization of the variable separation in the Hamilton-Jacobi equation. *J. Math. Phys.*, **38**, pp. 6578–6602, (1997).

BENENTI, S., CHANU, C. & RASTELLI, G. Remarks on the connection between the additive separation of the Hamilton-Jacobi equation and the multiplicative separation of the Schrödinger equation. I. The completeness and Robertson conditions. Tech. Rep. 22, Quaderni del Dipartimento di Matematica - Università di Torino, (2001).

BENETTIN, G., GALGANI, L. & GIORGILLI, A. A proof of Nekhoroshev's theorem for the stability times in nearly integrable Hamiltonian systems. *Celestial Mech.*, **37**, pp. 1–25, (1985).

BENETTIN, G., GALGANI, L., GIORGILLI, A. & STRELCYN, J.-M. A proof of Kolmogorov's theorem on invariant tori using canonical transformations defined by Lie method. *Il Nuovo Cimento*, **79 B**, pp. 201–223, (1984).

BOWMAN, F. *Introduction to Elliptic Functions with Applications.* Dover Publications, New York, (1961).

CHANDRASEKHARAN, K. *Elliptic Functions.* Springer-Verlag, Berlin Heidelberg, (1985).

CHEVALLEY, C. *Theory of Lie Groups I.* Princeton University Press, Princeton, N. J., (1946).

CHOQUET-BRUHAT, Y. *Géométrie Différentielle et Systèmes Extérieurs.* Dunod, Paris, (1968).

CODEGONI, S. *Funzioni Ellittiche e Integrazione di Problemi in Meccanica Analitica.* Master's thesis, Corso di laurea in Matematica dell'Università di Milano, (2000).

COFFEY, S. L., DEPRIT, A. & DEPRIT, E. Frozen orbits for satellites close to an Earth-like planet. *Celes. Mech. & Dyn. Astron.*, **59**, pp. 37–72, (1994).

COFFEY, S. L., DEPRIT, A. & MILLER, B. R. The critical inclination in artificial satellite theory. *Celestial Mech.*, **39**, pp. 365–406, (1986).

COLWELL, P. *Solving Kepler's Equation over Three Centuries.* Willmann-Bell, Inc., Richmond, Virginia, (1993).

CONLEY, C. On some new long periodic solutions of the plane restricted three body problem. *Commun. Pure App. Math.*, **16**, pp. 449–467, (1963).

CORDANI, B. Conformal regularization of the Kepler problem. *Commun. Math. Phys.*, **103**, pp. 403–413, (1986).

CORDANI, B. Kepler problem with a magnetic monopole. *J. Math. Phys.*, **27**, pp. 2920–2921, (1986b).

CORDANI, B. Quantization of the Kepler manifold. *Commun. Math. Phys.*, **113**, pp. 649–657, (1988).

CORDANI, B. On the Fock quantization of the hydrogen atom. *J. Phys. A: Math. Gen.*, **22**, pp. 2695–2707, (1989).

CORDANI, B. On the generalization of the Kustaanheimo-Stiefel transformation. *J. Phys. A: Math. Gen.*, **22**, pp. 2441–2446, (1989b).

CORDANI, B. Perturbations of the Kepler problem in global coordinates. *Celes. Mech. & Dyn. Astron.*, **77**, pp. 185–200, (2000).

CORDANI, B., FEHÉR, L. G. & HORVÁTHY, P. A. Kepler-type dynamical symmetries of long-range monopole interactions. *J. Math. Phys.*, **31**, pp. 202–211, (1990).

CORDANI, B. & MERLINI, G. Perturbations of the Kepler problem in global coordinates: a program. *Celes. Mech. & Dyn. Astron.*, **80**, pp. 313–319, (2001).

CORDANI, B. & REINA, C. Spinor regularization of the n-dimensional Kepler Problem. *Lett. Math. Phys.*, **13**, pp. 79–82, (1987).

CORNWELL, J. F. *Group Theory in Physics.* Academic Press, London, (1989).

COULSON, C. A. & ROBINSON, P. D. Wave function for the Hydrogen Atom in spheroidal coordinates I: The derivation and properties of the functions. *Proc. Phys. Soc.*, **71**, pp. 815–827, (1958).

CRAMPIN, M. & PIRANI, F. A. E. *Applicable Differential Geometry.* London Mathematical Society Lecture Notes Series. Cambridge University Press, Cambridge, (1986).

CROOM, F. H. *Principles of Topology.* Saunders College Publishing, Philadelphia, (1989).

CUSHMAN, R. Reduction, Brower's Hamiltonian, and the critical inclination. *Celestial Mech.*, **31**, pp. 401–429, (1983).

CUSHMAN, R. A survey of normalization techniques applied to perturbed Keplerian systems. In Jones, K., ed., *Dynamics Reported, Volume 1, New Series*, pp. 54–112. Springer-Verlag, New York, (1991).

CUSHMAN, R. H. & BATES, L. M. *Global Aspects of Classical Integrable Systems.* Birkhäuser, Basel, (1997).

CUSHMAN, R. H. & KNÖRRER, H. The momentum mapping of the Lagrange top. *Lect. Notes in Math.*, **1139**, pp. 12–24, (1985).

CUSHMAN, R. H. & SADOVSKIÍ, D. A. Monodromy in perturbed Kepler systems: Hydrogen atom in crossed fields. *Europhys. Lett.*, **47**, pp. 1–7, (1999).

CUSHMAN, R. H. & SADOVSKIÍ, D. A. Monodromy in the hydrogen atom in crossed fields. *Physica D*, **142**, pp. 166–196, (2000).

DEPRIT, A. Le problème des deux centres fixes. *Bull. Soc. Math. de Belgique*, pp. 12–45, (1962).

DEPRIT, A. Canonical transformations depending on a small parameter. *Celestial Mechanics*, **1**, pp. 12–30, (1969).

DEPRIT, A. The elimination of the parallax in satellite theory. *Celestial Mech*, **24**, pp. 111–153, (1981).

DERAHM, G. *Varietés Differentiables.* Hermann, Paris, (1960).

DIRAC, P. A. M. *The Principles of Quantum Mechanics.* Clarendon Press, Oxford, (1958).

DUBROVIN, B. A. Theta functions and non linear equations. *Russian Math. Surveys*, **36**, pp. 11–92, (1981).

DUBROVIN, B. A., NOVIKOV, S. & FOMENKO, A. *Géométrie Contemporaine.* MIR, Moskow, (1982–1987).

DUISTERMAAT, J. J. On global action-angle variables. *Comm. Pure Appl. Math.*, **33**, pp. 687–706, (1980).

EISENHART, L. P. Separation of the variables in the one-particle Schrödinger equation in 3-space. *Proc. Nat. Acad. Science*, **35**, pp. 412–418, (1949).

ENGLEFIELD, M. J. *Group Theory and the Coulomb Problem.* Wiley Interscience, New York, (1972).

FASANO, A. & MARMI, S. *Meccanica Analitica.* Bollati Boringhieri, Torino, (1994).

FASSÒ, F. The Euler-Poinsot top: A non-commutatively integrable system without global action-angle coordinates. *J. App. Math. Phys. (ZAMP)*, **47**, pp. 1-24, (1996).

FEHER, L. G. Dynamical O(4) symmetry in the asymptotic field of a Prasad-Sommerfield monopole. *J. Phys. A*, **19**, pp. 1259-1270, (1986).

FLECKENSTEIN, J. O. Notiz zur Lagrangeschen losung des Keplerschen problem. *Commentarii Mathematici Helvetici*, **13**, pp. 83-89, (1941).

FOCK, V. A. Zur theory des wasserstoffatoms. *Z. Phys.*, **98**, pp. 145-154, (1935).

GEL'FAND, I. M. & SHILOV, G. E. *Generalized Functions*, vol. 1. Academic Press, New York, (1964).

GIBBONS, G. W. & MANTON, N. S. Classical and quantum dynamics of BPS monopoles. *Nucl. Phys.*, **274B**, pp. 183-224, (1986).

GILMORE, R. *Lie Groups, Lie Algebras, and some of their Applications.* John Wiley & Sons, New York, (1974).

GINZBURG, V., GUILLEMIN, V. & KARSHON, Y. *Moment Maps, Cobordisms, and Hamiltonian Group Actions.* Preprint downloadable at http://www.ma.huji.ac.il/ karshon/monograph/, (2002).

GIORGILLI, A. New insights on the stability problem from recent results in classical perturbation theory. In Benest, D. & Froeschle, C., eds., *Modern Methods in Celestial Mechanics*, pp. 249-284. Editions Frontières, Gif-sur-Yvette, France, (1989).

GOLDSTEIN, H. *Classical Mechanics.* Addison-Wesley, Reading, Mass., (1950).

GOTAY, M. Constraints, reduction and quantization. *J. Math. Phys.*, **27**, pp. 2051-2066, (1986).

GOURSAT, É. *A Course in Mathematical Analysis*, vol. 1. Dover Publications, New York, (1959).

GROENEWOLD, H. J. On the principles of the elementary quantum mechanics. *Physica*, **12**, pp. 405-460, (1946).

GUILLEMIN, V. & STERNBERG, S. *Geometric Asymptotics.* AMS, Providence, RI, (1977).

GUILLEMIN, V. & STERNBERG, S. Geometric quantization and multiplicities of group representations. *Invent. Math.*, **67**, pp. 515-538, (1982).

GUILLEMIN, V. & STERNBERG, S. *Symplectic Techniques in Physics.* Cambridge University Press, Cambridge, (1984).

GUILLEMIN, V. & STERNBERG, S. *Variations on a Theme by Kepler.* American Mathematical Society, Providence, RI, (1990).

GYÖRGYI, G. Kepler's equation, Fock variables, Bacry's generators and Dirac brackets. *Nuovo Cimento,* **53A**, pp. 717–735, (1968).

HELGASON, S. *Differential Geometry, Lie Groups, and Symmetric Spaces.* Academic Press, New York, (1978).

HORI, G. Theory of general perturbations with unspecified canonical variables. *Publ. Astron. Soc. Japan,* **18**, pp. 287–296, (1966).

INCE, E. L. *Ordinary Differential Equations.* Dover Publications, New York, (1956).

IWAI, T. The geometry of the SU(2) Kepler problem. *JGP,* **7**, pp. 507–535, (1990).

IWAI, T. & UWANO, Y. The four-dimensional conformal Kepler problem reduces to three-dimensional Kepler problem with a centrifugal potential and Dirac's monopole field. Classical theory. *J. Math. Phys.,* **27**, pp. 1523–1529, (1986).

KALNINS, E. G. *Separation of Variables for Riemannian Spaces of Constant Curvature.* Longman Scientic & Technical, Burnt Mill, Harlow, Essex, England, (1986).

KALNINS, E. G. & MILLER, JR., W. Killing tensors and variable separation for Hamilton-Jacobi and Helmholtz equations. *SIAM J. Math. Anal.,* **11**, pp. 1011–1026, (1980).

KNAPP, A. W. *Representation Theory of Semisimple Groups.* Princeton University Press, Princeton, New Jersey, (1986).

KOBAYASHI, S. & NOMIZU, K. *Foundations of Differential Geometry,* vol. I–II. Interscience Publishers, John Wiley & Sons, New York, (1963–1968).

KOLMOGOROV, A. N. Preservation of conditionally periodic movements with small change in the Hamilton function. *Dokl. Acad. Nauk. SSSB,* **98**, pp. 527–530, (1954). Reprinted in *Lecture Notes in Physics* No. 93 (Berlin, 1979) pp. 51–56.

KOSTANT, B. Quantization and unitary representations. In *Lecture Notes in Mathematics,* vol. 170, pp. 87–208. Springer-Verlag, Berlin, (1970).

KUMMER, M. On the regularization of the Kepler problem. *Commun. Math. Phys.,* **84**, pp. 133–152, (1982).

KUMMER, M. On the three-dimensional Lunar problem and other perturbation problems of the Kepler problem. *J. Math. Anal. Appl.,* **93**, pp. 142–194, (1983).

LANCZOS, C. *The Variational Principles of Mechanics.* University of Toronto Press, Toronto, 4th edn., (1970).

LANDAU, L. & LIFCHITZ, E. *Mechanics.* Pergamon Press, Oxford, (1960).

LERMAN, L. M. & UMANSKII, Y. L. *Four-Dimensional Integrable Hamiltonian Systems with Simple Singular Points (Topological Aspect).* AMS, Providence, RI, (1991).

LEVI-CIVITA, T. Sur la résolution qualitative du probléme restreint des trois corps. In *Opere Matematiche, Vol Secondo.* Zanichelli, Bologna, (1956).

LIBERMANN, P. & MARLE, C.-M. *Symplectic Geometry and Analytical Mechanics.* D. Reidel Publishing Company, Dordrecht, (1987).

LIGON, T. & SCHAAF, M. On the global symmetry of the classical Kepler problem. *Rep. Math. Phys.*, **9**, pp. 281–300, (1976).

MACK, G. & TODOROV, I. Irreducibility of the ladder representations of U(2,2) when restricted to the Poincaré subgroup. *J. Math. Phys.*, **10**, pp. 2078–2085, (1969).

MALKIN, I. A. & MAN'KO, V. I. Symmetry of the hydrogen atom. *JETP Lett.*, **2**, pp. 146–148, (1963).

MARSDEN, J. E. & RATIU, T. Reduction of Poisson manifolds. *Lett. Math. Phys.*, **11**, pp. 161–169, (1986).

MARSDEN, J. E. & RATIU, T. S. *Introduction to Mechanics and Symmetry.* Springer-Verlag, New York, (1994).

MARSDEN, J. E. & WEINSTEIN, A. Reduction of symplectic manifolds with symmetry. *Rep. Math. Phys.*, **5**, pp. 121–130, (1974).

MASLOV, V. P. *Théorie Des Perturbations et Méthodes Asymptotiques.* Dunod, Paris, (1972).

MCINTOSH, H. & CISNEROS, A. Degeneracy in the presence of a magnetic monopole. *J. Math. Phys.*, **11**, pp. 896–916, (1970).

MILNOR, J. *Topology from Differentiable Viewpoint.* The Univ. Press of Virginia, Charlottesville, Va., (1965).

MLADENOV, I. M. Geometric quantization of the MIC-Kepler problem via extension of the phase space. *Ann. Inst. Henri Poincaré*, **50**, pp. 219–227, (1989).

MLADENOV, I. M. & TSANOV, V. Geometric quantization of the multidimensional Kepler problem. *JGP*, **2**, pp. 17–24, (1985).

MLADENOV, I. M. & TSANOV, V. V. Geometric quantisation of the MIC-Kepler problem. *J. Phys. A: Math. Gen.*, **20**, pp. 5865–5871, (1987).

MORBIDELLI, A. & GIORGILLI, A. On a connection between KAM and Nekhoroshev's theorems. *Phys. D*, **86**, pp. 514-516, (1995).

MORSE, P. M. & FESHBACH, H. *Methods of Theoretical Physics.* McGraw-Hill, New York, (1953).

MOSER, J. K. Stabilitätsverhalten kanonisher differentialgleichungssysteme. *Nachr. Akad. Wiss. Göttingen Math.-Phys. Kl. IIa*, pp. 87-120, (1955).

MOSER, J. K. On invariant curves of area-preserving mappings of an annulus. *Nachr. Akad. Wiss. Göttingen Math.-Phys. Kl. II*, pp. 1-20, (1962).

MOSER, J. K. On the volume elements on a manifold. *Trans. Am. Math. Soc.*, **120**, pp. 286-294, (1965).

MOSER, J. K. Convergent series expansions for quasi-periodic motions. *Math. Ann.*, **169**, pp. 136-176, (1967).

MOSER, J. K. Lectures on Hamiltonian systems. *Mem. Am. Math. Soc.*, **81**, pp. 1-60, (1968).

MOSER, J. K. Regularization of Kepler's problem and the averaging method in a manifold. *Commun. Pure Appl. Math.*, **23**, pp. 609-636, (1970).

NASH, C. & SEN, S. *Topology and Geometry for Physicists.* Academic Press Inc., London, (1983).

NEKHOROSHEV, N. Exponential estimates of the stability time for near-integrable Hamiltonian systems. *Russ. Math. Surveys*, **32**, pp. 1-65, (1977).

NEKHOROSHEV, N. Exponential estimates of the stability time for near-integrable Hamiltonian systems, II. *Trudy Sem. Petrovs.*, **5**, pp. 5-50, (1979).

NGUYÊN TIÊN, Z. A note on focus-focus singularities. *Diff. Geom. and Appl.*, **7**, pp. 123-130, (1997).

ONOFRI, E. Dynamical quantization of the Kepler manifold. *J. Math. Phys.*, **17**, pp. 401-408, (1976).

PARS, L. *A Treatise on Analytical Dynamics.* John Wiley & Sons, New York, (1965).

PASSONI, C. *Monodromia in Sistemi Hamiltoniani.* Master's thesis, Corso di laurea in Matematica dell'Università di Milano, (2000).

PAULI, W. Über das wasserstoffspektrum vom standpunkt der neuen Quantummechanic. *Z. Phys.*, **36**, pp. 336-363, (1926).

PENROSE, R. Twistor algebra. *J. Math. Phys.*, **8**, pp. 345-366, (1967).

PENROSE, R. Relativistic symmetry groups. In Barut, A. O., ed., *Group Theory in Non Linear Problems.* Reidel Publishing Company, Dordrecht, (1974).

PERKO, L. *Differential Equations and Dynamical Systems.* Springer-Verlag, New York, (1991).

POINCARÉ, H. *Les Méthodes Nouvelles de la Mecanique Céleste. Vol 1, 2, 3.* Gauthier-Villars, Paris, (1892-1893-1899).

PUTA, M. On the reduced phase space of a cotangent bundle. *Lett. Math. Phys.*, **8**, pp. 189-194, (1984).

ROBERTSON, H. P. Bemerkung über separierbare systeme in der wellenmeckanik. *Math. Ann.*, **98**, pp. 749-752, (1927).

RÜSSMANN, H. On optimal estimates for the solutions of linear partial differential equations of first order with constant coefficients on the torus. In *Dynamical systems, theory and applications (Rencontres, Battelle Res. Inst., Seattle, Wash., 1974)*, pp. 598-624. Lecture Notes in Phys., Vol. 38. Springer-Verlag, Berlin, (1975).

SIMMS, D. J. Geometric quantization of the energy levels in the Kepler problem. *Symp. Math.*, **14**, pp. 125-137, (1974).

SOMMERFELD, A. *Mechanics.* Academic Press, New York, London, (1964).

SOURIAU, J.-M. Sur la variété de Kepler. *Symp. Math.*, **14**, pp. 343-360, (1974).

SOURIAU, J.-M. Géometrie globale du problème à deux corps. *Proc. IUTAM-ISSIM Symp. on Mod. Devl. Anal. Mech., Atti Acad.Sci. Torino*, **Suppl. 117**, pp. 369-418, (1983).

SOURIAU, J.-M. *Structure of Dynamical Systems.* Birkhäuser, Boston, (1997).

STERNBERG, S. *Lectures on Differential Geometry.* Chelsea Publishing Company, New York, (1983).

STIEFEL, E. L. & SCHEIFELE, G. *Linear and Regular Celestial Mechanics.* Springer-Verlag, Berlin, (1971).

VAN DER WAERDEN, B. L. *Sources of Quantum Mechanics.* North Holland, Amsterdam, (1967).

VAN HOVE, L. Sur certaines representations unitaires d'un group infini de transformations. *Mem. Acad. Roy. Belg.*, **26**, pp. 61-102, (1951).

VILENKIN, N. Y. *Special Functions and the Theory of Group Representations.* Trans. Math. Monographs (AMS) vol 22, Providence, RI, (1968).

VIVARELLI, M. D. The KS-transformation in hypercomplex form. *Celestial Mech.*, **29**, pp. 45-50, (1983).

VOGAN, D. A. *Unitary Representations of Reductive Lie Groups.* Annals of Mathematical Studies. Princeton University Press, Princeton, New Jersey, (1987).

VON WESTENHOLZ, C. *Differential Forms in Mathematical Physics.* North Holland, Amsterdam, (1978).

WEINSTEIN, A. Symplectic manifolds and their Lagrangian submanifolds. *Adv. Math.*, **6**, pp. 329–345, (1971).

WHITTAKER, E. T. *A Treatise on the Analytical Dynamics of Particles and Rigid Bodies.* Cambridge University Press, Cambridge, 2nd edn., (1917).

WHITTAKER, E. T. & WATSON, G. N. *A Course of Modern Analysis.* Cambridge Univ. Press, Cambridge, England, (1952).

WIGNER, E. P. On unitary representations of the inhomogeneous Lorentz group. *Ann. Math.*, **40**, pp. 149–204, (1939).

WILLIAMSON, J. On an algebraic problem concerning the normal forms of linear dynamical systems. *Amer. J. Math.*, **58**, pp. 141–163, (1936).

WOLF, J. A. Remark on nilpotent orbits. *Proc. Amer. Math. Soc.*, **51**, pp. 213–216, (1975).

WOODHOUSE, N. *Geometric Quantization.* Clarendon Press, Oxford, (1980).

ZWANZIGER, D. Exactly soluble nonrelativistic model of particles with both electric and magnetic charges. *Phys. Rev.*, **176**, pp. 1480–1488, (1968).

Index

Z

Editors-in-Chief:
A. Boutet de Monvel, Université Paris VII Denis Diderot, France
G. Kaiser, The Virginia Center for Signals and Waves, Glen Allen, USA

Progress in Mathematical Physics

Cnops, J., Gent Polytechnic, Belgium

An Introduction to Dirac Operators on Manifolds

2002. 208 pages. Hardcover
ISBN 0-8176-4298-6
Vol. 24

In this work, the basic theories underlying the concept of Dirac operators are explored. Starting with the requisite material on Clifford algebras and differential geometry, the text focuses on the two main properties of Dirac operators: conformal invariance, which determines the local behavior of the operator, and the unique continuation property dominating global behavior. Spin groups and spin or bundles are covered, as well as the relations with their classical counterparts, orthogonal groups and Clifford bundles.

Kashiwara, M., Kyoto University, Japan / **Miwa, T.,** Kyoto University, Japan

MathPhys Odyssey 2001

Integrable Models and Beyond

In Honor of Barry McCoy

2002. 496 pages. Hardcover
ISBN 0-8176-4260-9
Vol. 23

Developments in mathematical physics during the second half of the 20th century influenced a number of mathematical areas, among the more significant being representation theory, differential equations, combinatorics, and algebraic geometry. In all areas, the dynamic role of integrable models has been central. This volume focuses on the ongoing importance of integrability in: conformal field theory, massive quantum field theory, solvable lattice models, quantum affine algebras, the Painlevé equations and combinatorics.

Progress in Mathematical Physics (formerly Progress in Physics) is a book series encompassing all areas of mathematical physics. It is intended for mathematicians, physicists, and other scientists, as well as graduate students in the above related areas. This distinguished collection of books includes authored monographs and textbooks, the latter primarily at the senior undergraduate and graduate levels. Edited collections of articles on important research developments or expositions of particular subject areas may also be included.

Cercignani, C., Politecnico di Milano, Italy / **Kremer, G. M.,** Universidade Federal do Paraná, Curitiba, Brazil

The Relativistic Boltzmann Equation: Theory and Applications

2002. 394 pages. Hardcover
ISBN 3-7643-6693-1
Vol. 22

The aim of this book is to present the theory and applications of the relativistic Boltzmann equation in a self-contained manner, even for those readers who have no familiarity with special and general relativity.

For orders originating from all over the world except USA and Canada:

Birkhäuser Verlag AG
c/o Springer GmbH & Co
Haberstrasse 7
D-69126 Heidelberg
Fax: ++49 / 6221 / 345 4229
e-mail: birkhauser@springer.de

For orders originating in the USA and Canada:

Birkhäuser
333 Meadowland Parkway
USA-Secaucus
NJ 07094-2491
Fax: ++1 201 348 4505
e-mail: orders@birkhauser.com

Birkhäuser

Progress in Mathematical Physics

Previously published titles

Petters, A.O., Duke University, Durham, USA / **Levine, H.**, Brandeis University, Waltham, USA / **Wambsganss, J.**, University of Potsdam, Germany

Singularity Theory and Gravitational Lensing

2001. 350 pages. Hardcover
ISBN 0-8176-3668-4
Vol. 21

First monograph to develop a mathematical theory of gravitational lensing. Introductory chapters present historical highlights and astrophysical aspects of the subject. Main part of the book employs ideas and results of singularity theory to put gravitational lensing on a rigorous mathematical foundation to solve certain key lensing problems. Many results published for the first time. Includes examples, illustrations, bibliography and index. For a wide audience of students, mathematicians, mathematical physicists, astrophysicists, and physicists.

Stollmann, P., Technical University of Chemnitz, Germany

Caught by Disorder

Bound States in Random Media

2001. 160 pages. Hardcover
ISBN 0-8176-4210-2
Vol. 20

Disorder is one of the central topics in science today. Over the past 15 years various aspects of the effects of disorder have changed a number of paradigms in mathematics and physics. One such effect is a phenomenon called localization, which describes the very strange behavior of waves in random media. Instead of traveling through space as they do in ordered environments, localized waves stay in a confined region and are caught by disorder. This work is the first treatment of the subject in monograph or textbook form.

Vol. 19: **Ryan, J. / Sprössig, W.**(Eds.) Clifford Algebras and their Applications in Mathematical Physics, Volume II. Clifford Analysis (2000) ISBN 0-8176-4183-1

Vol. 18: **Ablamowicz, R. / Fauser, B.**(Eds.) Clifford Algebras and their Applications in Mathematical Physics, Volume I. Algebra and Physics (2000) ISBN 0-8176-4182-3

Vol. 17: **Baylis, W.E.** Electrodynamics. A Modern Geometric Approach (1999) ISBN 0-8176-4025-8

Vol. 16: **Maslov, V.P.** The Complex WKB Method for Nonlinear Equations I. Linear Theory (1994) ISBN 3-7643-5088-1

Vol. 15: **Leznov, A.N. / Saveliev, M.V.** Group-Theoretical Method for Integration of Nonlinear Dynamical Systems (1992) ISBN 3-7643-2615-8

Vol. 14: **Smirnow, V.A.** Renormalization and Asymptotic Expansions (1991)

Vol. 13: **Haba, Z. / Sobczyk, J.** Functional Integration, Geometry and Strings. Proceedings of the XXV Karpacz Winter School of Theoretical Physics, Karpacz, 20 February–5 March 1989 (1989)

Vol. 12: **Piguet, O. / Sibold, K.** Renormalized Supersymmetry. The Perturbation of N=1 Supersymmetric Theories in Flat Space Time (1986)

Vol. 11: **Ceausescu / Costache Georgescu** Critical Phenomena 1983. Brasov School Conference (1998)

Vol. 10: **Fritz / Jaffe / Szasz** Statistical Physics and Dynamical Systems. Rigorous Results

Vol. 9: **Weldon / Langacker / Steinhardt** Fourth Workshop on Grand Unification

Vol. 8: **Milton K.A. / Samuel M.A.** Workshop on Non Pertubative Quantum Chromodynamics ISBN 0-8176-3127-5

Vol. 7: **Fröhlich** Scaling and Self-Similarity in Physics. Renormalization in Statistical Mechanics and Dynamics

Vol. 6: **Frampton P.H. / Glashow S.L. / Van Dam H.** Third Workshop on Grand Unification, 1982 ISBN 0-8176-3105-4

Vol. 5: **Dita / Georgescu / Purice** Gauge Theories. Fundamental Interactions and Rigorous Results

Vol. 4: **Atwood / Bjorken / Brodsky / Stroynowski** Lectures on Lepton Nucleon Scattering and Quantum Chromodynamics

Vol. 3: **Manin** Mathematics and Physics

Vol. 2: **Jaffe / Taubes** Vortices and Monopoles, Structure of Statistic Gauge Theories

Vol. 1: **Collet, P. / Eckmann, J.P.** Iterated Maps on the Interval as Dynamical Systems (1997) ISBN 0-8176-3026-0

For orders originating from all over the world except USA and Canada:

Birkhäuser Verlag AG
c/o Springer GmbH & Co
Haberstrasse 7
D-69126 Heidelberg
Fax: ++49 / 6221 / 345 4 229
e-mail: birkhauser@springer.de

For orders originating in the USA and Canada:

Birkhäuser
333 Meadowland Parkway
USA-Secaucus
NJ 07094-2491
Fax: ++1 201 348 4505
e-mail: orders@birkhauser.com

Birkhäuser

MATHEMATICAL PHYSICS WITH BIRKHAUSER

Singer, S.F., Haverford College, USA

Symmetry in Mechanics

A Gentle, Modern Introduction

2001. 206 pages. Softcover
ISBN 0-8176-4145-9

Recent years have seen the appearance of several books bridging the gap between mathematics and physics; most are aimed at the graduate level and above. Symmetry in Mechanics: A Gentle, Modern Introduction is aimed at anyone who has observed that symmetry yields simplification and wants to know why.

The monograph was written with two goals in mind: to chip away at the language barrier between physicists and mathematicians and to link the abstract constructions of symplectic and differential geometry to concrete, explicitly calculated examples. The vehicle is a careful exposition of the two-body problem, namely, the derivation of Kepler's Laws of planetary motion from Newton's laws of gravitation.

Beletsky, V.V., Russian Academy of Sciences, Moscow, Russia

Essays on the Motion of Celestial Bodies

2001. 384 pages. Hardcover
ISBN 3-7643-5866-1

From the review of the first edition in *Priroda* by V.I. Arnold and Ya.B. Zeldovich:

"...For a solid scientific monograph, V.V. Beletsky's book is out of the ordinary in many respects. Without exaggeration we can say that it marks the affirmation of a new style in the scientific literature. The author explains in a frank and detailed manner the reasons behind each calculation, its difficulties, and the psychological side of research.

...The general impression that the "Essays" make is not that of a boring lesson, but rather a discussion with a brilliant, knowledgeable and wise interlocutor."

For orders originating from all over the world except USA and Canada:

Birkhäuser Verlag AG
c/o Springer GmbH & Co
Haberstrasse 7
D-69126 Heidelberg
Fax: ++49 / 6221 / 345 4 229
e-mail: birkhauser@springer.de

For orders originating in the USA and Canada:

Birkhäuser
333 Meadowland Parkway
USA-Secaucus
NJ 07094-2491
Fax: +1 201 348 4505
e-mail: orders@birkhauser.com

http://www.birkhauser.ch

Birkhäuser

MIX
Papier aus verantwortungsvollen Quellen
Paper from responsible sources
FSC® C105338

If you have any concerns about our products,
you can contact us on
ProductSafety@springernature.com

In case Publisher is established outside the EU,
the EU authorized representative is:
Springer Nature Customer Service Center GmbH
Europaplatz 3, 69115 Heidelberg, Germany

Printed by Libri Plureos GmbH
in Hamburg, Germany